I0131605

An Invitation to Real Analysis

Adopting a student-centered approach, this book anticipates and addresses the common challenges that students face when learning abstract concepts like limits, continuity, and inequalities. The text introduces these concepts gradually, giving students a clear pathway to understanding the mathematical tools that underpin much of modern science and technology. In addition to its focus on accessibility, the book maintains a strong emphasis on mathematical rigor. It provides precise, careful definitions and explanations while avoiding common teaching pitfalls, ensuring that students gain a deep understanding of core concepts. Blending algebraic and geometric perspectives to help students see the full picture. The theoretical results presented in the book are consistently applied to practical problems. By providing a clear and supportive introduction to real analysis, the book equips students with the tools they need to confidently engage with both theoretical mathematics and its wide array of practical applications.

Features

- Student-Friendly Approach making abstract concepts relatable and engaging.
- Balanced Focus combining algebraic and geometric perspectives.
- Comprehensive Coverage: Covers a full range of topics, from real numbers and sequences to metric spaces and approximation theorems, while carefully building upon foundational concepts in a logical progression.
- Emphasis on Clarity: Provides precise explanations of key mathematical definitions and theorems, avoiding common pitfalls in traditional teaching.
- Perfect for a One-Semester Course: Tailored for a first course in real analysis.
- Problems, exercises and solutions.

Andrew D. Hwang earned his PhD in mathematics at the University of California, Berkeley. After a 30-year career as a mathematician in academia, he now pursues mathematical art, design, education, and outreach through his company, Differential Geometry (www.diffgeom.com).

Textbooks in Mathematics

Series editors:
Al Boggess, Kenneth H. Rosen

Quantitative Literacy Through Games and Gambling
Mark Hunacek

Measure and Integral
Theory and Practice
John Srdjan Petrovic

Contemporary Abstract Algebra, Eleventh Edition
Joseph A. Gallian

Student Solutions Manual for Gallian's Contemporary Abstract Algebra, Eleventh Edition
Joseph A. Gallian

Algebra, Second Edition
Groups, Rings, and Fields
Louis Halle Rowen and Uzi Vishne

Functional Analysis for the Applied Mathematician
Todd Arbogast and Jerry L. Bona

Exploring Linear Algebra, Second Edition
Labs and Projects with Mathematica®
Crista Arangala

Measure Theory and Fine Properties of Functions, Second Edition
Lawrence Craig Evans and Ronald F. Gariepy

Set Theory
An Introduction to Axiomatic Reasoning
Robert André

Introduction to Differential and Difference Equations Through Modeling
William P. Fox and Robert E. Burks

Abstract Algebra, Third Edition
An Interactive Approach
William Paulsen

Elements of Algebraic Topology, Second Edition
James R. Munkres, Steven G. Krantz, and Harold R. Parks

One Complex Variable from the Several Variable Point of View
Peter V. Dovbush and Steven G. Krantz

Math Anxiety How to Beat It
Brian Cafarella

Lectures on Differential Geometry with Maple
Mayer Humi

Numerical Analysis for Engineers
Methods and Applications
Bilal M. Ayyub and Richard H. McCuen

An Invitation to Real Analysis
Andrew D. Hwang

https://www.routledge.com/Textbooks-in-Mathematics/book-series/CANDHTEXBOOMTH

An Invitation to Real Analysis

Andrew D. Hwang

CRC Press
Taylor & Francis Group
Boca Raton London New York

CRC Press is an imprint of the
Taylor & Francis Group, an **informa** business

A CHAPMAN & HALL BOOK

Designed cover image: Andrew D. Hwang

First edition published 2026
by CRC Press
2385 NW Executive Center Drive, Suite 320, Boca Raton, FL 33431

and by CRC Press
4 Park Square, Milton Park, Abingdon, Oxon, OX14 4RN

CRC Press is an imprint of Taylor & Francis Group, LLC

© 2026 Andrew D. Hwang

Reasonable efforts have been made to publish reliable data and information, but the author and publisher cannot assume responsibility for the validity of all materials or the consequences of their use. The authors and publishers have attempted to trace the copyright holders of all material reproduced in this publication and apologize to copyright holders if permission to publish in this form has not been obtained. If any copyright material has not been acknowledged, please write and let us know so we may rectify in any future reprint.

Except as permitted under U.S. Copyright Law, no part of this book may be reprinted, reproduced, transmitted, or utilized in any form by any electronic, mechanical, or other means, now known or hereafter invented, including photocopying, microfilming, and recording, or in any information storage or retrieval system, without written permission from the publishers.

For permission to photocopy or use material electronically from this work, access www.copyright.com or contact the Copyright Clearance Center, Inc. (CCC), 222 Rosewood Drive, Danvers, MA 01923, 978-750-8400. For works that are not available on CCC, please contact mpkbookspermissions@tandf.co.uk

Trademark notice: Product or corporate names may be trademarks or registered trademarks and are used only for identification and explanation without intent to infringe.

ISBN: 978-1-032-98913-6 (hbk)
ISBN: 978-1-032-98714-9 (pbk)
ISBN: 978-1-003-60135-7 (ebk)

DOI: 10.1201/9781003601357

Typeset in Latin Modern font
by KnowledgeWorks Global Ltd.

Publisher's note: This book has been prepared from camera-ready copy provided by the authors.

*For everyone possessed of
curiosity, honesty,
and kindness.*

Contents

To the Instructor

An Invitation to Real Analysis exists to serve a variety of introductory courses in real analysis in a pedagogical interval, from a one-semester analysis-based introduction to proofs course near the left endpoint, to a one- or two-term course for more advanced students in mathematics and other quantitative fields at the right endpoint. The book is far from alone among analysis texts, so a few words are in order about the book's features, and prospective courses.

What Is Inviting About *An Invitation to Real Analysis*?

Two pedagogical principles stand out for shaping the book. First, presenting definitions of analysis as adversarial games, not as a passing remark but as a uniform stance, powerfully engages students' intuition. Second, real analysis is a *language* in addition to its substantial mathematical content. In particular, consistency of definitions is vital given the founding role analysis plays in so much of school mathematics, not to mention calculus and beyond. Explicitly acknowledging the language aspect, both for ourselves as instructors and for students, and structuring material in careful logical order, ultimately make everyone's life easier.

Atop these anchoring principles, the book takes special care: to be self-contained[1]; to use consistent and evocative terminology, notation, and proof idioms, stringently avoiding unstated edge cases; to phrase definitions and theorem statements for ease of comprehension and use; to structure arguments as simply and concretely as possible, and only directly or contrapositively, not by contradiction[2]; starting with axioms for the real numbers in Chapter 3, and earlier where feasible, to develop material in strict logical order, with proofs resting solidly on definitions and specific, established theorems; to present interesting examples as soon as possible; to provide students with comparable proofs before posing exercises, both as illustrations of proof-writing and as incentives to study and absorb existing proofs; to encapsulate *technical* points to avoid repetition; to cultivate *conceptual* revisiting, foreshadowing coming ideas and connecting material with what has already been developed.

---

[1]Except for the integer division algorithm.

[2]There is precisely one exception, in an exercise: The proof there exists no surjection from a set to its power set.

In my experience, real analysis tends to be the course of "maximum dynamic pressure" for math majors, future teachers, and other students, a blizzard of complicated and unfamiliar definitions, unstated idioms involving the use of arbitrary inequalities to establish equalities, and bewildering technicalities stated as tone poems written from Greek alphabet soup. In effect, students are asked to read and write literature while learning a foreign language. Neglecting to mention this often creates a mysterious layer of frustration: Students who enjoyed calculus suddenly find their day-to-day understanding of mathematics has sharply decreased for no apparent reason.

To the best of my ability, the book is a gradual, paved, well-marked path with multiple trailheads into the language and landscape of real analysis. It is written to be read, studied, taught from, referenced for many years, and enjoyed.

Sample Courses

Table 0.1 contains an approximate syllabus for a 10–15-week analysis-based proofs course for mathematics and education majors who have had a year of one-variable calculus, and possibly a semester of linear algebra and/or multi-variable calculus. Stars signify increasingly optional sections that can be omitted or partially combined with neighboring material as time demands. This course starts with statements and logic and makes its way to construction of the elementary functions, with emphases on the structure of the real number system and the definition and usage of sequential limits.

Table 0.2 contains an approximate syllabus for a 13–15-week introductory real analysis for upper-level undergraduate or beginning graduate students in mathematics, physics, computer science, statistics, engineering, and other quantitative fields, who have had a year of one-variable calculus, a semester of linear algebra, and possibly multi-variable calculus and/or a course in techniques of proof. This course starts with a brisk review of the real numbers and ends with about two or three weeks devoted to topics in metric spaces and functional analysis.

The book provides a solid foundation for a differential equations course devoted to techniques of solution for first- and second-order equations, or to qualitative theory of dynamical systems; for numerical analysis; for a theoretical course in multi-variable calculus; for mathematical probability and statistics; for measure theory; for spectral decomposition.

Each section comes with exercises. Across the book, these range from routine verification to challenging extended write-ups. Exercises that have hints (H), answers (A), or solutions (★) in the back are clearly marked. There is a complete solution manual available on request.

The book's essential thread is power series: Sections 7.4, 8.3, 9.4, and 11.1.

TABLE 0.1
Section coverage for an introductory course.

1. §1.1*	10. §4.1	19. §6.2	28. §8.3	37. §10.3
2. §1.2	11. §4.2	20. §6.3	29. §8.4	38. §11.1
3. §1.3	12. §4.2–3	21. §6.4	30. §8.5	39. §11.2*
4. §2.1	13. §5.1**	22. §7.1*	31. §9.1	40. §12.1
5. §2.1*	14. §5.2*	23. §7.2	32. §9.2	41. §12.2*
6. §2.2*	15. §5.3	24. §7.3*	33. §9.3	42. §13.1
7. §3.1	16. §5.4*	25. §7.4	34. §9.4	43. §13.2*
8. §3.2	17. §6.1	26. §8.1	35. §10.1	44. §13.2**
9. §3.3	18. §6.1*	27. §8.2	36. §10.2*	45. §14.2**

Power series illustrate material on sequential limits, uniform estimates, properties of functions, and order of approximation, among other topics. They permit easy construction and study of non-algebraic functions.

TABLE 0.2
Section coverage for an advanced course.

1. §3	10. §6.4	19. §9.1–2	28. §12.4*	37. §16.3
2. §4.1	11. §7.1*	20. §9.2–3	29. §13.1	38. §16.4
3. §4.2	12. §7.2	21. §9.4	30. §13.2	39. §16.5
4. §4.3	13. §7.3*	22. §10.1–2	31. §14.1–2	40. §17
5. §4.4*	14. §7.4	23. §10.3–4	32. §15.1	41. §17
6. §5.1–3*	15. §8.1–2	24. §11.1–2	33. §15.2	42. §17
7. §5.4	16. §8.3	25. §11.3	34. §15.3	43. §17
8. §6.1–2	17. §8.4	26. §12.1–2	35. §16.1	44. §17
9. §6.3*	18. §8.5	27. §12.2–3*	36. §16.2	45. §17

Inversely, Sections 2.4 (construction of natural, whole, and rational numbers) and 8.6 (discrete dynamical systems) are optional in the book's overall structure. Section 4.4 (topology) is recapitulated in Section 16.1 on metric spaces, so one or the other may be skimmed or omitted accordingly. The examples of functions in Section 5.1 are likely known to students, but are included for self-containment. Although the terms *countable* and *uncountable* occur in various places throughout the book, Section 5.4 (cardinality) contains more material than is strictly used. Series reordering is covered in Section 7.3, but is not used elsewhere. Section 12.3 (hyperbolic functions) contains material that could be covered only as needed. Finally, Chapter 17 contains selective applications. Aside from 17.2 and 3 (uniform convergence), these sections are mutually independent and optional.

What Is Distinctive About *An Invitation to Real Analysis*?

We mathematicians tend toward eponyms for concepts and theorems. These are opaque to students, and rarely withstand historical inspection. With two peripheral exceptions that are well-known in popular culture—Venn diagrams and Fibonacci numbers—this book instead uses descriptive names: the *ordered product* of sets, *condensing sequences* whose terms can be made as close to each other as we like, the *cross-term bound* for inner products, the *polar formula* $e^{i\theta} = \cos\theta + i\sin\theta$, *spectral decomposition* for periodic functions, and many others. Eponyms *are*, however, indexed for easy reference.

Students at this writing are not unlikely to have programming experience. A couple of salient expository choices reflect this reality. First, definitions, theorems, and exercises are formulated with an eye toward algorithmic implementation where appropriate. Sums and factorials, for example, are defined recursively. Generally, the book avoids using ellipses to connote "continuing patterns" except for illustration. Consequently, mathematical induction underpins the entire book. More than once, these recursive foundations simplified the "traditional" presentations and proofs I had used for many years in the classroom.

In this book, curly braces signify unordered sets while round parentheses connote ordered lists. Counting starts at 0 (the cardinal of the empty set) and, for a list of length n, ends at $(n-1)$, as in programming languages. In this book, $0^0 = 1$, the number of mappings from the empty set to itself. Again, being systematic with these choices pleasantly clarified edge cases and simplified at least a few calculations.

An Invitation to Real Analysis is written for the 21st century. I hope and expect both students and you will find the book especially friendly, supportive for your respective needs, internally consistent, mathematically substantive, and thought-provoking in positive ways.

To the Student

Welcome to *An Invitation to Real Analysis*: Not just this book, but to the prospect of a math course with a similar title and probably an intimidating reputation. Presumably, this book will be your close companion for a few months, and maybe your occasional companion for longer.

Real analysis places integral and differential calculus on a solid logical foundation. Calculus traffics in "infinitesimals" such as dt. Formally, an integral is "a sum of infinitely many infinitesimals" while a derivative is a ratio of infinitesimals. In the real numbers, however, there *are* no infinitesimals. As a calculus student, you may have wondered, "What exactly is dt?" or "Why do these calculations work?" This book contains one extended answer: Real analysis, the mathematics of arbitrarily close approximation, allows us to express and prove the computational rules of calculus using only real numbers.

Because infinitesimals extend to the roots of calculus, we must dig deeply into structure of the number line in order to establish results of calculus as theorems. As a result, the material does not look like calculus. Instead, analysis initially focuses on real numbers, sets of real numbers, and sequences. Later on as well, real analysis avoids infinitesimal notation, so in theoretical work, even integrals and derivatives do not look much like calculus.

Related to these items of unfamiliarity, real analysis is something like a *foreign language*. The story of limits you may have learned in calculus, anything that sounded like "approaching, but never reaching" or "closer and closer," is casually workable but lies somewhere between *misleading* and *technically wrong*; a large part of analysis is developing this story with logical consistency.

Further, in real analysis, we establish equalities via *arbitrarily close inequalities*: If x and x' are real numbers, and if $|x' - x| < \varepsilon$ (epsilon) for every positive ε, then $x = x'$. Computationally, you may be less familiar with inequalities than with equalities. Rest assured: This book will help you develop your expertise. Conceptually, for each particular ε, say $\varepsilon = 10^{-6}$, the assumption $|x' - x| < \varepsilon$ *does not* imply $x = x'$. On closer reflection, *no matter how small ε is*, the assumption $|x' - x| < \varepsilon$ does not imply $x = x'$. So, what gives? A direct answer involves "infinitely many hypotheses," but perhaps a clearer answer is, if $x \neq x'$, then $0 < |x' - x|$. So, if we pick $\varepsilon = |x' - x|/2$, then $|x' - x| < \varepsilon$ is false. This idea justifies much of analysis.

This book develops real analysis from little more than everyday common sense and mathematical properties of addition, multiplication, and comparison of real numbers. It is written to be read, actively and mindfully. Strive to connect what you read with what you already know. Browse chapters and sections your course skips. The book takes special care to develop material in logical order, to define all mathematical terms introduced, and to supply "sample" proofs before asking you to construct your own. Do use prior results in developing proofs, and use the book's index to locate unfamiliar terms.

Some exercises consist of an isolated statement or question, such as "$\sqrt{3}$ exists," or "Is $0.99\overline{9} = 1$?," or "$x^5 + 7x - 1 = y$ has a real solution x for every real y." (These are made-up examples.) Each is a request for an answer *backed up by proof, using only concepts and results established in the book so far*. The bare answer may be of interest, but providing incontrovertible justification within the framework of the material is an essential part of the subject.

You may ask yourself, "Why do we prove things?" To make an athletic analogy, why do we struggle up the precipice or hike the rocky trail instead of driving or taking the cable car? One partial answer is, "It depends what you want to do." There is looking at the view, and there is *the pleasure of earning* the view. Another partial answer is, "Maybe there *is* no road or cable car," and you are required to build one. *Someone* has to climb that cliff and map the territory for the first time. You are training for that possibility.

Mathematics leads to mind-stretching conclusions. Has every ordering of a deck of playing cards been seen at some point in history? Given an arbitrary infinite string s of 0s and 1s, is there a computer program that prints s in the idealized sense of "if we wait infinitely long"? If $B^n(1)$ denotes a ball of radius 1 in n-dimensional space, how rapidly does the volume of $B^n(1)$ grow with n? What does "$e^{i\pi} + 1 = 0$" even mean?

If you are pursuing mathematics purely for its own sake, give due respect to mathematics that sheds light on practical problems or opens technological frontiers. Even when a "simple" idea has useful applications, *the fact of noticing and acting* is itself an achievement. The world of experience is larger and more complicated than human mathematics encompasses, even if mathematics provides our best descriptive and predictive physical frameworks.

Inversely, if you are pursuing mathematics for its applications, take care to enjoy its intrinsic wonder, a profound aspect of the human experience independent of applications. Remember as well: Large pieces of "pure," "irrelevant" mathematics have found surprising technological applications decades or centuries after they were developed. None of us is omniscient.

Technical intricacy aside, doing mathematics has elements of play, discovery, and art. The more you seek and cultivate these pleasant aspects for yourself, the happier and more productive will be your time with this material.

Welcome again, and all good wishes with your journey!

1

Logic and Sets

Mathematics encompasses a language for expressing logical and quantitative propositions, and computational tools for using known truths to deduce new ones. In order to read, understand, and write mathematics fluently, you must learn and practice using the terminology, syntax, and idioms of this language. This chapter introduces essentials needed throughout the book.

As a body of recorded knowledge, mathematics proceeds via logical deduction from explicit but unproven hypotheses toward conclusions. The wish to present mathematics in logical order entails a pedagogical chicken and egg problem: Humans learn by example, but when material is developed in strict logical order, there are no familiar examples until considerable work has been accomplished. In this chapter, we "cheat" a little. Specifically, we'll give examples that refer to integers (whole numbers), natural numbers (for us this means non-negative integers), rational numbers (fractions of integers), and real numbers (which at this stage are difficult to describe precisely, but may be viewed as filling in gaps left by rational numbers on the number line).

1.1 Statements and Logical Connectives

Definition 1.1.1. A *statement* is a sentence having a *truth value*, T (True) or F (False).

Remark 1.1.2. Contact with the external world must be made via experience and is necessarily approximate. For example, no known phenomenon behaves exactly like integers. ◇

Example 1.1.3. -4 is an even integer. (True.)

The decimal representation of π contains the string "999999." (True) (The number π is defined in Chapter 13.)

$2 + 2 = 5$. (False) ◇

Example 1.1.4. Sentences that are *not* statements include "10^{1000} is a large number" ("large" has not been given a precise meaning), and "x is a positive real number" (whose truth value depends on x, so the sentence is not a statement unless x is specified). ◇

DOI: 10.1201/9781003601357-1

1

Conventionally, abstract statements are denoted P and Q.

Definition 1.1.5. The *negation* of a statement P is its logical opposite $\neg P$.

Remark 1.1.6. You may regard the negation as P preceded by the clause "It is not the case that...," but usually a more pleasant wording can be found. ⬦

Example 1.1.7. P: $2 + 2 = 4$. $\neg P$: $2 + 2 \neq 4$.   ◇

Assume P and Q are statements. New statements can be constructed using the "logical connectives" *and, or*, and *implies*.

Definition 1.1.8. The statement "P and Q," the *conjunction* of P and Q, has its ordinary meaning: The compound statement is true provided both P and Q are true, and is false otherwise.

Example 1.1.9. $2 + 2 = 4$ and $0 < 1$. (True)
    $2 + 2 = 5$ and $0 < 1$. (False)
    $2 + 2 = 5$ and $1 < 0$. (False)   ◇

Definition 1.1.10. The statement "P or Q," the *disjunction* of P and Q, always has the "inclusive" meaning in mathematics: P is true, or Q is true, *or both*.

Example 1.1.11. $2 + 2 = 4$ or $0 < 1$. (True)
    $2 + 2 = 5$ or $0 < 1$. (True)
    $2 + 2 = 5$ or $1 < 0$. (False)   ◇

Implication

"Logical implication" plays a central role in mathematics as our tool for deducing new true statements from known ones.

Definition 1.1.12. A *logical implication* is a statement of the form "If P then Q," also read "P *implies* Q." The statement P is called the *hypothesis*, and Q the *conclusion*. In logic, "P implies Q" is true unless the hypothesis P is true and the conclusion Q is false.

Example 1.1.13. If $1 \neq 0$, then $1^2 \neq 0$. (True)
    If $1 \neq 0$, then $1^2 = 0$. (False)
    If $1 = 0$, then $0 = 0$. (True)
    If $1 = 0$, then $1^2 = 0$. (True)   ◇

Remark 1.1.14. There are two potentially confusing consequences of logical implication. First, it is logically sound to deduce an arbitrary conclusion from a false hypothesis. An implication with false hypothesis is said to be *vacuous*. Humorous examples abound: "If $1 = 0$, then money grows on trees."

In Example 1.1.13, the third and fourth implications are vacuous. Note carefully that in each case, we can give a proof. If $1 = 0$, subtracting this equation from itself gives $0 = 0$, which proves the third statement. Similarly, squaring gives $1^2 = 0^2 = 0$, proving the fourth statement.

Second, an implication need not connect logically related statements. The implication "If $0 = 0$, then 2 is an even integer" is true because both the hypothesis and conclusion are true, but is effectively a *non sequitur*; the conclusion does not "follow" from the hypothesis in any obvious sense. A sequence of true implications does not, of itself, constitute a proof. In the example at hand, we know the implication is true only because there exists a valid proof, consisting of implications whose truth is apparent.

In these two senses, mathematicians are liberal in deeming an implication to be true. Truth of an implication is the weakest criterion that excludes the act of drawing a false conclusion from a true hypothesis.

In this book, and throughout mathematics in practice, implications do actually link "logically related" statements. Most implications involve classes of objects and assert that every object satisfying some condition also satisfies some other condition. ◇

Definition 1.1.15. An implication of the form "P implies P" is a *tautology*. A statement P such that "P implies $\neg P$ and $\neg P$ implies P" is a *contradiction*.

Remark 1.1.16. Existence of a contradiction is fatal for logic: If P is a contradiction and Q is an arbitrary statement, then either P implies Q or $\neg P$ implies Q is vacuously true. Thus, Q is provable, and there are no logical grounds for distinguishing truth and falsehood. Informally, "This sentence is false" is a contradiction. Example 1.3.11 describes a dramatic formal example. ◇

Truth Tables

The logical operators "not," "and," "or," and "implies" may be summarized with a *truth table*, Table 1.1. A truth table involving compounds of two statements contains four rows, one for each combination of truth values of P and Q. The value in the column of an expression represents the truth or falsity of that statement.

Remark 1.1.17. If P and Q are statements, the statement "P and Q" is false if *at least one* of P and Q is false. If someone assures you two statements are both true, only one has to be false for the assurance to be unfounded.

Analogously, if someone assures you at least one statement of two is true, then both must be false for the assurance to be unfounded.

Exercise 1.3.5 formalizes these observations. Loosely, the connectives "and" and "or" are interchanged by negation, perhaps contrary to first impression. Consequently, the order of negation and conjunction matters. ◇

TABLE 1.1
A truth table for not, and, or, and implies.

P	Q	$\neg P$	P and Q	P or Q	P implies Q
T	T	F	T	T	T
T	F	F	F	T	F
F	T	T	F	T	T
F	F	T	F	F	T

Example 1.1.18. The integers 1 and 0 are **not both** zero. (True.)
   The integers 1 and 0 are **both not** zero. (False.)   ◇

Logical Equivalence

Definition 1.1.19. Two statements P and Q are *logically equivalent* if each
implies the other: P implies Q and Q implies P.

Remark 1.1.20. A truth table shows P and Q are equivalent precisely when
they have the same truth value. In Table 1.2 we write P equiv Q to indicate
P and Q are logically equivalent.   ◇

TABLE 1.2
Logically equivalent statements have the same truth value.

P	Q	P implies Q	Q implies P	P equiv Q
T	T	T	T	T
T	F	F	T	F
F	T	T	F	F
F	F	T	T	T

   Every abstract logical implication P implies Q belongs to a quadruple of
implications. These are laid out as a "seating chart" in Table 1.3. We'll meet
each partner in more detail, then return to their collective relationships. You
do not need to memorize these names before reading further.

TABLE 1.3
The quadruple of an abstract implication.

Direct:	P implies Q	Contrapositive:	$\neg Q$ implies $\neg P$
Converse:	Q implies P	Inverse:	$\neg P$ implies $\neg Q$

TABLE 1.4
An implication is logically equivalent to its contrapositive.

P	Q	P implies Q	$\neg Q$	$\neg P$	$\neg Q$ implies $\neg P$
T	T	T	F	F	T
T	F	F	T	F	F
F	T	T	F	T	T
F	F	T	T	T	T

Definition 1.1.21. The implications "P implies Q" and "$\neg Q$ implies $\neg P$" are said to be *contrapositive* to each other.

Example 1.1.22. Direct: If x is a multiple of 4, then x is even.
   Contrapositive: If x is not even, then x is not a multiple of 4.   ◇

Remark 1.1.23. As shown in Table 1.4, an implication and its contrapositive are logically equivalent: "If it was raining when I left home this morning, then I am carrying an umbrella" is equivalent to "If I am not carrying an umbrella, then it was not raining when I left home this morning."   ◇

Remark 1.1.24. The implication "if P then Q" sometimes gets expressed "Q if P." This is a natural shortening of "Q (is true) if P (is true)."
   The same implication sometimes gets expressed "P only if Q." This paraphrases the contrapositive, "if Q is false, then P is false."   ◇

Example 1.1.25. Implications with multiple hypotheses are generally easier to understand and prove in contrapositive form. In each condition, x stands for an integer. Assume P is the condition "$x^2 - 1 \neq 0$" and Q is "$x \neq 1$."
   The implication P implies Q is true, but may require a few seconds' thought to see. By contrast, the logically equivalent contrapositive, "If $x = 1$, then $x^2 - 1 = 0$," is immediate.   ◇

Definition 1.1.26. The implications "P implies Q" and "Q implies P" are *converse* to each other.

Remark 1.1.27. Table 1.2 shows that an implication and its converse are not logically equivalent. Humans are particularly prone to conflating an implication and its converse: Everyone who works hard for years will become wealthy. "Therefore," everyone who is wealthy worked hard for years.   ◇

Remark 1.1.28. In Example 1.1.25, the converse implication, "If $x \neq 1$, then $x^2 - 1 \neq 0$" is false. The number $x = -1$ is a *counterexample*: It satisfies the converse hypothesis Q but not the converse conclusion P.   ◇

Remark 1.1.29. Logical equivalence of P and Q means "P if Q" (Q implies P) and "P only if Q" (equivalent to P implies Q). This naturally shortens to "P if and only if Q." The phrase "if and only if" pervades mathematics.   ◇

Definition 1.1.30. The implications "P implies Q" and "$\neg P$ implies $\neg Q$" are *inverse* to each other.

Remark 1.1.31. Exercise 1.1.2 shows that an implication and its inverse are not logically equivalent. Humans are particularly prone to conflating an implication and its inverse: Ordinary people die. "Therefore," because I am not ordinary, I will not die.   ◇

Remark 1.1.32. To summarize, every abstract implication P implies Q travels in an entourage of four. This entourage divides into two pairs of logically equivalent statements: direct-contrapositive, and converse-inverse. In general, a direct implication and its converse do not have the same truth value. Their truth values are the same precisely when P and Q are logically equivalent.

   Any two statements in the entourage are reciprocally related. For example, the direct implication is the converse of its converse.   ◇

Remark 1.1.33. In this book, a proven conclusion is a *theorem* if it merits a name, a *proposition* if noteworthy but not named, a *lemma* if idiomatic or useful in some other proof but not strictly immediate from definitions, and a *corollary* if it amounts to a useful observation on a theorem or proposition.   ◇

Remark 1.1.34. Mathematical theorems involve an ironclad guarantee "P implies Q." In many useful instances, P is some easily verified condition, while Q is onerous to check directly. In this situation the proof amounts to performing the onerous checking *one time in sufficient generality.*

   Do take care: While mathematics "tells the truth" it need not "tell the whole truth." For example, "If n is the square of an even integer, then n is a multiple of 4" is true. The inverse, "If n is not a square, then n is not a multiple of 4" is false.

   Theorems (etc.) of the form "P if and only if Q" *do* tell the whole truth in the preceding sense. For example, a theorem might guarantee that if easily-verified condition P holds, then useful-but-onerous conclusion Q also holds, and *in addition* they characterize Q by guaranteeing that if P is false, then Q is false.   ◇

Exercises for Section 1.1

Exercise 1.1.1. (★) Assume P and Q are statements. Use a truth table to prove "P implies Q" is equivalent to "$(\neg P)$ or Q."

Exercise 1.1.2. Assume P and Q are statements. Use a truth table to prove

(a) "P implies Q" is not equivalent to its inverse "$\neg P$ implies $\neg Q$."

(b) "$\neg P$ implies $\neg Q$" *is* equivalent to "Q implies P." (In words, the inverse and converse of an implication are logically equivalent.)

Exercise 1.1.3. (★) It is crucial to do this exercise yourself before reading the solution, which is a spoiler. Assume we have a set of cards bearing a letter, either D or N, on one side and a positive integer on the other. The question is to detect the logical condition, "If the card's letter is not N, then the integer on the reverse is greater than or equal to 21."

(a) Four cards are shown. Which cards satisfy the condition?

20	46	16	25
D	D	N	N

(b) Four more cards are shown on one side. Which cards must be turned over in order to check the condition?

18	35	D	N

These tasks are probably a little tricky. That is both normal and pedagogically optimal. This exercise and the solution in the back of the book illustrate how humans have at least two cognitive modes for logical deduction. We started here with the (slow, difficult) analytical mode needed to apprehend mathematical concepts.

Exercise 1.1.4. (★) The calculations you may have learned to associate with *solving* or *doing math* are really working backward from conclusion to hypothesis. This three-part question asks you to think carefully about the process and why it works.

(a) A question asks, "Find all real x such that $0 = x^3 - 4x^2 - 4x + 16$." Customarily, we factor the polynomial (which in general is hard), set each factor to 0, and read out values of x by inspection. Here, we might notice $(x - 4)$ is a factor. Thus $0 = (x - 4)(x^2 - 4) = (x - 4)(x - 2)(x + 2)$, so $x = 4$ or $x = 2$ or $x = -2$.

Explain carefully: What are the requested hypothesis and conclusion? What are the hypothesis and conclusion established by the calculation? Given that what was asked and what was shown are not logically equivalent in general, "Why does this work?" Specifically, why go through this process instead of being direct, and what cautions must we take when "solving equations" in this way?

(b) Analyze the following two-column "proof" that $-1 = 1$.

$$-1 = 1 \qquad \text{to be shown,}$$
$$(-1)^2 = 1^2 \qquad \text{square both sides,}$$
$$1 = 1 \qquad \text{true statement.}$$

Therefore, $-1 = 1$.

(c) Let a and b denote real numbers, and assume $a = b$. Analyze the following "proof" that $2 = 1$.

$$b^2 = ab \qquad a = b,$$
$$b^2 - a^2 = ab - a^2 \qquad \text{subtract } a^2,$$
$$(b + a)(b - a) = a(b - a) \qquad \text{factor each side,}$$
$$(b + a) = a \qquad \text{cancel common factor,}$$
$$2a = a \qquad a = b,$$
$$2 = 1 \qquad \text{cancel common factor.}$$

Exercise 1.1.5. Many types of logical error exist, including conflating an implication with its converse or inverse, assuming the conclusion, over-generalization (proof by example), using undefined terms or using the same name for different things, and logical disconnect (where a true statement is followed by "therefore the desired conclusion is true").

Recall that an integer p greater than 1 is *prime* if the only positive divisors of p are 1 and p. Analyze the following arguments. Determine whether the hypotheses and conclusions of each are true, and whether one of the errors of the preceding paragraph has been made, or if the hypotheses justify the conclusions.

(a) 3 is prime, 5 is prime, 7 is prime. Each is an odd integer greater than 1. Therefore, every odd integer greater than 1 is prime.

(b) Every even integer greater than 2 is not prime. Therefore, every prime greater than 2 is not even.

(c) Every even integer greater than 2 is not prime. Therefore, every odd integer greater than 2 is prime.

(d) It is repugnant to the nature of a prime to be even. Therefore, every prime greater than 2 is odd.

Exercise 1.1.6. (★) A truthful promotional flier proclaims: You are the guaranteed recipient of at least two of the following.

- Hotel Resort Platinum Getaway!

- $2,500.00 Instant Scratch Ticket!

- Home Theater System (retail value $500)!

- $1,000.00 Instant Scratch Ticket!

- $10,000 in cash!

What is the maximum value of the guaranteed prizes? What is the minimum value of the guaranteed prizes? If one or both questions cannot be answered using information given, what additional information is needed?

1.2 Quantification

To accommodate classes of objects in the framework of statements, we allow statements to contain *variables*, so long as each variable is "quantified," accompanied by the phrase "for every" or "there exists." The quantifiers are crucial; pay close attention to them while reading, and *do not omit them when thinking and writing*.

Example 1.2.1. For every integer n, $0 < 1 + n^2$. (True)
For every integer n, $0 < n^2$. (False)
For every integer n, $n^2 = 1$. (False) ◊

A variable preceded by "for every" or "for all" is said to be *universally quantified*. A universally quantified statement encapsulates multiple statements. For example, the first statement of the preceding example encapsulates an infinite collection of statements, one for each integer.

Example 1.2.2. There exists an integer n such that $0 < 1 + n^2$. (True)
There exists an integer n such that $0 < n^2$. (True)
There exists an integer n such that $n^2 = 1$. (True)
There exists an integer n such that $n = n + 1$. (False) ◊

A variable preceded by "there exists" or "for some" is *existentially quantified*. An existentially quantified statement encapsulates multiple statements, expressing that at least one truism is found among the statements. The fourth compound statement is false because *every* individual statement is false.

Remark 1.2.3. Statements contain only "bound" variables, namely, variables that are universally or existentially quantified.

Sentences containing "free" or "unbound" variables (such as "n is even" or "$x^2 + x - 2 = 0$") are not statements. However, sentences containing unbound variables play the useful role of *conditions* in mathematics, selecting objects (perhaps integers n or real numbers x) for which the condition is true.   ◇

Many mathematical conclusions take the universally quantified form "For every x satisfying $P(x)$, condition $Q(x)$ is true." For stylistic variety, such statements may be worded as implications involving *arbitrary* values of variables.

Example 1.2.4. If n is an integer such that $n^2 + n - 2 = 0$, then $n = 1$ or $n = -2$. (True)

If n is an arbitrary integer, then there exist unique integers q and r such that $n = 4q + r$ and $0 \le r < 4$. (True)

If a, b, and c are positive integers, then $a^3 + b^3 \ne c^3$. (True)   ◇

Quantifiers and Negation

The universal quantifier "for every" may be viewed as an enhancement of the "and" conjunction: "For every integer n, the condition $P(n)$ is true" means that the infinitely many statements $P(0)$, $P(1)$, $P(-1)$, and so forth, are *all* true.

The existential quantifier "there exists" may be viewed similarly as an enhancement of "or": "There exists an integer n such that the condition $P(n)$ is true" means that among the infinitely many statements $P(0)$, $P(1)$, $P(-1)$, ..., *at least one* is true.

Example 1.2.5. Logical negation "converts" a "for every" statement into a "there exists" statement of negations, and converts a "there exists" statement into a "for every" statement of negations:
 P: For every integer n, $n^2 \ge 0$.
 $\neg P$: There exists an integer n such that $n^2 < 0$.

 P: There exist integers m and n such that $m^2 + n^2 = 1$.
 $\neg P$: For all integers m and n, $m^2 + n^2 \ne 1$.   ◇

Remark 1.2.6. A statement "For every x, $P(x)$" can be disproved by finding a counterexample, but cannot be proved by exhibiting an example.   ◇

Remark 1.2.7. When the hypothesis of a logical implication contains a variable but no quantifier is explicitly present, the convention is to read "for every." For example, "If $n > 0$ then $n^2 > 0$" should be read "For every integer n, if $n > 0$ then $n^2 > 0$" (assuming context dictates n is an integer).

If an implicitly quantified statement is negated, the existential quantifier must be added explicitly: "There exists a positive n such that $n^2 \leq 0$."

To avoid confusion, including your own, include logical quantifiers explicitly. This book makes a special effort to set a good example. ◇

Quantifiers and Contraposition

Contraposition tends to clarify implications having multiple hypotheses. Particularly in analysis, an implication may have *infinitely many* hypotheses.

Example 1.2.8. Assume x denotes a real number and n a positive integer.
Direct implication: If $x < 1/n$ for every n, then $x = 0$.
Contrapositive: If $x > 0$, then there exists an n such that $1/n \leq x$.

The contrapositive is true: If $x > 0$ and n is an integer such that $n \geq 1/x$, then $1/n \leq x$. The direct implication is therefore true, since its contrapositive is true. However, the direct implication exhibits a new phenomenon: The hypothesis consists of infinitely many statements, $x < 1$, $x < 1/2$, $x < 1/3$, etc., but *no finite number of these statements implies the conclusion*. Indeed, if we assume only finitely many inequalities of the form $x < 1/n$, then there exists a largest denominator, say N, and our collection of inequalities is equivalent to the single inequality $x < 1/N$, which does not imply $x = 0$.

Incidentally, the converse "If $x = 0$, then $x < 1/n$ for every positive integer n" is easily seen to be true, even though the conclusion consists of infinitely many statements: $0 < 1$, $0 < 1/2$, $0 < 1/3$, etc. ◇

Multiple Quantifiers and Adversarial Games

Among the most subtle conditions in mathematics are those containing multiple quantifiers. The basic definitions of analysis entail multiple quantifiers. The good news is, multiply quantified definitions can be phrased as *adversarial two-player games*, and human intuition is remarkably strong in that arena, as we saw in Exercise 1.1.3.

Example 1.2.9. A joke the author often shared with students ran like this:

Let's play "Who can pick the larger integer?" I'll let you go first.

Students generally laughed immediately, even though the rules had not been specified, because human intuition is good at navigating potentially adversarial situations. The joke is, although letting someone play first in a new game is often a courtesy, here the second player can always win.

Consider the multiply quantified statement, "For every integer n, there exists an integer L such that $n < L$." Probably this looks forbidding, but its truth encapsulates the outcome of "Who can pick the larger integer?"

Let's give our players names: Player n and Player L (for "larger"). The game is, Player n picks an integer n. Then Player L picks an integer L. The player with the larger integer wins. Because there is no largest integer, it's better to pick second (assuming the goal is to win). For definiteness, Player L may pick $L = n + 1$. For every n (Player n's choice) there exists an integer L (Player L's choice, say $L = n + 1$) such that $n < L$ (Player L wins).

In a certain explanatory sense, the multiply quantified statement is true *because* Player L has a winning strategy against a perfect opponent. Player L's n-dependent strategy $L = n + 1$ *proves* the truth of "For every integer n, there exists an integer L such that $n < L$." $\diamond$

Example 1.2.10. What if the game is "Who can pick the smaller natural number?" (The players may not choose the same number.) Would you rather play first or second? Before reading further, take a minute to write a multiply quantified sentence that encapsulates the outcome.

Let's call the players n and S (for "smaller"). We can write a multiply quantified sentence just as before: "For every natural number n, there exists a natural number S such that $S < n$." The crucial mathematical difference is, there exists a smallest natural number, 0. So the truth of our multiply quantified statement implies "there exists a natural number S such that $S < 0$." Since no such S exists, our multiply quantified statement is false. And this proves Player n has a winning strategy against a perfect opponent. $\diamond$

Example 1.2.11. In the spirit of giving a "live" example, here is the type of condition analysts work with:

> For every (positive real number) ε (epsilon),
>   there exists a natural number N such that
>   if $k \geq N$, then $1/(k + 1) < \varepsilon$.

This is not the place to explain what this statement means, but when the time comes (Chapter 6), you'll have in mind an adversarial game and strategic tools to show the statement is true. $\diamond$

Remark 1.2.12. When you encounter multiply quantified statements, slow down, pay attention, and strive to understand the adversarial dependencies encoded in the *type* (universal or existential) and *ordering* of quantifiers. Changing word order can completely change meaning. The fundamental definitions of analysis, such as Definition 6.1.6, depend delicately on the precise wording. $\diamond$

Exercises for Section 1.2

Exercise 1.2.1. (★) The phrases "for every," "there exists," and "such that" require practice. Explain why each of the following is anomalous, determine whether or not the statement is true, and give the presumed meaning.

(a) There exists a real number x such that $2 + 2 = 4$.

(b) If $0 < x$ for every x such that $1 < x$, then $0 < 1 < x^2$.

(c) If $y = x^2$ for every $x > 0$, then $y > 0$.

Exercise 1.2.2. Negate the statement, "For every even integer $n \geq 6$, there exist primes p_1 and p_2 such that $n = p_1 + p_2$." (At this writing, the truth of this statement is unknown.)

Discuss the possibilities for resolving this statement by checking examples.

Exercise 1.2.3. (★) Let $[0, 1]$ be the set of real numbers x such that $0 \leq x \leq 1$. Analyze the two-player game, "Who can pick the larger number in $[0, 1]$?" Particularly, determine which player can force a win, and write a multiply-quantified sentence whose truth value expresses which player wins.

Exercise 1.2.4. Let $(0, 1)$ be the set of real numbers x such that $0 < x < 1$. Analyze the two-player game, "Who can pick the larger number in $(0, 1)$?" Particularly, determine which player can force a win, and write a multiply-quantified sentence whose truth value expresses which player wins.

Exercise 1.2.5. Consider the statement, "For every integer x, there exists an integer y greater than x such that for every integer z, if $z < y$ then $z \leq x$."

Describe a two-player adversarial game whose outcome is described by this statement, and analyze this game. Does your analysis change if we replace "integer" by "rational number"?

1.3 Sets

Modern mathematics is built on the concept of a "set," a collection of "elements." These primitive notions will serve in lieu of definitions.

Remark 1.3.1. Abstract sets will be denoted with capital letters, such as X or Y. Elements are normally denoted with lower case letters, such as x and y. We write $x \in X$ to mean "x is an element of (the set) X," and $y \notin X$ for the negation "y is not an element of X."

The set of natural numbers is denoted $\mathbf{N}$. The set of integers is denoted $\mathbf{Z}$ (from the German, *Zahl*, number). The set of rational numbers is denoted $\mathbf{Q}$ (for *quotients*). The set of real numbers is denoted $\mathbf{R}$.

We have $0 \in \mathbf{N}$, $-1 \notin \mathbf{N}$, and $1/2 \notin \mathbf{Z}$.   ◇

Definition 1.3.2. Assume X and Y are sets. We say X is a *subset* of Y, denoted $X \subseteq Y$, if $x \in X$ implies $x \in Y$: Every element of X is an element of Y. With the same meaning, Y is a *superset* of X, denoted $Y \supseteq X$.

Two sets X and Y are *equal* if $X \subseteq Y$ and $X \supseteq Y$, namely if they have exactly the same elements: $x \in X$ if and only if $x \in Y$. If $X \subseteq Y$ and $X \neq Y$, namely, Y contains an element not in X, we say X is a *proper subset* of Y.

Example 1.3.3. We have proper inclusions $\mathbf{N} \subseteq \mathbf{Z} \subseteq \mathbf{Q} \subseteq \mathbf{R}$.   ◇

Example 1.3.4. Let X be a set. For each element x in X, there is a *singleton* set $\{x\}$ contained in X. Take care to distinguish x and $\{x\}$; x is an object, while $\{x\}$ is a "package" containing exactly one object.   ◇

Example 1.3.5. There exists an *empty set* $\varnothing$ containing *no* elements. For all x, the clause $x \in \varnothing$ is false. In particular, for every set X the logical implication "$x \in \varnothing$ implies $x \in X$" is vacuous (has false hypothesis). Consequently, $\varnothing \subseteq X$ is true for all X.   ◇

Proposition 1.3.6. *The empty set is unique.*

Remark 1.3.7. In logic and mathematics, *unique* has its literal meaning: There is precisely one. The phrase "very unique" is nonsensical in mathematics, and best avoided in colloquial English since it implicitly redefines "unique."   ◇

Remark 1.3.8. Sometimes we want to introduce a name inline, such as calling the empty set X. The notation "$X := \varnothing$," read "X, by definition equal to $\varnothing$," signifies defining the symbol on the left to be the object on the right.   ◇

Proof of Proposition 1.3.6. The standard mathematical idiom for proving uniqueness is to assume two objects satisfy stated properties, and prove they are identical. Suppose $\varnothing$ and $\varnothing'$ are empty sets. To prove they are equal, it suffices to prove each is a subset of the other. But $X := \varnothing'$ is a set and $\varnothing$ is empty. By Example 1.3.5, $\varnothing \subseteq \varnothing'$.

Similarly, $X := \varnothing$ is a set and $\varnothing'$ is empty. By Example 1.3.5, $\varnothing' \subseteq \varnothing$.   □

Remark 1.3.9. Since the structure of the final paragraph mirrors the structure of the preceding argument with the roles of $\varnothing$ and $\varnothing'$ exchanged, in practice we often simply write, "Exchanging the roles of $\varnothing$ and $\varnothing'$, [conclusion follows, here $\varnothing' \subseteq \varnothing$]."   ◇

Example 1.3.10. What sets can we construct at this stage? In the spirit of providing examples and motivating the need to introduce additional tools, let's see what we can conjure.

We know about the empty set, which has no elements. It may help to think of the empty set as a box (set) with nothing in it (no elements).

Can we give a set having *one* element? We can, because a set can be an element of another set. This leads us to consider the (non-empty) set $\{\varnothing\}$, the singleton whose unique *element* is $\varnothing$. In our packing analogy, this set is a box (set) containing an empty box (an element, which happens to be a box).

Note carefully: We have both $\varnothing \subseteq \{\varnothing\}$ and $\varnothing \in \{\varnothing\}$, but the first is true for every set while the second is particular to this singleton.

What about a set with *two* elements? The set $\{\varnothing, \varnothing\}$ is equal to $\{\varnothing\}$ by Definition 1.3.2; multiply listed elements count only once. We have little choice but $\{\varnothing, \{\varnothing\}\}$. Inside *this* box we have two things (elements): One empty box (the empty set), and one box *containing an empty box* (the set $\{\varnothing\}$).

Perhaps you can see how we might continue, though our notation quickly becomes unwieldy. ◊

To proceed further in constructing sets, it will help to return to logical operations and express them in terms of sets. Before doing that, we note a contradiction that arises if we are careless about what constitutes a set.

Example 1.3.11 (Russell's paradox). A set X may be an element of itself $(X \in X)$ or not $(X \notin X)$, and for every set, precisely one of these is true. Define $R = \{X : X \notin X\}$, "the set of all sets that are not elements of themselves." Which statement is true, $R \in R$ or its negation $R \notin R$?

If $R \in R$, then R *satisfies the membership criterion* for R, which is to say $R \notin R$. But, if $R \notin R$, then R *fails to satisfy the membership criterion* for R, so $\neg(R \notin R)$, or $R \in R$. We have proven $R \in R$ if and only if $R \notin R$. This is a logical contradiction, see Remark 1.1.16. ◊

Remark 1.3.12. The problem was "the set of all sets" inside which we could speak of "the set of all sets that are not elements of themselves." To avoid Russell's paradox, mathematicians always restrict attention, at least implicitly, to sets that are contained in a fixed set $\mathcal{U}$, called a *universe*.

Specific subsets of $\mathcal{U}$ are conveniently described using *set-builder notation*, in which elements are selected according to logical conditions formally known as a *predicates*. The expression $\{x \text{ in } \mathcal{U} : P(x)\}$ is read "the set of all x in $\mathcal{U}$ such that $P(x)$."

If $\mathcal{U}$ is a population whose elements are individuals, a subset X of $\mathcal{U}$ is a club or organization, and the predicate defining X is a membership card. We screen individuals x for membership in X by checking whether or not x carries the membership card for X, namely whether $P(x)$ is true. ◊

Example 1.3.13. The set of *even integers* is $\{n$ in $\mathbf{Z} : n = 2k$ for some k in $\mathbf{Z}\}$. The set of *odd integers* is $\{n$ in $\mathbf{Z} : n = 2k + 1$ for some k in $\mathbf{Z}\}$. ◊

Remark 1.3.14. For brevity, we sometimes write, for example, the set of even integers as $\{2n : n \in \mathbf{Z}\}$, read "the set of $2n$ such that n is an integer." This way of writing a set is convenient, and the meaning is generally clear, but it isn't technically proper grammar. To define a set formally, first give the universe, then specify the predicate. ◊

Sets and Logic

Remark 1.3.15. Assume $\mathcal{U}$ is a universe, and X, Y are subsets of $\mathcal{U}$. The predicates $x \in X$ and $x \in Y$ may be viewed as conditions P and Q on elements of $\mathcal{U}$. By definition, the logical implication "$x \in X$ implies $x \in Y$" corresponds to "being a subset": $X \subseteq Y$. ◊

Logical negation, disjunction (or), and conjunction (and) similarly have natural interpretations in terms of X and Y.

Definition 1.3.16. Assume $\mathcal{U}$ is a universe, X and Y subsets of $\mathcal{U}$.
The *complement* of X is the set $X^c = \{x$ in $\mathcal{U} : x \notin X\}$.
The *union* of X and Y is the set $X \cup Y = \{x$ in $\mathcal{U} : x \in X$ or $x \in Y\}$.
The *intersection* of X and Y is $X \cap Y = \{x$ in $\mathcal{U} : x \in X$ and $x \in Y\}$.

Remark 1.3.17. A *Venn diagram* represents subsets of a universe $\mathcal{U}$ pictorially. The universe is depicted as a rectangle, and subsets are disks or, if necessary, more complicated shapes. The complement of X, or the union and intersection of two sets X and Y, might be drawn as indicated in Figure 1.1. ◊

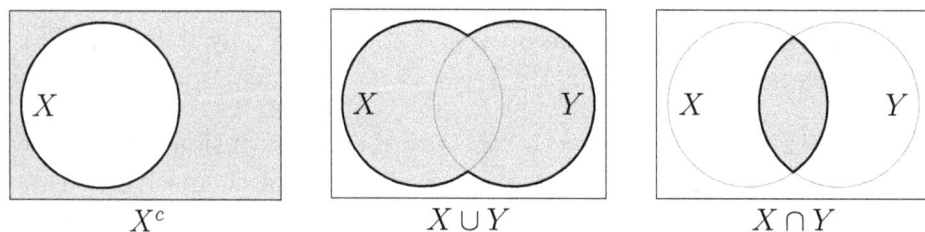

$X^c \qquad\qquad X \cup Y \qquad\qquad X \cap Y$

FIGURE 1.1
The complement of a set; union and intersection of two sets.

Definition 1.3.18. Assume X and Y are subsets of $\mathcal{U}$. Their *difference* is defined to be $X \setminus Y = \{x$ in $X : x \notin Y\} = X \cap Y^c$.

Definition 1.3.19. We say X and Y are *disjoint* if $X \cap Y = \varnothing$, namely if X and Y have no elements in common.

Families of Sets

Definition 1.3.20. Let $\mathscr{I}$ (script-I) be a non-empty set, possibly containing infinitely many elements, and suppose we have a collection of sets X_i, one for each i in $\mathscr{I}$, with each X_i contained in some fixed universe $\mathcal{U}$. We denote the collection of sets by $\{X_i\}_{i \in \mathscr{I}}$, and say the collection is *indexed by* $\mathscr{I}$.

Definition 1.3.21. Assume $\{X_i\}_{i \in \mathscr{I}}$ is a collection of subsets of $\mathcal{U}$. We define the *union* and *intersection* of the collection to be the following subsets of $\mathcal{U}$:

$$\bigcup_{i \in \mathscr{I}} X_i = \{x \text{ in } \mathcal{U} : x \in X_i \text{ for } \textit{some } i \text{ in } \mathscr{I}\},$$

$$\bigcap_{i \in \mathscr{I}} X_i = \{x \text{ in } \mathcal{U} : x \in X_i \text{ for } \textit{all } i \text{ in } \mathscr{I}\}.$$

Remark 1.3.22. If $\mathscr{I} = \{0, 1, 2, \ldots, n-1\}$, a collection of sets indexed by $\mathscr{I}$ is often denoted $\{X_i\}_{i=0}^{n-1}$. The union and intersection are denoted

$$\bigcup_{i \in \mathscr{I}} X_i = \bigcup_{i=0}^{n-1} X_i, \qquad\qquad \bigcap_{i \in \mathscr{I}} X_i = \bigcap_{i=0}^{n-1} X_i.$$

By convention, the collection, union, and intersection are empty if $n = 0$. $\quad\diamond$

Theorem 1.3.23 (The complement laws and distributive laws). *Assume $\{X_i\}_{i \in \mathscr{I}}$ is a collection of subsets of some set $\mathcal{U}$. If $Y \subseteq \mathcal{U}$, then*

(i) $Y \setminus (\bigcup_{i \in \mathscr{I}} X_i) = \bigcap_{i \in \mathscr{I}} (Y \setminus X_i)$ *and* $Y \setminus (\bigcap_{i \in \mathscr{I}} X_i) = \bigcup_{i \in \mathscr{I}} (Y \setminus X_i)$.

(ii) $(\bigcup_{i \in \mathscr{I}} X_i) \cap Y = \bigcup_{i \in \mathscr{I}} (X_i \cap Y)$ *and* $(\bigcap_{i \in \mathscr{I}} X_i) \cup Y = \bigcap_{i \in \mathscr{I}} (X_i \cup Y)$.

Proof. See Exercise 1.3.5. $\qquad\qquad\qquad\qquad\qquad\qquad\qquad\qquad\qquad \square$

Definition 1.3.24. Assume $\{X_k\}_{k \in \mathbf{N}}$ is a collection of sets. We say $\{X_k\}_{k \in \mathbf{N}}$ is *nested outward* if $X_k \subseteq X_{k+1}$ for every k in $\mathbf{N}$.

We say $\{X_k\}_{k \in \mathbf{N}}$ is *nested inward* if $X_k \supseteq X_{k+1}$ for every k in $\mathbf{N}$.

Definition 1.3.25. The *power set* of X, $\mathscr{P}(X)$, is the set of all subsets of X.

Example 1.3.26. If $X = \{0, 1\}$ has two elements, the power set $\mathscr{P}(X)$ has four elements: $\mathscr{P}(X) = \{\varnothing, \{0\}, \{1\}, X\}$. The empty set and X itself are always subsets of X, a.k.a., *elements* of $\mathscr{P}(X)$, so a power set is never empty. For example, $\mathscr{P}(\varnothing) = \{\varnothing\}$ is the singleton we met earlier. $\quad\diamond$

Definition 1.3.27. If X is a set, a *partition* of X is a collection of *non-empty* sets $\{X_i\}_{i \in \mathscr{I}}$ indexed by a set $\mathscr{I}$ such that $X_i \cap X_j = \varnothing$ if $i \neq j$, and $X = \bigcup_{i \in \mathscr{I}} X_i$.

Remark 1.3.28. In words, a partition of a set is a way of "dividing X into disjoint, non-empty subsets": Each element of X is contained in *precisely one* of the X_i.

A partition of X is a particular type of *subset of the power set* $\mathscr{P}(X)$. $\diamond$

Example 1.3.29. The sets

$$2\mathbf{Z} := \{x \text{ in } \mathbf{Z} : x = 2n \text{ for some integer } n\},$$
$$2\mathbf{Z} + 1 := \{x \text{ in } \mathbf{Z} : x = 2n + 1 \text{ for some integer } n\}$$

of even and odd integers are a partition of $\mathbf{Z}$: If we put $X_0 := 2\mathbf{Z}$ and $X_1 := 2\mathbf{Z} + 1$, then every integer x is even ($x \in X_0$) or odd ($x \in X_1$), and no integer is both: X_0 and X_1 are disjoint. The index set here is $\mathscr{I} = \{0, 1\}$. $\Diamond$

Example 1.3.30. For each integer n, define $X_n = \{x \text{ in } \mathbf{R} : n \le x < n+1\}$. The sets $\{X_n\}_{n \in \mathbf{Z}}$ are a partition of $\mathbf{R}$: Every real number x may be written uniquely as an integer n plus a real number x' such that $0 \le x' < 1$, see Corollary 4.3.2. The index set here is $\mathscr{I} = \mathbf{Z}$. $\Diamond$

Exercises for Section 1.3

Exercise 1.3.1. ($\bigstar$) Assume X and Y are subsets of $\mathcal{U}$. Prove $X \subseteq Y$ if and only if $Y^c \subseteq X^c$, and illustrate with a Venn diagram. How is this result related to contrapositives?

Exercise 1.3.2. Let $\mathcal{U}$ be a universe. Prove that if $A \subseteq \mathcal{U}$, then $A \subseteq A^c = \mathcal{U} \setminus A$ if and only if $A = \varnothing$.

Exercise 1.3.3. ($\bigstar$) Let P and Q denote arbitrary statements. Use truth tables to prove the indicated pairs of statements are logically equivalent.

(a) "$\neg(P \text{ or } Q)$" and "$\neg P \text{ and } \neg Q$."

(b) "$\neg(P \text{ and } Q)$" and "$\neg P \text{ or } \neg Q$."

(c) Let X and Y be arbitrary subsets of a universe $\mathcal{U}$. Prove the *complement laws* $(X \cup Y)^c = X^c \cap Y^c$ and $(X \cap Y)^c = X^c \cup Y^c$.

Exercise 1.3.4. ($\bigstar$) Assume X, Y and Z are subsets of a universe $\mathcal{U}$. Prove the *distributive laws*

$$(X \cup Y) \cap Z = (X \cap Z) \cup (Y \cap Z),$$
$$(X \cap Y) \cup Z = (X \cup Z) \cap (Y \cup Z).$$

Exercise 1.3.5. Prove Theorem 1.3.23.

Exercise 1.3.6.(★) If $X = \{0, 1, 2\}$, list the elements of the power set $\mathscr{P}(X)$.

Exercise 1.3.7. Assume n is a natural number and X is a set of n elements. How many elements does the power set $\mathscr{P}(X)$ have? Argue as formally as you can.

1.4   Mappings and Relations

"Mappings" and "binary relations" can be defined in a couple of paragraphs, but they underlie this entire book, as well as large swaths of mathematics and computer science. Ramifications of these brief definitions take years to explore.

Definition 1.4.1. Assume X and Y are sets, not necessarily different. If $x \in X$ and $y \in Y$, the *ordered pair* (x, y) is a set such that for all x' in X and all y' in Y, we have $(x, y) = (x', y')$ if and only if $x = x'$ and $y = y'$.

The set of all ordered pairs (x, y) such that $x \in X$ and $y \in Y$ is the *ordered product* $X \times Y$, often read "X cross Y."

Remark 1.4.2. One construction is $(x, y) = \{\{x\}, \{x, y\}\}$, see Exercise 1.4.5.   ◇

Example 1.4.3. If $\mathbf{R}$ is the set of real numbers, the ordered product $\mathbf{R} \times \mathbf{R}$ is denoted $\mathbf{R}^2$, read "are two," and is called the *plane*. The real numbers x and y are the *rectangular coordinates* of the point (x, y). The set of (x, y) such that x and y are positive is the (*open*) *first quadrant*.   ◇

Definition 1.4.4. Assume X and Y are sets. A *mapping* $f : X \to Y$ (read "f from X to Y") is a subset $f \subseteq X \times Y$ with the following property: For every x in X, there exists a unique y in Y such that $(x, y) \in f$.

If $(x, y) \in f$, we write $y = f(x)$, and call y the *value* of f at *input* x.

Remark 1.4.5. The ordered product $X \times Y$ may be viewed schematically as a rectangle with base X and side Y. A mapping $f : X \to Y$ is then a set satisfying the vertical line test, see Figure 1.2.   ◇

Definition 1.4.6. Let $\mathbf{N}$ denote the set of natural numbers, and let X be a non-empty set. A *sequence* in X is a mapping $\mathbf{a} : \mathbf{N} \to X$. The value $a_k := \mathbf{a}(k)$ is called the *kth term*. The *set of terms* is $\{x$ in $X : x = a_k$ for some $k\}$.

Remark 1.4.7. Sequences are among the most important concepts in this book. We usually view a sequence as an *ordered list* of elements of X, in which case we denote the sequence by $(a_k)_{k \in \mathbf{N}}$, and its set of terms by $\{a_k\}_{k \in \mathbf{N}}$.   ◇

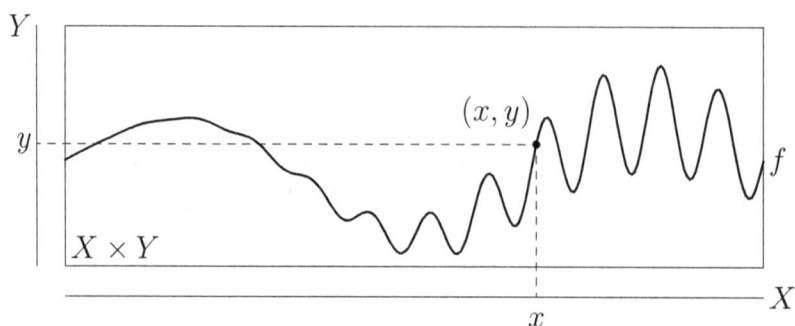

FIGURE 1.2
A mapping $f : X \to Y$.

Definition 1.4.8. Assume X is a non-empty set. A *binary relation* on X is a subset $R \subseteq X \times X$. If x and x' are arbitrary elements of X, we write "$x \, R \, x'$" and say "x is R-related to x'," if $(x, x') \in R$.

Example 1.4.9. In the set $X = \mathbf{N}$ of natural numbers, equality is a binary relation R comprising ordered pairs (x, x') such that $x = x'$. Inequality, the logical negation, is the complement $(\mathbf{N} \times \mathbf{N}) \setminus R$ of pairs such that $x \neq x'$.

   Less-than ($<$) and less-than-or-equal ($\leq$) are binary relations on $\mathbf{N}$. $\diamond$

Equivalence Relations and Partitions

Definition 1.4.10. Assume X is a non-empty set and R a binary relation on X. We say R is:

 (i) *Reflexive* if for all x in X, $x \, R \, x$;

 (ii) *Symmetric* if for all x, y in X, $x \, R \, y$ implies $y \, R \, x$;

 (iii) *Transitive* if for all x, y, z in X, $x \, R \, y$ and $y \, R \, z$ imply $x \, R \, z$.

Remark 1.4.11. To personify, imagine X is a set of people, and $x \, R \, y$ means "x is a friend of y." Reflexivity means everyone is their own friend. Symmetry means all friendships are mutual. Transitivity means "a friend of a friend is a friend." $\diamond$

Definition 1.4.12. Assume X is a non-empty set and R a binary relation on X. If R is reflexive, symmetric, and transitive, we say R is an *equivalence relation* on X.

   If R is an equivalence relation on X, the set $[x] := \{x' \text{ in } X : x \, R \, x'\}$ is the *equivalence class* of x, and an element of $[x]$ is called a *representative*.

Remark 1.4.13. In a set X where "is a friend of" is an equivalence relation, each person belongs to a "clique" of mutual friends. No one outside a clique is a friend of anyone in that clique. The set X is partitioned into cliques. $\diamond$

Proposition 1.4.14. *If X is a non-empty set and $\equiv$ an equivalence relation on X, then the distinct equivalence classes are a partition of X.*

Proof. By reflexivity, $x \in [x]$ for all x, so every equivalence class is non-empty, and the union of all equivalence classes is X. It suffices to check that distinct equivalence classes are disjoint: For all x and z in X, $[x] \neq [z]$ as sets implies $[x] \cap [z] = \varnothing$.

We prove the contrapositive: If $y \in [x] \cap [z]$ for some y, then $[x] = [z]$. Our friendship example guides the proof: If y is a friend of both x and z, then x and z are friends by transitivity through y, so every friend of x is a friend of z by transitivity through x, and similarly every friend of z is a friend of x.

Formally, $y \in [x]$ means $x \equiv y$. Similarly, $y \in [z]$ means $z \equiv y$, so $y \equiv z$ by symmetry. Since $x \equiv y$ and $y \equiv z$, transitivity implies $x \equiv z$, so $z \in [x]$. Now assume $z' \in [z]$. By assumption, $z \equiv z'$. Transitivity through z implies $x \equiv z'$, that is, $z' \in [x]$. We have shown $[z] \subseteq [x]$. Reversing the roles of x and z, $[x] \subseteq [z]$. $\qquad\square$

Remark 1.4.15. Conversely, by Exercise 1.4.8, a partition $\{X_i\}_{i \in \mathscr{I}}$ of a set X defines an equivalence relation whose equivalence classes are precisely the sets of the partition. Conceptually, equivalence relations on X and partitions of X are "the same thing." $\qquad\qquad\diamond$

Induced Mappings

Mathematicians use equivalence relations to "ignore 'irrelevant' distinctions," with "irrelevant" depending on context.

Example 1.4.16. Suppose a room has a single on-off light switch. When we left, the lights were off. Later when we arrived, the lights were on. How many times was the switch flipped while we were gone? There is no way to know, but we *can* be sure it was an odd number.

To a mathematician in this situation, natural numbers (counting flips of the light switch) come in two types, state-preserving (on remains on, off remains off) and state-reversing (on becomes off, off becomes on). Every number is one or the other, and no number is both: These sets partition the natural numbers. By Remark 1.4.15, they define an equivalence relation. When a mathematician "looks at" the natural numbers through the "glasses" of this relation, they see a set with two elements: state-preserving and state-reversing, or $\{[0], [1]\}$. $\quad\diamond$

Two definitions formalize this example: the quotient of a set by an equivalence relation, and induced mappings.

Definition 1.4.17. Assume X is a non-empty set and R an equivalence relation on X. The *quotient* of X by R is the partition into equivalence classes, viewed as a set: $X/R := \{X_i\}_{i \in \mathscr{I}}$.

Definition 1.4.18. Assume X is a non-empty set, R an equivalence relation on X, and $f : X \to Y$ a mapping. We say f is *constant on equivalence classes of R*, or *well-defined mod R*, if for all x and x' in X, $x \, R \, x'$ implies $f(x) = f(x')$.

If f is well-defined mod R, we define the *induced mapping* $\overline{f} : X/R \to Y$ by $\overline{f}([x]) = f(x)$ for all x in X, and we say f *factors through the quotient* X/R.

Example 1.4.19. In the light switch example, $X = \mathbf{N}$ is the set of natural numbers, counting times the switch might have been flipped. The relation R distinguishes two types of natural number: the equivalence class $[0] = 2\mathbf{N}$ of even numbers, and the equivalence class $[1] = 2\mathbf{N} + 1$ of odd numbers.

Now let $Y = \{F, T\}$ be the set of truth values, representing state-preserving (F) and state-reversing (T), and let f be the mapping $f : \mathbf{N} \to Y$ defined by $f(n) = F$ if n flips preserve the state, and $f(n) = T$ if n flips reverse the state. Because f is constant on equivalence classes of R, there is an induced mapping $\overline{f} : X/R \to Y$. The induced mapping is defined by $\overline{f}([0]) = F$ and $\overline{f}([1]) = T$. Existence of an induced mapping amounts to the claim that although we cannot tell how many times the switch was flipped, we *can* tell whether it was flipped an even or odd number of times.   ◇

Example 1.4.20. A 52-card deck of playing cards has four "suits" and 13 "denominations" in each suit. The suits form a set $S = \{\clubsuit, \diamondsuit, \heartsuit, \spadesuit\}$: clubs, diamonds, hearts, and spades. Clubs and spades are *black*, diamonds and hearts are *red*. The denominations form a set $D = \{A, 2, 3, 4, 5, 6, 7, 8, 9, 10, J, Q, K\}$: ace, two through ten, jack, queen, and king. The entire deck is the ordered product $C = D \times S$. Typical elements (cards) are the *ace of diamonds* $(A, \diamondsuit)$ and the *jack of clubs* $(J, \clubsuit)$.

Let $Y = \{\text{black}, \text{red}\}$ be the set of colors, and let $f : C \to Y$ be the mapping that sends a card to its color. Before continuing with the next paragraph, take a minute to ponder and answer: If R is the relation "same suit as," is f well-defined mod R? And if R' is the relation "same denomination as," is f well-defined mod R'? If the answer is "yes" in either case, describe the induced mapping.

Intuitively we're asking: If we know a card's suit, do we know its color? And, if we know a card's denomination, do we know its color?

Since each suit has a single color, f *is* well-defined mod R. The set of equivalence classes of R may be viewed as S, the set of suits. The induced mapping $\overline{f} : S \to Y$ sends each suit to its color.

By contrast, the denomination of a card does not determine its color; f is *not* well-defined mod R', and there is no induced mapping. More explicitly, the hallmark of a mapping is having a unique value for each input. But while, for example, $[(4, \diamondsuit)]' = [(4, \spadesuit)]'$ as R'-equivalence classes, at least one of the following equalities must be false:

$$\text{red} = f\big((4, \diamondsuit)\big) = \overline{f}\big([(4, \diamondsuit)]'\big) = \overline{f}\big([(4, \spadesuit)]'\big) = f\big((4, \spadesuit)\big) = \text{black}.$$

The first and last are true, so either the defining condition of an induced mapping fails (the second or fourth equality is false) or $\overline{f}$ is not well-defined (takes two or more values at a single input). ◊

Exercises for Section 1.4

Exercise 1.4.1.(★) Assume $X = \{0, 1, 2\}$.

(a) How many distinct mappings $f : X \to X$ are there?

(b) How many mappings $f : X \to X$ take three distinct values? List all of them. Suggestion: Such a mapping is uniquely determined by its values $\left(f(0), f(1), f(2)\right)$.

Exercise 1.4.2.(★) Assume X is a set of 2 elements and Y a set of 5 elements.

(a) How many distinct mappings $f : X \to Y$ are there? How many of these take distinct values at the two inputs?

(b) How many partitions of Y are there into two subsets?

Exercise 1.4.3. Assume X is a set of 3 elements and Y a set of 5 elements.

(a) How many distinct mappings $f : X \to Y$ are there? How many of these take three distinct values?

(b) How many partitions of Y are there into three subsets?

Exercise 1.4.4. Suppose k and n are positive integers, X is a set of k elements, and Y is a set of n elements.

(a) How many distinct mappings $f : X \to Y$ are there?

(b) If $k \leq n$, how many mappings $f : X \to Y$ take k distinct values?

Exercise 1.4.5.(★) Assume X and Y are sets. Prove that if x, x' are elements of X and y, y' are elements of Y, then $\left\{\{x\}, \{x, y\}\right\} = \left\{\{x'\}, \{x', y'\}\right\}$ if and only if $x = x'$ and $y = y'$. (See Remark 1.4.2.)

Exercise 1.4.6.(H) Let $Y = \{T, F\}$ be the set of *truth values* (true and false). A *truth function on* X is a mapping $f : X \to Y$.

Prove that if X is a set, subsets of X correspond to truth functions on X. Conclude that binary relations on X correspond to truth functions on $X \times X$.

Exercise 1.4.7. (A) Let $X = \mathbf{Z}$ be the set of integers. In the table, five binary relations R on $\mathbf{Z}$ are given by the condition $m\ R\ n$. Determine with justification whether each R is reflexive, symmetric, and/or transitive.

	$m \neq n$	$m < n$	$m \leq n$	$0 < mn$	$0 \leq mn$
Reflexive					
Symmetric					
Transitive					

Exercise 1.4.8. Conversely to Proposition 1.4.14, suppose $\{X_i\}_{i \in \mathscr{I}}$ is a partition of X, and define a binary relation $\sim$ on X by declaring $x \sim y$ if and only if there exists an index i such that $\{x, y\} \subseteq X_i$. Prove $\sim$ is an equivalence relation, and the equivalence classes of $\sim$ are the sets X_i.

Exercise 1.4.9. (★) Let C denote a deck of playing cards, S the set of suits, and R the "same color" equivalence relation on C.

(a) Describe the equivalence classes of R in S.

(b) Is the mapping $f : C \to S$ that sends a card to its suit well-defined mod R?

Exercise 1.4.10. Assume $f : X \to Y$ is a mapping and $\{X_i\}_{i \in \mathscr{I}}$ a partition. If x and x' are elements of X, write $x \equiv x'$ if there exists an i in $\mathscr{I}$ such that x and x' are both elements of X_i.

Prove that "$x \equiv x'$ implies $f(x) = f(x')$" if and only if "there exists a mapping $F : \mathscr{I} \to Y$ defined by $F(i) = f(x)$ for all x in X_i."

Exercise 1.4.11. Assume X is a set and R a binary relation on X. Someone claims that in defining an equivalence relation, reflexivity is redundant: "Suppose x and y are arbitrary elements of X. If $x\ R\ y$, then $y\ R\ x$ by symmetry. But now taking $x = z$, transitivity implies $x\ R\ x$." Is this correct? If not, where is the error?

Exercise 1.4.12. (★) Let $X = \mathbf{Z}$ be the set of integers.

(a) Let R be the binary relation $x\ R\ x'$ if and only if $x - x'$ is even. Determine whether R is an equivalence relation. If so, determine its equivalence classes.

(b) Let R be the binary relation $x\ R\ x'$ if and only if $x - x'$ is odd. Determine whether R is an equivalence relation. If so, determine its equivalence classes.

Exercise 1.4.13. In this exercise, assume that for every integer n, there exist *unique* integers d and r such that $n = 3d + r$ and $0 \leq r < 3$.

(a) Define a binary relation $\equiv_3$ on $\mathbf{Z}$ by $n \equiv_3 n'$ if and only if $n - n' = 3d$ for some integer d. Prove $\equiv_3$ is an equivalence relation.

(b) Describe the equivalence classes of $\equiv_3$.

(c) If $m \equiv_3 m'$ and $n \equiv_3 n'$, prove $m + n \equiv_3 m' + n'$ and $m \cdot n \equiv_3 m' \cdot n'$.

(d) By part (c), *addition of equivalence classes is well-defined*: If we add two classes by picking representatives, adding, and taking the equivalence class of the sum, we get the same value regardless of representatives. It makes sense to write $[r] + [s] = [r + s]$ for all r and s in $\{0, 1, 2\}$.

Similarly, $[r] \cdot [s] = [r \cdot s]$ for all r and s in $\{0, 1, 2\}$.

Make a 3×3 "addition table" and "multiplication table" for equivalence classes.

2

Natural Numbers and Induction

The set $\mathbf{N}$ of natural numbers is a mathematician's prototype of an infinite list. There is an "initial element" 0 and a concept of "successorship." Every natural number arises by starting with 0 and taking successors, and every natural number except 0 is the successor of a unique natural number. We denote natural numbers with their familiar Hindu-Arabic symbols, $\mathbf{N} = \{0, 1, 2, 3, 4, 5, \dots\}$.

2.1 Induction and Recursion

We first give axioms for the natural numbers. To avoid "lengthy start-up," however, for now we'll assume the familiar structure of the natural numbers, including positional notation, addition, multiplication, and ordering.

Definition 2.1.1. There exists a set $\mathbf{N}$, whose elements are called *natural numbers*, with an element 0 and *successorship* mapping $S : \mathbf{N} \to \mathbf{N}$ satisfying the following conditions.

(i) 0 is not the successor of any natural number.

(ii) Every natural number other than 0 is the successor of a unique natural number.

(iii) (The inductive property) If $I \subseteq \mathbf{N}$ is a subset such that $0 \in I$, and for every natural number k, $k \in I$ implies $k + 1 \in I$, then $I = \mathbf{N}$.

The inductive property lets us prove infinitely many statements that are logically structured in a list indexed by the natural numbers. This technique is one of our primary tools in this book.

Theorem 2.1.2 (Mathematical induction). *Assume $P(n)$ is a collection of statements, one for each natural number n. If $P(0)$ is true, and if $P(k)$ implies $P(k+1)$ for each k, then $P(n)$ is true for all n.*

Proof. Put $I = \{n \text{ in } \mathbf{N} : P(n) \text{ is true}\}$. By hypothesis, $0 \in I$ (the statement $P(0)$ is true) and if $k \in I$, then $k + 1 \in I$ (for each natural number k, $P(k)$ implies $P(k+1)$). By the inductive property, $I = \mathbf{N}$, namely, $P(n)$ is true for every natural number n. $\qquad\square$

DOI: 10.1201/9781003601357-2

Remark 2.1.3. In Theorem 2.1.2, $P(0)$ is called the *base case* and "$P(k)$ implies $P(k+1)$ for every natural number k" is the *inductive step*.

It is sometimes convenient to take a positive index n_0 for the base case. In this event, if $P(n_0)$ is true, and $P(k)$ implies $P(k+1)$ for k greater than or equal to n_0, then $P(n)$ is true for all n greater than or equal to n_0. ◇

Theorem 2.1.4 (Well-ordering). *If $A \subseteq \mathbf{N}$ is non-empty, then there exists a "smallest element" of A, namely, a natural number n_0 such that every element of A occurs in the chain of successorship starting at n_0.*

Proof. We will prove the contrapositive: If $A \subseteq \mathbf{N}$ has no smallest element, then A is empty. For each natural number n, let $P(n)$ be the statement "If m is a natural number such that $m \leq n$, then $m \notin A$."

The base case $P(0)$ is true, since if $0 \in A$ then 0 is the smallest element of A by the inductive property. Assume inductively that $P(k)$ is true for some natural number k, namely, A contains no natural number m such that $m \leq k$. If $k + 1 \in A$, then $k + 1$ is the smallest element of A. Since A has no smallest element by hypothesis, $k + 1 \notin A$. Consequently, A contains no natural number m such that $m \leq k + 1$; thus $P(k+1)$ is true. Since the base case $P(0)$ is true and the inductive step $P(k)$ implies $P(k+1)$ for every k is true, $P(n)$ is true for all n. Particularly, $n \notin A$ for all n, so A is empty. □

Before looking at "concrete" examples of induction, we'll give recursive definitions of summation and exponentiation. These will give us material to work with inductively.

Definition 2.1.5. If $(a_k)_{k \in \mathbf{N}}$ is a sequence of numbers (integers, rationals, reals, ...), the sequence $(s_n)_{n \in \mathbf{N}}$ of *partial sums* is defined recursively by

$$s_0 = 0, \qquad s_{n+1} = s_n + a_n \quad \text{if } n \geq 0.$$

Example 2.1.6. Expanding this definition if $n = 3$ gives

$$\begin{aligned}
s_3 &= s_2 + a_2 \\
&= (s_1 + a_1) + a_2 \\
&= (s_0 + a_0) + a_1 + a_2 \\
&= a_0 + a_1 + a_2,
\end{aligned}$$

the sum of the terms a_k such that $0 \leq k < n = 3$. ◇

Remark 2.1.7. Partial sums arise frequently enough to get special notation:

$$s_n = \sum_{k=0}^{n-1} a_k = a_0 + a_1 + a_2 + \cdots + a_{n-1},$$

read, "the sum from $k = 0$ to $n - 1$ of a sub k." Note that the sum is 0 if $n = 0$. The recursion relation is written

$$\sum_{k=0}^{n} a_k = \left[\sum_{k=0}^{n-1} a_k \right] + a_n,$$

which says the sum over $0 \leq k \leq n$ is obtained by adding the nth term to the sum over $0 \leq k \leq n - 1$. The *summation sign* is a stylized Σ (Sigma), for "sum."   ◇

Definition 2.1.8. If m is a natural number, define $m^0 = 1$. (Note: We *define* $0^0 = 1$.) If n is a natural number, recursively define

$$m^{n+1} = m^n \cdot m \qquad \text{for all } m \text{ in } \mathbf{N}.$$

We read m^n as "m to the nth power" and call this operation *exponentiation* of natural numbers.

Example 2.1.9. Expanding the definition if $m = 2$ and $n = 3$, we have

$$2^3 = 2^2 \cdot 2 = 2^1 \cdot 2 \cdot 2 = 2^0 \cdot 2 \cdot 2 \cdot 2 = 1 \cdot 2 \cdot 2 \cdot 2 = 2 \cdot 2 \cdot 2 = 8.$$

After $2^0 = 1$, the next twenty powers of 2 are:

$2^1 = 2$	$2^6 = 64$	$2^{11} = 2048$	$2^{16} = 65,536$
$2^2 = 4$	$2^7 = 128$	$2^{12} = 4096$	$2^{17} = 131,072$
$2^3 = 8$	$2^8 = 256$	$2^{13} = 8192$	$2^{18} = 262,144$
$2^4 = 16$	$2^9 = 512$	$2^{14} = 16,384$	$2^{19} = 524,288$
$2^5 = 32$	$2^{10} = 1024$	$2^{15} = 32,768$	$2^{20} = 1,048,576.$

Note that $2^{10} = 1024 \approx 1000 = 10^3$ and $2^{20} = 1,048,576 \approx 10^6$.   ◇

A useful but less common recursive operation related to both sums and powers is general products.

Definition 2.1.10. If $(a_k)_{k \in \mathbf{N}}$ is a sequence of numbers, the sequence $(p_n)_{n \in \mathbf{N}}$ of *partial products* is defined recursively by

$$p_0 = 1, \qquad p_{n+1} = p_n \cdot a_n \quad \text{if } n \geq 0.$$

Remark 2.1.11. As with sums, partial products get special notation:

$$p_n = \prod_{k=0}^{n-1} a_k = a_0 \cdot a_1 \cdot a_2 \cdots a_{n-1},$$

with the product equal to 1 if $n = 0$. The *product sign* is a stylized Π (Pi), for "product."   ◇

Example 2.1.12. If m is a natural number, then $m^n = \prod\limits_{k=0}^{n-1} m$. ◇

We now turn to examples of mathematical induction.

Example 2.1.13. Suppose someone tells us the sum of the first n positive odd integers is equal to n^2. What basis do we have for believing this claim?

As a start, we might verify instances by hand with the implicit hope of finding a counterexample. For example, $1 + 3 + 5 = 9 = 3^2$, so the claim is true when $n = 3$. Perhaps skeptical, we add the first ten odd integers, or the first twenty, each time verifying the claim. Perhaps we are starting to believe.

Logically, however, testing special cases leaves us no closer to mathematical certainty. Finding a single counterexample would prove the claim false, but no matter how many cases we verify there remain infinitely many unverified cases.

Induction allows us to resolve such questions with a finite proof. The idea is to structure the statement "For every natural number n, the sum of the first n odd positive integers is equal to n^2" as an infinite list of statements indexed by n. Here, we let $P(n)$ be the statement

$$\sum_{j=0}^{n-1} (2j + 1) = 1 + 3 + 5 + \cdots + (2n - 1) = n^2.$$

To say $P(100)$ is true, for example, means the sum of the first hundred odd integers is equal to $10,000 = 100^2$.

The original statement may be rephrased "For every natural number n, $P(n)$ is true." This single statement P encapsulates the infinite list of statements: $P(0)$ is true, $P(1)$ is true, $P(2)$ is true, etc. To establish the truth of P by mathematical induction, it suffices to prove $P(0)$, and to prove $P(k)$ implies $P(k + 1)$ for all k in $\mathbf{N}$.

The base case $P(0)$ reads "$0 = 0^2$," which is true.

Next, "assume inductively" that $P(k)$ is true for some (fixed but arbitrary) k. By the recursive definition of summation, the sum of the first $k + 1$ odd positive integers is equal to the sum of the first k plus the $(k + 1)$th. By hypothesis, the sum of the first k is equal to k^2. We therefore deduce

$$\underbrace{1 + 3 + 5 + \cdots + (2k - 1)}_{= \, k^2 \text{ by } P(k)} + (2k + 1) = k^2 + (2k + 1) = (k + 1)^2$$

by algebra. This equation says the sum of the first $(k + 1)$ odd positive integers is equal to $(k + 1)^2$. By assuming $P(k)$, we proved $P(k + 1)$.

To summarize, the base case $P(0)$ is true, and the inductive step, "$P(k)$ implies $P(k + 1)$ for every k" is true. By Theorem 2.1.2, $P(n)$ is true for all n. ◇

Remark 2.1.14. The preceding discussion contains explanation needed mostly because it was a first example. Here is a "final draft" argument.

For each natural number n, let $P(n)$ be the statement

$$\sum_{j=0}^{n-1}(2j+1) = n^2.$$

We will prove by mathematical induction that $P(n)$ is true for every n in $\mathbf{N}$.

The base case $P(0)$ reads $0 = 0^2$, which is true.

Assume inductively that $P(k)$ is true for some natural number k. By the recursive definition of summation, the inductive hypothesis, and algebra,

$$\sum_{j=0}^{k}(2j+1) = \left[\sum_{j=0}^{k-1}(2j+1)\right] + (2k+1)$$
$$= k^2 + (2k+1) = (k+1)^2;$$

that is, $P(k+1)$ follows from $P(k)$. We have proven $P(0)$ is true, and $P(k)$ implies $P(k+1)$ for all k. By mathematical induction, $P(n)$ is true for all n. $\diamond$

Remark 2.1.15. Our use of n or k to denote an arbitrary natural number in an inductive proof signifies a subtle but important conceptual distinction. In this book, $P(n)$ connotes the general statement of an inductive list, whose truth value is to be established. By contrast, $P(k)$ connotes a general statement that is "inductively true": We *grant for the sake of argument* that $P(k)$ is true for some (particular but arbitrary) k, and try to deduce $P(k+1)$.

To emphasize, in an inductive proof we never assume $P(k)$ is true without qualification. To do so (for arbitrary k) would be to assume the conclusion we wish to establish, namely that $P(k)$ is true for every k. $\diamond$

Induction can be used to establish families of statements that are not formulas, such as inequalities.

Example 2.1.16. Prove there exists a natural number n_0 such that $2^{n+2} \leq 3^n$ if $n \geq n_0$. For reasons to be explained, we start with the inductive step.

Assume inductively that $P(k)$ is true, namely, that $2^{k+2} \leq 3^k$, for some k. By definition of powers, properties of inequalities, and the inductive hypothesis, we have

$$2^{(k+1)+2} = 2 \cdot 2^{k+2} \leq 2 \cdot 3^k \leq 3 \cdot 3^k = 3^{k+1}.$$

That is, $P(k)$ implies $P(k+1)$ for all k.

There is a curious and important feature here: Although the inductive step "$P(k)$ implies $P(k+1)$ for all k" is true, the statements $P(n)$ themselves are initially false; the first few implications are vacuous. This highlights the need to prove both a base case and the inductive step, and emphasizes that "inductively true" does not mean "true."

As for a base case, $P(3)$ reads $2^5 \le 3^3$, or $32 \le 27$, which is false, while $P(4)$ reads $2^6 \le 3^4$, or $64 \le 81$, which is true. We conclude that $2^{n+2} \le 3^n$ if $n \ge 4$, so we may take $n_0 = 4$ (or any larger integer). $\diamondsuit$

Exercises for Section 2.1

Exercise 2.1.1. Prove $\displaystyle\sum_{j=0}^{n-1} j^2 = \frac{n(n-1)(2n-1)}{6}$ for all natural numbers n.

Exercise 2.1.2. Prove that for all natural numbers n:

(a) $\displaystyle\sum_{j=0}^{n-1} j = \frac{n(n-1)}{2}.$

(b) $\displaystyle\sum_{j=0}^{n-1} j^3 = \left[\frac{n(n-1)}{2}\right]^2.$

Exercise 2.1.3. If $(a_k)_{k \in \mathbf{N}}$ is a sequence of numbers, prove the *telescoping sum* identity:

$$a_n - a_0 = \sum_{k=0}^{n-1}(a_{k+1} - a_k) \quad \text{for every } n \text{ in } \mathbf{N}.$$

Exercise 2.1.4.($\bigstar$) If n is a natural number, define $n! := \prod_{j=0}^{n-1}(j+1)$, and let $P(n)$ be the statement $2^n < n!$. Following Example 2.1.16, prove $P(k)$ implies $P(k+1)$ if $k \ge 1$, and prove $P(n)$ is true if $n \ge 4$.

Exercise 2.1.5. If n is a natural number and X is a set of n elements, use induction to prove $\mathscr{P}(X)$ has 2^n elements.

Exercise 2.1.6. Assume n is a positive integer. The *Tower of Hanoi* puzzle consists of n disks of decreasing size, stacked on one of three spindles. The object is to move the entire stack to one of the other spindles, moving only one disk at a time, and never placing a larger disk atop a smaller one. The initial configuration (with seven disks) is shown in Figure 2.1.

Determine how many individual transfers are required to "solve" the Tower of Hanoi, and prove your guess is correct.

2.2 Counting Subsets

Definition 2.2.1. If m and n are integers, the number of m-element subsets of a set of n elements is called the *binomial coefficient* $\binom{n}{m}$, read "n choose m."

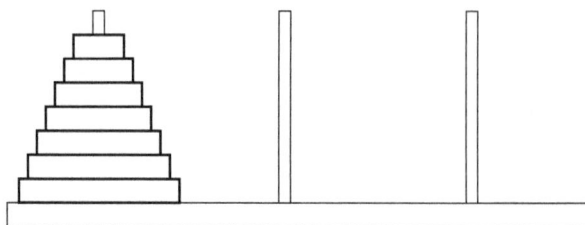

FIGURE 2.1
The Tower of Hanoi, initial configuration.

Remark 2.2.2. Each binomial coefficient is a non-negative integer, and $\binom{n}{m} = 0$ unless $0 \leq m \leq n$. Further, the binomial coefficients satisfy the symmetry $\binom{n}{m} = \binom{n}{n-m}$: If X is a set of n elements, then each m-element subset A of X has a unique complement $X \setminus A$ containing $(n-m)$ elements.   ◇

Proposition 2.2.3. *If m and n are integers, then $\binom{n+1}{m} = \binom{n}{m} + \binom{n}{m-1}$.*

Proof. If $m < 0$ or $n + 1 < m$, each term is 0. Assume, therefore, that $0 \leq m \leq n+1$. Consider the set $X = \{0, 1, 2, \ldots, n\}$ containing $(n+1)$ elements, and assume A is a subset having precisely m elements.

If $0 \notin A$, then $A \subseteq \{1, 2, \ldots, n\}$. By definition, there exist $\binom{n}{m}$ distinct sets of this type.

If $0 \in A$, then $A \setminus \{0\} \subseteq \{1, 2, \ldots, n\}$. By definition, there exist $\binom{n}{m-1}$ distinct sets of this type.

Since every m-element set A falls into exactly one of these categories, $\binom{n+1}{m} = \binom{n}{m} + \binom{n}{m-1}$.   □

Remark 2.2.4. Proposition 2.2.3 leads us to construct the *binomial table*, Table 2.1, in which the top row contains a single 1 among infinitely many 0s, and each entry in subsequent rows is the sum of its "parents" in the preceding row.   ◇

Remark 2.2.5. In addition to the binomial table, we seek a *closed formula* for the binomial coefficients: One that allows us to calculate $\binom{100}{50}$, for example, without computing the first 100 rows of the table. To this end, we'll informally count subsets in two ways, which gives a prospective formula. Then we'll prove the formula reproduces the first row of the binomial table and satisfies the recursion rule of Proposition 2.2.3.

The informal counting argument separates into two steps. First, there are $n \cdot (n-1) \cdot (n-2) \cdots (n-m+1)$ ways to pick an ordered list of m distinct elements in X: We have n choices for the first element, and then $(n-1)$ choices for the second (because we cannot pick the same element twice), and then $(n-2)$ choices for the third, etc., until at the mth step we have $n - (m-1) = n - m + 1$

TABLE 2.1
The binomial table, "classic" format (top) and tabular.

0	0	0	1	0	0	0	0	...
0	0	0	1	1	0	0	0	...
0	0	1	2	1	0	0	0	...
0	0	1	3	3	1	0	0	...
0	1	4	6	4	1	0	0	...
0	1	5	10	10	5	1	0	...
⋮	⋮	⋮	⋮	⋮	⋮	⋮	⋮	

	$m=-1$	$m=0$	$m=1$	$m=2$	$m=3$	$m=4$	$m=5$	$\cdots$
$n=0$	0	1	0	0	0	0	0	$\cdots$
$n=1$	0	1	1	0	0	0	0	$\cdots$
$n=2$	0	1	2	1	0	0	0	$\cdots$
$n=3$	0	1	3	3	1	0	0	$\cdots$
$n=4$	0	1	4	6	4	1	0	$\cdots$
$n=5$	0	1	5	10	10	5	1	$\cdots$
⋮	⋮	⋮	⋮	⋮	⋮	⋮	⋮	

choices. Since these choices are independent, the total number of choices is the product of the individual numbers of choices at each step.

Second, to emphasize, the preceding paragraph does not count subsets of m elements, but *ordered* subsets of m elements. There are $m \cdot (m-1) \cdots 3 \cdot 2 \cdot 1$ ways to order the elements of a m-element set: We have m choices for the first element, and then $(m-1)$ choices for the second element, and then $(m-2)$ choices for the third element, and so on.

Each of the $\binom{n}{m}$ m-element subsets of X gives rise to $m \cdot (m-1) \cdots 3 \cdot 2 \cdot 1$ ordered m-element subsets. We have found two ways to count ordered m-element subsets of X:

$$n \cdot (n-1) \cdot (n-2) \cdots (n-m+1) = \Big(m \cdot (m-1) \cdots 3 \cdot 2 \cdot 1 \Big) \binom{n}{m}.$$

Our informal argument has given us a formula we can try to verify formally by induction. Since products of consecutive integers arise repeatedly, it makes sense to introduce special terminology and notation. ◇

Definition 2.2.6. If n is a natural number, the *factorial* $n!$ of n is defined recursively by

$$0! = 1, \qquad (n+1)! = (n+1) \cdot n! \quad \text{if } n \geq 0.$$

Remark 2.2.7. The convention $0! = 1$ may be viewed as "The empty set has a unique ordering."   ◇

Example 2.2.8. Expanding the recursive definition if $n = 5$ gives

$$5! = 5 \cdot 4! = 5 \cdot 4 \cdot 3! = 5 \cdot 4 \cdot 3 \cdot 2! = 5 \cdot 4 \cdot 3 \cdot 2 \cdot 1! = 5 \cdot 4 \cdot 3 \cdot 2 \cdot 1 \cdot 0!.$$

Since $0! = 1$, we have $5! = 5 \cdot 4 \cdot 3 \cdot 2 \cdot 1 = 120$.

Generally, if n is a *positive* integer, then $n! = \prod_{k=1}^{n} k$ is the product of the integers from 1 to n. The first several factorials are worth memorizing, and their approximate sizes worth remembering:

n	$n!$	n	$n!$	n	$n!$	n	$n!$
1	1	5	120	9	362 880	13	6 227 020 800
2	2	6	720	10	3 628 800	14	87 178 291 200
3	6	7	5 040	11	39 916 800	15	1 307 674 368 000
4	24	8	40 320	12	479 001 600	16	20 922 789 888 000

Since each successive factorial results from multiplying by an ever-larger number, factorials grow faster than exponentially.   ◇

We are now ready to formally justify our informal counting argument.

Proposition 2.2.9. *If m and n are integers, then*

$$\binom{n}{m} = \begin{cases} \dfrac{n!}{m!(n-m)!} & \text{if } 0 \le m \le n, \\ 0 & \text{otherwise.} \end{cases}$$

Proof. If $m < 0$ or $n < m$, there are no m-subsets of an n-element set, and the stated formula is correct. It therefore suffices to establish the formula in the proposition for all natural numbers m and n. We proceed by induction on n. For each natural number n, let $P(n)$ be the statement

For every natural number m, $\quad \binom{n}{m} = \begin{cases} \dfrac{n!}{m!(n-m)!} & \text{if } 0 \le m \le n, \\ 0 & \text{otherwise.} \end{cases}$

The base case $P(0)$ reads "$\binom{0}{m} = \frac{0!}{0!\,0!} = 1$ if $m = 0$, and 0 if $m \ge 1$." This is true: The empty set has a unique empty subset, and no subsets of m elements if $m \ge 1$.

Assume inductively that $P(k)$ is true for some k. By Proposition 2.2.3,

$$\binom{k+1}{m} = \binom{k}{m} + \binom{k}{m-1} \quad \text{for all } m.$$

If $0 < m < k+1$, the inductive hypothesis gives

$$\binom{k+1}{m} = \binom{k}{m} + \binom{k}{m-1} = \frac{k!}{m!\,(k-m)!} + \frac{k!}{(m-1)!\,(k-m+1)!}$$

$$= \frac{k!}{(m-1)!\,(k-m)!}\left[\frac{1}{m} + \frac{1}{k-m+1}\right]$$

$$= \frac{k!}{(m-1)!\,(k-m)!}\left[\frac{(k-m+1)+m}{m(k-m+1)}\right]$$

$$= \frac{(k+1)!}{m!\,(k-m+1)!}.$$

If $m = 0$, the second summand is 0, and the inductive hypothesis gives the true statement $\binom{k+1}{0} = \binom{k}{0} = 1$ (every set has a unique empty subset). Similarly, if $m = k+1$, the first summand is 0 and the inductive hypothesis gives $\binom{k+1}{k+1} = \binom{k}{k} = 1$. In each case, $P(k)$ implies $P(k+1)$. By mathematical induction, $P(n)$ is true for all n. □

Exercises for Section 2.2

Exercise 2.2.1. (★) Without a calculator, determine how many ways 5 cards can be selected from a deck of 52.

Exercise 2.2.2. Assume m and n are natural numbers. How many paths are there from $(0,0)$ to (m,n) that consist solely of unit steps to the right or up?

Exercise 2.2.3. (★) In the history of the world, counting every time a deck of 52 playing cards has been shuffled, has every possible ordering occurred? If so, about how many times? If not, what fraction of possible orderings have been seen? The point is to make order-of-magnitude estimates.

Exercise 2.2.4. (A) The *double factorial* $n!!$ of a natural number n is defined inductively by

$$0!! = 1!! = 1 \qquad (n+2)!! = (n+2)n!! \quad \text{if } n \geq 0.$$

Without a calculator, compute the double factorials up to $10!!$. Use induction to prove

$$\left.\begin{aligned}(2n+1)!!(2n)!! &= (2n+1)!, \\ (2n+2)!!(2n+1)!! &= (2n+2)!,\end{aligned}\right\} \quad \text{for every } n \text{ in } \mathbf{N},$$

and find formulas for $(2n)!!$ and $(2n+1)!!$ in terms of factorials. (That is, informally find formulas, and use induction to prove them correct.)

Exercise 2.2.5. ($\bigstar$) Assume $B = \{0,1\}$ is a two-element set. Define $B^0 = \{0\}$, $B^1 = B$, and recursively define $B^{n+1} = B^n \times B$ if $n \geq 1$. The set B^n is the *n-fold ordered product of B*, and an element of $B^n = \{0,1\}^n$ is an *ordered n-tuple* $(b_k)_{k=0}^{n-1}$ such that $b_k \in B$ for each k.

If b_0 and b_0' are elements of B, define their *distance* to be $|b_0' - b_0| = 0$ if $b_0 = b_0'$ and 1 if $b_0 \neq b_0'$. If n is a natural number and b and b' are in B^n, define the *distance* from b to b' to be

$$|b' - b| = \sum_{k=0}^{n-1} |b_k' - b_k|.$$

(a) Sketch the sets B^1, B^2, and B^3, and label all their points as ordered tuples. Suggestion: After B^1, each product amounts to two copies of the preceding. Use the recursive structure to be systematic about labeling.

Use induction on n to prove that for all b and b' in B^n, $|b' - b|$ is the number of components in which b and b' differ.

Conclude that for all b, b', and b'' in B^n, we have $|b' - b| = 0$ if and only if $b = b'$ (*positivity*), $|b - b'| = |b' - b|$ (*symmetry*), and $|b'' - b| \leq |b'' - b'| + |b' - b|$ (the *triangle inequality*).

(b) If n is a natural number, $0 \leq m \leq n$, and $b \in B^n$, how many elements of B^n are at distance m from b? Confirm this in your sketches. Write the number of elements in B^n in two different ways.

Exercise 2.2.6. In probability, the set B^n of Exercise 2.2.5 models the *sample space for n binary trials*; think enumerating the set of possible outcomes if a two-sided coin is tossed n times. A *probability density* on B^n is an assignment of a non-negative real number, or *probability*, to each point of B^n so that the sum of all the probabilities is 1. This mathematical structure models a situation where the experiment "n binary trials" is performed many times, and the number attached to a point of B^n is the fraction of times that outcome is achieved. One interprets this idealized fraction as a quantitative, if "fuzzy," prediction or description of what occurs if we perform n binary trials.

Assume p is a real number such that $0 \leq p \leq 1$, outcome 0 has probability p, and (therefore) outcome 1 has probability $(1 - p)$. If trials are *independent*, intuitively if the outcome of any particular trial is not affected by prior trials, or "the process is memoryless," the probability of an element $(b_k)_{k=0}^{n-1}$ in B^n is defined to be the product of the individual probabilities. In symbols,

$$P\left((b_k)_{k=0}^{n-1}\right) = \prod_{k=0}^{n-1} P(b_k).$$

Under these assumptions, find the probability of $b = (b_k)_{k=0}^{n-1}$ if $b_k = 0$ for precisely m indices k. Find the probability, in n binary trials, of *some* outcome with precisely m trials equal to 0. If $p = 1$, does your formula make intuitive sense?

Exercise 2.2.7. Assume $I = [0, 1]$ is the set of real numbers x such that $0 \leq x \leq 1$, and $I^\circ = (0, 1)$ the set of real numbers x such that $0 < x < 1$. Define $[0, 1]^0 = \{0\}$ and $[0, 1]^1 = [0, 1]$, and recursively define $[0, 1]^{n+1} = [0, 1]^n \times [0, 1]$ if $n \geq 1$. Call $[0, 1]^n$ the *unit n-cube*.

The partition $I = \{0\} \cup I^\circ \cup \{1\}$ induces a partition of the unit n-cube into ordered n-tuples $(I_k)_{k=0}^{n-1}$ for which each I_k is either $\{0\}$, I°, or $\{1\}$.

For each natural number m, we say a tuple $(I_k)_{k=0}^{n-1}$ is an *m-face* if $I_k = I^\circ$ for precisely m indices k.

Write out the m-faces of the unit 2-cube if $0 \leq m \leq 2$, and sketch them. If n is a natural number and $0 \leq m \leq n$, determine with justification how many m-faces the unit n-cube has. Check your formulas if $n = 2$ and 3, and calculate the result if $n = 4$.

2.3 Binary Operations

Addition of natural numbers assigns a unique natural number $m + n$, the *sum*, to each ordered pair (m, n) of natural numbers, the *summands*. In other words, addition may be viewed as a mapping $+ : \mathbf{N} \times \mathbf{N} \to \mathbf{N}$. The same is true for multiplication and exponentiation of natural numbers. The underlying concept is both simple and widely occurring, and earns a special name.

Definition 2.3.1. Assume X is a non-empty set. A *binary operation* on X is a mapping $* : X \times X \to X$. The pair $(X, *)$ is called a *magma*.

Remark 2.3.2. If $(X, *)$ is a magma, and if x, x' are elements of X, we use *infix notation* $x * x'$ instead of *prefix* (function) notation $*(x, x')$.

Sometimes no operator symbol is used at all. For example, if (m, n) is a pair of natural numbers, the result of multiplication is often denoted mn rather than, say, $m \cdot n$. Similarly, exponentiation is denoted m^n; there is no standard infix notation at all. ◇

Example 2.3.3. If X is a set and $\mathscr{P}(X)$ its power set, then union is a binary operation on $\mathscr{P}(X)$: To each pair of subsets A and B in X we associate $A \cup B$. Intersection $A \cap B$ and difference $A \setminus B$ are binary operations on $\mathscr{P}(X)$. ◇

Definition 2.3.4. Assume $(X, *)$ is a magma.
We say $*$ is *associative* if $x * (x' * x'') = (x * x') * x''$ for all x, x', x'' in X.
We say $*$ is *commutative* if $x' * x = x * x'$ for all x, x' in X.

If there exists an element e of X such that $e * x = x$ and $x * e = x$ for all x in X, we call e an *identity element* for $*$.

If $*$ has an identity element e and $x \in X$, we say an element y of X is an *inverse* of x if $x * y = e$ and $y * x = e$.

Remark 2.3.5. If $(X, *)$ is a magma and $*$ is associative, we say $(X, *)$ is an *associative magma*. If $*$ is commutative, $(X, *)$ is a *commutative magma*.   ◇

Example 2.3.6. The magma $(\mathbf{N}, +)$ is both associative and commutative, and has 0 as identity element. The only natural number with an inverse is 0.

The magma $(\mathbf{N}, \cdot)$ is both associative and commutative, and has 1 as identity element. The only natural number with an inverse is 1.

Exponentiation in $\mathbf{N}$ is neither associative nor commutative, and there is no identity element for exponentiation. We do have $n^1 = n$ for all n, but do not have $1^n = n$ for all n.   ◇

Proposition 2.3.7. *Assume $(X, *)$ is a magma with identity element e.*

(i) *The identity element for $*$ is unique.*

(ii) *If $*$ is associative, each x in X has at most one inverse.*

Proof. See Exercise 2.3.5.   □

Exercises for Section 2.3

Exercise 2.3.1. (H) Assume n is a natural number and $X = \{j\}_{j=0}^{n-1}$ is a set of n elements. Use informal counting (not induction) to answer the following.

(a) How many binary relations are there on X? How many are reflexive? How many are both reflexive and symmetric? How many are symmetric?

(b) How many binary operations are there on X? How many have 0 as identity element? How many have 0 as identity element and are commutative? How many are commutative?

Exercise 2.3.2. (★) On the set $\mathbf{Z}$ of integers, define a binary operation by $a * b = a + b - 1$.

(a) Prove $*$ is associative and commutative.

(b) Prove $*$ has an identity element, and every integer has an inverse.

Exercise 2.3.3. On the set $\mathbf{Z}$ of integers, define a binary operation by $a * b = ab + a + b$.

(a) Prove $*$ is associative and commutative.

(b) Determine whether $*$ has an identity element. If so, which integers have an inverse?

Exercise 2.3.4. (★) Assume X is a set, and $\mathscr{P}(X)$ its power set.

(a) Prove $\cup$ is commutative and associative.

(b) Determine whether $\cup$ has an identity element. If so, which subsets of X have an inverse?

Exercise 2.3.5. Prove Proposition 2.3.7. Hint: As needed, re-read the proof of Proposition 1.3.6.

Exercise 2.3.6. (H) Assume $(X, *)$ is an *associative* magma. Use induction to prove that if $\{x_j\}_{j=0}^{n}$ are elements of X, then the value $x_0 * x_1 * \cdots * x_n$ is independent of how consecutive operands are grouped. (By definition, a binary operation takes precisely two operands. If we are literal, no product of more than two operands is syntactically correct.)

2.4 Construction of Numbers

In this optional section, we construct the natural numbers recursively as sets, define addition and multiplication of natural numbers recursively, and show how to establish familiar laws of algebra with induction. Then we sketch constructions of the integers and rational numbers. The preceding material, especially mathematical induction, supplies motivation.

Definition 2.4.1 (Construction of the natural numbers). Define $0 = \varnothing$. If n is a set, define $S(n) = n \cup \{n\} = \{n, \{n\}\}$. Let $\mathbf{N}$ be the smallest set containing 0 and *closed under successorship*: If $n \in \mathbf{N}$ then $S(n) \in \mathbf{N}$.

Remark 2.4.2. This recapitulates and generalizes Example 1.3.10.   ◇

Definition 2.4.3 (Addition). We define $1 = S(0)$. If n is a natural number, we define $n + 0 = n$. If n and m are natural numbers, we define $n + S(m) = S(n+m)$, namely, $n + (m + 1) = (n + m) + 1$. Particularly, $n + 1 = S(n)$.
   We call $n + m$ the *sum* of n and m, and call this operation *addition*.

Definition 2.4.4 (Ordering). If m and n are natural numbers, we say m *is less than or equal to* n, and write $m \leq n$, if there exists a natural number k such that $m + k = n$. If $k \neq 0$, we say m *is less than* n, and write $m < n$.

Remark 2.4.5. If m and n are natural numbers, precisely one of the following is true: $m < n$, $m = n$, or $n < m$.   ◇

Definition 2.4.6 (Multiplication). For all natural numbers n and m, we define $n \cdot 0 = 0$ and $n \cdot (m + 1) = (n \cdot m) + n$.

   We call $n \cdot m$ the *product* of n and m, and call this operation *multiplication*.

Proposition 2.4.7. *Addition and multiplication of natural numbers are associative and commutative, and multiplication distributes over addition. That is, for all natural numbers ℓ, m, and n, we have*

$$(\ell + m) + n = \ell + (m + n) \qquad \text{associativity of } +,$$
$$m + n = n + m \qquad \text{commutativity of } +,$$
$$(\ell \cdot m) \cdot n = \ell \cdot (m \cdot n) \qquad \text{associativity of } \cdot,$$
$$m \cdot n = n \cdot m \qquad \text{commutativity of } \cdot,$$
$$\ell \cdot (m + n) = (\ell \cdot m) + (\ell \cdot n) \qquad \text{distributivity of } \cdot \text{ over } +.$$

Proof. We prove addition is associative for illustration. Throughout, let k, ℓ, m, and n denote natural numbers. For each n, let $P(n)$ be the statement

$$(\ell + m) + n = \ell + (m + n) \quad \text{for all } \ell \text{ and } m.$$

The single statement $P(1)$ says $(\ell + m) + 1 = \ell + (m + 1)$ for all ℓ and m, which is true by the recursive definition of addition. (Incidentally, $P(0)$ is immediate from the "base case" of addition.)

   Now assume inductively that $P(k)$ is true for some k, that is,

$$(\ell + m) + k = \ell + (m + k) \quad \text{for all } \ell \text{ and } m.$$

Then, for all ℓ and m, we have

$$
\begin{aligned}
(\ell + m) + (k + 1) &= \big((\ell + m) + k\big) + 1 && \text{(recursive definition of } +) \\
&= \big(\ell + (m + k)\big) + 1 && \text{(inductive hypothesis } P(k)) \\
&= \ell + \big((m + k) + 1\big) && \text{(recursive definition of } +) \\
&= \ell + \big(m + (k + 1)\big) && \text{(recursive definition of } +);
\end{aligned}
$$

thus $P(k + 1)$ follows. Since $P(0)$ is true and $P(k)$ implies $P(k + 1)$ for all k, $P(n)$ is true for all n. The remaining parts are Exercise 2.4.1.   □

Integers

To define integers in terms of natural numbers, let's start with the school intuition that an integer is a "difference" of natural numbers, say $m - n$. For example, $-3 = 0 - 3$. We might therefore try to represent the integer $m - n$ by the ordered pair (m, n) of natural numbers.

There is a snag: Infinitely many distinct pairs correspond to the same integer. For example, $(0, 3)$, $(42, 45)$, and $(1965, 1968)$ all represent -3. A pair is not an integer, but an "avatar" of an integer. In mathematical terms, the set $\mathbf{N} \times \mathbf{N}$ comprises representatives of integers. We'd like to construct an equivalence relation on $\mathbf{N} \times \mathbf{N}$ whose equivalence classes are integers.

Suppose (m, n) and (m', n') are integer avatars. These avatars are equal if and only if $m = m'$ and $n = n'$. On the other hand, using our school knowledge, $m - n = m' - n'$ if and only if $m + n' = m' + n$. This motivates a binary relation on $\mathbf{N} \times \mathbf{N}$:

Definition 2.4.8. On the set $\mathbf{N} \times \mathbf{N}$, define a binary relation $\equiv$ by $(m, n) \equiv (m', n')$ if and only if $m + n' = m' + n$.

Lemma 2.4.9. $\equiv$ *is an equivalence relation on* $\mathbf{N} \times \mathbf{N}$.

Proof. See Exercise 2.4.3. □

Definition 2.4.10 (Construction of the integers). Assume m and n are natural numbers. The *integer* $[(m, n)]$ is the set of all ordered pairs (m', n') such that $m + n' = m' + n$, namely, the $\equiv$ equivalence class of the avatar (m, n).

To motivate addition and multiplication, let's add and multiply differences using laws of algebra and see what we get:

$$(m_1 - n_1) + (m_2 - n_2) = (m_1 + m_2) - (n_1 + n_2),$$
$$(m_1 - n_1) \cdot (m_2 - n_2) = (m_1 \cdot m_2) + (n_1 \cdot n_2) - \big((m_1 \cdot n_2) + (n_1 \cdot m_2)\big).$$

These formulas suggest we should define

$$[(m_1, n_2)] + [(m_1, n_2)] = [(m_1 + m_2, n_1 + n_2)],$$
$$[(m_1, n_2)] \cdot [(m_1, n_2)] = [(m_1 \cdot m_2) + (n_1 \cdot n_2), (m_1 \cdot n_2) + (n_1 \cdot m_2)].$$

As we saw in Chapter 1, however, there is a potential snag: These formulas are defined for *integer avatars* (ordered pairs), but an integer is an *equivalence class*. We need to check that the "sum" and "product" depend only on the integers, not on the specific avatars used. In other words, we need to prove these formulas are *well-defined* mod $\equiv$.

Explicitly, we need to check that if $a_1 = [(m_1, n_1)]$ and $a_1' = [(m_1', n_1')]$ are equal as integers, and if $a_2 = [(m_2, n_2)]$ and $a_2' = [(m_2', n_2')]$ are equal as integers, then $a_1 + a_2 = a_1' + a_2'$ as integers, and similarly that $a_1 \cdot a_2 = a_1' \cdot a_2'$. The details (with substantial hints for multiplication) are left to you, Exercises 2.4.4 and 2.4.5. In light of these exercises:

Definition 2.4.11. If $a = [(m, n)]$ and $a' = [(m', n')]$ are integers, we define their *sum* and *product* to be the integers

$$a + a' = [(m + m', n + n')],$$
$$a \cdot a' = [(m \cdot m' + n \cdot n', m \cdot n' + n \cdot m')].$$

Exercise 2.4.6 establishes that integer addition and multiplication satisfy the associative, commutative, and distributive laws.

Rational Numbers

Conceptually, the construction of rational numbers from integers runs parallel to the construction of integers from natural numbers. In school, a rational number (or "fraction") is an expression $r = p/q$ with p and q integers and $q \neq 0$. This numerator/denominator representation is not unique: $p/q = p'/q'$ if and only if $pq' = qp'$.

The sum and product of fractions may be expressed in terms of numerators and denominators using only integer addition and multiplication:

$$\frac{p}{q} + \frac{p'}{q'} = \frac{pq' + qp'}{qq'}, \qquad\qquad \frac{p}{q} \cdot \frac{p'}{q'} = \frac{pp'}{qq'}.$$

Definition 2.4.12. A "rational avatar" is an ordered pair (p, q) of integers such that $q \neq 0$. Two rational avatars (p, q) and (p', q') are *equivalent* if $pq' = p'q$. A *rational number* $p/q := [(p, q)]$ is the set of rational avatars equivalent to (p, q).

If $r_1 = [(p_1, q_1)]$ and $r_2 = (p_2, q_2)$ are rational numbers, we define

$$r_1 + r_2 = (p_1, q_1) + (p_2, q_2) = (p_1 q_2 + p_2 q_1, q_1 q_2),$$
$$r_1 \cdot r_2 = (p_1, q_1) \cdot (p_2, q_2) = (p_1 p_2, q_1 q_2).$$

As with integers, the sum and product are well-defined: Sums of equivalent avatars are equivalent, and similarly for products. And again, this must be checked, see Exercises 2.4.8 and 2.4.9. It must also be checked that rational addition and multiplication satisfy the associative, commutative, and distributive laws, Exercise 2.4.10.

Real Numbers

Constructing the real numbers from the rationals turns out to be substantially more technical than constructing the rational numbers in set theory. Particularly, infinite sets of rationals are needed, see Corollary 17.1.7. Our perspective in Chapter 3 is to give axioms characterizing "the real number system": the set of real numbers, the operations of addition and multiplication, and the relation of order (less-than).

In these axioms, the sets $\mathbf{N}$ of natural numbers, $\mathbf{Z}$ of integers, and $\mathbf{Q}$ of rational numbers are readily constructed inside the set $\mathbf{R}$ of real numbers.

Exercises for Section 2.4

Exercise 2.4.1. In each part, use induction to establish part of Proposition 2.4.7. Earlier parts may be needed to do later parts, and one-variable induction may be needed for the base case of a multivariable induction.

(a) Prove addition of natural numbers is commutative.

(b) Prove multiplication of natural numbers distributes over addition. Hint: Do separate inductions on n using "$\ell(m+n) = \ell m + \ell n$ for all ℓ and m" and "$(\ell + m)n = \ell n + mn$ for all ℓ and m."

(c) Prove multiplication of natural numbers is associative.

(d) Prove multiplication of natural numbers is commutative.

Exercise 2.4.2. (★) If ℓ, m, and n are natural numbers, and if $m < n$, prove $\ell + m < \ell + n$. Prove $\ell \cdot m \leq \ell \cdot n$, with equality if and only if $\ell = 0$.

Exercise 2.4.3. (H) Prove Lemma 2.4.9.

Exercise 2.4.4. (H) Prove that addition of integers is well-defined.

Exercise 2.4.5. Prove multiplication of integers is well-defined. Systematic notation, such as $a = [(m,n)]$, $a_1 = [(m_1, n_1)]$, $a'_1 = [(m'_1, n'_1)]$, etc., will help keep details organized.

(a) Prove that avatar multiplication is commutative, namely, $a_1 \cdot a_2 = a_2 \cdot a_1$.

(b) Prove that if $a_1 \equiv a'_1$, then $a \cdot a_1 \equiv a \cdot a'_1$ and $a_1 \cdot a \equiv a'_1 \cdot a$.

(c) Use part (b) to prove that if $a_1 \equiv a'_1$ and $a_2 \equiv a'_2$, then $a_1 \cdot a_2 \equiv a'_1 \cdot a'_2$.

Exercise 2.4.6. For the integers as constructed in the text, prove:

(a) Addition is associative.

(b) There exists an identity element e for addition. (We denote e by 0. This does not cause ambiguity with the natural number 0, see Exercise 2.4.7.)

(c) Every integer a has an additive inverse b. (We denote b by the symbol $-a$, and call it the *negative* of a.)

(d) Multiplication is associative and distributes over addition.

Exercise 2.4.7. (★) If n is a natural number, write $\phi(n) = (n, 0)$.

(a) Prove that for all natural numbers m and n, $\phi(m) + \phi(n) = \phi(m+n)$ and $\phi(m) \cdot \phi(n) = \phi(m \cdot n)$.

(b) For every integer a, precisely one of the following is true: (i) $a = 0$. (ii) $a \equiv (k, 0)$ for some non-zero natural number k. In this case we say a is *positive*. (iii) $a \equiv (0, k)$ for some non-zero natural number k. In this case we say a is *negative*.

Exercise 2.4.8. Prove that addition of rational numbers is well-defined and commutative.

Exercise 2.4.9. (★) Prove that multiplication of rational numbers is well-defined and commutative.

Exercise 2.4.10. Establish the following properties of the rational numbers as constructed in the text.

(a) Addition is associative.

(b) Addition has an identity element e. (We denote e by 0. This does not cause ambiguity with the integer 0, see Exercise 2.4.11.)

(c) Every rational number a has an additive inverse b. (We denote b by the symbol $-a$, and call it the *negative* of a.)

(d) Multiplication is associative, distributes over addition, and has an identity element e. (From now on we denote e by 1.)

(e) Every non-zero rational number a has a multiplicative inverse b. (We denote b by the symbol a^{-1} or $1/a$, and call it the *reciprocal* of a.)

Exercise 2.4.11.

(a) If p is an integer, write $\phi(p) = (p, 1)$. Prove that for all integers m and n, $\phi(m) + \phi(n) = \phi(m+n)$ and $\phi(m) \cdot \phi(n) = \phi(m \cdot n)$. (Note carefully: On the left the operations act on rational avatars, while on the right the operations act on integers.)

(b) For every rational number a, precisely one of the following is true: (i) $a = 0$. (ii) $a = [(p, q)]$ for some positive integers p and q. In this case we say a is *positive*. (iii) $a = [(-p, q)]$ for some positive integers p and q. In this case we say a is *negative*.

On Writing Proofs

This completes the book's introductions to logic, sets, and induction. Now that you have an idea what this book means by *mathematics*, a few general words may be in order about proofs.

As we have seen, a mathematical theorem is an idealized contract: *If* certain conditions are met (the hypotheses), *then* other conditions are guaranteed (the conclusion). No theorem has even one exception. To guarantee a theorem's correctness, a mathematical proof must leave no logical possibility unexamined, no contingency unresolved.

Most theorems make infinitely many guarantees, one for each assignment of values to variables. A proof often amounts to a calculation or other argument with *arbitrary* (unconstrained) values of variables. Any particular set of values in a theorem statement implicitly expands to a set of values throughout the proof. Exercise 1.1.4, and its solution in the back, illustrate this principle when locating errors.

Avoid pronouns when thinking, speaking, and writing, especially "it." In the middle of even a simple proof, two or three objects tend to be under consideration, and "it" can often refer to any of them. If you're unable to decide exactly what "it" refers to, you've located something you don't fully understand.

There are times, in this book and beyond, where you are asked to prove two quantities are equal. Keep in mind the *one moving target* principle: **It is easier to prove the difference is** 0. The tools of analysis, based on inequalities, are particularly well-adapted to proving quantities are arbitrarily small, and therefore 0. (In analysis, *small* means *close to* 0 and *large* means *far from* 0. Thus we speak of *large negative numbers* like -10^6.)

Although the book illustrates proof organization, writing, and strategizing, first-hand experience is essential. Read actively: As you develop familiarity with idioms of real analysis, think about theorem statements when you first read them and try to prove them yourself, or sketch out a viable approach. Many exercises ask for proofs. In the book's earlier chapters, you will be able to find a comparable proof worked out, either in the text or in the back, and possibly advice on how you might have thought of the proof yourself.

Find your own writing style. Do write accurately and precisely, but don't be pedantic, excessively wordy, or terse. Declarative sentences expressing one idea are generally effective at conveying ideas. Additional suggestions may be found in [17].

Though it may feel awkward at first, read your proofs aloud, either to yourself or someone else. Speaking and listening engage different parts of the brain than writing. Lapses of grammar, narrative continuity, and logic can be more apparent when spoken aloud, partly because the more times you silently re-read your own prose and calculations, the more you skim.

Mathematical writing, like all formal communication, is a craft developed over a lifetime, not a toolbox skill picked up in an afternoon or a few weeks. "Good writing" is subjective and context-dependent. That said, in mathematics there *are* expository desiderata: clarity at levels from notation to word choice to organization, and effectiveness at conveying the structure and beauty of ideas.

3

Real Numbers

"Real analysis" originates with differential and integral calculus, systematic formal rules for working with rates of change, and total change. The "changing" objects are real-valued functions of one real variable, deterministic relationships between numerical quantities. "Numerical quantities" here are real numbers.

What exactly *are* real numbers? We usually depict the set of real numbers as a number line, and write individual real numbers as decimal expressions, such as $3.14159265358979\ldots$. The number line is a visual metaphor and guide. Infinite decimals are symbolic representations. Neither perspective really addresses the essence of "being a real number."

There are two complementary philosophical viewpoints about the nature of real numbers (or any mathematical objects). The first specifies what "structure" and "properties" real numbers have, the types of formal rules we need to work with "generic" or "arbitrary" real numbers symbolically. If our properties "abstractly characterize" the real numbers, we may take our properties as *axioms*. In this book, axioms for the real numbers are our foundation.

The second viewpoint is *construction*: defining real numbers in terms of set theory or some other framework of primitive concepts, and proving that the objects constructed satisfy a set of axioms.

3.1 Axioms for the Real Numbers

Our legal contract for the real number system is a list of axioms. The remainder of the chapter is devoted to (mostly familiar) consequences of the algebraic and order axioms. Completeness is discussed in Chapter 4. Our axioms for the real number system fall into three categories:

Algebraic Properties. (A1.–A4., M1.–M4., D.) The algebraic axioms concern the operations of addition and multiplication: What properties each operation has (associativity, commutativity, existence of an identity element and inverses), and how the two operations interact (the distributive law).

Order Properties. (O1.–O3.) These three axioms formalize the idea of one real number being "greater than" or "less than" another, and the fact that every pair of real numbers is "comparable." Geometrically, the order axioms

DOI: 10.1201/9781003601357-3

guarantee that the real number system can be visualized on a line. "Less than" means "to the left of," and "greater than" means "to the right of."

Completeness. This axiom formalizes the geometric intuition that the real number system "has no gaps," or that "any quantity that can be approximated by real numbers is itself a real number." Geometrically, if A is a non-empty set on the number line, and if *some* point M lies to the right of every point of A, then there exists a *leftmost* point (i.e., smallest number) lying on or to the right of every element of A.

Definition 3.1.1. The *real number system* consists of a non-empty set **R**, two binary operations, $+$ and $\cdot$, and a subset P of **R** (the set of "positive" real numbers) satisfying the following thirteen axioms:

A1. Addition is associative: For all x, y, z in **R**, $x + (y + z) = (x + y) + z$.

A2. Additive identity: There exists a unique element 0 in **R** such that for all x in **R**, $0 + x = x + 0 = x$.

A3. Additive inverses: For every x in **R**, there exists a unique $-x$ in **R** such that $x + (-x) = (-x) + x = 0$.

A4. Addition is commutative: For all x, y, in **R**, $x + y = y + x$.

M1. Multiplication is associative: For all x, y, z in **R**, $x \cdot (y \cdot z) = (x \cdot y) \cdot z$.

M2. Multiplicative identity: There exists a unique element 1 in **R**, distinct from 0, such that for all x in **R**, $1 \cdot x = x \cdot 1 = x$.

M3. Multiplicative inverses: For every non-zero x in **R**, there exists a unique x^{-1} in **R** such that $x \cdot x^{-1} = x^{-1} \cdot x = 1$.

M4. Multiplication is commutative: For all x, y, in **R**, $x \cdot y = y \cdot x$.

D. Multiplication (on the left) distributes over addition: For all x, y, z in **R**, $x \cdot (y + z) = (x \cdot y) + (x \cdot z)$.

O1. Trichotomy: For every real number x, *precisely one* of the following holds: $x \in P$, $-x \in P$, or $x = 0$.

O2. Closure under addition: If x and y are in P, then $x + y$ is in P.

O3. Closure under multiplication: If x and y are in P, then $x \cdot y$ is in P.

C. Completeness: If A is a non-empty subset of **R** that is bounded above, then A has a least upper bound in **R**.

Remark 3.1.2. The multiplication dot is often omitted: $x \cdot y = xy$.   ◇

Remark 3.1.3. These axioms have minor redundancies built in for convenience. For example, if $x + y = 0$, then by the commutative Axiom A4., $y + x = 0$ as well; there is no need to assume both equations in A3. Further, the uniqueness conditions in the axioms for identity elements and inverses can be deduced from the other axioms, see the proof of Lemma 3.1.14.

To reiterate, the real number system, including the operations of addition and multiplication and the set of positive numbers, can be constructed from a much smaller number of axioms. Our axioms above would then be proven as theorems. ⋄

Auxiliary Concepts

We define the operations of *subtraction* and *division* in terms of addition and multiplication.

Definition 3.1.4. If x and y are real numbers, we define their *difference* x minus y to be $x - y := x + (-y)$.

If $y \neq 0$, the *reciprocal* of y is the real number $1/y := y^{-1}$. The *quotient* x over y is defined to be $x/y := x \cdot y^{-1} = y^{-1} \cdot x$.

Remark 3.1.5. Subtraction and division are neither associative nor commutative, as you should check. ⋄

We define the concepts of *positive* and *negative* numbers, and the relations *greater-than* and *less-than*, using Axioms O1.–O3.

Definition 3.1.6. Assume x, y and z are real numbers.

If $z \in P$, we say z is *positive*, or that 0 *is less than* z, and write $0 < z$.

If $y - x \in P$, we say x *is less than* y and write $x < y$. If $x < y$ or $x = y$, we say x *is less than or equal to* y, and write $x \leq y$.

Remark 3.1.7. If $0 < z$, we also say z *is greater than* 0 and write $z > 0$.

If $x < y$, we also say y *is greater than* x and write $y > x$. If $x \leq y$ we also say y *is greater than or equal to* x and write $y \geq x$. ⋄

Whole numbers and fractions constitute some of the most important classes of real numbers.

Definition 3.1.8. The set **N** of *natural numbers* is the smallest subset of **R** satisfying the following conditions: (i) $0 \in \mathbf{N}$; (ii) If $x \in \mathbf{N}$, then $x + 1 \in \mathbf{N}$.

A real number x is an *integer* or *whole number* if either x or $-x$ is a natural number. The set of integers is denoted **Z**, from the German *Zahl*.

A real number x is a *rational number* if there exist integers p and q such that $q > 0$ and $x = p/q$. The set of rational numbers (a.k.a. *ratios* or *quotients*) is denoted **Q**.

A real number x is *irrational* if x is not a rational number.

Remark 3.1.9. In this book, 0 is a natural number. This convention makes natural numbers correspond to "cardinalities" of finite sets, but is not universal. The set of positive integers is denoted $\mathbf{Z}^{+}$.   ◇

Definition 3.1.10. A set $\mathbf{F}$ of two or more elements, together with two binary operations $+$ and $\cdot$ that satisfy Axioms A1.–A4., M1.–M4., and D, is called a *field*. A field containing a subset P satisfying Axioms O1.–O3. is called an *ordered field*.

Remark 3.1.11. The real number system is an ordered field and is abstractly characterized by Axiom C. The rational number system is a non-complete ordered field. The integers are not a field; Axiom M3 fails.   ◇

Algebraic Properties

Though the axioms for the real numbers are numerous, we must still establish "familiar," "elementary" properties needed for routine calculation. Doing so gives us a chance to see the axioms at work.

We first collect a few useful consequences of Axioms A1.–D. These properties hold in every field.

Lemma 3.1.12. *Multiplication on the right distributes over addition: For all* x, y, z *in* $\mathbf{R}$, $(y + z) \cdot x = (y \cdot x) + (z \cdot x)$.

Proof. If x, y, and z are arbitrary real numbers, then

$$
\begin{aligned}
(y + z) \cdot x &= x \cdot (y + z) && \text{Axiom M4.,} \\
&= (x \cdot y) + (x \cdot z) && \text{Axiom D.,} \\
&= (y \cdot x) + (z \cdot x) && \text{Axiom M4.}
\end{aligned}
$$
□

Remark 3.1.13. Because of the commutativity axioms, any identity involving addition or multiplication on the left has a corresponding version with the operation on the right.   ◇

Lemma 3.1.14. *Assume* x, y, *and* z *are real numbers.*

(i) *If* $x + y = x + z$, *then* $y = z$.

(ii) *If* $x \neq 0$ *and* $xy = xz$, *then* $y = z$.

Proof. (i). Assume $x + y = x + z$, and $-x$ is the additive inverse of x. Then

$$
\begin{aligned}
y &= 0 + y && \text{Axiom A2.,} \\
&= \left((-x) + x \right) + y && \text{Axiom A3.,} \\
&= (-x) + (x + y) && \text{Axiom A1.,}
\end{aligned}
$$

$$= (-x) + (x + z) \qquad\qquad \text{hypothesis,}$$
$$= \big((-x) + x\big) + z \qquad\qquad \text{Axiom A1.,}$$
$$= 0 + z \qquad\qquad \text{Axiom A3.,}$$
$$= z \qquad\qquad \text{Axiom A2.}$$

(ii). Mimic the proof of (i), replacing addition by multiplication and replacing the negative $-x$ by the reciprocal x^{-1}. $\qquad\square$

Remark 3.1.15. These conclusions are called the "left cancellation laws." There are corresponding "right cancellations laws." For practice, state and prove these identities for yourself. $\qquad\diamond$

Lemma 3.1.16. *If $x \in \mathbf{R}$, then $-(-x) = x$. If $x \neq 0$, then $(x^{-1})^{-1} = x$, or $1/(1/x) = x$.*

Proof. The condition $x + (-x) = 0$ asserts that the (unique) additive inverse of $-x$ is x itself: $-(-x) = x$. Similarly, if $x \neq 0$, then $x(x^{-1}) = 1$ asserts that $(x^{-1})^{-1} = x$. $\qquad\square$

Proposition 3.1.17. *For all real numbers x and y:*

(i) $x \cdot 0 = 0 \cdot x = 0$.

(ii) *If $xy = 0$, then $x = 0$ or $y = 0$.*

(iii) $-x = x \cdot (-1) = (-1) \cdot x$.

(iv) $(-x)(-y) = xy$. *Particularly, $(-1)(-1) = 1$.*

Proof. Assume x and y are arbitrary real numbers.

(i) Taking $x = 0$ in A2, we have $0 = 0 + 0$. Now let x denote an arbitrary real number. Multiplying on the left by x and using the distributive law, we have
$$0 + (x \cdot 0) = (x \cdot 0) = x \cdot (0 + 0) = (x \cdot 0) + (x \cdot 0).$$
By cancellation, $0 = x \cdot 0$. By the commutative Axiom M4., $0 = 0 \cdot x$ as well.

(ii) Suppose $xy = 0$. If $x = 0$, there is nothing to prove. If $x \neq 0$,
$$y = (x^{-1}x)y = x^{-1}(xy) = x^{-1}(0) = 0.$$

(iii) Multiply $0 = \big(1 + (-1)\big)$ on the left by x:
$$x + (-x) = 0 = x \cdot 0 = x \cdot \big(1 + (-1)\big) = x \cdot 1 + x \cdot (-1) = x + x \cdot (-1).$$
By cancellation, $-x = x \cdot (-1)$. By commutativity, $-x = (-1) \cdot x$.

(iv) Taking $x = -1$ in (iii), we have $(-1)(-1) = -(-1) = 1$ by Lemma 3.1.16. Thus
$$(-x)(-y) = \big((-1)x\big)\big((-1)y\big) = (-1)(-1)(xy) = xy. \qquad\square$$

Continued Fractions

Definition 3.1.18. If $(a_k)_{k \in \mathbf{N}}$ is a real sequence and $a_k > 0$ if $k \geq 1$, then for each n, the expression

$$[a_0, a_1, a_2, \ldots, a_n] := a_0 + \cfrac{1}{a_1 + \cfrac{1}{a_2 + \cfrac{1}{\ddots + \cfrac{1}{a_n}}}}$$

is called the *finite continued fraction* with *coefficients* $(a_k)_{k=0}^n$. A continued fraction is a term of the sequence defined recursively by

$$[a_0] = a_0, \qquad [a_0, a_1, a_2, \ldots, a_n, a_{n+1}] = [a_0, a_1, a_2, \ldots, a_n + (1/a_{n+1})].$$

Remark 3.1.19. Continued fractions having all numerators equal to 1 are called *simple* continued fractions, see for example Krishnan, [19]. In this book we consider only simple continued fractions, and omit "simple" for brevity.   ◇

Example 3.1.20. Using the recursion repeatedly, we have

$$[1, 2, 3, 4] = [1, 2, 3 + (1/4)] = [1, 2, 13/4]$$
$$= [1, 2 + (4/13)] = [1, 30/13] = [1 + 13/30] = 43/30.$$

Inversely, successive long division, terminating when the division is even, gives

$$17/12 = [1 + (5/12)] = [1, 12/5] = [1, 2 + (2/5)] = [1, 2, 5/2] = [1, 2, 2, 2].   ◇$$

Remark 3.1.21. In practice, as in Example 3.1.20, continued fractions usually arise from *integer* sequences. The recursion rule itself, however, involves non-integers.   ◇

Remark 3.1.22. Continued fractions can be grouped from the right. Specifically, for all positive integers k and m we have

$$[a_0, a_1, \ldots, a_k, a_{k+1}, \ldots, a_{k+m}] = \Big[a_0, a_1, \ldots, a_k, [a_{k+1}, \ldots, a_{k+m}]\Big].   ◇$$

Proposition 3.1.23. *Assume $(a_k)_{k \in \mathbf{N}}$ is a real sequence such that a_k is positive if $k \geq 1$. Recursively define real sequences (p_n) and (q_n) by*

$$p_{-2} = 0, \qquad p_{-1} = 1, \qquad\qquad p_n = a_n p_{n-1} + p_{n-2},$$
$$q_{-2} = 1, \qquad q_{-1} = 0, \qquad\qquad q_n = a_n q_{n-1} + q_{n-2}.$$

For every natural number n, we have $p_n/q_n = [a_0, a_1, a_2, \ldots, a_n]$.

Proof. See Exercise 3.1.6.   □

Exercises for Section 3.1

Exercise 3.1.1. (★) Assume x and y are real numbers. Using one axiom per step, prove that $(x+y)^2 = x^2 + 2xy + y^2$. Notes: By definition, $2 = 1+1$; at each step but the last, there should be enough parentheses that no sub-expression contains more than two operands; this takes about ten steps.

Exercise 3.1.2. (H) Assume a, b, c, and d are real numbers such that $bd \neq 0$. Use the axioms to prove

(a) $(bd)^{-1} = b^{-1}d^{-1}$. (b) $\dfrac{a}{b} \cdot \dfrac{c}{d} = \dfrac{ac}{bd}$. (c) $\dfrac{a}{b} + \dfrac{c}{d} = \dfrac{ad + bc}{bd}$.

Exercise 3.1.3. (★) If x, y, u, and v are real, prove that

$$\begin{matrix} 2u = x + y, \\ 2v = x - y, \end{matrix} \quad \text{if and only if} \quad \begin{matrix} x = u + v, \\ y = u - v. \end{matrix}$$

(Use algebra "normally" rather than one axiom per step.)

Exercise 3.1.4. Assume x and y are real. Prove that $x^2 = y^2$ if and only if $x = y$ or $x = -y$. (Use the axioms and results from the text. Do not use properties of square roots, which have not yet been established.)

Exercise 3.1.5. Consider the set $\mathbf{Z}^2 := \mathbf{Z} \times \mathbf{Z} \subseteq \mathbf{R}^2$ of *integer points*, whose rectangular coordinates are integers. If α is real, does the equation $y = \alpha x$ always have a non-zero solution in $\mathbf{Z}^2$? If so, give a proof; if not, characterize α for which $y = \alpha x$ has a non-zero integer point solution. Does any equation $y = \alpha x$ have a *unique* non-zero integer point solution?

Exercise 3.1.6. (★) Prove Proposition 3.1.23.

Exercise 3.1.7. In this exercise, assume the *integer division algorithm*: If p and q are integers and $q > 0$, there exist unique integers a and r such that $p = aq + r$ and $0 \leq q < r$.

Use repeated integer division to represent $5/7$, $-8/5$, and $355/113$ as continued fractions. Prove every rational number is a finite continued fraction $[a_0, a_1, \ldots, a_n]$ with a_k an integer for all k, a_k positive if $k \geq 1$, and $a_n > 1$.

3.2 Order Properties

Mathematical inequalities are the bread and jam of analysis. The following properties of inequalities are among our most basic and common tools. A good habit upon encountering such a proposition or theorem is to express each part in procedural form, what each rule for manipulating inequalities "allows" us to do. For example, (i) says the less-than relation is transitive: less-than inequalities can be "daisy-chained" to obtain new less-than inequalities; (iii) says that multiplying an inequality by a positive number preserves the inequality.

Proposition 3.2.1. *For all real numbers x, y, and z:*

(i) *If $x < y$ and $y < z$, then $x < z$.*

(ii) *If $x < y$, then $x + z < y + z$.*

(iii) *If $x < y$ and $0 < z$, then $xz < yz$.*

(iv) *If $x < y$ and $z < 0$, then $yz < xz$.*

(v) *If $0 < x < y$, then $0 < 1/y < 1/x$.*

Two additional extensions are useful occasionally:

Corollary 3.2.2. *Assume x, y, z, and w are real numbers.*

(i) *If $x < y$ and $w < z$, then $x + w < y + z$.*

(ii) *If $0 < x < y$ and $0 < w < z$, then $0 < xw < yz$.*

In particular, $0 \le x \cdot x$, with equality if and only if $x = 0$.

Remark 3.2.3. The set P of numbers mentioned by Axioms O1.–O3. is uniquely characterized by these axioms. Specifically, P turns out to be precisely the set

$$\{y \text{ in } \mathbf{R} : y = x^2 \text{ for some non-zero real } x\}$$

of squares of non-zero real numbers. Corollary 3.2.2 (ii) guarantees every non-zero square is positive. The converse implication, that every positive real number is the square of some real number, is Theorem 4.2.11.   ◇

To reduce the properties in Proposition 3.2.1 to Axioms O1.–O3., it is convenient first to establish special cases where one comparand is zero:

Lemma 3.2.4. *For all real numbers x and y,*

(i) *If $0 < x$ and $y < 0$, then $xy < 0$.*

(ii) *If $x < 0$ and $y < 0$, then $0 < xy$.*

(iii) *If $0 < x$, then $0 < 1/x$.*

Proof. (i). By trichotomy, $y < 0$ if and only if $-y \in P$. By Axiom O3. and Proposition 3.1.17 (iii), $-(xy) = x(-y) \in P$; that is, $xy < 0$.

(ii). If $-x$ and $-y \in P$, then by Axiom O3. and Proposition 3.1.17 (iv), $0 < (-x)(-y) = xy$.

(iii). Since $1 \neq 0$ by Axiom M2., trichotomy implies $1 \in P$ or $-1 \in P$. Since P is closed under multiplication, either $1 \cdot 1 = 1$ is positive, or $(-1) \cdot (-1) = 1$ is positive. Under either alternative, we conclude 1 is positive, or $0 < 1$.

Suppose $0 < x$. If $x^{-1} < 0$, then by (i), we have $1 = x(x^{-1}) < 0$, which we have just seen is false. Contrapositively, if $0 < x$, then $0 < x^{-1}$. $\square$

Proof of Proposition 3.2.1. (i). By definition, $x < y$ if and only if $y - x \in P$, and $y < z$ if and only if $z - y \in P$. By Axiom O2.,

$$z - x = (z - y) + (y - x) \in P,$$

which is equivalent to $x < z$.

(ii). For all x, y, and z, we have $y - x = (y + z) - (x + z)$. The claim follows immediately.

(iii). If $x < y$ and $0 < z$, then $y - x \in P$ and $z \in P$. By Axiom O3., the product $(y - x)z = yz - xz$ is in P. That is, $xz < yz$.

(iv). Since $x < y$ and $0 < -z$, part (iii) gives

$$0 < (y - x)(-z) = (x - y)z = xz - yz,$$

or, $yz < xz$.

(v). By Lemma 3.2.4 (iii), if $0 < x < y$, then $0 < 1/x$ and $0 < 1/y$. Algebra gives $0 < (y - x)/(xy) = 1/x - 1/y$, or, $1/y < 1/x$. $\square$

Proof of Corollary 3.2.2. (i). By Proposition 3.2.1 (ii), adding x to $w < z$ and adding z to $x < y$ gives $x + w < x + z < y + z$.

(ii). follows similarly from Proposition 3.2.1 (iii).

For the assertion about squares, Proposition 3.1.17 (i) implies $0 = 0 \cdot 0$, while we have just shown that if $x \neq 0$, then $0 < x \cdot x$. $\square$

Definition 3.2.5. Assume x and y are real numbers. If t is real and $0 \leq t \leq 1$, the expression $(1 - t)x + ty$ is called a *convex linear combination* of x and y.

Proposition 3.2.6. *Assume x and y are real numbers such that $x < y$.*

(i) *If $0 < t < 1$, then $x < (1 - t)x + ty < y$.*

(ii) *If $s < t$ are real, then $(1 - s)x + sy < (1 - t)x + ty$.*

Proof. See Exercise 3.2.8. $\square$

The Triangle Inequalities

Definition 3.2.7. If x is a real number, the *absolute value* of x is

$$|x| = \begin{cases} x & \text{if } x \geq 0, \\ -x & \text{if } x < 0. \end{cases}$$

Lemma 3.2.8. *If x and y are real, then $|xy| = |x|\,|y|$.*

Proof. Immediate from Proposition 3.1.17 (iii) and (iv). □

Lemma 3.2.9. *If x is real, then $-|x| \leq x \leq |x|$.*

Proof. If $x \geq 0$, then $|x| = x$ and $-|x| = -x \leq 0$. Combining these inequalities, $-|x| \leq 0 \leq x \leq |x|$, as claimed.

If instead $x < 0$, then $|x| = -x$ and $0 < -x = |x|$. Consequently, we have $-|x| \leq x < 0 < |x|$. □

Proposition 3.2.10. *Assume x and b are real numbers.*

(i) $|x| < b$ *if and only if* $-b < x < b$.

(ii) *If $|x| \leq b$ for all positive b, then $x = 0$.*

Proof. (i). For all real x, we have $0 \leq |x|$. If $|x| < b$, then Proposition 3.2.1 (i) implies $0 \leq |x| < b$, while Proposition 3.2.1 (iv) implies $-b < -|x|$. Combining with Lemma 3.2.9, we have $-b < -|x| \leq x \leq |x| < b$.

Conversely, assume $-b < x < b$. If $0 \leq x$, then $x = |x|$, so in particular $|x| < b$. If instead $x < 0$, then $x = -|x|$. By hypothesis, $-b < x = -|x|$. Multiplying by -1 implies $|x| < b$ in this case as well.

(ii). The hypothesis involves infinitely many statements, one for each positive real number b, and *no finite number of these hypotheses implies the conclusion.* It is therefore more natural to consider the contrapositive: To assume the conclusion is false ($x \neq 0$) and prove the hypothesis is false (there exists a positive real number b such that $b < |x|$). But if $x \neq 0$, then $0 < |x|$. Setting $b = |x|/2$, we have $0 < b < |x|$. □

We now state two of the most ubiquitous inequalities in analysis, the *triangle inequality* and *reverse triangle inequality*.

Proposition 3.2.11. *If x and y are real numbers, then*

(i) $|x + y| \leq |x| + |y|$.

(ii) $\bigl||x| - |y|\bigr| \leq |x - y|$.

Proof. See Exercise 3.2.4. □

Corollary 3.2.12. *For all real x and y, we have*

$$\Big||x| - |y|\Big| \le |x \pm y| \le |x| + |y|.$$

Proof. See Exercise 3.2.5. □

Remark 3.2.13. Conceptually, there are *a priori* lower and upper bounds on $|x \pm y|$, the absolute value of a sum or difference, in terms of $|x|$ and $|y|$. ◇

Definition 3.2.14. If x and y are real numbers, the *number line distance* between x and y is

$$|y - x| = |x - y| = \begin{cases} y - x & \text{if } x \le y, \\ x - y & \text{if } y < x. \end{cases}$$

Corollary 3.2.15. *If x, y and z are real numbers, then*

(i) $|x - z| \le |x - y| + |y - z|$.

(ii) $\Big||x - y| - |z - y|\Big| \le |x - z|$.

Remark 3.2.16. These claims follow by applying the corresponding part of Proposition 3.2.11 to the identity

$$(x - z) = (x - y) + (y - z) = (x - y) - (z - y).$$

We may view real numbers x, y, and z as vertices of a (degenerate) triangle on the number line. Part (i) of the corollary says the length of a side of a triangle does not exceed the sum of the lengths of the other two sides. Part (ii) says the length of a side of a triangle is no shorter than the difference of the lengths of the other two sides. ◇

Minimum and Maximum

Definition 3.2.17. Assume A is a non-empty set of real numbers. A real number β (beta) is a *maximum* of A, or a *largest element* of A, if

(i) $\beta \in A$, and

(ii) $x \le \beta$ for all x in A.

A real number α (alpha) is a *minimum* of A, or a *smallest element* of A, if

(i) $\alpha \in A$, and

(ii) $\alpha \le x$ for all x in A.

Example 3.2.18. If a is a real number and $A = \{a\}$, then a itself is both the largest and the smallest element of A, both in the ordinary English sense and (check this) according to Definition 3.2.17. $\diamond$

Lemma 3.2.19. *Assume A is a non-empty set of real numbers. If β and β' are largest elements of A, then $\beta = \beta'$. In words, a largest element, if it exists, is unique.*

Proof. Suppose β and β' are largest elements of A. Since $x \leq \beta$ for all x in A and $\beta' \in A$, we have $\beta' \leq \beta$. The same argument with the roles reversed shows $\beta \leq \beta'$. Consequently, $\beta = \beta'$. $\qquad\qquad\qquad\qquad\qquad\qquad\qquad\qquad\quad$ $\square$

Remark 3.2.20. An analogous claim is true for smallest elements. Exercise 3.2.9 asks you to give a formal statement, and to prove your statement by modifying the proof above. $\diamond$

Remark 3.2.21. A non-empty set of real numbers may have both a largest and a smallest element, or it may have one but not the other, or it may have neither.

For example, the set of positive real numbers has neither: If x is a positive real number, then $x/2$ is smaller and positive, so x is not the smallest positive real. Similarly, $2x$ is larger and positive, so x is not the largest positive real.

Proposition 3.2.25 guarantees that every non-empty *finite* set of real numbers has both a smallest and a largest element. $\diamond$

Definition 3.2.22. If a and b are real numbers, we define their *maximum* and *minimum* by

$$\max(a, b) = \begin{cases} b & \text{if } a \leq b, \\ a & \text{if } b < a, \end{cases} \qquad \min(a, b) = \begin{cases} a & \text{if } a \leq b, \\ b & \text{if } b < a. \end{cases}$$

Remark 3.2.23. If a and b are real and $A = \{a, b\}$, then $\max(a, b)$ is the largest element of A and $\min(a, b)$ is the smallest element of A according to Definition 3.2.17. Take a minute to check this. We will give a separate proof shortly. $\diamond$

Proposition 3.2.24. *If a and b are real numbers, then*

$$\max(a, b) = \frac{a + b + |a - b|}{2}, \qquad \min(a, b) = \frac{a + b - |a - b|}{2}.$$

Proof. Adding and subtracting the formulas for $\max(a, b)$ and $\min(a, b)$ in Definition 3.2.22 gives

$$\max(a, b) + \min(a, b) = a + b$$

$$\max(a, b) - \min(a, b) = |a - b| = \begin{cases} b - a & \text{if } a \leq b, \\ a - b & \text{if } b < a. \end{cases}$$

The proposition follows by adding and subtracting *these* and dividing by 2, compare Exercise 3.1.3. □

Proposition 3.2.25. *If $A = \{a_1, a_2, \ldots, a_n\}$ is a* finite, *non-empty set of real numbers, there exist unique elements* $\max A$ *and* $\min A$ *in A such that*

$$\min A \leq x \leq \max A \quad \text{for all } x \text{ in } A.$$

Proof. Uniqueness of largest elements was established in Lemma 3.2.19, while Exercise 3.2.9 handles smallest elements. It therefore suffices to establish *existence* of largest elements.

We proceed by induction on the number of elements. For each positive integer n, let $P(n)$ be the statement,

Every set of n distinct real numbers has a largest element.

A set of one real number has a largest element; the base case $P(1)$ is true.

Assume inductively that $P(m)$ is true for some positive integer m. Assume $A_{m+1} = \{a_k\}_{k=1}^{m+1}$ is a set of $(m+1)$ distinct numbers, and write $A_m = \{a_k\}_{k=1}^{m}$, so that $A_{m+1} = A_m \cup \{a_{m+1}\}$. By the inductive hypothesis, A_m has a largest element $\max A_m$. Define

$$\beta_{m+1} = \max\left(\max A_m, a_{m+1}\right).$$

We first show β_{m+1} is an element of A_{m+1}: Either $\beta_{m+1} = \max A_m$, which is an element of $A_m \subseteq A_{m+1}$, or $\beta_{m+1} = a_{m+1}$, which is an element of A_{m+1}.

It remains to prove $x \leq \beta_{m+1}$ for all x in A_{m+1}. Assume x is an arbitrary element of A_{m+1}. If $x \in A_m$, then $x \leq \max A_m \leq \beta_{m+1}$. If instead $x \notin A_m$, then $x = a_{m+1} \leq \beta_{m+1}$.

We have shown that $\beta_{m+1} \in A_{m+1}$ and $x \leq \beta_{m+1}$ for all x in A_{m+1}. By definition, β_{m+1} is a largest element of A_{m+1}. Since A_{m+1} was an arbitrary set of $(m+1)$ elements, we have established the inductive step, "$P(m)$ implies $P(m+1)$ for every m." By mathematical induction, $P(n)$ is true for every positive integer n; every finite set of real numbers has a largest element.

The proof that every non-empty finite set has a smallest element is entirely analogous, see Exercise 3.2.10. □

Exercises for Section 3.2

Exercise 3.2.1. In each part, assume n is a positive integer, and x, y are real numbers such that $x < y$. Use induction to prove:

(a) If $0 \leq x$, then $0 \leq x^n < y^n$.   (b) If n is odd, then $x^n < y^n$.

Exercise 3.2.2. Assume x and y are real numbers. Prove that

(a) $2|xy| \leq x^2 + y^2$, and the inequality is strict unless $|x| = |y|$.

(b) $0 \leq x^2 + xy + y^2$, and the inequality is strict unless $x = y = 0$.

Exercise 3.2.3. (H) Assume a, b, and c are real numbers.

(a) If a is positive, prove $ax^2 + bx + c \geq (4ac - b^2)/(4a)$ for all real x, with equality if and only if $x = -b/(2a)$. Suggestion: Complete the square.

(b) If $a \neq 0$, prove there exist two real numbers x such that $ax^2 + bx + c = 0$ if and only if there exists a non-zero real number r such that $r^2 = b^2 - 4ac$ (if and only if $0 < b^2 - 4ac$ by Theorem 4.2.11).

(c) (A) In (b), write $r = \sqrt{b^2 - 4ac}$. If x is real and $ax^2 + bx + c = 0$, find a formula for x in terms of a, b, and c.

Exercise 3.2.4. (H) Prove Proposition 3.2.11.

Exercise 3.2.5. ($\bigstar$) Prove Corollary 3.2.12.

Exercise 3.2.6. Give a formal proof by mathematical induction that if $n \geq 2$ and $\{x_j\}_{j=0}^{n-1}$ are real numbers, then

$$\left| \sum_{j=0}^{n-1} x_j \right| \leq \sum_{j=0}^{n-1} |x_j|.$$

Exercise 3.2.7. Assume x_0 and r are real numbers, and $r > 0$. Prove, for all real x,

(a) $|x - x_0| < r$ if and only if $x_0 - r < x < x_0 + r$.

(b) $0 < |x - x_0| < r$ if and only if $x_0 - r < x < x_0$ or $x_0 < x < x_0 + r$.

Exercise 3.2.8. Assume x and y are real numbers such that $x < y$.

(a) Prove that $x < \frac{1}{2}(x + y) < y$ and $x < \frac{1}{3}(2x + y) < \frac{1}{3}(x + 2y) < y$.

(b) Prove Proposition 3.2.6.

(c) Interpret the conclusions of Proposition 3.2.6 geometrically.

Exercise 3.2.9. Give a formal statement for smallest elements analogous to Lemma 3.2.19, and prove your statement.

Exercise 3.2.10. (H) Complete the proof of Proposition 3.2.25 by showing every non-empty finite set of real numbers has a smallest element.

Exercise 3.2.11. (H) If x_0, y_0, x, and y are real, prove that

$$|xy - x_0 y_0| \leq |x|\,|y - y_0| + |y_0|\,|x - x_0|$$

and give a geometric interpretation.

Exercise 3.2.12. (H) If $x_0 \neq 0$, and if $|r| < |x_0|/2$, prove $1/|x_0 + r| \leq 2/|x_0|$.

Exercise 3.2.13. ($\bigstar$) Use induction to prove that if n and m are natural numbers, then $(n+1)^m n! \leq (n+m)!$ and the inequality is strict if $m > 1$.

Exercise 3.2.14. Assume m is an integer greater than 1. Prove there exists an integer N such that if $n \geq N$, then $m^n \leq n!$.

Exercise 3.2.15. This five-part exercise introduces the complex field. Although formally a complex number is an expression $a + bi$ with a and b real and i a formal symbol satisfying $i^2 = -1$ and commuting with real numbers, the modern definition refers to nothing but real numbers and set-theoretic constructions: A *complex number* is an ordered pair (a, b) of real numbers. We call a the *real part* of (a, b) and b the *imaginary part*. If $b = 0$, we say $(a, b) = (a, 0)$ is *real*.

If $z = (x, y)$, $w = (u, v)$ are complex, define their *sum* and *product* to be

$$z + w = (x, y) + (u, v) = (x + u, y + v),$$
$$zw = (x, y) \cdot (u, v) = (xu - yv, xv + yu).$$

These operations are called *complex addition* and *complex multiplication*.

(a) Prove complex addition is commutative and associative, has an identity element, and every complex number has an additive inverse.

(b) Prove complex multiplication is commutative, associative, and distributes over complex addition.

(c) Prove $(1, 0)$ is an identity element for complex multiplication. If x and y are real, calculate $(x, y)(x, -y)$. Prove that every non-zero complex number $z = (x, y)$ has a complex reciprocal: There exists a complex number w such that $zw = (1, 0)$.

(d) Find all complex numbers $z = (x, y)$ such that $z^2 = z \cdot z = (-1, 0)$.

(e) If $z = (x, y)$, the number $\overline{z} = (x, -y)$ is called the *complex conjugate* of z. Prove that if z and w are complex, then $\overline{z + w} = \overline{z} + \overline{w}$ and $\overline{zw} = \overline{z}\,\overline{w}$.

3.3   Powers and Sums

Definition 3.3.1. If x is a real number, we define *powers* of x recursively by

$$x^0 = 1, \qquad x^{k+1} = x^k \cdot x \quad \text{if } k \text{ is a natural number.}$$

If x is non-zero and k is a natural number, we define $x^{-k} = (x^k)^{-1}$.

Remark 3.3.2. Particularly, we define $0^0 = 1$.   ◇

Remark 3.3.3. The expression "x^k," read "x to the k," is called "the kth power of x." Intuitively, x^k is the result of multiplying k factors of x.   ◇

Theorem 3.3.4 (The law of exponents). *If x and y are non-zero real numbers, then*

$$\left.\begin{array}{ll} \text{(i)} & (xy)^n = (x^n)(y^n), \\ \text{(ii)} & x^{m+n} = x^m \cdot x^n, \\ \text{(iii)} & x^{nm} = (x^n)^m, \end{array}\right\} \quad \text{for all integers } m \text{ and } n.$$

In particular, $x^{-n} = (x^{-1})^n$ for all non-zero real x and all integers n.

Proof. (i). For each natural number n, consider the statement $P(n)$:

$$(xy)^n = (x^n)(y^n).$$

The base case $P(0)$ reduces to $1 = 1$, which is true. Assume inductively that $P(k)$ is true for some natural number k. We have

$$\begin{array}{ll} (xy)^{k+1} = (xy)^k(xy) & \text{definition of exponentiation,} \\ = (x^k y^k)(xy) & \text{inductive hypothesis,} \\ = x^k(y^k \cdot x)y & \text{associativity,} \\ = x^k(x \cdot y^k)y & \text{commutativity,} \\ = (x^k \cdot x)(y^k \cdot y) & \text{associativity,} \\ = (x^{k+1})(y^{k+1}) & \text{definition of exponentiation.} \end{array}$$

Since $P(0)$ is true and $P(k)$ implies $P(k+1)$ for all k, the statement $P(n)$ is true for all n by mathematical induction.

   To prove $(xy)^{-n} = (x^{-n})(y^{-n})$ if $n > 0$, recall $x^{-n} = (x^n)^{-1}$ by definition. The preceding argument and Exercise 3.1.2 (a) imply

$$(xy)^{-n} = [(xy)^n]^{-1} = [(x^n)(y^n)]^{-1} = (x^n)^{-1}(y^n)^{-1} = (x^{-n})(y^{-n}).$$

Finally, $x^n(x^{-n}) = 1 = (x \cdot x^{-1})^n = x^n(x^{-1})^n$. By cancellation, $x^{-n} = (x^{-1})^n$ for all n.

   The proofs of (ii) and (iii) are Exercises 3.3.1 and 3.3.2.   □

The "limiting behavior" of x^n as n grows without bound plays a central role in analysis. We will use the following estimates repeatedly.

Proposition 3.3.5. *If u is a positive real number, then*

$$1 + nu \leq (1 + u)^n \quad \text{if } n \geq 0.$$

Particularly, $n < 1 + n \leq 2^n$ if $n \geq 0$.

Proof. We proceed by induction on n. Let $P(n)$ denote the inequality in the theorem. The base case $P(0)$ asserts $1 \leq 1$, which is true. Assume inductively that $P(k)$ is true for some natural number k. We have

$$
\begin{aligned}
1 + (k+1)u &\leq 1 + (k+1)u + ku^2 && 0 \leq ku^2 \\
&= (1 + ku)(1 + u) && \text{algebra} \\
&\leq (1 + u)^k (1 + u) && \text{inductive hypothesis} \\
&= (1 + u)^{k+1} && \text{definition of exponentiation}
\end{aligned}
$$

so that $P(k)$ implies $P(k+1)$. $\qquad\square$

Corollary 3.3.6. *If x is a real number such that $0 < x < 1$, there exists a positive real number u such that $x = 1/(1 + u)$, and*

$$0 < x^n \leq \frac{1}{1 + nu} \quad \text{if } n \geq 0.$$

Proof. If $0 < x < 1$, then $1 < 1/x$ by Proposition 3.2.1 (v), so we may write $1/x = 1 + u$ for some positive real number u. By Proposition 3.3.5, we have $0 < 1 + nu \leq (1/x)^n = 1/(x^n)$ if $n \geq 0$. Taking reciprocals again establishes the corollary. $\qquad\square$

Geometric Sums

Finite sums whose terms form a geometric progression (consecutive terms all have the same ratio) play an important role in real analysis.

Definition 3.3.7. If a and r are real numbers, and n is a natural number, the *geometric sum* with *first term a*, *ratio r*, and n terms is

$$\sum_{k=0}^{n-1} ar^k = a + ar + ar^2 + \cdots + ar^{n-1}.$$

Proposition 3.3.8. *The geometric sum with first term a, ratio 1, and n terms is equal to an. If instead $r \neq 1$, then*

$$\sum_{k=0}^{n-1} ar^k = a\frac{r^n - 1}{r - 1} = a\frac{1 - r^n}{1 - r}.$$

Proof. See Exercise 3.3.4. $\qquad\square$

Remark 3.3.9. The form with positive denominator tends to be most useful. $\quad\diamond$

The Binomial Theorem

If x and y are real numbers, then $(x+y)^2 = x^2 + 2xy + y^2$. The *binomial theorem* generalizes to arbitrary positive integer powers $(x+y)^n$.

Theorem 3.3.10 (The binomial theorem). *For all real numbers x and y and every non-negative integer n,*

$$(x+y)^n = \sum_{k=0}^{n} \binom{n}{k} x^{n-k} y^k$$
$$= \binom{n}{0} x^n + \binom{n}{1} x^{n-1} y + \binom{n}{2} x^{n-2} y^2 + \cdots + \binom{n}{n} y^n.$$

Example 3.3.11. From the binomial table, Table 2.1, if x and y are real, then

$$(x+y)^3 = x^3 + 3x^2 y + 3xy^2 + y^3,$$
$$(x+y)^4 = x^4 + 4x^3 y + 6x^2 y^2 + 4xy^3 + y^4. \qquad \diamond$$

Example 3.3.12. The binomial theorem can be used with specific numbers. For example,

$$11^3 = (10+1)^3 = 10^3 \quad + 3 \cdot 10^2 \cdot 1 + 3 \cdot 10 \cdot 1^2 + 1^3$$
$$= 1000 + 300 \qquad + 30 \qquad + 1 = 1331.$$

The digits comprise the fourth row of the binomial table. $\qquad \diamond$

Proof of Theorem 3.3.10. Conceptually, the n-fold product

$$(x+y)^n = (x+y)(x+y) \cdots (x+y)$$

is expanded by the following procedure:

1. Pick an arbitrary integer k such that $0 \le k \le n$;

2. Distribute k check marks among the n copies of $(x+y)$;

3. If a copy of the factor $(x+y)$ is unchecked, choose x from that copy; otherwise choose y. Multiply the resulting n factors to get $x^{n-k} y^k$.

4. Sum over all k and all ways of distributing k check marks.

By Points 1. and 3., the expanded product has the form

$$(x+y)^n = \underline{\quad} x^n + \underline{\quad} x^{n-1} y + \underline{\quad} x^{n-2} y^2 + \cdots + \underline{\quad} xy^{n-1} + \underline{\quad} y^n$$

for some coefficients. By Points 2. and 4., the coefficient of $x^{n-k} y^k$ is $\binom{n}{k}$, the number of distinct ways of distributing k check marks among n parenthesized binomials. This completes the proof. $\qquad \square$

Exercises for Section 3.3

Exercise 3.3.1. (H) Prove Theorem 3.3.4 (ii).

Exercise 3.3.2. Prove Theorem 3.3.4 (iii).

Exercise 3.3.3. Assume $u > 0$. Use induction to prove

$$1 + nu + \frac{n(n-1)}{2}\, u^2 \le (1+u)^n \quad \text{for all } n \text{ in } \mathbf{N}.$$

Exercise 3.3.4. (H) Use induction on n to establish the geometric sum formula, Proposition 3.3.8.

Exercise 3.3.5. ($\bigstar$) If x and y are real numbers, find a closed formula for

$$\sum_{k=0}^{n-1} x^{n-k-1} y^k = x^{n-1} + x^{n-2}y + x^{n-3}y^2 + \cdots + xy^{n-2} + y^{n-1}.$$

Exercise 3.3.6. ($\bigstar$) Assume n is a natural number, h and x real numbers such that $h \neq 0$. Use the binomial theorem to give a *polynomial formula* for $\big((x+h)^n - x^n\big)/h$. To what does this formula reduce when $h = 0$?

Exercise 3.3.7. ($\bigstar$) Assume x is real and n a natural number. Express

$$\tfrac{1}{2}\big((1+x)^n + (1-x)^n\big),$$
$$\tfrac{1}{2}\big((1+x)^n - (1-x)^n\big)$$

as polynomials in x. Write out the results explicitly if $n = 2$, 3, and 4.

Exercise 3.3.8. Use the binomial theorem to expand:

(a) $(x+y)^3$, $(x-y)^3$, and $\tfrac{1}{2}\big((x+y)^3 \pm (x-y)^3\big)$.

(b) $(x+y)^4$, $(x-y)^4$, and $\tfrac{1}{2}\big((x+y)^4 \pm (x-y)^4\big)$.

(c) $(x+y)^6$.

Exercise 3.3.9. Assume x, y, and z are real. State and prove a *trinomial theorem* for $(x+y+z)^n$.

Exercise 3.3.10. ($\bigstar$) For each n in $\mathbf{N}$, let $P(n)$ be the inequality $3^n < n!$.

(a) Prove that $P(k)$ implies $P(k+1)$ if $k \ge 2$.

(b) Prove there exists an n_0 such that $P(n_0)$ is true.

Exercise 3.3.11. If m and n are positive integers such that $m < n$, then

$$\left[\frac{n+1}{n}\right]^m = \left[1+\frac{1}{n}\right]^m < 1+\frac{m^2}{n}.$$

Exercise 3.3.12. Assume m is an integer greater than 1. Prove there exists an integer N such that if $n \geq N$, then $n^m \leq 2^n$. Hint: Use Exercise 3.3.11.

Exercise 3.3.13.(H) Prove the binomial theorem by induction on the exponent, taking $P(n)$ to be

$$(x+y)^n = \sum_{k\in\mathbf{Z}} \binom{n}{k} x^{n-k}y^k.$$

The sum is taken over all integers k, but the summands are 0 unless $0 \leq k \leq n$.

Exercise 3.3.14. (Binary, or base 2, representation of natural numbers) For purposes of this question, a *bit string* is a real sequence $(b_k) = (b_k)_{k\in\mathbf{N}}$ such that $b_k = 0$ or 1 for each k, and $b_k = 0$ except for finitely many k. We say two bit strings (b_k) and (b'_k) are *equal* if $b'_k = b_k$ for all k.

Suppose N is a natural number. If (b_k) is a bit string such that $k \geq N$ implies $b_k = 0$, define its *bitrep* to be the natural number

$$\sum_{k\in\mathbf{N}} b_k \cdot 2^k = \sum_{k=0}^{N-1} b_k \cdot 2^k$$

$$= b_0 + b_1 \cdot 2 + b_2 \cdot 2^2 + b_3 \cdot 2^3 + \cdots + b_{N-1} \cdot 2^{N-1}.$$

Prove that every natural number is uniquely written as a bitrep.

4

The Real Number Line

Fundamental concepts of real analysis, especially limits, depend not on individual numbers, but on infinite objects, especially sequences (Chapter 6) and intervals. This chapter introduces notation, definitions, theorems, and important examples related to infinite sets and the completeness axiom. Unless stated otherwise, a and b denote real numbers such that $a < b$.

4.1 Sets of Real Numbers

Definition 4.1.1. The sets

$$(a, b) = \{x \text{ in } \mathbf{R} : a < x < b\}$$
$$[a, b] = \{x \text{ in } \mathbf{R} : a \le x \le b\}$$

are called the *open interval* and the *closed interval* with endpoints a and b.

Remark 4.1.2. There are also *half-open* intervals $[a, b)$ and $(a, b]$, though these play only passing roles in this book. ◇

Definition 4.1.3. An interval of real numbers with endpoints a and b has *center* $x_0 = \frac{1}{2}(b + a)$, *radius* $r = \frac{1}{2}|b - a|$, and *length* $|b - a| = 2r$, Figure 4.1.

Lemma 4.1.4. *Assume* $a < b$. *If* $x_0 = \frac{1}{2}(b + a)$ *and* $r = \frac{1}{2}|b - a|$, *then*

$$(a, b) = \{x \text{ in } \mathbf{R} : |x - x_0| < r\}, \qquad [a, b] = \{x \text{ in } \mathbf{R} : |x - x_0| \le r\}.$$

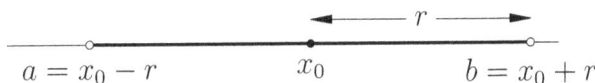

FIGURE 4.1
An interval represented using its center and radius.

DOI: 10.1201/9781003601357-4

Proof. By Proposition 3.2.24, $a = \min(a, b) = x_0 - r$ and $b = \max(a, b) = x_0 + r$. If x is real, then $x \in (a, b)$ if and only if $a < x < b$, namely, $x_0 - r < x < x_0 + r$. Subtracting x_0 from each term gives $-r < x - x_0 < r$. By Proposition 3.2.10 this is equivalent to $|x - x_0| < r$. Replacing the "$<$"s with "$\leq$" proves the assertion about the closed interval. $\qquad\qquad\qquad\qquad\qquad\qquad\qquad\qquad\qquad\qquad\quad\square$

Definition 4.1.5. Assume x_0 and r are real numbers, and $r > 0$. The *open ball of radius r about x_0* is the open interval

$$B_r(x_0) = (x_0 - r, x_0 + r).$$

Remark 4.1.6. Open balls are nested by transitivity of inequality: If $r' < r$, then $|x - x_0| < r'$ implies $|x - x_0| < r$, so $B_{r'}(x_0) \subseteq B_r(x_0)$. $\diamond$

Definition 4.1.7. Assume x_0 and r are real numbers, and $r > 0$. The *punctured ball* of radius r about x_0 is the set $B_r^\times(x_0) = B_r(x_0) \setminus \{x_0\}$, Figure 4.2.

Remark 4.1.8. That is, $B_r^\times(x_0)$ is the open ball $B_r(x_0)$ from which the center x_0 has been removed, namely the set

$$\{x \text{ in } \mathbf{R} : 0 < |x - x_0| < r\} = (x_0 - r, x_0) \cup (x_0, x_0 + r).   \qquad \diamond$$

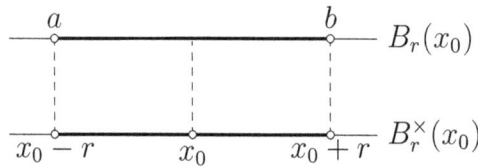

FIGURE 4.2
The open ball and punctured ball of radius r about x_0.

Remark 4.1.9. Distinct real numbers are not "neighbors": If $x \neq x_0$, there is a "gulf" separating x and x_0. In order to bridge this gulf, we consider "neighbors" of x_0 not as individual numbers, but as open balls and punctured balls. $\diamond$

Definition 4.1.10. If x_0 is real, a *neighborhood* of x_0 is a set containing $B_r(x_0)$ for some positive real r. A *punctured neighborhood* of x_0 is a neighborhood from which x_0 has been removed.

Operations on Sets

We can build useful, surprisingly complicated sets of real numbers by recursively applying simple operations to simple sets.

Definition 4.1.11. Assume $A \subseteq \mathbf{R}$ and k is real. The *translate* of A by k is the set

$$k + A = \{x \text{ in } \mathbf{R} : x = k + a \text{ for some } a \text{ in } A\}.$$

The *scaling* of A by k is the set

$$kA = \{x \text{ in } \mathbf{R} : x = ka \text{ for some } a \text{ in } A\}.$$

In particular, $-A = \{-a : a \in A\}$ is the reflection of A across the origin.

Remark 4.1.12. If $k = 0$, then $k + A = A$ and $kA = \{0\}$. In practice, we usually assume $k \neq 0$. ◇

Example 4.1.13. If q is a positive integer, the scaling of $\mathbf{Z}$ by $1/q$,

$$(1/q)\mathbf{Z} = \{x \text{ in } \mathbf{R} : x = p/q \text{ for some integer } p\},$$

consists of all rational numbers that can be represented as a (possibly improper and/or non-reduced) fraction whose denominator is *precisely* q. The elements of $(1/q)\mathbf{Z}$ are spaced regularly along the number line, with adjacent elements separated by a distance of $1/q$, Figure 4.3.

If q and q' are positive, then $(1/q)\mathbf{Z} \subseteq (1/q')\mathbf{Z}$ if and only if q divides q'. ◇

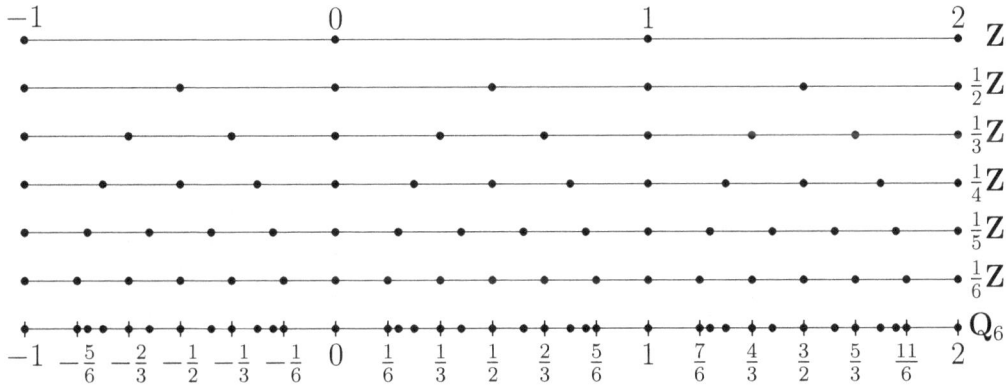

FIGURE 4.3
The sets $\frac{1}{q}\mathbf{Z}$ and $\mathbf{Q}_N$.

Example 4.1.14. For each positive integer N, the set

$$\mathbf{Q}_N := \bigcup_{q=1}^{N} (1/q)\mathbf{Z}$$

consists of all rational numbers that can be represented as a fraction whose denominator is *no larger than* N, Figure 4.3.

The sets $\mathbf{Q}_N$ are nested outward: $\mathbf{Q}_N \subseteq \mathbf{Q}_{N+1} = \mathbf{Q}_N \cup [1/(N+1)]\mathbf{Z}$. The set of rational numbers can be expressed as

$$\mathbf{Q} = \bigcup_{q \in \mathbf{Z}^+} (1/q)\mathbf{Z} = \bigcup_{N \in \mathbf{Z}^+} \mathbf{Q}_N. \qquad \Diamond$$

Definition 4.1.15. A *splitting* of the closed interval $I = [a, b]$ is a finite set $\Pi \subseteq [a, b]$ (Pi) containing both endpoints. Every splitting may be written uniquely in the form $\{t_i\}_{i=0}^n$ with $t_0 = a$, $t_n = b$, and $t_i < t_{i+1}$ if $0 \le i < n$. With this understanding, $I_i := [t_i, t_{i+1}]$ is called the *ith piece* of the splitting; its length is $\Delta t_i := t_{i+1} - t_i$, see Figure 4.4.

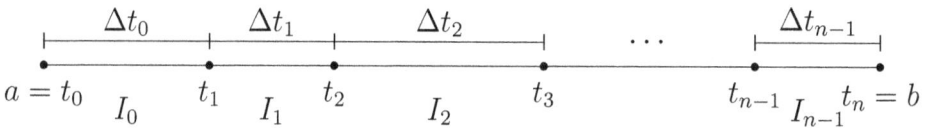

FIGURE 4.4
Splitting a closed interval $[a, b]$ into n pieces.

Example 4.1.16. Assume n is a positive integer, and put $\Delta t = (b - a)/n$. The *equal-length splitting* of $[a, b]$ with n pieces is $\Pi_n = \{a + i\,\Delta t\}_{t=0}^n$. $\qquad \Diamond$

Example 4.1.17. Assume $[a, b]$ is an arbitrary closed interval and Π_3 is the equal-length splitting with 3 pieces. By definition, *removing the open middle third of* $[a, b]$ gives the set

$$[a, b]^{\vee} := [a, b] \setminus (t_1, t_2) = [t_0, t_1] \cup [t_2, t_3]. \qquad \Diamond$$

Definition 4.1.18. If $\{I_i\}_{i=0}^n$ is a finite collection of *disjoint* closed intervals, we call each I_i a *(connected) component* of the union $\bigcup_i I_i$.

Example 4.1.19 (The ternary set). Recursively construct sets $(K_n)_{n \in \mathbf{N}}$ as follows. Let $K_0 = [0, 1]$ be the closed unit interval. Inductively, if n is a natural number and $K_n = \bigcup_i I_i$ is a finite union of 2^n disjoint closed intervals each of length 3^{-n}, define $K_{n+1} = \bigcup_i I_i^{\vee}$, the result of removing the open middle third of each component of K_n, Figure 4.5. The sets K_n are nested inward: $K_n \supseteq K_{n+1}$ for each n. The *ternary set* is defined to be the intersection,

$$K = \bigcap_{n \in \mathbf{N}} K_n.$$

Each endpoint of K_n is an element of K_{n+1}, and consequently "persists" in K. The ternary set therefore contains the union of the endpoints of the

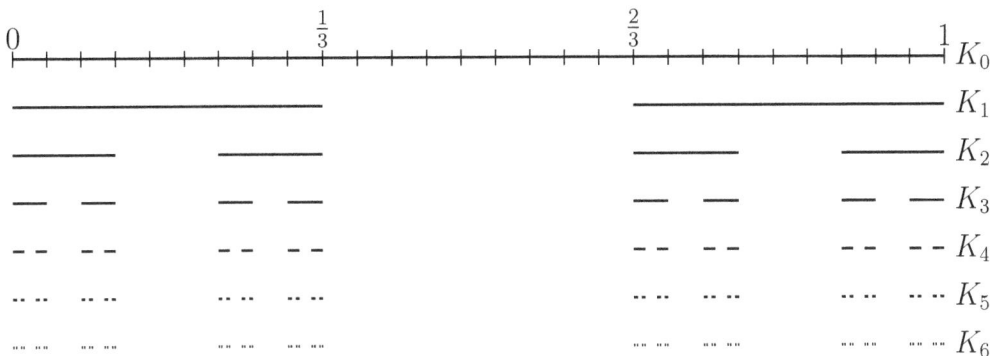

FIGURE 4.5
Approximations to the ternary set.

sets K_n. Since K_n has 2^{n+1} endpoints, K has at least 2^{n+1} elements for each natural number n, and therefore has infinitely many elements.

The set K also contains "non-endpoint" elements. In a well-defined sense, and with terrific understatement, "most" elements of the ternary set are not endpoints of any K_n, see Remark 5.4.22.

The ternary set is *self-similar*. Specifically, $K = (1/3)K \cup (1/3)(K+2)$ is a union of two disjoint subsets, each a scaled copy one-third the size of the entire set. $\diamond$

Upper and Lower Bounds

In Chapter 3 we saw "largest" and "smallest" elements of a set. These useful concepts are, unfortunately, too restrictive for real analysis. The limitation is already seen with an open interval such as $(0,1)$. Loosely, the right endpoint 1 "should be" the largest element, but fails to be since 1 is not in the interval. Similarly, 0 should be the smallest element, but fails to be.

In Section 4.2, we isolate a suitable generalization of "largest element": the "least upper bound" or "supremum" of a set A, the smallest real number greater than or equal to every element of A. The corresponding generalization for a "smallest element" is the "greatest lower bound" or "infimum" of a set. We approach these definitions in stages, first discussing "bounds" on a set.

Definition 4.1.20. Assume A is a set of real numbers. A real number U is an *upper bound* of A if $x \leq U$ for every x in A. If there exists an upper bound of A, we say A is *bounded above* (in **R**).

A real number L is a *lower bound* of A if $L \leq x$ for every x in A. If there exists a lower bound of A, we say A is *bounded below* (in **R**).

The set A is *bounded* if A is bounded above and bounded below.

Remark 4.1.21. If U is an upper bound of A and if $U < U'$, then U' is an upper bound of A by transitivity of inequality. Similarly, if L is a lower bound of A and if $L' < L$, then L' is also a lower bound of A, Figure 4.6.   ◇

FIGURE 4.6
Upper and lower bounds for an interval.

Example 4.1.22. Every interval $[a, b]$ is bounded. The left-hand endpoint a is a lower bound, and the right-hand endpoint b is an upper bound.

Every subset of a bounded set is bounded. (Why?) For example, the open interval (a, b) and the ternary set $K \subseteq [0, 1]$ are bounded.   ◇

Remark 4.1.23. If $A \subseteq \mathbf{R}$ is non-empty, then the following are equivalent:

(i) L is a lower bound of A and U is an upper bound of A.

(ii) $A \subseteq [L, U]$.   ◇

Example 4.1.24. The set $\mathbf{N}$ of natural numbers is bounded below; 0 is a lower bound. In Section 4.3, see Theorem 4.3.1, we will prove $\mathbf{N}$ is not bounded above in $\mathbf{R}$: For every real number x, there exists a natural number n such that $x < n$.

The set $\mathbf{Z}$ of integers is not bounded above or below in $\mathbf{R}$, nor is any set that contains $\mathbf{Z}$; thus $\mathbf{Q}$ and $\mathbf{R}$ are not bounded above or below in $\mathbf{R}$.   ◇

Lemma 4.1.25. *A subset A of $\mathbf{R}$ is bounded if and only if there exists a positive real number M such that $|x| \leq M$ for all x in A.*

Remark 4.1.26. Practically speaking, when a set is bounded we may as well work with "symmetric" upper and lower bounds.   ◇

Proof. If there exists an M such that $|x| \leq M$ for all x in A, then $-M \leq x \leq M$ for all x in A; that is, $L = -M$ is a lower bound of A, and $U = M$ is an upper bound of A.

Conversely, if $L \leq x \leq U$ for all x in A, take $M = \max\big(|L|, |U|\big)$. For all x in A, we have $-M \leq -|L| \leq L \leq x \leq U \leq |U| \leq M$.   □

Exercises for Section 4.1

Exercise 4.1.1. (★) If $r > 0$ and x_0 is real, prove that $B_r^\times(x_0) = x_0 + B_r^\times(0)$.

Exercise 4.1.2. Assume A is a non-empty set of real numbers, the real number L is a lower bound of A, and the real number U is an upper bound of A. If k is an arbitrary real number, find lower and upper bounds on $k + A$. If $k > 0$, find lower and upper bounds on the sets kA and $(-k)A$.

Exercise 4.1.3. Assume b is a non-zero real number and $0 < r \leq |b|/2$. If $x \in B_r(|b|)$, namely, if $\big| x - |b| \big| < r \leq |b|/2$, then $|b|/2 < |x|$ and $1/|x| < 2/|b|$.

Exercise 4.1.4. Use the given strategies to prove that for every x in (a, b), there exists a positive real ε such that $B_\varepsilon(x) \subseteq (a, b)$.

(a) If $x \in (a, b)$, show $\varepsilon = \min(x - a, b - x)$ is positive, and use properties of real inequalities to prove $B_\varepsilon(x) \subseteq (a, b)$.

(b) Write the interval (a, b) as an open ball with center x_0 and radius r. If $x \in B_r(x_0)$, show $\varepsilon := r - |x - x_0|$ is positive, and use the triangle inequality to prove $B_\varepsilon(x) \subseteq B_r(x_0) = (a, b)$.

Exercise 4.1.5. (★) Suppose $J = [a, b]$ and $J' = [a', b']$ are closed intervals whose union is an interval. Prove $J \cap J'$ is non-empty.

Exercise 4.1.6. (H) Suppose $J = (a, b)$ and $J' = (a', b')$ are open intervals whose union is an open interval. Prove $J \cap J'$ is an open interval.

Exercise 4.1.7. Assume a and b are real numbers such that $a < b$. Prove that for every positive real r, the intersection $B_r(b) \cap (a, b)$ is non-empty.

4.2 The Completeness Axiom

A non-empty set of real numbers may be bounded (above and below) yet have neither a maximum nor a minimum.

Example 4.2.1. The open interval (a, b) is bounded below by a and bounded above by b, but contains no smallest or largest element: If $x \in (a, b)$, namely, if $a < x < b$, then by Proposition 3.2.6,

$$a < \tfrac{1}{2}(a + x) < x < \tfrac{1}{2}(x + b) < b.$$

In words, $\tfrac{1}{2}(a + x) \in (a, b)$ and is smaller than x (so x is not the minimum of (a, b)), while $\tfrac{1}{2}(x + b) \in (a, b)$ and is larger (so x is not the maximum). ◇

Definition 4.2.2. Assume A is a set of real numbers that is bounded above. A real number β is called a *least upper bound* or *supremum* of A if

(i) $x \leq \beta$ for all x in A, namely, β is an upper bound of A.

(ii) For every upper bound U of A, we have $\beta \leq U$.

Lemma 4.2.3. *If $A \subseteq \mathbf{R}$, and β and β' are suprema of A, then $\beta = \beta'$.*

Proof. By hypothesis, β' is an upper bound of A, so $\beta \leq \beta'$ by condition (ii). Reversing roles, $\beta' \leq \beta$. $\qquad\qquad\qquad\qquad\qquad\qquad\qquad\qquad\qquad\qquad$ $\square$

Remark 4.2.4. Lemma 4.2.3 guarantees we may speak of *the* supremum of A rather than *a* supremum of A. We are further justified in writing $\sup A$ to denote the supremum of A. For a given set A, the symbol $\sup A$ may signify no real number at all, but it never signifies more than one. $\qquad\qquad\quad$ ◇

Remark 4.2.5. The completeness axiom for the real number system says: If A is a *non-empty* set of real numbers that is bounded above, then A has a real supremum. Loosely, every set of real numbers that "should" have a real supremum *does* have a real supremum.

Completeness underpins all of real analysis. $\qquad\qquad\qquad\qquad\qquad\qquad\quad$ ◇

Remark 4.2.6. If A has a largest element, then $\sup A = \max A$. In this sense, suprema generalize maxima.

For every non-empty set that is bounded above, $\sup A$ is the leftmost number lying on or to the right of every element of A, Figure 4.7. $\qquad\quad$ ◇

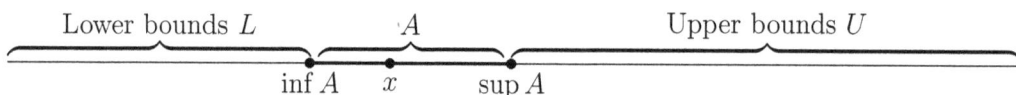

FIGURE 4.7
The supremum and infimum of an interval.

An adversarial game-like formulation of a supremum is useful in practice:

Lemma 4.2.7. *If A is a set of real numbers having a supremum β, then β is the unique real number satisfying the conditions*

(i) *If $x \in A$, then $x \leq \beta$.*

(ii)' *If $\varepsilon > 0$, then there exists an x in A such that $\beta - \varepsilon < x$.*

Proof. Conceptually, the conditions (ii) and (ii)' are contrapositives.

In other words, β is the supremum of A if and only if every upper bound U of A satisfies $\beta \leq U$. The contrapositive says every real M such that $M < \beta$ *fails* to be an upper bound of A. That is, if $\varepsilon > 0$, there exists an x in A such that $M = \beta - \varepsilon < x$. $\qquad\qquad\qquad\qquad\qquad\qquad\qquad\qquad\qquad$ $\square$

Everything said above for upper bounds has a corresponding concept or statement for lower bounds.

Definition 4.2.8. A real number α is called a *greatest lower bound* or *infimum* of A if

(i) α is a lower bound of A, and

(ii) If L is a real lower bound of A, then $L \leq \alpha$.

Remark 4.2.9. Geometrically, an infimum of A lies to the left of A, but is the rightmost such point. Infima are unique (if they exist), so we are justified in writing inf A. For practice, write out an alternative characterization of infima corresponding to Lemma 4.2.7. ◇

Lemma 4.2.10 (Constriction). *Assume A and B are non-empty sets of real numbers. If $A \subseteq B$, then* inf $B \leq$ inf $A \leq$ sup $A \leq$ sup B.

Proof. Every lower bound of B is *a fortiori* a lower bound of A. In particular, inf $B \leq$ inf A since inf B is a lower bound of B. Similarly, sup $A \leq$ sup B. Finally, if $a \in A$, then inf $A \leq a \leq$ sup A. □

To illustrate the power of suprema, we'll establish that every positive real number b has a positive real *square root* $\sqrt{b}$.

Theorem 4.2.11 (Real square roots). *If b is a positive real number, there exists a unique positive real number $\sqrt{b}$ such that $(\sqrt{b})^2 = b$.*

Proof. We'll prove existence if $b = 2$. Existence in general and uniqueness are Exercise 4.2.4. Consider the set $A = \{x \text{ in } \mathbf{R} : x^2 < 2\}$. It suffices to prove A is non-empty, bounded above, and $(\sup A)^2 = 2$. In the course of the argument, we'll actually prove more: If U is a positive real number, then A is bounded above by U *if and only if* $2 \leq U^2$.

Since $1^2 = 1 < 2$, we have $1 \in A$, so A is non-empty.

Next, assume U is positive and $2 \leq U^2$. By Proposition 3.2.1 (iii), if $U \leq U'$ then $2 \leq U^2 \leq U'^2$, so $U' \notin A$. Contrapositively, if $x \in A$, then $x < U$; in other words, A is bounded above by every positive real U satisfying $2 \leq U^2$.

Since A is non-empty and bounded above, $\sqrt{2} := \sup A$ exists by completeness, and $1 \leq \sqrt{2}$. It suffices to prove $(\sqrt{2})^2 = 2$. We'll prove, contrapositively:

1. If $0 < U$ and $2 < U^2$, there exists a smaller upper bound of A, so $U \neq \sqrt{2}$;

2. If $0 < V$ and $V^2 < 2$, then V is not an upper bound of A, so $V \neq \sqrt{2}$.

Trichotomy implies $(\sqrt{2})^2 = 2$.

(Proof of 1). If $0 < U$ and $2 < U^2$, set $U' = (1/2)(U + (2/U))$. Since

$$0 < \frac{U^2 - 2}{2U} = \frac{1}{2}\left[U - \frac{2}{U}\right] = U - \frac{1}{2}\left[U + \frac{2}{U}\right] = U - U',$$

we have $U' < U$. Further,

$$0 < \frac{(U^2 - 2)^2}{(2U)^2} = \frac{(U^2 + 2)^2}{(2U)^2} - 2 = U'^2 - 2,$$

so $2 < U'^2$. This establishes 1.

(Proof of 2). By Proposition 3.2.1, if $0 < V$ and $V^2 < 2$, then $U := 2/V$ is positive and $2 < U^2$. If we define U' as above and set $V' := 2/U'$, then $V < V'$ and $V'^2 < 2$, so $V' \in A$. Consequently, V is not an upper bound of A.   □

Getting comfortable with suprema takes practice. To illustrate the definition further, we'll establish three useful formal properties.

Proposition 4.2.12. *Assume A is a bounded, non-empty set of real numbers, and k is real.*

(i) $\sup(k + A) = k + \sup A$.

(ii) *If $k \geq 0$, then* $\sup(kA) = k \sup A$.

(iii) *If $k > 0$, then* $\sup(-kA) = -k \inf A$.

Example 4.2.13. If $A = [a, b]$ is an interval, then $k + A = [k+a, k+b]$. If $k \geq 0$, then $kA = [ka, kb]$ and $-kA = [-kb, -ka]$. In each case, the supremum can be read off by inspection. This is a useful way of remembering the conclusion of Proposition 4.2.12.   ◇

Proof. The strategy in each part is to show that the right-hand side satisfies the two conditions for a supremum in Definition 4.2.2.

(i). Since $k + A = \{x' \text{ in } \mathbf{R} : x' = k + x \text{ for some } x \text{ in } A\}$, a real number U is an upper bound of A if and only if $x \leq U$ for all x in A,

if and only if $x' := k + x \leq k + U$ for all x' in $k + A$,

if and only if $k + U$ is an upper bound of $k + A$.

But $\sup A$ is an upper bound of A, so $k + \sup A$ is an upper bound of $k + A$. By Definition 4.2.2 (ii), $\sup(k + A) \leq k + \sup A$.

Conversely, $\sup(k + A)$ is an upper bound of $k + A$, so $\sup(k + A) - k$ is an upper bound of A. By Definition 4.2.2 (ii), $\sup A \leq \sup(k + A) - k$, or $k + \sup A \leq \sup(k + A)$. This completes the proof. (Why?)

(ii). If $k = 0$, then $kA = \{0\}$ and the claim is immediate. Assume, therefore, that $k > 0$. The proof in this case is a mechanical modification of the preceding argument. The main idea is, multiplication or division by k preserves the sense of an inequality since k is positive, so a real number U is an upper bound of A if and only if kU is an upper bound of kA. Just as in the preceding argument, $k \sup A$ is an upper bound of kA, so $\sup(kA) \leq k \sup A$, and $\sup(kA)/k$ is an upper bound of A, so $k \sup A \leq \sup(kA)$.

(iii). Since $-k < 0$, multiplication or division by $-k$ *reverses* inequalities. Consequently, L is a *lower* bound of A if and only if $U' = -kL$ is an *upper* bound of $-kA$. Particularly, $-k \inf A$ is an upper bound of $-kA$, so $\sup(-kA) \leq -k \inf A$. Conversely, $\sup(-kA)/(-k)$ is a lower bound of A, so $\sup(-kA)/(-k) \leq \inf A$, or $-k \inf A \leq \sup(-kA)$.

The proofs of the last two parts are only sketched. One hope is to convey the conceptual frame of the proofs uncluttered by details. A second, complementary aim is to coax you to study part (i) carefully, then fill in details for (ii) and (iii) as practice working with suprema yourself. □

Finally, we prove a technical result that recurs in multiple guises.

Theorem 4.2.14 (Interval induction). *Assume $[a,b]$ is a closed, bounded interval of real numbers, $I \subseteq \mathbf{R}$, and $J := \{t \text{ in } [a,b] : [a,t] \subseteq I\}$. Assume J satisfies*

(i) (*Priming*) $a \in J$.

(ii) (*Climbing*) *If $t \in J$, then $B_r(t) \cap [a,b] \subseteq J$ for some positive real r.*

(iii) (*Capping*) $\sup J \in J$.

Then $[a,b] \subseteq J$.

Proof. Since $a \in J \subseteq [a,b]$, climbing implies $[a, a+r) \cap [a,b] \subseteq J$ for some positive real r. Thus J is non-empty, so $\sup J$ is real and $a < \sup J \leq b$.

By construction, $t \in J$ if and only if $[a,t] \subseteq I$, if and only if $[a,s] \subseteq I$ for all s such that $a \leq s \leq t$, if and only if $[a,t] \subseteq J$.

Since $\sup J \in J$ by capping, $[a, \sup J] \subseteq J$. It suffices to prove $\sup J = b$. Contrapositively, if $t \in J$ and $t < b$, then climbing implies $B_r(t) \cap [a,b] \subseteq J$ for some positive real r. Particularly, $t' := \min(t + \frac{1}{2}r, b)$ is in J and is greater than t, so t is not an upper bound of J. Since $\sup J$ is an upper bound of J, we have $\sup J = b$ and $[a,b] \subseteq J$. □

Exercises for Section 4.2

Exercise 4.2.1.(★) Prove $\sup(a,b) = b$. Suggestion: Use condition (ii)′.

Exercise 4.2.2. Suppose A and B are non-empty sets of real numbers. State the converse of constriction, and determine whether the converse is true.

Exercise 4.2.3. Assume $C \subseteq (a,b)$ is a finite set of real numbers and that $A = (a,b) \setminus C$. Prove $\sup A = b$.

Exercise 4.2.4. Assume $b > 0$.

(a) Finish the proof of Theorem 4.2.11, that b has a positive real square root.

(b) Use Exercise 3.1.4 to prove b has a *unique* positive square root.

(c) Prove every positive real number has a unique positive fourth root.

Exercise 4.2.5. Assume A is a bounded, non-empty set of real numbers, and k is real. Formulate and prove a version of Proposition 4.2.12 for infima.

Exercise 4.2.6. (H) Assume $I_n = [a_n, b_n]$ is a *closed*, bounded interval of real numbers for each natural number n. Prove that if the intervals are nested inward, namely, $I_n \supseteq I_{n+1}$ for every n, then the intersection $\bigcap_n I_n$ is non-empty.

Exercise 4.2.7. (H) Assume $[a,b]$ is a closed, bounded interval of real numbers, and that U and V are disjoint, non-empty open intervals whose union contains $[a,b]$. Prove that either $[a,b] \subseteq U$ or $[a,b] \subseteq V$.

Exercise 4.2.8. This exercise continues (and finishes) the introduction to complex numbers in Exercise 3.2.15, and uses the same notation. If z is complex, the non-negative real number $|z| := \sqrt{z\bar{z}}$ is the *magnitude* of z. This exercise establishes properties of magnitude formally identical to the triangle and reverse triangle inequalities.

(a) Prove that if z and w are complex, then $|zw| = |z|\,|w|$ and $|z+w| \le |z|+|w|$. Hint for the second: Square both sides. Note that $z\bar{w} + \bar{z}w$, is twice the real part of $z\bar{w}$, and thus no larger than $2|z|\,|w|$.

(b) Prove that if z and w are complex, then $\big||z| - |w|\big| \le |z \pm w|$.

4.3   Finitude of Real Numbers

Although the set $\mathbf{N}$ of natural numbers is infinite, each specific natural number is finite. The completeness axiom implies, analogously, that there are no real infinities: Every real number is finite in the following sense.

Theorem 4.3.1 (Finitude). *For every number x, there exists a natural number n such that $x < n$.*

Proof. Suppose ω (omega, the last letter of the Greek alphabet) is a real number and $\omega - 1$ is *not* an upper bound of $\mathbf{N}$, that is, there exists a natural number n_0 such that $\omega - 1 < n_0$. Adding 1 to both sides shows $\omega < n_0 + 1$. Since $n_0 + 1$ is a natural number, ω is not an upper bound of $\mathbf{N}$, either.

Contrapositively, if ω is a real upper bound of $\mathbf{N}$, then $\omega - 1$ is also a real upper bound of $\mathbf{N}$. But this means $\mathbf{N}$ has no real supremum: For every real upper bound of $\mathbf{N}$ there is a strictly smaller real upper bound. By the completeness axiom, $\mathbf{N}$ is not bounded above in $\mathbf{R}$. In other words, for every real number x, there exists a natural number n such that $x < n$. □

In school, real numbers are introduced as having an "integer part" and a "decimal part." This useful representation of real numbers is not immediate from the definition, but a consequence of finitude.

Corollary 4.3.2. *For every real number x, there exist a unique integer $\lfloor x \rfloor$ and a unique real number x' such that $x = \lfloor x \rfloor + x'$ and $0 \le x' < 1$.*

Proof. Exercise 4.3.5. □

Definition 4.3.3. If x is a real number, the *floor* of x is the integer $\lfloor x \rfloor$ in Corollary 4.3.2.
The *ceiling* of x is the integer

$$\lceil x \rceil = \begin{cases} \lfloor x \rfloor + 1 & \text{if } x \text{ is not an integer,} \\ \lfloor x \rfloor & \text{if } x \text{ is an integer.} \end{cases}$$

Example 4.3.4. If $x = 42$, then $\lfloor x \rfloor = \lceil x \rceil = x = 42$.
If $x = 3.14159$, then $\lfloor x \rfloor = 3$, $x' = 0.14159$, and $\lceil x \rceil = 4$.
If $x = -3.14159$, then $\lfloor x \rfloor = -4$, $x' = 0.85841$, and $\lceil x \rceil = -3$.
Generally, every real number x is either an integer (and $\lfloor x \rfloor = \lceil x \rceil = x$), or is strictly between two consecutive integers ($\lfloor x \rfloor < x < \lceil x \rceil = \lfloor x \rfloor + 1$). ◇

Corollary 4.3.5 (The accretion principle). *For every real number M and every positive ε, there exists a positive integer n such that $M < n\varepsilon$.*

Remark 4.3.6. Metaphorically, a journey of 1000 miles (M) can be accomplished one step (ε) at a time, no matter how small the steps. ◇

Proof. Since M/ε is real, finitude implies there exists a positive integer n such that $M/\varepsilon < n$. Multiplying by ε gives $M < n\varepsilon$. □

Corollary 4.3.7 (Reciprocal finitude). *For every positive real number ε, there exists a positive integer n such that $1/n < \varepsilon$. Contrapositively, if $\varepsilon \le 1/n$ for every n, then $\varepsilon \le 0$.*

Proof. By finitude, there exists a positive integer n such that $1/\varepsilon < n$. Proposition 3.2.1 (v) implies $1/n < \varepsilon$. $\qquad\qquad\qquad\qquad\qquad\qquad\qquad\qquad\square$

Example 4.3.8. In this example, we'll *denote* the set $\{x \text{ in } \mathbf{R} : 0 < x\}$ of positive real numbers by $(0, \cdot)$. (We'll introduce more suggestive notation shortly.) For each positive integer n, consider the intervals $A_n = (0, 2^n)$. We will use finitude to prove

$$\bigcup_{n \in \mathbf{N}} A_n = \bigcup_{n \in \mathbf{N}} (0, 2^n) = (0, \cdot).$$

In particular, a union of bounded intervals may be unbounded.

Let's establish this equality of sets from the definitions. A real number x is in the union of the A_n if and only if x is an element of $(0, 2^n)$ for some positive integer n. On the other hand, by definition x is an element of $(0, \cdot)$ if and only if $0 < x$. We must show each of these conditions implies the other.

If $0 < x < 2^n$ for some n, then $0 < x$ in particular; $(0, 2^n) \subseteq (0, \cdot)$ for all n. Consequently, the union of $(0, 2^n)$ is contained in $(0, \cdot)$. In symbols, $\bigcup_n A_n \subseteq (0, \cdot)$.

Conversely, assume $0 < x$. By finitude, there exists a natural number n such that $x < n$. By Proposition 3.3.5, $n < 2^n$. Thus $x \in A_n$ for this n, and therefore x is in the union of the A_n. In symbols, $\bigcup_n A_n \supseteq (0, \cdot)$. $\qquad\qquad\Diamond$

Example 4.3.9. For each positive integer n, let $B_n = (0, 2^{-n})$. The intervals $\{B_n\}_{n \in \mathbf{N}}$ are nested inward: $B_n \supseteq B_{n+1}$. We will use reciprocal finitude to prove

$$\bigcap_{n \in \mathbf{N}} B_n = \bigcap_{n \in \mathbf{N}} (0, 2^{-n}) = \varnothing.$$

In particular, an intersection of nested, non-empty sets may be empty.

Again, let's establish this equality of sets from the definitions. The inclusion $\supseteq$ is vacuous, so it suffices to show $(\subseteq)$ that no real number is contained in the intersection.

Assume $x \in \mathbf{R}$. If $x \leq 0$, then $x \notin B_0 = (0, 1)$, so $x \notin \bigcap_n B_n$. If instead $0 < x$, Corollaries 3.3.6 and 4.3.7 imply there exists a positive integer n such that $1/2^n < 1/n < x$. For this n we have $x \notin B_n$, which implies $x \notin \bigcap_n B_n$. We have shown that every real number fails to be in the intersection, implying the intersection is empty. $\qquad\qquad\qquad\qquad\qquad\qquad\qquad\qquad\qquad\qquad\qquad\qquad\Diamond$

Density of the Rational Numbers

Definition 4.3.10. A set A of real numbers is *dense* (in $\mathbf{R}$) if the complement $\mathbf{R} \setminus A$ contains no open interval.

Remark 4.3.11. Equivalently, for every real number x and every positive ε, there exists a number x_0 in A such that $|x - x_0| < \varepsilon$.

If A is dense in $\mathbf{R}$, then every real number can be approximated arbitrarily closely by elements of A. ◇

Theorem 4.3.12 (Density of the rationals). *If x and y are real numbers and $x < y$, then there exists a non-zero rational number r such that $x < r < y$. Particularly, the set $\mathbf{Q}$ of rational numbers is dense in $\mathbf{R}$.*

Proof. Assume first that $0 \leq x < y$. The real number $\varepsilon := y - x$ is positive because $x < y$. By reciprocal finitude, there exists a positive integer n such that $1/n < \varepsilon$. It suffices to prove that some integer multiple $r = m/n$ lies between x and y. Geometrically this is plausible: By taking steps of size $1/n$ across the interval between x and y, we must step in the interior at least once.

Rigorously, consider the set $S = \{p \text{ in } \mathbf{Z}^+ : y \leq p/n\}$. By the accretion principle, there is a positive integer p such that $y < p/n$; that is, the set S is non-empty. By well-ordering, a non-empty set of positive integers has a smallest element, say $m + 1$. By definition of the set S, we have $m/n < y \leq (m+1)/n$. To complete the proof, it suffices to prove $x < m/n$. But

$$y - \frac{m}{n} \leq \frac{m+1}{n} - \frac{m}{n} = \frac{1}{n} < y - x.$$

Rearranging the inequality gives $x < m/n$. We have proven that if $0 \leq x < y$, then there exists a non-zero rational number $r = m/n$ such that $x < r < y$.

If instead $x < y \leq 0$, then $0 \leq -y < -x$. The preceding argument guarantees there exists a non-zero rational number r such that $-y < r < -x$. The non-zero rational number $-r$ satisfies $x < -r < y$.

Finally, if neither alternative holds, then $x < 0 < y$; by the first part of the proof there is a rational number $x < 0 < r < y$. □

Corollary 4.3.13. *The set $\mathbf{R} \setminus \mathbf{Q}$ of irrational numbers is dense in $\mathbf{R}$.*

Proof. See Exercise 4.3.6. □

Remark 4.3.14. It may be tempting to conclude that rational and irrational numbers "alternate" along the number line, as if by painting rational numbers blue and irrational numbers red, the number line would consist of alternating red and blue points. Unfortunately, this picture is utterly incorrect. Distinct real numbers are never adjacent to each other, but instead are endpoints of an interval containing infinitely many rational and irrational numbers. ◇

The Extended Real Numbers

Although every real number is finite, number-like notation for referring to arbitrarily large positive and negative numbers is convenient. Earlier, for example, we wrote $(0, \cdot)$ to denote the set of positive real numbers.

Definition 4.3.15. Let $+\infty$ and $-\infty$ denote objects that are not real numbers, and extend the ordering on the real numbers by declaring $-\infty < x < +\infty$ for every real number x. The set $\overline{\mathbf{R}} = \mathbf{R} \cup \{-\infty, +\infty\}$ is called the *extended real number system*.

Definition 4.3.16. If $A \subseteq \mathbf{R}$, we write $\sup A = +\infty$ if and only if A is not bounded above, and write $\inf A = -\infty$ if and only if A is not bounded below.

Remark 4.3.17. *Every* set A of real numbers has an infimum and a supremum in the extended real number system.

If $A = \varnothing$, then $\sup A = -\infty$ (because every extended real number is an upper bound) and $\inf A = +\infty$. The empty set is the only set whose supremum is smaller than its infimum. If $a \in A$, then

$$-\infty \le \inf A \le a \le \sup A \le +\infty. \qquad \diamond$$

Definition 4.3.18. If a is a real number, we define *unbounded* open and closed intervals by

$$(-\infty, a) = \{x \text{ in } \mathbf{R} : x < a\} \qquad (a, \infty) = \{x \text{ in } \mathbf{R} : a < x\},$$
$$(-\infty, a] = \{x \text{ in } \mathbf{R} : x \le a\} \qquad [a, \infty) = \{x \text{ in } \mathbf{R} : a \le x\}.$$

Example 4.3.19. If $a < b$, then $(a, b) = (-\infty, b) \cap (a, \infty)$. $\qquad \diamond$

Example 4.3.20. For every real a, we have $(-\infty, a) = \mathbf{R} \setminus [a, \infty)$. $\qquad \diamond$

Remark 4.3.21. To emphasize, the symbols $-\infty$ and ∞ do not denote real numbers, but are place-holders for an omitted inequality.

In the same spirit, a sequence $(a_k)_{k \in \mathbf{N}}$ may be denoted $(a_k)_{k=0}^{\infty}$, read, "a sub k for $k = 0$ to infinity." This flexibly handles sequences whose initial index is not naturally 0, such as $(1/k)_{k=1}^{\infty}$. $\qquad \diamond$

Exercises for Section 4.3

Exercise 4.3.1. ($\bigstar$) Assume N is a positive integer. Prove that for every real x, there exists an element x' in $\mathbf{Q}_N$ such that $|x - x'| \le 1/N$. Is there a smaller upper bound?

Exercise 4.3.2. Prove that if $A \subseteq \mathbf{R}$ is dense, then for all real a and b such that $a < b$, we have $\inf[A \cap (a, b)] = a$ and $\sup[A \cap (a, b)] = b$.

Exercise 4.3.3. Assume $A = \{x \text{ in } \mathbf{R} : x = 1/n \text{ for some } n \text{ in } \mathbf{Z}^+\}$. Find $\sup A$ and $\inf A$. (A sketch is not a proof, but may help you guess the answers and guide your construction of a proof.)

Exercise 4.3.4. Assume K is the ternary set, and $A = [0,1] \setminus K$ its complement in the closed unit interval $[0,1]$. Find $\inf A$ and $\sup A$. Does A have a smallest element?

Exercise 4.3.5. (H) Prove Corollary 4.3.2.

Exercise 4.3.6. (H) Irrational numbers exist by Theorem 4.2.11.

(a) If r is a non-zero rational and α is irrational, prove $r\alpha$ is irrational.

(b) Prove the set $\mathbf{R} \setminus \mathbf{Q}$ of irrational numbers is dense in $\mathbf{R}$.

Exercise 4.3.7. An element of the set $\mathbf{Z}[\frac{1}{2}] := \bigcup_{n\in\mathbf{N}} 2^{-n}\mathbf{Z}$ is called a *dyadic rational*. In other words, a dyadic rational has the form $m \cdot 2^{-n}$ for some integer m and some natural number n.

(a) Determine whether $\mathbf{Z}[\frac{1}{2}]$ is closed under addition, multiplication, and/or taking negatives. Is $\mathbf{Z}[\frac{1}{2}]$ a field?

(b) Prove $\mathbf{Z}[\frac{1}{2}]$ is dense in $\mathbf{R}$.

(c) A set $A \subseteq \mathbf{R}$ is *midpoint-closed* if $\{a,b\} \subseteq A$ implies $\frac{1}{2}(a+b) \in A$. Prove that if $\{0,1\} \subseteq A$ and A is closed under addition, taking negatives, and is midpoint-closed, then $\mathbf{Z}[\frac{1}{2}] \subseteq A$.

Exercise 4.3.8. (★) One basic goal of real analysis, which we'll pursue systematically in Chapter 7, is to assign useful numerical values to "infinite sums." Among the prototypical examples is the "geometric series with first term 1 and ratio $1/2$":

$$\sum_{k=0}^{\infty} \frac{1}{2^k} = 1 + \frac{1}{2} + \frac{1}{2^2} + \frac{1}{2^3} + \frac{1}{2^4} + \cdots,$$

which we define to be

$$\sup_{n\in\mathbf{N}} \sum_{k=0}^{n-1} \frac{1}{2^k} = \sup_{n\in\mathbf{N}} \left[1 + \frac{1}{2} + \frac{1}{2^2} + \cdots + \frac{1}{2^{n-1}}\right].$$

Evaluate the supremum with proof.

Exercise 4.3.9. Assume $0 < r < 1$.

(a) Consider the set of powers $A = \{r^n : n \in \mathbf{N}\}$, which is bounded below by 0. Find $\inf A$ with proof.

(b) The *geometric series with first term* 1 *and ratio* r is the formal expression

$$\sum_{k=0}^{\infty} r^k = 1 + r + r^2 + r^3 + \cdots,$$

which we define to be

$$\sup_{n \in \mathbf{N}} \sum_{k=0}^{n-1} r^k = \sup_{n \in \mathbf{N}} \left[1 + r + r^2 + \cdots + r^{n-1} \right].$$

Evaluate the supremum in terms of r.

Exercise 4.3.10. (★) For each natural number m, define

$$H_m = \sum_{k=1}^{2^m} \frac{1}{k} = 1 + \frac{1}{2} + \frac{1}{3} + \cdots + \frac{1}{2^m}.$$

Let $H = \{H_m\}_{m=0}^{\infty}$ denote the set of all these sums. Prove H is not bounded above. Hint: How many summands comprise the difference $H_{m+1} - H_m$, and what is the least each summand could be?

Exercise 4.3.11. Assume p is an integer and $p > 1$. For each positive integer m, define

$$H_m = \sum_{k=1}^{2^m - 1} \frac{1}{k^p} = 1 + \frac{1}{2^p} + \frac{1}{3^p} + \cdots + \frac{1}{(2^m - 1)^p}.$$

Let $H = \{H_m\}_{m=0}^{\infty}$ denote the set of all these sums. Prove H is bounded above. Hint: How many summands comprise the difference $H_{m+1} - H_m$, and what is the largest each summand could be?

Caution: Do not attempt to evaluate the supremum. The suprema are known for p even (each is a rational multiple of π^p—yes, *that* π; the first few are $\pi^2/6$, $\pi^4/90$, $\pi^6/945$), but at this writing all are unknown for p odd.

4.4   Topology

How is a real number x_0 situated relative to a set A of real numbers? If we look at x_0 in isolation, either $x_0 \in A$ or $x_0 \notin A$. If we use open balls and punctured balls in the spirit of Remark 4.1.9, however, we find four mutually exclusive logical conditions. We start with open balls. Throughout, ε connotes a *positive* real number.

Definition 4.4.1. Assume $A \subseteq \mathbf{R}$, and recall that $A^c = \mathbf{R} \setminus A$. For each x_0 in $\mathbf{R}$, precisely one of the following three conditions holds:

(i) There exists an ε such that $B_\varepsilon(x_0) \subseteq A$. In this case we say x_0 is an *interior point* of A.

(ii) There exists an ε such that $B_\varepsilon(x_0) \subseteq A^c$. In this case we say x_0 is an *exterior point* of A.

(iii) For every ε, the sets $B_\varepsilon(x_0) \cap A$ and $B_\varepsilon(x_0) \cap A^c$ are both non-empty. In this case we say x_0 is a *boundary point* of A.

The *interior* of A is the set of interior points of A. The *exterior* and the *boundary* ∂A of A are defined analogously.

Example 4.4.2. Assume a and b are real numbers such that $a < b$.

A	Interior	Exterior	Boundary
$\mathbf{Z}$	$\varnothing$	$\mathbf{R} \setminus \mathbf{Z}$	$\mathbf{Z}$
$\mathbf{Q}$	$\varnothing$	$\varnothing$	$\mathbf{R}$
$\mathbf{R}$	$\mathbf{R}$	$\varnothing$	$\varnothing$
$(-\infty, a)$	$(-\infty, a)$	(a, ∞)	$\{a\}$
$[a, b)$	(a, b)	$(-\infty, a) \cup (b, \infty)$	$\{a, b\}$
$[a, b) \cap \mathbf{Q}$	$\varnothing$	$(-\infty, a) \cup (b, \infty)$	$[a, b]$

$\diamondsuit$

Remark 4.4.3. The exterior of A is the interior of A^c. The boundary of A is the boundary of A^c. In symbols, $\partial A = \partial(A^c)$. $\diamond$

Remark 4.4.4. An interior point of A is an element of A, since $x_0 \in B_\varepsilon(x_0)$ regardless of ε. Similarly, an exterior point of A is not an element of A. However, a boundary point of A may lie in either A or its complement. $\diamond$

Now we turn to punctured open balls, which detect whether or not $A \setminus \{x_0\}$ "neighbors" x_0.

Definition 4.4.5. Assume A is a set of real numbers. If $B_\varepsilon^\times(x_0) \cap A$ is non-empty for every ε, we say x_0 is a *limit point* of A.

The *closure* $\overline{A}$ of A is the union of A and its set of limit points.

Remark 4.4.6. Contrapositively, x_0 is *not* a limit point of A if and only if there exists an ε such that $B_\varepsilon^\times(x_0) \cap A = \varnothing$. $\diamond$

Definition 4.4.7. We say x_0 is an *isolated point of A* if there exists an ε such that $B_\varepsilon(x_0) \cap A = \{x_0\}$, namely, $x_0 \in A$ and x_0 is not a limit point of A.

We say x_0 is a *border point* of A if x_0 is both a limit point of A and boundary point of A.

Proposition 4.4.8. *Assume $A \subseteq \mathbf{R}$. For every real number x_0, precisely one of the following conditions holds:*

	Boundary $= F$	Boundary $= T$
Limit $= F$	*Exterior*	*Isolated*
Limit $= T$	*Interior*	*Border*

Proof. First assume x_0 is not a boundary point of A. By Definition 4.4.1, x_0 is either an interior or exterior point of A. If x_0 is not a limit point, namely, if $B_\varepsilon^\times(x_0) \cap A = \varnothing$ for some positive ε, then x_0 is an exterior point of A. If instead x_0 *is* a limit point, then x_0 is an interior point of A.

Next assume x_0 is a boundary point of A: Every ε-ball about x_0 contains points of A and points of A^c. If x_0 is not a limit point of A, then $B_\varepsilon^\times(x_0) \cap A = \varnothing$ for some ε. That is, $x_0 \in A$, and x_0 is an isolated point of A by definition. Otherwise, x_0 is a border point of A by definition.   □

Remark 4.4.9. By inspection, the boundary of A is the disjoint union of the isolated and border points of A. The set of limit points of A is the disjoint union of the interior and border points of A. The closure of A is the disjoint union of the interior, isolated, and border points of A, namely, the complement of the exterior of A.   ◇

Example 4.4.10. Assume a and b are real numbers such that $a < b$.

A	Interior	Boundary	Isolated	Border	Closure
$\mathbf{Q}$	$\varnothing$	$\mathbf{R}$	$\varnothing$	$\mathbf{R}$	$\mathbf{R}$
$(-\infty, 0) \cup \mathbf{N}$	$(-\infty, 0)$	$\mathbf{N}$	$\mathbf{Z}^+$	$\{0\}$	$(-\infty, 0) \cup \mathbf{N}$
$[a, b) \cup \{b+1\}$	(a, b)	$\{a, b, b+1\}$	$\{b+1\}$	$\{a, b\}$	$[a, b] \cup \{b+1\}$
$[a, b) \cap \mathbf{Q}$	$\varnothing$	$[a, b]$	$\varnothing$	$[a, b]$	$[a, b]$

◇

Definition 4.4.11. Assume $A \subseteq \mathbf{R}$. We say A is an *open set* if every element of A is an interior point of A, namely, if A contains *none* of its boundary points.

We say A is a *closed set* if A contains all of its limit points; that is, $\overline{A} = A$, or A contains *all* of its boundary points.

Remark 4.4.12. Generally, a set contains some, but not all, of its boundary points, and is therefore neither open nor closed.   ◇

Proposition 4.4.13. *If $A \subseteq \mathbf{R}$, the following are equivalent:*

(i) *A is closed.*   (ii) *$\overline{A} = A$.*   (iii) *$\partial A \subseteq A$.*   (iv) *A^c is open.*

Proof. Interior and isolated points are in A. Exterior points are not in A. Each condition is equivalent to "A contains all its border points."   □

Proposition 4.4.14. *An arbitrary open interval (a, b) is an open set. An arbitrary closed interval $[a, b]$ is a closed set.*

Proof. Two proofs of the first claim were requested in Exercise 4.1.4.

For the second, let $x_0 = \frac{1}{2}(b + a)$ be the midpoint and $r = \frac{1}{2}|b - a|$ the radius. It suffices to prove the complement of $[a, b] = [x_0 - r, x_0 + r]$ is open.

Assume x is an arbitrary point of the complement. By hypothesis, $|x - x_0| > r$, so $\varepsilon := |x - x_0| - r > 0$. It suffices to prove $B_\varepsilon(x) \cap [a, b]$ is empty. But if x' is an arbitrary point of $B_\varepsilon(x)$, then $|x' - x| < \varepsilon$. The reverse triangle inequality implies

$$|x' - x_0| = |(x' - x) + (x - x_0)| \geq |x - x_0| - |x' - x| > |x - x_0| - \varepsilon = r.$$

Since $|x' - x_0| > r$, $x' \notin [a, b]$. □

Proposition 4.4.15. *A union of open sets is open, and a finite intersection of open sets is open. Precisely:*

(i) *If $\{O_i\}_{i \in \mathscr{I}}$ is a collection of open subsets of $\mathbf{R}$, then $\bigcup_i O_i$ is open.*

(ii) *If $\{O_i\}_{i=0}^{n-1}$ is a finite collection of open subsets of $\mathbf{R}$, then $\bigcap_i O_i$ is open.*

Proof. (i). Assume $\{O_i\}_{i \in \mathscr{I}}$ is an arbitrary collection of open sets, and let O denote the union $\bigcup_i O_i$. If $x \in O$, then $x \in O_i$ for some i. Since O_i is open, there exists an ε such that $B_\varepsilon(x) \subseteq O_i$. But $O_i \subseteq O$, so x is an interior point of O; since x was arbitrary, the union O is open.

(ii). Assume $\{O_i\}_{i=0}^{n-1}$ is a finite collection of open subsets of $\mathbf{R}$, and let O denote the intersection $\bigcap_i O_i$. If $x \in O$, then $x \in O_i$ for all i, so there exist positive numbers $\{\varepsilon_i\}_{i=0}^{n-1}$ such that $B_{\varepsilon_i}(x) \subseteq O_i$. The minimum $\varepsilon = \min\{\varepsilon_i\}_{i=0}^{n-1}$ is positive, and for all i we have $B_\varepsilon(x) \subseteq B_{\varepsilon_i}(x) \subseteq O_i$, so $B_\varepsilon(x) \subseteq O$. □

Remark 4.4.16. By Proposition 4.4.13 and the complement laws, an intersection of closed sets is closed, and a finite union of closed sets is closed. ◇

Proposition 4.4.17. *If O is a non-empty open set of real numbers, there exists a unique partition $\{O_i\}_{i \in \mathscr{I}}$ of O into open intervals.*

Proof. See Exercise 4.4.8. □

Definition 4.4.18. The sets $\{O_i\}_{i \in \mathscr{I}}$ are the *(connected) components* of O.

Proposition 4.4.19. *The ternary set K is closed, contains no interval, and contains no isolated points.*

Proof. See Exercise 4.4.5. □

Exercises for Section 4.4

Exercise 4.4.1. (★) For each positive integer n, the set $O_n = (-1/n, 1/n)$ is open. Is the intersection $\bigcap_{n=1}^\infty O_n$ open?

Exercise 4.4.2. For each positive integer n, the set $F_n = [0, n/(n+1)]$ is closed. Is the union closed?

Exercise 4.4.3. (★) Assume $(A_k)_{k \in \mathbf{N}}$ is a sequence of non-empty sets of real numbers, each having no limit points.

(a) Prove that $A_0 \cup A_1$ has no limit points.

(b) Prove that $\bigcup_{k=0}^{n-1} A_k$ has no limit points.

(c) Show by example that $\bigcup_{k \in \mathbf{N}} A_k$ may have no limit points, or may be dense in $\mathbf{R}$. (These are far from the only possibilities.)

Exercise 4.4.4. (H) Assume $A = \{x \text{ in } \mathbf{R} : x = 1/n \text{ for some } n \text{ in } \mathbf{Z}^+\}$. Determine with proof which real numbers (if any) are interior points of A, exterior points of A, boundary points of A, isolated points of A, limit points of A. Is A closed?

Exercise 4.4.5. Let K be the ternary set, Example 4.1.19.

(a) Prove K is closed.

(b) Prove K contains no intervals.

(c) Prove K contains no isolated points.

Exercise 4.4.6. If A is a set of real numbers, let $P(A)$ be the statement, "The closure of the interior of A is equal to $\overline{A}$." Is $P(A)$ true for all A? If so, give a proof. If not, give a counterexample, and find necessary conditions on A for $P(A)$ to be true, and sufficient conditions on A for $P(A)$ to be true.

Exercise 4.4.7. (H) Assume $\{O_i\}_{i \in \mathscr{I}}$ is a collection of open sets of real numbers whose union contains $[a, b]$. Prove that $[a, b]$ is *finitely covered* from this collection: There exist finitely many of these sets, say $\{O_{i_j}\}_{j=1}^{N}$ whose union $\bigcup_{j=1}^{N} O_{i_j}$ contains $[a, b]$.

Exercise 4.4.8. Prove Proposition 4.4.17. A suggested outline is provided.

(a) (Existence.) Define a binary relation $\sim$ on O by declaring $x \sim y$ if there exists an open interval $(a, b) \subseteq O$ such that $\{x, y\} \subseteq (a, b)$. Prove $\sim$ is an equivalence relation, and its equivalence classes are disjoint open intervals whose union is O.

(b) (Uniqueness.) Assume $\{O_i\}$ is a partition of O into open intervals. If $(a, b) \subseteq O$ is an arbitrary open interval, then $(a, b) \subseteq O_i$ for some i.

Consequently, if $\{O'_{i'}\}$ is a partition of O into open intervals, then for every i' there exists an i such that $O'_{i'} = O_i$ and conversely, so $\{O_i\} = \{O'_{i'}\}$.

Exercise 4.4.9. (★) Let O be $\{x \text{ in } \mathbf{R} : x \neq 0,\ x \neq 1/n \text{ for every } n \text{ in } \mathbf{Z}^+\}$. Sketch the set O, show O is open, and find the components of O, Exercise 4.4.8.

Rational Numbers in Real Analysis

Loosely, rational numbers are "arithmetically explicit" from the real axioms and irrational numbers are not. To substantiate this breezy assertion, let's briefly review what real numbers we can construct from which axioms.

The only real numbers explicitly mentioned by the axioms are 0 and 1. The only binary operations (taking two numbers and returning a number) mentioned in the real axioms are addition and multiplication. Each comes with its unary inversion operator; negatives for addition, reciprocals for multiplication.

The real number 0 forms a self-contained algebraic universe, in that the singleton set $\{0\}$ is closed under addition and under multiplication.

A set S of real numbers containing 1 and also closed under addition contains the set $\mathbf{Z}^+$ of positive integers, finite sums of 1s. If further S is closed under taking negatives, then S contains the set $\mathbf{Z}$ of integers.

Although the set of integers is closed under addition and multiplication, it is not closed under taking reciprocals of non-zero numbers. Any set of reals containing 1, closed under addition and taking negatives, and closed under multiplication and taking reciprocals, contains the set $\mathbf{Q}$ of rational numbers. The rational numbers constitute the universe generated from the identity elements for addition and multiplication, the arithmetic operations themselves, and taking inverses. In this sense, the rational numbers, and no others, are arithmetically explicit.

To define irrational real numbers, we need more. Suprema, encoded in the order and completeness axioms, turn out to suffice: Every real number is the supremum of some subset of $\mathbf{Q}$. In Chapter 6 we'll develop "limits" of real sequences, which serve, in part, as a "user-friendly front end" to suprema.

5

Functions

Real numbers provide a mathematical model for "continuous quantities" such as position, length, duration, and mass. When scientists speak of the natural world being "predictable," they generally mean natural phenomena are accurately described by numerical quantities satisfying deterministic mathematical models, such as "equations of motion."

Functions, the subject of this chapter, are mathematical idealizations of deterministic relationships, the means by which we extrapolate information about unknown quantities from known quantities.

Our starting point is the simple, general, and widely useful concept of a "mapping," introduced at the end of Chapter 1 and repeated below with a few extra features. Functions are a special case of mappings.

5.1 Mappings

Definition 5.1.1. Assume X and Y are sets. A *mapping* $f : X \to Y$ (read "f from X to Y") is a set f of ordered pairs (x, y) with the property that for each x in X, there exists *exactly one* y in Y such that $(x, y) \in f$, see Figure 5.1.

If $(x, y) \in f$, we write $y = f(x)$ and call y the *value of f at x*. The set X is the *domain* of f; its elements are the "inputs" of f. The set Y is the *codomain* of f; its elements are "potential outputs" of f.

Remark 5.1.2. If X is a set and Y is a subset of $\mathbf{R}$, a mapping $f : X \to Y$ is called a *real-valued function* on X, or often simply a "function on X."

If in addition X is a union of non-empty intervals of real numbers, a mapping $f : X \to Y$ is called a *function of one variable*, or simply a *function*.   ◇

Remark 5.1.3. You may have seen a function defined as a "rule" sending each point of X to a unique point of Y. Our set-theoretic definition does precisely this, without our having to say what a "rule" is. The set f itself, namely the set of ordered pairs (x, y) such that $y = f(x)$, is also called the *graph of f*.   ◇

Remark 5.1.4. A function is not merely a procedure such as "squaring" or formula such as $f(x) = x^2$: To define a function, we must explicitly specify

DOI: 10.1201/9781003601357-5

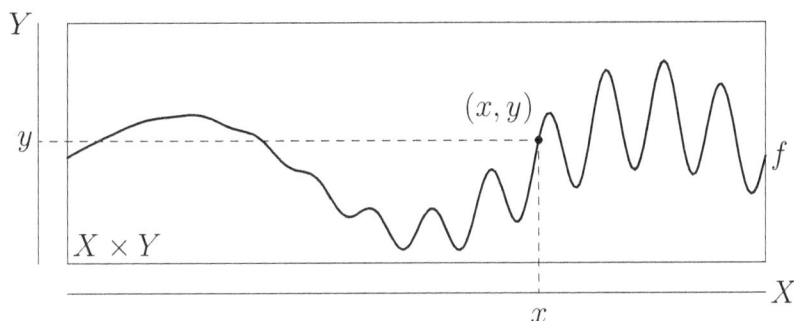

FIGURE 5.1
A mapping $f : X \to Y$, and a point where $y = f(x)$.

the domain X and the codomain Y. For example, changing the domain by removing a single point specifies a *different function*. ◇

To emphasize the preceding remark, we introduce two recurring concepts:

Definition 5.1.5. Let $f : X \to Y$ be a mapping.

If $A \subseteq X$, the *restriction* of f to A is the mapping $f|_A : A \to Y$ satisfying $f|_A(x) = f(x)$ for every x in A.

Dually, if $X \subseteq X'$, an *extension* of f to X' is a mapping $F : X' \to Y$ whose restriction to X is f.

Polynomial Functions

Loosely, a "polynomial" in one variable x is an expression involving x and finitely many constants that can be evaluated using only addition and multiplication. Although polynomials are simple and explicit, they are surprisingly versatile. Every finite numerical sequence is generated by some polynomial, see Theorem 5.1.20. Polynomials are the basis for "power series" in Chapter 7. In Chapter 17 we prove every "continuous" function on a closed, bounded interval can be approximated as closely as we like by polynomials.

Definition 5.1.6. Let X be a set of real numbers. For each real number c, there is a *constant function* $c : X \to \mathbf{R}$ defined by $c(x) = c$ for all x in X.

Definition 5.1.7. If X is a set of real numbers, the function $\iota_X : X \to X$ defined by $\iota_X(x) = x$ for all x in X is called the *identity function* on X.

Remark 5.1.8. Here, in practice, we may be lax about codomains. With the same notation, the (technically different) function $\iota_X : X \to \mathbf{R}$ defined by $\iota_X(x) = x$ for all x in X is also sometimes called the *identity function* on X. ◇

Definition 5.1.9. If f and g are real-valued functions with domain X, then the functions $f \pm g$ and fg with domain X are defined by the formulas

$$(f \pm g)(x) = f(x) \pm g(x), \qquad (fg)(x) = f(x) \cdot g(x)$$

for each x in X. These functions are called the *pointwise* sum, difference, and product of f and g.

Definition 5.1.10. If $(a_j)_{j=0}^n$ is a finite real sequence, the expression

$$p(x) = \sum_{j=0}^n a_j x^j = a_0 + a_1 x + a_2 x^2 + \cdots + a_n x^n$$
$$= a_0 + x\Big(a_1 + x(a_2 + \cdots + x(a_n)\ldots)\Big)$$

is the *polynomial* (*in one variable*) with *coefficients* (a_j).

A real number x_0 is a *root* of p if $p(x_0) = 0$.

If $a_n \neq 0$, we say p has *degree* n, denoted $\deg p = n$. In this case, we write $p(x) = a_n x^n + \cdots$, and call $a_n x^n$ the *top-degree term*. The ellipsis signifies *lower-degree terms*. If $a_n = 1$, namely $p(x) = x^n + \cdots$, we say p is *monic*.

If $a_j = 0$ for all j we call $p(x)$ the *zero polynomial* and define $\deg p = -\infty$.

If X is a non-empty set of real numbers and p is a polynomial, the function $f : X \to \mathbf{R}$ defined by $f(x) = p(x)$ is called a *polynomial function* on X.

Remark 5.1.11. The final expression for $p(x)$ in Definition 5.1.10 entails at most n multiplications and n additions. "Backward inductively," we define $p_{n+1}(x) = 0$, and $p_m(x) = a_m + x p_{m+1}(x)$ if $n \geq m \geq 0$, so that $p_0 = p$. $\diamond$

Example 5.1.12. A polynomial function of degree at most 1, defined by a formula $f(x) = a_0 + a_1 x$, is called an *affine function*. An affine function is constant if and only if $a_1 = 0$. $\diamond$

Remark 5.1.13. The term "linear" is often used because the graph of an affine function is (part of) a line. In real analysis and elsewhere in mathematics, however, "linear" has connotations that are not satisfied unless $a_0 = 0$. $\diamond$

Example 5.1.14. A polynomial function of degree at most 2, defined by a formula $f(x) = a_0 + a_1 x + a_2 x^2$, is called a *quadratic function*. A quadratic function is affine if and only if $a_2 = 0$. $\diamond$

Proposition 5.1.15. *If p and q are polynomials, then $\deg(pq) = \deg p + \deg q$.*

Proof. If neither p nor q is the zero polynomial, we may write $p(x) = a_n x^n + \cdots$ and $q(x) = b_m x^m + \cdots$. Since a_n and b_m are non-zero, $(pq)(x) = a_n b_m x^{n+m} + \cdots$, and consequently $\deg(pq) = n + m = \deg p + \deg q$.

If instead either is the zero polynomial, then pq is the zero polynomial, and $\deg(pq) = -\infty = \deg p + \deg q$. $\square$

Theorem 5.1.16 (Polynomial division). *If p and q are polynomials and q is not the zero polynomial, then there exist unique polynomials $d(x)$ and $r(x)$ such that $p(x) = d(x)q(x) + r(x)$ and $\deg r < \deg q$.*

Proof. See Exercise 5.1.4. □

Definition 5.1.17. If p and q are polynomials, we say q *divides* p if there exists a polynomial d such that $p = qd$.

Example 5.1.18. The polynomial $q(x) = x - 1$ divides $p(x) = x^3 - 1$ since $x^3 - 1 = (x - 1)(x^2 + x + 1)$.
 The polynomial $q(x) = x^2 + 2x + 2$ divides $p(x) = x^4 + 4$ since

$$x^4 + 4 = (x^2 + 2)^2 - 4x^2 = (x^2 + 2x + 2)(x^2 - 2x + 2). \qquad \Diamond$$

Corollary 5.1.19. *If p is a polynomial and x_0 is real, then $p(x_0) = 0$ if and only if $(x - x_0)$ divides $p(x)$.*

Proof. By polynomial division with $q(x) = x - x_0$, there exist unique polynomials d and r such that $p(x) = (x - x_0)d(x) + r(x)$ and $\deg r < \deg q = 1$; thus r is a constant polynomial. By definition, q divides p if and only if $r(x) = 0$, if and only if $r(x_0) = p(x_0) = 0$. □

Theorem 5.1.20 (Polynomial interpolation). *Assume $\{(x_i, y_i)\}_{i=0}^n$ is a finite set of points in the plane, and $x_i < x_{i+1}$ if $0 \leq i < n$. There exists a unique polynomial p of degree at most n such that $y_i = p(x_i)$ for each i.*

Proof. See Exercise 5.1.9. □

Rational Functions

Definition 5.1.21. Assume X is a non-empty set of real numbers. If f is a function on X, the *zero set* of f is the set $Z(f) = \{x \text{ in } X : f(x) = 0\}$.

Definition 5.1.22. If X is a non-empty set of real numbers, f and g are functions on X, and $Z = Z(g)$, then the formula $(f/g)(x) = f(x)/g(x)$ defines the *(pointwise) quotient* $f/g : X \setminus Z \to \mathbf{R}$.

Definition 5.1.23. Assume $p(x) = \sum_{j=0}^n a_j x^j$ and $q(x) = \sum_{i=0}^m b_i x^i$ are polynomials, and Z is the zero set of q on $\mathbf{R}$. The formula

$$f(x) = \frac{p(x)}{q(x)} = \frac{a_0 + a_1 x + a_2 x^2 + \cdots + a_n x^n}{b_0 + b_1 x + b_2 x^2 + \cdots + b_m x^m}$$

defines the *rational function* $f = p/q : \mathbf{R} \setminus Z \to \mathbf{R}$. The set $\mathbf{R} \setminus Z$, on which $q(x) \neq 0$, is called the *natural domain* of f.

Example 5.1.24. The formulas

$$f(x) = \frac{x}{1 - x^2}, \qquad\qquad g(x) = \frac{x^4 + 1}{x^2 + 1}$$

define rational functions. The natural domain of f is $\mathbf{R} \setminus \{\pm 1\}$. The natural domain of g is $\mathbf{R}$ ◊

Example 5.1.25. The formulas

$$f(x) = \frac{1 - x}{1 - x^2}, \qquad\qquad g(x) = \frac{1}{1 + x}$$

define distinct rational functions that are equal everywhere both are defined. The natural domain of f is $\mathbf{R} \setminus \{\pm 1\}$. The natural domain of g is $\mathbf{R} \setminus \{-1\}$. Particularly, f is a restriction of g, and g is an extension of f.   ◊

Example 5.1.26. A finite sum of "pure singularities" $c/(x - x_0)^m$ defines a rational function. For example, putting terms over a common denominator gives

$$\frac{1}{x^3} + \left[\frac{3}{(x - 1)^2} + \frac{1}{x - 1}\right] = \frac{1}{x^3} + \frac{x + 2}{(x - 1)^2} = \frac{1 - 2x + x^2 + 2x^3 + x^4}{x^3(x - 1)^2}. \quad ◊$$

Conversely, a rational function whose denominator factors completely may be decomposed into a sum of this form. We first give a precise statement, then work through the preceding example "backward" to suggest how to prove the theorem.

Theorem 5.1.27 (Partial fractions). *Assume p and q are polynomials having no common root, and $q(x) = \prod_{j=0}^{n-1}(x - x_j)^{m_j}$ for distinct real numbers $\{x_j\}_{j=0}^{n-1}$, and for some positive integers $\{m_j\}_{j=0}^{n-1}$. There exist real numbers $c_{j,k}$, $0 \le j < n$ and $1 \le k \le m_j$, and a polynomial d, such that*

$$\frac{p(x)}{q(x)} = d(x) + \sum_{j=0}^{n-1}\left[\sum_{k=1}^{m_j} \frac{c_{j,k}}{(x - x_j)^k}\right] \quad \text{if } x \notin \{x_j\}_{j=0}^{n-1}.$$

Proof. See Exercise 5.1.6.   □

Remark 5.1.28. Not every real polynomial factors completely over the reals. Every real polynomial *does* factor completely over the complex numbers, the so-called *fundamental theorem of algebra*. A proof of this, and details of the resulting partial fractions decomposition, lie beyond the scope of this book.   ◊

Example 5.1.29. Suppose we wished, ignorant of Example 5.1.26, to decompose in partial fractions the rational function

$$\frac{p(x)}{q(x)} = \frac{1 - 2x + x^2 + 2x^3 + x^4}{x^3(x-1)^2}.$$

Theorem 5.1.27 guarantees this rational function may be written

$$\left[\frac{c_{0,3}}{x^3} + \frac{c_{0,2}}{x^2} + \frac{c_{0,1}}{x}\right] + \left[\frac{c_{1,2}}{(x-1)^2} + \frac{c_{1,1}}{(x-1)}\right].$$

In special cases there may be speedy tricks for evaluating the coefficients. Here, we'll proceed in a way that suggests how Theorem 5.1.27 might be proven.

First check for possible common factors. Since $p(0) = 1$, Corollary 5.1.19 implies x does not divide $p(x)$. Similarly, $p(1) = 1 - 2 + 1 + 2 + 1 = 3$, so $(x-1)$ does not divide $p(x)$. (If either were 0, polynomial division would allow us to manually cancel common factors before proceeding.)

The strategy is to subtract off a highest-order singularity to reduce the degree of the denominator. Here we have two choices: c/x^3 or $c/(x-1)^2$. We pick the second, whose denominator has lower degree, since typically this involves less computation. Writing

$$\frac{p(x)}{q(x)} = \frac{1 - 2x + x^2 + 2x^3 + x^4}{x^3(x-1)^2} = \frac{(1 - 2x + x^2 + 2x^3 + x^4)/x^3}{(x-1)^2},$$

we evaluate the numerator to be $p(1)/1^3 = 3$. This motivates us to subtract $3/(x-1)^2$, obtaining

$$\frac{p(x)}{q(x)} - \frac{3}{(x-1)^2} = \frac{(1 - 2x + x^2 + 2x^3 + x^4) - 3x^3}{x^3(x-1)^2}$$

$$= \frac{1 - 2x + x^2 - x^3 + x^4}{x^3(x-1)^2}.$$

This numerator (necessarily) vanishes at 1. Polynomial division gives

$$\frac{1 - 2x + x^2 - x^3 + x^4}{x^3(x-1)^2} = \frac{(x-1)(-1 + x + x^3)}{x^3(x-1)^2} = \frac{-1 + x + x^3}{x^3(x-1)}.$$

Having reduced the degree of the denominator by subtracting a suitable pure singularity, we can repeat the process inductively, stopping when we have a sum of pure singularities. Writing

$$\frac{-1 + x + x^3}{x^3(x-1)} = \frac{(-1 + x + x^3)/x^3}{x - 1}$$

and setting $x = 1$ in the numerator leads us to subtract $1/(x-1)$:

$$\frac{-1 + x + x^3}{x^3(x-1)} - \frac{1}{x-1} = \frac{(-1 + x + x^3) - x^3}{x^3(x-1)} = \frac{-1 + x}{x^3(x-1)} = \frac{1}{x^3}.$$

What remains is a pure singularity, so we are done:

$$\frac{1 - 2x + x^2 + 2x^3 + x^4}{x^3(x-1)^2} - \frac{3}{(x-1)^2} - \frac{1}{x-1} = \frac{1}{x^3},$$

which is visibly equivalent to the initial expression in Example 5.1.26.   ◊

Secants

Definition 5.1.30. Assume X is an *interval* of real numbers and f a function on X. Two numbers a and b in X such that $a < b$ determine two points $\big(a, f(a)\big)$ and $\big(b, f(b)\big)$ of f. There is a unique affine function $f_{a,b} : X \to \mathbf{R}$ whose graph passes through these points, defined for all x in X by

$$f_{a,b}(x) = f(a) + \frac{f(b) - f(a)}{b - a}(x - a).$$

We call $f_{a,b}$ the *affine interpolation* of f on $[a, b]$, and the graph of $f_{a,b}$ the *secant line* of f on $[a, b]$, Figure 5.2.

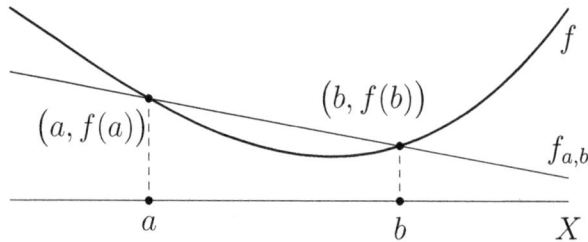

FIGURE 5.2
Affine interpolation of f on $[a, b]$.

If $f(x) \le f_{a,b}(x)$ for all x in (a, b), we say f *lies below its secant on* $[a, b]$. If this condition holds for every $[a, b]$ contained in X, we say f is *convex* on X.

If $f < f_{a,b}$ on (a, b) we say f *lies strictly below its secant on* $[a, b]$. If this condition holds for every $[a, b]$ contained in X, we say f is *strictly convex* on X.

Remark 5.1.31. We say f is *(strictly) concave* on X if $-f$ is (strictly) convex. This may be expressed in terms of "f lying above its secant lines."   ◊

Remark 5.1.32. The function in Figure 5.2 is strictly convex on X: For each interval $[a, b]$, we look only at the segment over $[a, b]$, not the whole secant line.

The function in Figure 5.3 is neither convex nor concave.   ◊

Functions Defined by Multiple Formulas

A function may be defined by multiple formulas, each holding on part of the domain.

Definition 5.1.33. Assume f is a function on some closed interval $[a, b]$. Each splitting $\{x_i\}_{i=0}^{n}$ of $[a, b]$ defines a *piecewise-affine interpolation* of f: The unique function whose graph over $I_i = [x_i, x_{i+1}]$, for each i, is the affine interpolation over I_i, Figure 5.3.

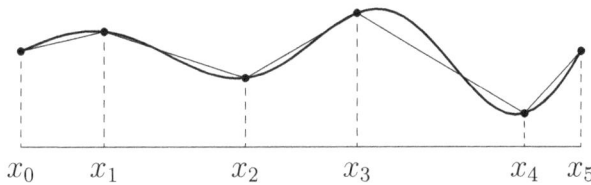

FIGURE 5.3
A piecewise-affine interpolation of f on $[a, b]$.

Definition 5.1.34. A function f on $[a, b]$ is a *step function* if there is a splitting $\{x_i\}_{i=0}^{n}$ such that for each i, f is constant on the open interval (x_i, x_{i+1}), namely if there exist real numbers $\{y_i\}_{i=0}^{n-1}$, such that $f(x) = y_i$ if $x_i < x < x_{i+1}$.

Example 5.1.35. The functions f, $g : [a, b] \to \mathbf{R}$ defined by $f(x) = \lfloor x^2 \rfloor$ and $g(x) = \lceil x^2 \rceil$ are step functions. Figure 5.4 shows the graphs, with compressed vertical scale, if $[a, b] = [0, 3]$. $\diamond$

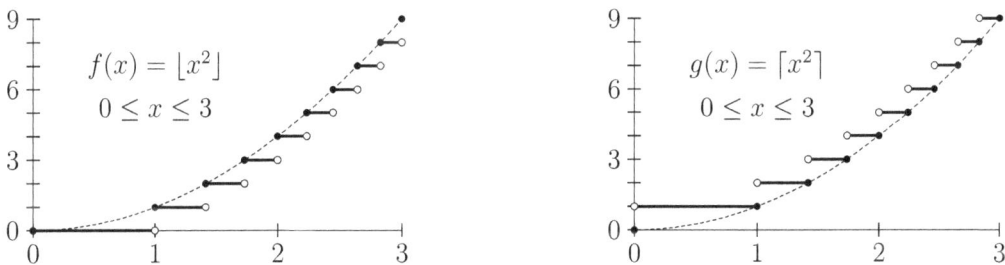

FIGURE 5.4
The step functions $f(x) = \lfloor x^2 \rfloor$ and $g(x) = \lceil x^2 \rceil$ on $[0, 3]$.

Example 5.1.36. The function $f : [0, 1] \to \mathbf{R}$ defined by

$$f(x) = \begin{cases} 1/2^n & \text{if } 1/2^{n+1} < x \leq 1/2^n, \\ 0 & \text{if } x = 0, \end{cases}$$

is not a step function because the domain has been divided into *infinitely many* pieces. For every δ (delta) in $(0,1)$, however, the restriction of f to $[\delta,1]$ *is* a step function. (Why?)   ◇

Definition 5.1.37. Let X be a set of real numbers. If $A \subseteq X$, the *indicator* of A (on X) is the function $\chi_A : X \to \mathbf{R}$ (chi sub A) defined by

$$\chi_A(x) = \begin{cases} 1 & \text{if } x \in A, \\ 0 & \text{if } x \notin A. \end{cases}$$

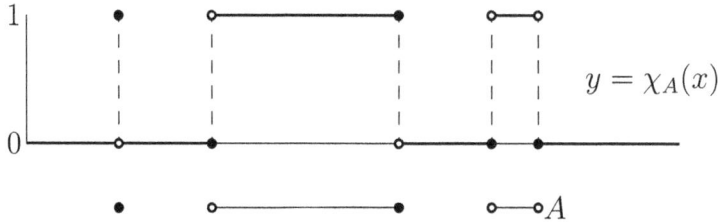

FIGURE 5.5
The indicator of a set.

Remark 5.1.38. The indicator χ_A inquires of each real number x in X, "Are you an element of A?" and returns 1 (yes) or 0 (no) accordingly, see Figure 5.5.   ◇

Example 5.1.39. The indicator of $\mathbf{Q}$, the set of rational numbers, is stylistically portrayed in Figure 5.6.   ◇

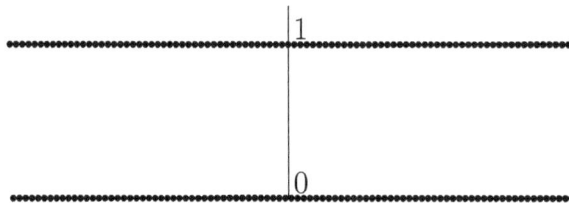

FIGURE 5.6
The indicator $\chi_{\mathbf{Q}}$ of $\mathbf{Q}$.

Definition 5.1.40. If X is a non-empty set of real numbers and $f : X \to \mathbf{R}$ is a function, we define the *positive part* f^+ and *negative part* f^- of f by the formulas

$$f^+(x) = \max\big(f(x),0\big) = \frac{|f(x)| + f(x)}{2},$$
$$f^-(x) = -\min\big(f(x),0\big) = \frac{|f(x)| - f(x)}{2}.$$

Each function is non-negative, and we have $f = f_+ - f_-$ and $|f| = f_+ + f_-$.

Exercises for Section 5.1

Exercise 5.1.1.(★) Assume p and q are polynomials of degree at most n.

(a) Prove that $p + q$ is a polynomial of degree at most n. If p and q both have degree n, does $p + q$ have degree n?

(b) Prove that if c is a real number, then cp is a polynomial of degree at most n. Is the degree of cp always equal to n?

(c) Suppose $\{x_j\}_{j=0}^n$ are real numbers, and $p(x_j) = 0$ for all j. Prove that $p(x) = 0$ for all real x.

Exercise 5.1.2.(★) Find the positive and negative parts of $p(x) = x(x-1)^2$; express your answer as a piecewise-polynomial function.

Exercise 5.1.3. For each function $f : \mathbf{R} \to \mathbf{R}$, sketch the graphs $y = f(x)$, $y = f^+(x)$, and $f^-(x)$ on the same set of axes.

(a) $f(x) = x$.   (b) $f(x) = 1 - |x|$.   (c) $f(x) = x^3 - x$.

Exercise 5.1.4.(H) Assume p and q are polynomials, q not the zero polynomial. Prove Theorem 5.1.16: There exist unique polynomials $d(x)$ and $r(x)$ such that $p(x) = d(x)q(x) + r(x)$ and $\deg r < \deg q$.

Exercise 5.1.5.(★) Decompose $\dfrac{x^4 - x^3 + 3x^2 - 6x + 1}{x^3 - 2x^2 + x}$ in partial fractions.

Exercise 5.1.6.(★) Use induction on the degree of q to prove Theorem 5.1.27.

Exercise 5.1.7.(A) Assume $a \neq b$. Decompose the following into partial fractions:

(a) $\dfrac{1}{(x-a)(x-b)}$.   (b) $\dfrac{1}{x^4 - a^2 x^2}$.   (c) $\dfrac{1}{x^2(x-a)(x-b)}$.

Exercise 5.1.8.(★) In this question we construct a polynomial of degree at most 2 whose graph (a line or parabola) passes through specified points (x_0, y_0), (x_1, y_1), and (x_2, y_2) such that $x_0 < x_1 < x_2$.

(a) The following formulas define polynomials e_i of degree 2:

$$
\begin{aligned}
f_0(x) &= (x - x_1)(x - x_2), & e_0(x) &= f_0(x)/f_0(x_0), \\
f_1(x) &= (x - x_0)(x - x_2), & e_1(x) &= f_1(x)/f_1(x_1), \\
f_2(x) &= (x - x_0)(x - x_1), & e_2(x) &= f_2(x)/f_2(x_2),
\end{aligned}
$$

Prove $e_i(x_j) = 1$ if $i = j$ and $e_i(x_j) = 0$ if $i \neq j$.

(b) Prove the polynomial $p(x) = y_0 e_0(x) + y_1 e_1(x) + y_2 e_2(x)$ has degree at most 2, and the points (x_0, y_0), (x_1, y_1), and (x_2, y_2) satisfy $y = p(x)$.

(c) Use parts (a) and (b) to find the unique quadratic polynomial $p(x)$ whose graph passes through the points $(-1, y_0)$, $(0, y_1)$, and $(1, y_2)$. Under what conditions does $p(x)$ have degree less than 2?

Exercise 5.1.9. (H) Suppose $\{(x_i, y_i)\}_{i=0}^n$ is a finite set of points in the plane, and $x_i < x_{i+1}$ if $0 \leq i < n$.

(a) For each i, define a monic polynomial e_i of degree n by the formula

$$f_i(x) = \prod_{j \neq i} (x - x_j), \qquad e_i(x) = \frac{f_i(x)}{f_i(x_i)},$$

with the product taken over j such that $0 \leq j \leq n$ and $j \neq i$, compare Exercise 5.1.8. Prove that $e_i(x_j) = 1$ if $i = j$ and $e_i(x_j) = 0$ if $i \neq j$.

(b) Prove the polynomial $p(x) = \sum_{j=0}^n y_j e_j(x)$ has degree at most n, and the points (x_i, y_i) satisfy $y = p(x)$.

(c) (H) Assume $y = q(x)$ is a polynomial of degree at most n and $y_i = q(x_i)$ if $0 \leq i \leq n$. Prove $q(x) = p(x)$ for all real x. (In words, the interpolation polynomial in (b) is unique.)

Exercise 5.1.10. Suppose A and B are subsets of $\mathbf{R}$. Establish the following:

(a) $1 - \chi_A = \chi_{A^c}$, the indicator of A^c.

(b) $\min(\chi_A, \chi_B) = \chi_A \cdot \chi_B = \chi_{A \cap B}$.

(c) $\max(\chi_A, \chi_B) = \chi_A + \chi_B - \min(\chi_A, \chi_B) = \chi_{A \cup B}$.

Exercise 5.1.11. (H) Assume I is an interval of real numbers and $f : I \to \mathbf{R}$ satisfies $|f(x') - f(x)| \leq |x' - x|^2$ for all x, x' in I. Prove f is constant.

5.2 Composition

Applying a function may be viewed as "acting on" on a quantity. In this section we study a useful auxiliary concept: Acting with functions successively, plugging the output of one function into another as input.

Definition 5.2.1. Assume X, Y, Y', and Z are sets of real numbers, and $f : X \to Y$, $g : Y' \to Z$ are mappings. If $f(x) \in Y'$ for every x in X, we say g is *composable* with f. Particularly, if $Y \subseteq Y'$ then g is composable with f.

If g is composable with f, the *composition* $g \circ f : X \to Z$, read "g of f," is defined by

$$(g \circ f)(x) = g\bigl(f(x)\bigr) \quad \text{for all } x \text{ in } X.$$

Example 5.2.2. The polynomial functions $f(x) = x + 1$ ("adding one") and $g(x) = x^2$ ("squaring") are composable in either order, and

$$
\begin{aligned}
(g \circ f)(x) &= g(x+1) = (x+1)^2 = x^2 + 2x + 1, \\
(f \circ g)(x) &= f(x^2) = x^2 + 1.
\end{aligned}
$$
◇

Remark 5.2.3. Note carefully that $g \circ f$ and $f \circ g$ are *different functions*: Mapping composition is not commutative. The same is true of everyday operations: Putting on your socks and then your shoes is not the same as putting on your shoes and then your socks.

By contrast, composition is associative: $(h \circ g) \circ f = h \circ (g \circ f)$ if the compositions are defined, see Exercise 5.2.2. ◇

Symmetries of Functions

In this section we'll refer to the Greek letters ρ (rho), a small r here suggesting "reflection," and τ (tau), a small t here suggesting "translation."

Example 5.2.4. Assume $g : \mathbf{R} \to \mathbf{R}$. The function $Rg : \mathbf{R} \to \mathbf{R}$ defined by $Rg(x) = g(-x)$, whose graph results from "reflecting" the graph of g across the vertical axis, may be written as a composition: If we define $\rho(x) = -x$ for all x, then $Rg = g \circ \rho$.

More generally, if the domain of g is a set X of real numbers, we may view $\rho : -X \to X$, and $Rg(x) = g(-x)$ is the composition $Rg = g \circ \rho : -X \to \mathbf{R}$. (The notation $-X$ is introduced in Definition 4.1.11.) ◇

Remark 5.2.5. If X is an interval $[-a, a]$, or generally a non-empty set of real numbers satisfying $-X = X$, namely "symmetric with respect to 0," then the function Rg of Example 5.2.4 has the same domain as g itself.

This situation invites a useful abstraction: Let $\mathscr{F}(X)$ denote the set (or "space") of *all functions* on X. The function g is an *element* of $\mathscr{F}(X)$. The function $Rg = g \circ \rho$ is another *element* of $\mathscr{F}(X)$, and is uniquely determined by g. In mathematics we have a name for this: *mapping*. We may view R as a *mapping on functions*. In symbols, $R : \mathscr{F}(X) \to \mathscr{F}(X)$.

A mapping on functions is often called an *operator*. The operator R is called the *domain reflection operator* on $\mathscr{F}(X)$.

Incidentally, the operator R can be composed with itself since its domain and codomain are the same set, $\mathscr{F}(X)$. Because "reflecting twice is the identity," or $-(-x) = x$ for all x, we have

$$R(Rg)(x) = Rg(-x) = g\big(-(-x)\big) = g(x) \quad \text{for all } x.$$

Composing R with itself is the identity operator on $\mathscr{F}(X)$.   ◇

Definition 5.2.6. Assume X is a non-empty set of real numbers such that $-X = X$. A function $g : X \to \mathbf{R}$ is said to be *even* if $g(-x) = g(x)$ for all x in X, and is said to be *odd* if $g(-x) = -g(x)$ for all x in X.

Remark 5.2.7. In our operator formalism, g is even if and only if $Rg = g$: The graph of g is invariant (unchanged) under reflection about the vertical axis. Similarly, g is odd if and only if $Rg = -g$: The graph of g is invariant by reflection in both axes, namely under a half-turn rotation about the origin.   ◇

Lemma 5.2.8. *Assume X is a non-empty set of real numbers such that $-X = X$, and $f : X \to \mathbf{R}$ is a function. The functions*

$$f_{\text{even}}(x) = \tfrac{1}{2}\big(f(x) + f(-x)\big),$$
$$f_{\text{odd}}(x) = \tfrac{1}{2}\big(f(x) - f(-x)\big),$$

are even and odd respectively, and satisfy $f_{\text{even}}(x) + f_{\text{odd}}(x) = f(x)$.

Proof. See Exercise 5.2.5.   □

Definition 5.2.9. The functions f_{even} and f_{odd} of Lemma 5.2.8 are called the *even part* of f and the *odd part* of f.

From reflection, we now turn to a different domain symmetry, translation.

Example 5.2.10. Assume $g : \mathbf{R} \to \mathbf{R}$ is a function. For each real number c, the function $T_c g : \mathbf{R} \to \mathbf{R}$ defined by $T_c g(x) = g(x - c)$ may be written as a composition: If we define $\tau_c(x) = x - c$, then $T_c g = g \circ \tau_c$. The graph of $T_c g$ results from translating the graph of g horizontally by c, to the right if c is positive and to the left if c is negative, Figure 5.7.

More generally, if X is an arbitrary non-empty set of real numbers, then the formula $\tau_c(x) = x - c$ defines a mapping $\tau_c : c + X \to X$. If $g : X \to \mathbf{R}$ is a function, then $T_c g(x) = g(x-c)$ is the composition $T_c g = g \circ \tau_c : c + X \to \mathbf{R}$.   ◇

Definition 5.2.11. Assume ℓ is a positive real number. A function $f : \mathbf{R} \to \mathbf{R}$ is *ℓ-periodic* if $f(x - \ell) = f(x)$ for all real x.

Remark 5.2.12. A function $f : \mathbf{R} \to \mathbf{R}$ is ℓ-periodic if and only if $T_\ell f = f$.   ◇

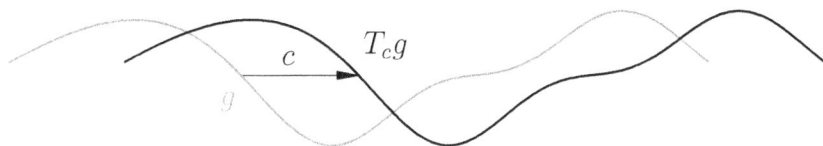

FIGURE 5.7
The domain translation operator T_c acting on a function g.

Example 5.2.13. The indicator $\chi_{\mathbf{Z}}$ is 1-periodic. ◇

Example 5.2.14. The indicator $\chi_{\mathbf{Q}}$ of $\mathbf{Q}$ is ℓ-periodic for every positive rational ℓ. ◇

Lemma 5.2.15. *If f is ℓ-periodic, then for every integer n, we have*

$$f(x + n\ell) = f(x) \quad \text{for all real } x.$$

Proof. See Exercise 5.2.11. □

Example 5.2.16. No non-constant polynomial or rational function f is periodic: If y is real, the equation $y = f(x)$ is equivalent to a non-constant polynomial equation in x, and therefore has at most finitely many solutions x. If f is periodic, however, Lemma 5.2.15 implies each equation $y = f(x)$ has either no solutions, or infinitely many. ◇

Proposition 5.2.17. *Assume $a < b$, and put $\ell = b - a$. If $f : [a, b) \to \mathbf{R}$ is a function, there exists a unique ℓ-periodic extension of f.*

Proof. See Exercise 5.2.12. □

Invertibility

Suppose $f : X \to Y$ is a mapping. The equation $y = f(x)$ may be viewed as a "true/false condition" (or "truth function") on the ordered product $X \times Y$: true if $(x, y) \in f$ and false otherwise.

If x is given, the definition of a mapping guarantees there is a unique y such that $y = f(x)$. Conceptually, we can solve $y = f(x)$ uniquely for y, though here the term "solve" sounds strange: we are merely evaluating f at x.

What about the reverse: If y is given, can we solve $y = f(x)$ for x? More precisely, if we are given an arbitrary y in Y, does there exist a unique x in X such that $y = f(x)$? If "yes," we may view x as a function of y and write the relation $y = f(x)$ in the form $x = g(y)$ for some mapping $g : Y \to X$.

Definition 5.2.18. If $f : X \to Y$ and $g : Y \to X$ are mappings, we say f and g are *inverses* if $(g \circ f)(x) = x$ for every x in X and $(f \circ g)(y) = y$ for every y in Y.

A mapping $f : X \to Y$ is *invertible* if there exists an inverse $g : Y \to X$.

Remark 5.2.19. Recall that ι_X denotes the identity mapping on a set X. A mapping $f : X \to Y$ is therefore invertible if and only if there exists a mapping $g : Y \to X$ such that $g \circ f = \iota_X$ and $f \circ g = \iota_Y$. $\diamond$

Example 5.2.20. Assume c is a non-zero real number. The scaling function $\mu_c : \mathbf{R} \to \mathbf{R}$ (mu sub c) defined by $\mu_c(x) = cx$ is invertible. The inverse function is $\mu_{1/c}$. $\diamond$

Example 5.2.21. If $X = \mathbf{R} \setminus \{0\}$ is the set of non-zero real numbers, then the reciprocal function $f : X \to X$ defined by $f(x) = 1/x$ is invertible. In fact, this mapping is its own inverse: $1/(1/x) = x$ for all non-zero x. $\diamond$

Remark 5.2.22. A mapping has at most one inverse: If f is invertible, then $g(y) = x$ if and only if $y = f(x)$, so the values of f uniquely determine the values of g. If f is invertible, we denote g by the symbol f^{-1}, read "f inverse." The inverse "undoes" the action of f. An equation $y = f(x)$ is "solved" for x by applying f^{-1} to both sides. $\diamond$

Exercises for Section 5.2

Exercise 5.2.1. (★) Consider the rational function $f(x) = (1 + x)/(1 - x)$. Calculate the compositions $f \circ f$, $f \circ f \circ f$, and $f \circ f \circ f \circ f$, giving the domain of each, and simplifying formulas as much as possible.

Exercise 5.2.2. (★) Mapping composition is associative, see Remark 5.2.3.

Exercise 5.2.3. (★) Find the even and odd parts of $p(x) = x(x - 1)^2$.

Exercise 5.2.4. Use the definition to find the even and odd parts of:

(a) $f(x) = 3 - 2x + 5x^2 + x^5$.   (b) $g(x) = 2\chi_{[0,1]}(x)$.

Exercise 5.2.5. Assume X is a non-empty set of real numbers satisfying $-X = X$, and f is a function on X. Prove Lemma 5.2.8, and prove f is written *uniquely* as the sum of an even function and an odd function.

Exercise 5.2.6. If n is an integer, define $f : \mathbf{R} \setminus \{0\} \to \mathbf{R}$ by $f(x) = x^n$. Prove f is an even function if and only if n is an even integer, and $f(x)$ is odd if and only if n is an odd integer.

Exercise 5.2.7. Assume $p(x)$ is a polynomial function.

(a) Prove p is an even function if and only if every term has even degree, if and only if there exists a polynomial q such that $p(x) = q(x^2)$ for all real x.

(b) Prove p is an odd function if and only if every term has odd degree, if and only if there exists a polynomial q such that $p(x) = xq(x^2)$ for all real x.

Exercise 5.2.8. Assume X is a non-empty set of real numbers such that $-X = X$, and f, g are functions on X.

(a) Prove that if f and g are both even or both odd, then fg is even.

(b) Prove that if f is even and g is odd, then fg is odd.

(c) Suppose f is even, and h is an arbitrary function defined on all of $\mathbf{R}$. Is either $f \circ h$ or $h \circ f$ necessarily even? Give a proof or counterexample.

Exercise 5.2.9. Assume $f : \mathbf{R} \to \mathbf{R}$ is ℓ-periodic, and that $g : \mathbf{R} \to \mathbf{R}$. Must $f \circ g$ be ℓ-periodic? Must $g \circ f$ be ℓ-periodic?

Exercise 5.2.10. Assume f and g are ℓ-periodic functions.

(a) Prove that $f + g$ and fg are ℓ-periodic.

(b) Prove that the even and odd parts of f are ℓ-periodic.

(c) Prove that the positive and negative parts of f are ℓ-periodic.

Exercise 5.2.11.($\bigstar$) Prove Lemma 5.2.15.

Exercise 5.2.12. Prove Proposition 5.2.17.

Exercise 5.2.13. Assume $\ell > 0$. We say a function $f : \mathbf{R} \to \mathbf{R}$ is ℓ-*anti-periodic* if $f(x + \ell) = -f(x)$ for all real x. Prove that such a function is 2ℓ-periodic.

5.3   Injectivity and Surjectivity

There are two reasons a mapping $f : X \to Y$ might not be invertible. First, for some y there might exist distinct points x_1, x_2 in X such that $y = f(x_1) = f(x_2)$. There might even exist infinitely many such points, for example if f is periodic.

Second, for some y, there might exist *no* points x in X such that $y = f(x)$. For example, if f is constant, say $f(x) = y_0$ for all x, then the equation $y = f(x)$ has no solutions if $y \neq y_0$.

Each of these alternatives is important enough to deserve a name. As we develop tools, especially continuity and derivatives, we will establish powerful theorems for checking whether or not functions satisfy these properties.

Definition 5.3.1. Assume $f : X \to Y$ is a mapping and $\{x_1, x_2\} \subseteq X$. If $x_1 \neq x_2$ and $f(x_1) = f(x_2)$, we say f *identifies* x_1 and x_2.

Remark 5.3.2. If X is a collection of people, Y a set of names, and f an assignment of names, then two people are identified by f (namely, are *made identical* by f) if they have the same name. Knowing only a name (value of f), we are unable to uniquely determine an individual (input of f).   ◇

Reiterating, to say $f : X \to Y$ is invertible means that for every y in Y, there exists *exactly one* x in X such that $y = f(x)$. Invertibility may therefore be separated into two useful criteria:

Definition 5.3.3. A mapping $f : X \to Y$ is *injective* if $f(x_1) = f(x_2)$ implies $x_1 = x_2$. Contrapositively, f is injective if $x_1 \neq x_2$ implies $f(x_1) \neq f(x_2)$, namely, if f does not identify any points, or f takes distinct values at distinct inputs. An injective mapping is called an *injection*.

Definition 5.3.4. A mapping $f : X \to Y$ is *surjective* if for every y in Y, there exists an x in X such that $f(x) = y$. A surjective mapping is called a *surjection*.

Definition 5.3.5. A mapping $f : X \to Y$ is *bijective* if f is both injective and surjective. A bijective mapping is called a *bijection*.

Example 5.3.6. Define $f : \mathbf{R} \to \mathbf{R}$ by $f(x) = x^2$. We wish to determine whether or not f is injective and/or surjective.

For injectivity, the general strategy is to assume $f(x_1) = f(x_2)$ and either deduce that $x_1 = x_2$, or find a specific pair of distinct inputs having equal output values. Here, $x_1^2 = x_2^2$ if and only if

$$0 = x_1^2 - x_2^2 = (x_1 - x_2)(x_1 + x_2),$$

if and only if $x_1 - x_2 = 0$ or $x_1 + x_2 = 0$. The second equation has solutions for which $x_1 \neq x_2$, such as $x_1 = 1 = -x_2$, so f is not injective.

For surjectivity, assume y is an arbitrary element of the codomain, and attempt to solve the equation $y = f(x)$ for x in the domain of f.

Here, we wish to solve $x^2 = y$ for x, with y an arbitrary real number. By Lemma 3.2.4 (ii), $x^2 \geq 0$ for all real x. Consequently, the equation $x^2 = -1$ has no real solution; that is, $y = -1$ is not a value of f, so f is not surjective.   ◇

Example 5.3.7. Define $g : (0, \infty) \to \mathbf{R}$ by $g(x) = x^2$. Since the domain of g differs from the domain of f in the preceding example, g and f are *different functions* even though they are defined by the same formula.

Though g is not surjective for the same reasons as f above, g *is* injective: If $g(x_1) = g(x_2)$ for some *positive* real numbers x_1 and x_2, then either $x_1 - x_2 = 0$ or $x_1 + x_2 = 0$. But since x_1 and x_2 are positive, their sum is positive; thus $x_1 - x_2 = 0$, or $x_1 = x_2$.   ◇

Example 5.3.8. Define $h : (0, \infty) \to (0, \infty)$ by $g(x) = x^2$. Since the codomain of h differs from the codomain of g in the preceding example, h and g are *different functions*.

The function h is injective for the same reasons as g. By Theorem 4.2.11, h is surjective. In words, squaring is a bijection on $(0, \infty)$. ◇

Images and Preimages

A mapping $f : X \to Y$ acts on subsets of X and on subsets of Y.

Definition 5.3.9. Let $f : X \to Y$ be a mapping. If $A \subseteq X$, the set

$$f(A) = \{y \text{ in } Y : y = f(x) \text{ for some } x \text{ in } A\}$$

is called the *image of A under f*, Figure 5.8a. In particular, the set $f(X)$ of values of f is called the *image* of f.

Remark 5.3.10. A mapping $f : X \to Y$ is surjective if and only if the image of f is the entire codomain, $f(X) = Y$.

By contrast with the domain of a mapping, the codomain of a general mapping is relatively unimportant. Specifically, if the codomain is not fixed by context, we may harmlessly replace a mapping $f : X \to Y$ with the surjection $f : X \to f(X)$, in effect "assuming f is surjective without loss of generality." Sometimes, however, context *does* fix the codomain, see for example [15]. ◇

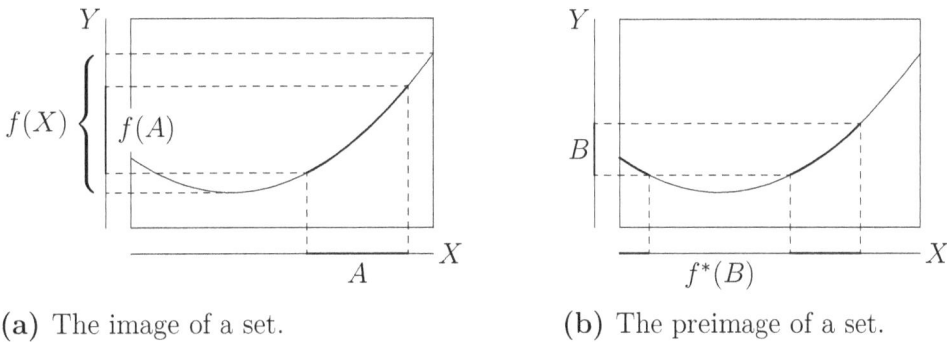

(a) The image of a set.   (b) The preimage of a set.

FIGURE 5.8
The forward and backward actions of a mapping $f : X \to Y$.

Definition 5.3.11. Let $f : X \to Y$ be a mapping. If $B \subseteq Y$, the set

$$f^*(B) = \{x \text{ in } X : f(x) \in B\}$$

is called the *preimage of B under f*, Figure 5.8b.

Proposition 5.3.12. *Let $f : X \to Y$ be a mapping. For all subsets S_1 and S_2 of X, and all subsets T_1 and T_2 of Y, we have*

(i) $f(S_1 \cup S_2) = f(S_1) \cup f(S_2)$.   (iii) $f^*(T_1 \cup T_2) = f^*(T_1) \cup f^*(T_2)$.

(ii) $f(S_1 \cap S_2) \subseteq f(S_1) \cap f(S_2)$.   (iv) $f^*(T_1 \cap T_2) = f^*(T_1) \cap f^*(T_2)$.

Proof. For illustration we prove (i). Suppose y is an arbitrary element of $f(S_1 \cup S_2)$. By definition, there exists an x in $S_1 \cup S_2$ such that $y = f(x)$. By definition of a union, either $x \in S_1$ and $y = f(x) \in f(S_1)$, or $x \in S_2$ and $y = f(x) \in f(S_2)$. In either case, $y \in f(S_1) \cup f(S_2)$. Since y was arbitrary, $f(S_1 \cup S_2) \subseteq f(S_1) \cup f(S_2)$.

Conversely, suppose $y \in f(S_1) \cup f(S_2)$. By definition of a union, either $y \in f(S_1)$ and there exists an x in S_1 such that $y = f(x)$, or $y \in f(S_2)$ and there exists an x in S_2 such that $y = f(x)$. In either case, $x \in S_1 \cup S_2$, so $y \in f(S_1 \cup S_2)$. Since y was arbitrary, $f(S_1 \cup S_2) \supseteq f(S_1) \cup f(S_2)$.

The remaining parts are Exercise 5.3.8. $\square$

Remark 5.3.13. Assume $f : X \to Y$ is a mapping, and $y \in Y$.

If f is injective, the equation $y = f(x)$ has *at most one* solution: The preimage $f^*(\{y\})$ contains at most one element for each y.

If f is surjective, $y = f(x)$ has *at least one* solution: The preimage $f^*(\{y\})$ is non-empty for each y.

Thus, f is bijective if and only if $y = f(x)$ has exactly one solution for every y in Y, if and only if f is invertible. $\diamond$

Remark 5.3.14. We may apply a function to both sides of an equation, obtaining a new equation. We do this when we "square both sides." Inversely, if f is injective, we may "cancel" f from both sides of an equation. That is, if f is injective and $f(x_1) = f(x_2)$ we may deduce $x_1 = x_2$, see also [8]. $\diamond$

Monotone Functions

Analysis deals in inequalities as well as equations. In this section we'll introduce names for functions that can be applied to inequalities, obtaining new inequalities. There are (unfortunately) four useful criteria, depending on (i) whether applying f preserves *strict* inequalities or possibly not, and (ii) whether f preserves or reverses the sense of inequalities.

Definition 5.3.15. Let X be a non-empty set of real numbers. A function $f : X \to \mathbf{R}$ is *strictly increasing* on X if for all x and x' in X, $x < x'$ implies $f(x) < f(x')$.

Lemma 5.3.16. *If n is a positive integer, then the function $f : (0, \infty) \to \mathbf{R}$ defined by $f(x) = x^n$, is strictly increasing: If $0 < x_1 < x_2$, then $0 < x_1^n < x_2^n$.*

Proof. Use Proposition 3.2.1 (iii) and induction on n. $\square$

Lemma 5.3.17. *A strictly increasing function is injective.*

Proof. If $x_1 \neq x_2$, then $x_1 < x_2$ without loss of generality (swapping names if necessary), so $f(x_1) < f(x_2)$, and therefore $f(x_1) \neq f(x_2)$. $\square$

Lemma 5.3.18. *If $f : X \to Y$ is strictly increasing and surjective, then the inverse function $f^{-1} : Y \to X$ is strictly increasing.*

Proof. Assume y and y' are arbitrary elements of Y, and write $x = f^{-1}(y)$ and $x' = f^{-1}(y')$. If $x' < x$, then $y' < y$ because f is increasing. Contrapositively, if $y < y'$, then $x < x'$, or $f^{-1}(y) < f^{-1}(y')$. Since y and y' were arbitrary, f^{-1} is increasing. $\square$

Definition 5.3.19. Assume X is a non-empty set of real numbers. A function $f : X \to \mathbf{R}$ is *non-decreasing* on X if for all x and x' in X, $x < x'$ implies $f(x) \leq f(x')$.

Example 5.3.20. The floor and ceiling functions are non-decreasing on $\mathbf{R}$. Neither is injective. $\diamond$

Remark 5.3.21. Non-decreasing functions preserve non-strict inequalities. Since the conclusion $f(x) \leq f(x')$ is automatic if $x = x'$, the hypothesis may be expressed freely as $x < x'$ or $x \leq x'$, whichever is convenient. $\diamond$

Remark 5.3.22. The corresponding conditions for functions that reverse the sense of inequalities are "strictly decreasing" and "non-increasing." For brevity, we generally omit explicit discussion of these conditions.

Properties of strictly increasing or non-decreasing functions have analogs for strictly decreasing or non-increasing functions. Writing out formal statements and proofs may provide beneficial practice. $\diamond$

Definition 5.3.23. A function that is either non-decreasing or non-increasing is *monotone*. A function that is either strictly increasing or strictly decreasing is *strictly monotone*.

Lemma 5.3.24. *A composition of strictly decreasing functions is strictly increasing. Generally, a composition of monotone functions is monotone.*

Proof. See Exercise 5.3.4 for the first part. (There are sixteen contingencies in general, but the idea is identical for each.) $\square$

Example 5.3.25. The restriction of the reciprocal function $f(x) = 1/x$ to $(0, \infty)$ is strictly decreasing by Proposition 3.2.1 (v). The composition with itself is the identity, which is strictly increasing.

The reciprocal function itself, defined on $\mathbf{R} \setminus \{0\}$, is invertible (indeed, $f^{-1} = f$), hence injective. Note carefully, however, that f is not monotone: $-1 < 1$ but $f(-1) < f(1)$, so f is not non-increasing; and $1 < 2$ but $f(2) < f(1)$, so f is not non-decreasing. $\diamond$

Exercises for Section 5.3

Exercise 5.3.1. (★) Assume $a < b$, and $X = (a, b) \setminus \{0\}$ is the set of non-zero real numbers between a and b. Find, with justification, the image of X under the reciprocal function $f(x) = 1/x$, assuming:

(a) $0 \leq a$.   (b) $b \leq 0$.   (c) $a < 0 < b$.

Exercise 5.3.2. Assume a and b are real and $a < b$. Prove there exists a unique affine bijection $f : (0, 1) \to (a, b)$, and find its inverse.

Exercise 5.3.3. Prove that $f(x) = \dfrac{x}{1 - |x|}$ defines a bijection $f : (-1, 1) \to \mathbf{R}$.

Exercise 5.3.4. Prove the first statement of Lemma 5.3.24.

Exercise 5.3.5. (★) Assume $f : X \to Y$ and $g : Y \to Z$ are mappings.

(a) If f and g are bijections, prove that the composition $g \circ f : X \to Z$ is a bijection. Express the inverse $(g \circ f)^{-1}$ in terms of f^{-1} and g^{-1}.

(b) Conversely, assume the composition $g \circ f : X \to Z$ is a bijection. Must f be injective? Surjective? What can you deduce about g?

Exercise 5.3.6. If f is an injective, non-vanishing real function, it has an inverse mapping f^{-1} and a reciprocal $1/f$. One might ask if these can be equal. Prove that if $f : X \to Y$ is bijective and $f^{-1} = 1/f$, then successive application of f sends each point x to $f(x)$, to $1/x$, to $f(1/x) = 1/f(x)$ to x. What constraints does this put on the domain and codomain?

Exercise 5.3.7. (A) Let X be the set of points (x, y) in the plane satisfying $x^2 + y^2 = 1$ and $(x, y) \neq (0, 1)$. This exercise introduces *inverse stereographic projection*, the mapping $\Pi : \mathbf{R} \to X$ (Pi, for projection) defined by joining $(0, 1)$ to $(t, 0)$ by a ray and letting $\Pi(t) = (x, y)$ be the point of intersection with the circle, Figure 5.9.

(a) Use similar triangles to find a formula for (x, y) in terms of t.

(b) Prove Π is bijective, both geometrically and by finding a formula for Π^{-1}, that is, expressing t in terms of x and y.

(c) Prove t is rational if and only if x and y are rational. Conclude that if A, B, and C are positive coprime integers satisfying $A^2 + B^2 = C^2$, then there exist positive integers p and q such that $C = p^2 + q^2$, and $\{A, B\} = \{2pq, p^2 - q^2\}$.

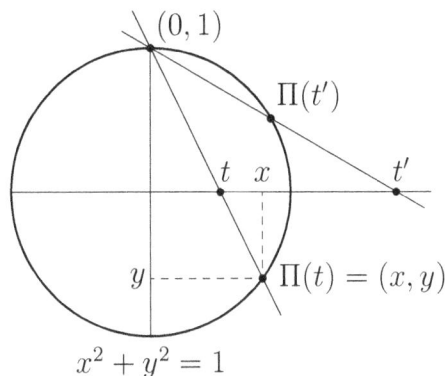

FIGURE 5.9
Inverse stereographic projection of the unit circle.

(d) Assume $f(t) = 1/t$ for non-zero t. Show that $\Pi \circ f \circ \Pi^{-1}(x, y) = (x, -y)$, reflection of the circle across the x-axis.

(e) Assume $f(t) = (1 + t)/(1 - t)$ if $t \notin \{-1, 0, 1\}$, compare Exercise 5.2.1. Show similarly that $\Pi \circ f \circ \Pi^{-1}(x, y) = (-y, x)$, the counterclockwise quarter turn of the circle.

Exercise 5.3.8. Prove parts (ii)–(iv) of Proposition 5.3.12.

Exercise 5.3.9. If $f : X \to Y$ is a mapping, prove f is injective if and only if $f(S_1 \cap S_2) \supseteq f(S_1) \cap f(S_2)$ for all subsets S_1 and S_2 of X. (Compare Proposition 5.3.12 (ii).)

Exercise 5.3.10. Assume α is irrational, and define $f : \mathbf{Z} \times \mathbf{Z} \to \mathbf{R}$ by $f(m, n) = m + n\alpha$. Prove f is injective, and the image $A = f(\mathbf{Z} \times \mathbf{Z})$ is dense.

5.4   Cardinality

For each natural number n there exists a set of n elements, for example, the initial segment $\underline{\mathbf{n}} = \{j\}_{j=0}^{n-1} \subseteq \mathbf{N}$. (By convention $\underline{\mathbf{0}} = \varnothing$.) In everyday life, "counting" a set X establishes a bijection from X to some initial segment $\underline{\mathbf{n}}$.

In mathematics, there are also "infinite" sets. In the late 19th century, German mathematician Georg Cantor extended the concept of counting to infinite sets, and showed there are "different sizes" of infinity. We can only splash a bit in the shallows of this deep ocean, but different sizes of infinity have bearing on real analysis, and we cannot avoid getting our feet wet.

Definition 5.4.1. Two sets X and Y are said to have the *same cardinality* if there exists a bijection $f : X \to Y$.

Remark 5.4.2. A finite set X does not have the same cardinality as any proper subset Y. This property *characterizes* finite sets, namely, is false for infinite sets.   ◇

Example 5.4.3. If n is an arbitrary natural number, the set $\mathbf{N}$ has the same cardinality as $n + \mathbf{N} = \{m \text{ in } \mathbf{N} : m \geq n\} = \{n, n+1, n+2, \dots\}$. The function $f : \mathbf{N} \to n + \mathbf{N}$ defined by $f(k) = n + k$ is a bijection.   ◇

Lemma 5.4.4. *If a and b are extended real numbers such that $a < b$, there exists an increasing bijection from $(0,1)$ to (a,b).*

Proof. See Exercise 5.4.1   □

Definition 5.4.5. A set X is *countable* if X has the same cardinality as $\mathbf{N}$, the set of natural numbers. We say X is *at most countable* if X is either finite or countable. If X is not at most countable, we say X is *uncountable*.

Lemma 5.4.6. *Every subset of $\mathbf{N}$ is at most countable.*

Proof. Assume $X \subseteq \mathbf{N}$. If X is empty or has a largest element, X is finite.

Otherwise, X is non-empty and unbounded. Define $f : \mathbf{N} \to X$ recursively as follows. Let $f(0)$ be the smallest element of X, which exists by well-ordering, Theorem 2.1.4, and define $X_1 = X \setminus \{f(0)\}$. For each positive integer m, define $f(m)$ to be the smallest element of X_m, and define $X_{m+1} = X_m \setminus \{f(m)\}$. The mapping f is bijective, so X is countable.   □

Corollary 5.4.7. *Every subset of a countable set is at most countable.*

Proof. Assume X is countable and $A \subseteq X$. Fix a bijection $f : \mathbf{N} \to X$. The preimage $f^*(A)$ is a subset of $\mathbf{N}$, hence at most countable by Lemma 5.4.6, and the restriction of f to the preimage is a bijection to A.   □

Corollary 5.4.8. *If Y is a set and $f : \mathbf{N} \to Y$ is a surjection, then Y is at most countable.*

Proof. See Exercise 5.4.2.   □

Proposition 5.4.9. *The ordered product $\mathbf{N} \times \mathbf{N}$ is countable.*

Proof. See Exercise 5.4.3.   □

Corollary 5.4.10. *If X_k is a countable set for each natural number k, then the union $\bigcup_k X_k$ is countable.*

Proof. See Exercise 5.4.4.   □

Corollary 5.4.11. *The set of polynomials with rational coefficients is countable.*

Proof. See Exercise 5.4.6. □

Definition 5.4.12. A real number x is *algebraic* if there exists a polynomial f with integer coefficients such that $f(x) = 0$. A non-algebraic real number is *transcendental*.

Example 5.4.13. Every rational number is algebraic: If $r = p/q$ is rational, the affine polynomial $f(x) = qx - p$ has r as a root.

Square roots of positive rational numbers are algebraic; if $f(x) = qx^2 - p$, then $r = \sqrt{p/q}$ is a root (in the sense of polynomials) of f.

Since a polynomial of degree n has at most n distinct real roots and the set of polynomials with integer coefficients is countable, there are only countably many algebraic numbers.

Could *every* real number be algebraic? And if not, can we hope to describe a single transcendental number? Without jumping too far ahead in the story, we will see momentarily that "most" real numbers are *not* algebraic.

In Chapter 12 we will carefully study the specific real number known as e, which is transcendental according to Theorem 12.4.12. ◇

Uncountable Sets

Despite looking at larger and larger sets of real numbers, including the set of algebraic numbers, rife with irrationals, we have not escaped the realm of the countable. Remarkably, the set of real numbers is uncountable; there exists no surjection $f : \mathbf{N} \to \mathbf{R}$. We'll first establish a more modest-looking result.

Definition 5.4.14. A mapping $\mathbf{a} : \mathbf{N} \to \{0, 1\}$ is a *binary sequence*.

Remark 5.4.15. Recall that the value $a_k = \mathbf{a}(k)$ is the *kth term*. We write $(a_k)_{k=0}^{\infty}$ to denote a sequence as an ordered list of terms. ◇

Proposition 5.4.16. *The set X of binary sequences is uncountable.*

Proof. Our method of proof is the *diagonal argument*. We will show that if f is an arbitrary mapping from $\mathbf{N}$ to X, then f is not surjective; there exists a sequence not in the image of f.

If $f : \mathbf{N} \to X$, the value $f(n) = (a_{n,k})_{k=0}^{\infty}$ is a binary sequence for each natural number n. Define a sequence $\mathbf{b}$ in X by the formula $b_k = 1 - a_{k,k}$; that is, take the kth term of $\mathbf{b}$ to be the "opposite" of the kth term of the sequence $f(k)$. By construction, $\mathbf{b} \neq f(n)$ for every n; these sequences do not have the same nth term. In other words, $\mathbf{b}$ is not in the image of f. But $f : \mathbf{N} \to X$ was an arbitrary mapping. Consequently, no mapping from $\mathbf{N}$ to X is surjective. □

Remark 5.4.17. If this proof leaves you with nagging thoughts similar to, "Why can't we just prepend the 'missing' sequence to our list of values?" you are not alone. To reiterate, however, the argument proves that if $f : \mathbf{N} \to X$ is an *arbitrary* mapping, then f is not surjective. In the adversarial game *Who can construct a surjection $f : \mathbf{N} \to X$?*, Player f loses. The "missing" sequence constructed is not the only binary sequence "missed" by f; there are others, necessarily uncountably many. ◇

Corollary 5.4.18. *The set $\mathbf{R}$ of real numbers is uncountable.*

Proof. Let X be the set of binary sequences, and define a mapping $f : X \to \mathbf{R}$ by sending each binary sequence to the corresponding infinite decimal whose digits are all either 0 or 1:

$$f\left((a_k)_{k=0}^{\infty}\right) = \sum_{k=0}^{\infty} a_k \cdot 10^{-k} := \sup_{n \in \mathbf{N}} \sum_{k=0}^{n-1} a_k \cdot 10^{-k}.$$

This mapping is injective, see Exercise 5.4.7, hence bijective to its image. Since $\mathbf{R}$ has an uncountable subset, $\mathbf{R}$ itself is uncountable. □

Remark 5.4.19. In Chapter 4 we saw that the set $\mathbf{Q}$ of rationals is dense in $\mathbf{R}$: Between any two real numbers, there is a rational number. We can now see a bit more deeply the complexity with which the rationals sit inside the reals, since the set of irrational numbers has larger cardinality than the set of rationals. ◇

Example 5.4.20. The set of binary sequences is in bijective correspondence with the ternary set K. The geometric idea is natural. Recall that the ternary set is the intersection of sets K_n, each obtained by removing the middle third of each component of K_{n-1}. Assume $x \in K$, and imagine "homing in" on x by answering an infinite number of "yes-no" questions. The answer to each question tells us whether to pass to a left-hand or right-hand piece when we remove the middle third.

In more detail, we know initially that $x \in K_0 = [0,1]$. If x is in $[0, 1/3]$, set $a_0 = 0$; if x is in $[2/3, 1]$, set $a_0 = 1$. Inductively, assume we have constructed a finite sequence $(a_j)_{j=0}^{n-1}$, and we have located x in a particular component of K_n. This component is split into two pieces in K_{n+1}. If x lies in the left piece, take $a_n = 0$; otherwise take $a_n = 1$, Figure 5.10.

Each x in K gives rise to a unique binary sequence, each binary sequence determines a unique element of K, and these associations are inverse to each other. It follows that K is uncountable. ◇

Remark 5.4.21. Exercise 7.1.15 outlines an algebraic/analytic proof of these claims. Exercise 8.3.8 constructs a non-decreasing surjection from the ternary set to the unit interval $[0,1]$. ◇

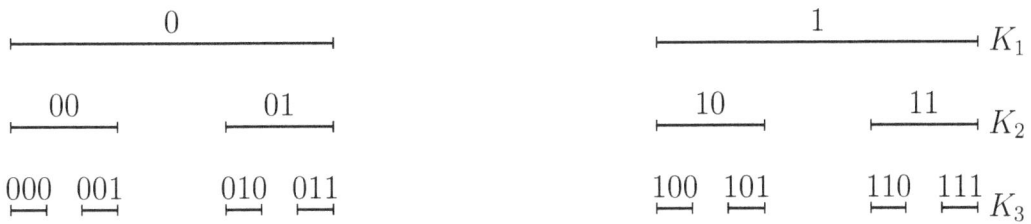

FIGURE 5.10
Binary sequences and intervals in the ternary set.

Remark 5.4.22. The ternary set may appear to consist entirely of endpoints of the approximating intervals. In fact, nearly the opposite is true: Only countably many elements of K are endpoints of some K_n. Each set K_n has only finitely many endpoints. The set of all endpoints in K is therefore a countable union of finite sets. The endpoints of K_n correspond to sequences that are constant after the nth term, see also Exercise 7.1.15, and there are only countably many such sequences. ◇

Remark 5.4.23. In Alan Turing's mathematical formalization of computation, there exist only countably many algorithms, countably many computer programs. There are more real numbers than there are computer programs. Only countably many *computable real numbers* have a decimal representation that can be output by a computer program.

This is not an issue of decimals having infinitely many digits: There *do* exist algorithms that output the infinite, non-repeating digits of algebraic irrational numbers such as $\sqrt{2}$, or of transcendental numbers we will meet later, such as e or π. The blunt fact is, there exist *individual binary sequences* that are not output by any computer program.

Computability of a binary sequence is not a property of any initial finite sequence. No matter how many terms we examine, the resulting sequence *is* computable. If we attempt to search for a non-computable binary sequence, we enter the exponentially growing labyrinth of finite binary sequences, knowing that the property we seek *is not determined by the portion we have examined*, no matter how far into the sequence we look.

If instead we ask an oracle to show us a non-computable binary sequence, we have no algorithmic way to verify the oracle's claim. This is awkward: We have proven that uncountably many binary sequences exist, but only countably are output by some computer program. In any practical sense, cannot exhibit a single non-computable sequence or real number.

Pondering seriously how this mathematical theorem interacts with the physical world, we eventually founder on our ignorance about the natures of space and time. Jorge Luis Borges' *The Library of Babel*, [4, 2], and Richard Preston's *The Mountains of Pi*, [23], provide thought-provoking reading. ◇

Exercises for Section 5.4

Exercise 5.4.1. Prove Lemma 5.4.4.

Exercise 5.4.2. Prove Corollary 5.4.8.

Exercise 5.4.3. (★) Prove that the function $f : \mathbf{N} \times \mathbf{N} \to \mathbf{N}$ defined by

$$f(m, n) = \tfrac{1}{2}\big((m + n)^2 + 3n + m\big)$$

is a bijection. (This establishes Proposition 5.4.9.)

Exercise 5.4.4. (★) Prove Corollary 5.4.10.

Exercise 5.4.5. Prove that the set of *finite subsets* of $\mathbf{N}$ is countable.

Exercise 5.4.6. Prove Corollary 5.4.11.

Exercise 5.4.7. Use Exercise 4.3.9 to prove that the mapping f in the proof of Corollary 5.4.18 is injective.

Exercise 5.4.8. (H) Show that the set of mappings $f : \mathbf{N} \to \{0, 1\}$ is in bijective correspondence with the power set $\mathscr{P}(\mathbf{N})$.

Exercise 5.4.9. Assume X is an arbitrary set and $f : X \to \mathscr{P}(X)$ a mapping. Prove f is not surjective by considering $N_f := \{x \text{ in } X : x \notin f(x)\}$. (This is the only proof by contradiction in the book.)

6

Sequences

Assume x_0 is a real number. Every real number distinct from x_0, and therefore (by induction) every finite set not containing x_0, is separated from x_0 by a gap of positive length. To "approach" x_0 we need some type of infinite set.

In Chapter 4, we saw how every punctured interval $B_r^\times(x_0)$ neighbors x_0. A punctured interval, however, is uncountable. In this chapter, we use real sequences, which are countable objects and therefore "closer to finite." The primary concept is a limit of a sequence, which represents "large-index behavior." Every real number is the limit of some rational sequence. In effect, limits provide a user-friendly front end to suprema.

6.1 Convergence and Limits

Remark 6.1.1. By definition, a real sequence is a mapping $\mathbf{a} : \mathbf{N} \to \mathbf{R}$. We usually denote a sequence by $(a_k)_{k \in \mathbf{N}}$ or $(a_k)_{k=0}^\infty$, emphasizing the interpretation as an *ordered list* of real numbers. The ordering is crucial; a sequence must be carefully distinguished from its set of terms $\{a_k\}_{k \in \mathbf{N}} = \{a_k\}_{k=0}^\infty$.

If k_0 is an integer and a_k is defined for k greater than or equal to k_0, we usually write $(a_k)_{k=k_0}^\infty$—rather than the literally correct $(a_{k+k_0})_{k=0}^\infty$—to signify "the sequence starting at k_0." We have done this a few times already in prior chapters, as you may have noticed.

If the starting index k_0 is unimportant we may simply write (a_k). ◇

Example 6.1.2. The formulas $a_k = 1/(k+1)$ and $b_k = (-1)^k$ define real sequences. The respective sets of terms are $A = \{a_k\}_{k=0}^\infty = \{1/k\}_{k=1}^\infty$, and the finite set $B = \{b_k\}_{k=0}^\infty = \{1, -1\}$. ◇

Example 6.1.3. For each real number x, the formula $a_k = x^k$ defines a real sequence whose set of terms is $\{a_k\}_{k=0}^\infty = \{1, x, x^2, x^3, \dots\}$. ◇

Example 6.1.4. The recursive rule $a_0 = 2$, $a_{k+1} = \frac{1}{2}\left(a_k + (2/a_k)\right)$ defines a real sequence. The next three terms are $a_1 = 3/2$, $a_2 = 17/12$, and $a_3 = 577/408$. ◇

Remark 6.1.5. The definition of limits refers a *positive* real number ε (epsilon), the small Greek e. To say ε is *arbitrary*, or to say *for every* ε, means there is no positive lower bound on the choice. ◇

DOI: 10.1201/9781003601357-6

Definition 6.1.6. Assume (a_k) is a real sequence, and a_∞ is real. We say (a_k) *converges* to a_∞, and write $(a_k) \to a_\infty$, if the following condition is true:

> For every ε,
> > there exists an index N such that
> > > if $k \geq N$, then $|a_k - a_\infty| < \varepsilon$.

Remark 6.1.7. Operationally, $(a_k) \to a_\infty$ if we can make the terms a_k as close as we like to a_∞ (within a distance ε, for arbitrary positive ε) by considering only terms with sufficiently large index ($k \geq N$ for some N). ◇

Remark 6.1.8. Geometrically, $(a_k) \to a_\infty$ if *every open ball $B_\varepsilon(a_\infty)$ contains a_k for all but finitely k.* ◇

Proposition 6.1.9. *If (a_k) is a real sequence that converges to a_∞ and to a'_∞, then $a_\infty = a'_\infty$.*

Proof. We will prove $|a'_\infty - a_\infty| < \varepsilon$ for every positive ε, which implies $a_\infty = a'_\infty$. (Why?) Assume ε is arbitrary. Since $(a_k) \to a_\infty$, there exists an index N_1 such that if $k \geq N_1$, then $|a_k - a_\infty| < \varepsilon/2$. Similarly, since $(a_k) \to a'_\infty$, there exists an index N'_1 such that if $k \geq N'_1$, then $|a_k - a'_\infty| < \varepsilon/2$.

Put $N = \max(N_1, N'_1)$. Since $N \geq N_1$ and $N \geq N'_1$, the triangle inequality implies

$$|a'_\infty - a_\infty| = |(a'_\infty - a_N) + (a_N - a_\infty)|$$
$$\leq |a'_\infty - a_N| + |a_N - a_\infty| < \varepsilon/2 + \varepsilon/2 = \varepsilon.$$

Since ε was arbitrary, $a'_\infty - a_\infty = 0$. □

Remark 6.1.10. Proposition 6.1.9 guarantees uniqueness of limits: A real sequence converges to at most one number. If $(a_k) \to a_\infty$, we call a_∞ the *limit* of (a_k), and write $a_\infty = \lim\limits_{k \to \infty} a_k$. ◇

Remark 6.1.11. Informally, we say "a_k approaches a_∞ as $k \to \infty$." Indeed, the notation a_∞ is meant to suggest "setting $k = \infty$ in the limit." Logically, however, this is not what convergence means. The terms a_k of a sequence (a_k) are merely individual real numbers. Convergence only makes sense for a *sequence* $\mathbf{a} = (a_k)$, an *ordered list of terms*. Further, convergence either happens or does not; we do not need to "wait," possibly forever, for k to run from 0 to ∞. ◇

The ε-N Game

Convergence of a sequence may be understood as an adversarial "challenge-response" game. A sequence (a_k) and a putative limit a_∞ are specified in advance. Player ε chooses a positive "tolerance" ε, which defines a target, the interval $B_\varepsilon(a_\infty) = (a_\infty - \varepsilon, a_\infty + \varepsilon)$.

Player N now tries to "hit the target," namely, to ensure $|a_k - a_\infty| < \varepsilon$, solely by taking k to be sufficiently large. A "successful response" to the "challenge" of Player ε is an index N such that if $k \geq N$, then $|a_k - a_\infty| < \varepsilon$.

If Player N is able to respond successfully to a particular ε, they "win the round." Otherwise Player ε wins the round.

To say $(a_k) \to a_\infty$ means that Player N *has a winning strategy against a perfect opponent*: No matter how "skillful" Player ε is (if ε is arbitrary, no matter how small ε is chosen), Player N can issue a successful response.

The ε-N game and its variants studied later are the essence of analysis. If you read proofs elsewhere, you may find that elaborate, seemingly magical choices of N are made. To make these choices, the author imagined an arbitrary ε was given, and formulated a strategy for choosing a "winning" index N, making idiomatic choices. The proof itself is merely a demonstration that Player N wins.

Example 6.1.12. Assume $c \in \mathbf{R}$. The sequence $a_k = c$ is called a *constant sequence*. A constant sequence converges to $a_\infty = c$: For every ε and for every k, we have $|a_k - a_\infty| = |c - c| = 0 < \varepsilon$. That is, Player N *cannot lose against a perfect opponent* when playing with a constant sequence!   ◇

Example 6.1.13. The sequence $(a_k) = (1/k)_{k=1}^\infty$ converges to 0. Before giving a proof, we'll play a few rounds of the ε-N game.

If $\varepsilon = 100$, Player N *cannot lose*, namely, may take $N = 1$: Indeed, $|a_k - 0| = 1/k \leq 1 < \varepsilon$ for every positive integer k. Note carefully: This fact does not prove $(a_k) \to 0$; Player N must be able to win against an *arbitrary* ε.

If $\varepsilon = 0.01 = 1/100$, Player N may take $N = 101$: If $k \geq 101$, then $|a_k - 0| = 1/k \leq 1/101 < 1/100 = \varepsilon$.

If $\varepsilon = 1/\sqrt{200}$, Player N's goal is to ensure $1/N < \varepsilon = 1/\sqrt{200}$, or after rearranging, $200 < N^2$. The smallest choice, $N = 15$, is easy to find in this example, but there is no harm in taking, say, $N = 200$.

To prove $(a_k) \to 0$, it suffices to construct a winning strategy for Player N. Fix ε arbitrarily. By reciprocal finitude, Corollary 4.3.7, there exists a positive integer N such that $1/N < \varepsilon$. This N is a winning response. If $k \geq N$, then $|a_k - 0| = 1/k \leq 1/N < \varepsilon$.   ◇

Remark 6.1.14. If an index N wins against some challenge ε, then *every larger integer $N' \geq N$ also wins*, because $k \geq N'$ implies $k \geq N$. To reiterate, it is not necessary (or always desirable) to pick the smallest winning N.

Correspondingly, making ε smaller makes the target smaller, which makes the condition $|a_k - a_\infty| < \varepsilon$ "harder to meet," and generally forces N to be larger. But note carefully: Player ε has no "optimal" choice: There exists no smallest positive real number.   ◇

Example 6.1.15. The sequence defined by $a_k = (-1)^k$ has terms that are alternately 1 and -1: The "even" terms $a_{2\ell}$ are all 1 and the "odd" terms $a_{2\ell+1}$ are -1. Intuitively, (a_k) does not approach a real limit.

Using the definition, we will prove (a_k) does not converge. That is, for every real number a_∞, the statement "$(a_k) \to a_\infty$" is false. To establish this, we fix a putative limit a_∞ arbitrarily, then take the side of Player ε and look for a winning strategy:

> No matter what N is, $k = 2N$ is even, and $k = 2N + 1$ is odd, and both are greater than N. We are therefore assured that $|a_k - a_\infty|$ takes both values $|1 - a_\infty|$ and $|-1 - a_\infty| = |1 + a_\infty|$ for some k greater than N. In order to win, Player N must make *both* of these quantities smaller than ε. But in that event, the triangle inequality implies $2 \leq |1 - a_\infty| + |1 + a_\infty| < \varepsilon + \varepsilon = 2\varepsilon$, or $2 < 2\varepsilon$. If this inequality is not satisfied, Player N loses.

Having reasoned thusly, Player ε chooses any ε such that $0 < \varepsilon \leq 1$, say $\varepsilon = 1$. There does not exist an N such that if $k \geq N$, then $|a_k - a_\infty| < \varepsilon = 1$; if such an N exists, then

$$2 \leq |1 - a_\infty| + |1 + a_\infty| = |a_{2N} - a_\infty| + |a_{2N+1} + a_\infty| < \varepsilon + \varepsilon = 2,$$

or $2 < 2$, which is false. Since "$(a_k) \to a_\infty$" is false for every real number a_∞, the sequence defined by $a_k = (-1)^k$ does not converge. $\Diamond$

Proposition 6.1.16. *If x is a real number such that $-1 < x \leq 1$, then the sequence $a_k = x^k$ converges; the limit is 1 if $x = 1$, and is 0 if $-1 < x < 1$.*

Proof. If $x = 1$, the sequence (a_k) is constant (since $a_k = 1^k = 1$ for all k), and therefore converges to 1 by Example 6.1.12. Similarly, if $x = 0$, then $a_k = 0$ for all k.

For the remainder of the proof, assume $0 < |x| < 1$, and set $a_\infty = 0$. By Corollary 3.3.6, if we write $|x| = 1/(1 + u)$, then $|x^k| \leq 1/(1 + ku)$ for all k.

Fix ε arbitrarily. By the accretion principle, Corollary 4.3.5, there exists a natural number N such that $1/\varepsilon < Nu$, or $1/(Nu) < \varepsilon$. If $k \geq N$, then

$$|x^k - a_\infty| = |x^k| \leq \frac{1}{1 + ku} < \frac{1}{ku} \leq \frac{1}{Nu} < \varepsilon.$$

Since ε was arbitrary, $(x^k) \to a_\infty = 0$ if $0 < |x| < 1$. $\square$

Example 6.1.17. For $x = 1/2$ or $x = -4/5$, say, the conclusion of Proposition 6.1.16 is not intuitively surprising. However, for a number such as $x = 0.99999999999999999999 = 1 - 10^{-20}$, a "fairly large" exponent n may be needed to make the power x^n "small," compare Exercise 6.1.6. $\Diamond$

Boundedness

Definition 6.1.18. A real sequence (a_k) is *bounded above* if its set of terms is bounded above, namely, if there exists a real number M such that $a_k \leq M$ for all k. Any such M is called an *upper bound* for the sequence.

Similarly, we say (a_k) is *bounded below* if there exists a real number m such that $m \leq a_k$ for all k.

We say the real sequence (a_k) is *bounded* if (a_k) is bounded above and bounded below, namely (Lemma 4.1.25), if there exists a positive real number M such that $-M \leq a_k \leq M$, or (Proposition 3.2.10) $|a_k| \leq M$, for all k.

Proposition 6.1.19. *If (a_k) is a convergent real sequence with limit a_∞, then*

(i) (a_k) *is bounded.*

(ii) *The sequence $(|a_k|)$ converges to $|a_\infty|$.*

Proof. By hypothesis, for every ε, there exists an N such that if $k \geq N$, then $|a_k - a_\infty| < \varepsilon$.

(i). Particularly, if $\varepsilon = 1$, there exists an N such that $|a_k - a_\infty| < 1$ if $k \geq N$. It suffices to prove $|a_k| \leq M := 2 + \max(\{|a_j|\}_{j=0}^N)$ for all k.

If $0 \leq k \leq N$, then $|a_k| < M$ by construction. If instead $k \geq N$, the triangle inequality applied to $a_k = (a_k - a_\infty) + (a_\infty - a_N) + a_N$ implies

$$|a_k| \leq |a_k - a_\infty| + |a_\infty - a_N| + |a_N| < 1 + 1 + |a_N| \leq M.$$

This completes the proof that $|a_k| \leq M$ for all k.

(ii). Assume $\varepsilon > 0$, and choose N such that if $k \geq N$, then $|a_k - a_\infty| < \varepsilon$. By the reverse triangle inequality, $k \geq N$ implies $\big||a_k| - |a_\infty|\big| \leq |a_k - a_\infty| < \varepsilon$. Since ε was arbitrary, $(|a_k|) \to |a_\infty|$. $\square$

Remark 6.1.20. Contrapositively, an unbounded sequence does not converge. In words, boundedness is *necessary* for convergence.

Example 6.1.15 shows that a bounded sequence may fail to converge. In words, boundedness is *not sufficient* for convergence. $\diamond$

Exercises for Section 6.1

Exercise 6.1.1. (★) Using the ε-N definition, prove $\displaystyle\lim_{k \to \infty} \frac{4k - 5}{3k + 2} = \frac{4}{3}$.

Exercise 6.1.2. Using the ε-N definition, evaluate $\displaystyle\lim_{k \to \infty} \frac{2 - 7k}{2k - 5}$.

Exercise 6.1.3. (★) Someone tells you a certain sequence (a_k) is *eventually constant*. Give a precise definition for this condition.

Exercise 6.1.4. (H) Assume (a_k) is a real sequence and a_∞ a real number. Consider the following conditions:

(i) For every ε, there exists an N such that if $k \geq N$ then $|a_k - a_\infty| < \varepsilon$.

(ii) There exists an N such that for every ε, if $k \geq N$ then $|a_k - a_\infty| < \varepsilon$.

Are these conditions logically equivalent? If so, give a proof. If not, find a "familiar" condition equivalent to (ii), and give an example of a sequence satisfying (i) but not (ii).

Exercise 6.1.5. Assume (a_k) is a real sequence and a_∞ a real number. Consider the following conditions:

(i) There exists an ε such that for every N, if $k \geq N$ then $|a_k - a_\infty| < \varepsilon$.

(ii) For every N, there exists an ε such that if $k \geq N$ then $|a_k - a_\infty| < \varepsilon$.

Are these conditions logically equivalent? If so, give a proof. If not, find a "familiar" condition equivalent to (ii), and give an example of a sequence satisfying (i) but not (ii).

Exercise 6.1.6. Assume $x = 1 - 10^{-20}$. In the proof of Proposition 6.1.16, what is the smallest N that makes the upper bound at most $1/2$?

Exercise 6.1.7. In calculus, limits of sequences are sometimes investigated by examining terms. For instance, if $x_0 = 1.75$ and $x_{n+1} = \frac{1}{2}\left(x_n + (3/x_n)\right)$, the next two terms in decimal are

$$97/56 \approx 1.732142857\ldots, \qquad 18817/10864 \approx 1.73205081\ldots,$$

while the mathematical limit, to the accuracy of a calculator display, is $1.732050808\ldots$. Discuss what we learn about convergence of a sequence by examining terms. Can we formulate convergence using numerical evidence of initial terms? If so, describe how; if not, explain why not in detail.

Exercise 6.1.8. If (a_k) and (b_k) are real sequences, define a new sequence (c_k) by "shuffling" the terms, setting $c_{2k} = a_k$ and $c_{2k+1} = b_k$ for every natural number k.

Write out the first six terms of (c_k). Prove that (c_k) converges if and only if (a_k) and (b_k) converge to the same limit.

Exercise 6.1.9. Assume (a_k) is an integer sequence, namely, a_k is an integer for all k. Prove that (a_k) converges if and only if (a_k) is eventually constant.

6.2   Algebraic Properties of Limits

The ε-N definition of convergence puts sequential limits on a firm logical foundation, but the definition is a little onerous to use. In this section, we establish formal rules for algebraically manipulating limits in equations.

Proposition 6.2.1. *Assume (a_k) is a real sequence converging to a_∞, and (b_k) is a real sequence converging to b_∞.*

(i) *The sequence $(a_k + b_k)$ converges to $a_\infty + b_\infty$.*

(ii) *The sequence $(a_k b_k)$ converges to $a_\infty b_\infty$.*

(iii) *If $b_k \neq 0$ for all k and if $b_\infty \neq 0$, then (a_k/b_k) converges to a_∞/b_∞.*

Proof. (i). We'll start with "scratch work," adopting the perspective of Player N: Strategizing a response is how analysis gets done, and our object is not just to prove statements, but to explain *how to develop estimates*.

> By hypothesis, we can make $|a_k - a_\infty|$ and $|b_k - b_\infty|$ as small as we like. Our goal is to ensure $|(a_k + b_k) - (a_\infty + b_\infty)|$ is smaller than ε. Algebra and the triangle inequality give
>
> $$|(a_k + b_k) - (a_\infty + b_\infty)| = |(a_k - a_\infty) + (b_k - b_\infty)| \leq |a_k - a_\infty| + |b_k - b_\infty|$$
>
> We have a total "error budget" ε. Idiomatically, we'll ensure each summand is smaller than $\varepsilon/2$.
>
> We can ensure $|a_k - a_\infty| < \varepsilon/2$ if $k \geq N_1$, and ensure $|b_k - b_\infty| < \varepsilon/2$ if $k \geq N_2$. To get a *single index* satisfying *both conditions*, idiomatically we pick $N = \max(N_1, N_2)$. We're ready to face Player ε.

The "conventional" proof follows.

Assume $\varepsilon > 0$. Since $(a_k) \to a_\infty$ by hypothesis, there exists an integer N_1 such that if $k \geq N_1$, then $|a_k - a_\infty| < \varepsilon/2$. Similarly, there exists an integer N_2 such that if $k \geq N_2$, then $|b_k - b_\infty| < \varepsilon/2$. Let $N = \max(N_1, N_2)$. If $k \geq N$, the triangle inequality implies

$$|(a_k + b_k) - (a_\infty + b_\infty)| \leq |a_k - a_\infty| + |b_k - b_\infty| < \varepsilon/2 + \varepsilon/2 = \varepsilon.$$

Since ε was arbitrary, $(a_k + b_k) \to a_\infty + b_\infty$.

(ii). Again we'll start by strategizing as Player N:

> By hypothesis, we can make $|a_k - a_\infty|$ and $|b_k - b_\infty|$ as small as we like. Our goal is to ensure $|a_k b_k - a_\infty b_\infty|$ is arbitrarily small. Operationally, we are trying to control the change in the product of two numbers, but we only have direct

control over each number separately. This suggests we "vary one factor at a time," see also Exercise 3.2.11. Doing so amounts to subtracting and adding a product where only one factor changes, say $a_\infty b_k$:

$$
\begin{aligned}
|a_k b_k - a_\infty b_\infty| &= |a_k b_k - a_\infty b_k + a_\infty b_k - a_\infty b_\infty| \\
&\leq |a_k b_k - a_\infty b_k| + |a_\infty b_k - a_\infty b_\infty| \\
&= |a_k - a_\infty|\,|b_k| + |a_\infty|\,|b_k - b_\infty|.
\end{aligned}
$$

The triangle inequality gives us two summands, each a product having one "small" factor. If we bound the "other" factor in each summand, we can make each summand, and therefore their sum, as small as we like. The factor $|a_\infty|$ is just a number, already bounded. The factor $|b_k|$ might range over infinitely many numbers, but those numbers are terms of a convergent sequence, hence are bounded by Proposition 6.1.19.

Let L and M be positive numbers such that $|a_\infty| < L$ and $|b_k| < M$ for all k. To bring our estimate in under budget, it suffices to ensure each summand is smaller than $\varepsilon/2$, or $|a_k - a_\infty| < \varepsilon/(2M)$ and $|b_k - b_\infty| < \varepsilon/(2L)$. We're ready to face Player ε.

The "conventional" proof follows.

Assume $\varepsilon > 0$. Put $L = 1 + |a_\infty|$ (to ensure $L > 0$). By Proposition 6.1.19, there exists a positive real number M such that $1 + |b_k| \leq M$ for all k. Put $r = \varepsilon/(2M)$ and $s = \varepsilon/(2L)$. Since $(a_k) \to a_\infty$, there exists an integer N_1 such that if $k \geq N_1$, then $|a_k - a_\infty| < r$. Since $(b_k) \to b_\infty$, there exists an integer N_2 such that if $k \geq N_2$, then $|b_k - b_\infty| < s$. Let $N = \max(N_1, N_2)$. If $k \geq N$, then

$$
\begin{aligned}
|a_k b_k - a_\infty b_\infty| &= |a_k b_k - a_\infty b_k + a_\infty b_k - a_\infty b_\infty| \\
&\leq |a_k b_k - a_\infty b_k| + |a_\infty b_k - a_\infty b_\infty| \\
&= |a_k - a_\infty|\,|b_k| + |a_\infty|\,|b_k - b_\infty| \\
&< \varepsilon/(2M) \cdot M + L \cdot \varepsilon/(2L) = \varepsilon.
\end{aligned}
$$

Since ε was arbitrary, $(a_k b_k) \to a_\infty b_\infty$.

(iii). We'll prove the special case $(1/b_k) \to 1/b_\infty$, then use (ii) to deduce the full claim. Again, we adopt the viewpoint of Player N:

By hypothesis, we can make $|b_k - b_\infty|$ as small as we like. Our goal is to ensure

$$
\left| \frac{1}{b_k} - \frac{1}{b_\infty} \right| = \frac{|b_\infty - b_k|}{|b_k|\,|b_\infty|} = |b_k - b_\infty| \cdot \frac{1}{|b_k|\,|b_\infty|}
$$

is arbitrarily small. The factor $|b_\infty|$ is just a positive real number. The obstacle is to bound $1/|b_k|$ for large k, namely, to *bound b_k away from zero*, or find a positive lower bound for $|b_k|$.

Figure 6.1 depicts the situation geometrically, and suggests an estimate: If b_k is no further than $r = |b_\infty|/2$ from b_∞, then $|b_\infty|/2 \leq |b_k|$, and $1/|b_k| \leq 2/|b_\infty|$.

It therefore suffices in addition to ensure $|b_k - b_\infty| < r = \varepsilon \cdot |b_\infty|^2/2$.

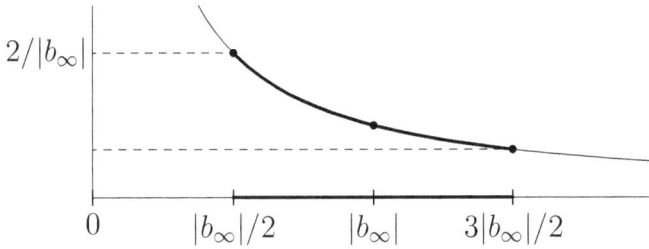

FIGURE 6.1
Bounding the reciprocal on a neighborhood of a non-zero number.

The conventional proof follows. To avoid a technical digression mid-estimate, we'll start with bounding $|b_k|$ away from 0.

Lemma 6.2.2. *Assume b is a non-zero real number and $0 < r \leq |b|/2$. If $|x - b| < r$, namely, if $x \in B_r(b)$, then $|b|/2 < |x|$ and $1/|x| < 2/|b|$.*

Proof. The reverse triangle inequality applied to $x = b + (x - b)$ implies

$$|x| \geq \Big||b| - |x - b|\Big| \geq |b| - |x - b| > |b| - r \geq |b| - |b|/2 = |b|/2,$$

or $|b|/2 < |x|$. Taking reciprocals implies $1/|x| < 2/|b|$. $\qquad\square$

Assume $\varepsilon > 0$. Put $r = \min(|b_\infty|/2, \varepsilon \cdot |b_\infty|^2/2)$. By construction, $r \leq |b_\infty|/2$, so Lemma 6.2.2 guarantees that if $|b_k - b_\infty| < r$, then $1/|b_k| < 2/|b_\infty|$.

Since $0 < r$ and $(b_k) \to b_\infty$, there exists an integer N such that if $k \geq N$, then $|b_k - b_\infty| < r \leq \varepsilon \cdot |b_\infty|^2/2$, and consequently

$$\left|\frac{1}{b_k} - \frac{1}{b_\infty}\right| = \frac{|b_k - b_\infty|}{|b_k| \, |b_\infty|} < \frac{2r}{|b_\infty|^2} \leq \varepsilon.$$

We have shown: For every ε, there exists an index N such that if $k \geq N$, then $|1/b_k - 1/b_\infty| < \varepsilon$. Since ε was arbitrary, $(1/b_k) \to 1/b_\infty$.

(iii) now follows from (ii) by writing $a_k/b_k = a_k \cdot (1/b_k)$. $\qquad\square$

Example 6.2.3. If m is a positive integer, the sequence $(a_k)_{k=1}^\infty$ defined by $a_k = k^{-m} = 1/k^m$ converges to 0. If $m = 1$, this is Example 6.1.13. For larger m, Proposition 6.2.1 (ii) establishes the inductive step. $\qquad\diamond$

Example 6.2.4. In practice, Proposition 6.2.1 grants license to move a limit into or out of an arithmetic expression:

$$\lim_{k \to \infty} \frac{k-1}{k+1} = \lim_{k \to \infty} \frac{1 - (1/k)}{1 + (1/k)} = \frac{1 - \lim_k (1/k)}{1 + \lim_k (1/k)} = \frac{1}{1} = 1;$$

$$\lim_{k \to \infty} \frac{2k}{k^2+1} = \lim_{k \to \infty} \frac{(2k)/k^2}{1 + (1/k^2)} = \frac{2\lim_k (1/k)}{1 + \lim_k (1/k^2)} = \frac{0}{1} = 0;$$

$$\lim_{k \to \infty} \left[\frac{k+1}{k} \right]^m = \left[\lim_{k \to \infty} \frac{k+1}{k} \right]^m = (1)^m = 1; \quad \text{etc.}$$

In the third example, the integer exponent $m \geq 0$ is arbitrary but "fixed," independent of k. Note carefully: Proposition 6.2.1 does *not* imply

$$\lim_{k \to \infty} \left[\frac{k+1}{k} \right]^k$$

exists. As we will see in Chapter 12, the limit does exist, but is not 1. Instead, this limit is $e \approx 2.718281828\ldots$, a transcendental number that pervades pure and applied mathematics. For now, see Exercise 6.2.6.   ◇

Exercises for Section 6.2

Exercise 6.2.1. (★) With justification, evaluate the limits, or show the limit does not exist.

(a) $\displaystyle\lim_{k \to \infty} \frac{4k^2 - 5k + 7}{5k^2 + 1}$.

(b) $\displaystyle\lim_{k \to \infty} \frac{4k^5 - 5k + 7}{5k^4 + 1}$.

Exercise 6.2.2. (H) Prove that if p is a polynomial function, and if (x_k) is a convergent real sequence, then $\lim_k p(x_k) = p(\lim_k x_k)$.

Exercise 6.2.3. (★) If $k \geq 1$, define $f_k : [0,1] \to \mathbf{R}$ by $f_k(x) = x^k$.

(a) Prove that for each x in $[0,1]$, $f(x) := \lim_k f_k(x)$ exists.

(b) Prove that $\sup\{|f_k(x) - f(x)| : 0 \leq x \leq 1\} \not\to 0$.

Exercise 6.2.4. Suppose $\phi : \mathbf{R} \to \mathbf{R}$ is defined by $\phi(x) = x/(1+x^2)$, and put $f_k(x) = \phi(kx)/k$ for positive integer k. Prove that $\sup\{|f_k(x)| : x \in \mathbf{R}\} \to 0$.

Exercise 6.2.5. (H) Find the limit of the sequence whose general term is:

(a) $a_k = k!/k^k$.

(b) $b_k = 10^k/k!$.

(c) $c_k = (k!)^2/(2k)!$.

Exercise 6.2.6. If $k \geq 1$, prove $2 \leq [(k+1)/k]^k \leq \sum_{j=0}^{k} 1/j!$.

6.3   Limits and Inequalities

Concepts of monotonicity from Chapter 5 make sense for real sequences. Since convergence depends only on "arbitrary tails" of a sequence, it is convenient in practice to "forget" finitely many initial terms and consider "eventually monotone" sequences. In this section we'll see how limits and suprema (or infima) amount to the same thing for eventually monotone sequences, and show that limits respect non-strict inequalities.

Definition 6.3.1. A real sequence (a_k) is *eventually non-decreasing* if there exists an index N_0 such that $a_k \leq a_{k'}$ whenever $N_0 \leq k \leq k'$.

Lemma 6.3.2. *A real sequence (a_k) is eventually non-decreasing if and only if there exists an N_0 such that $k \geq N_0$ implies $a_k \leq a_{k+1}$.*

Proof. Since $k < k' := k + 1$, an eventually non-decreasing sequence satisfies the condition in the lemma. Conversely, write $k = N_0 + k_0$, $k' = k + n$, and argue by induction on n, see also Exercise 6.3.1. $\square$

Example 6.3.3. The real sequence defined by $a_k = k^2/(k^2 - 500)$ is eventually non-decreasing, since

$$\frac{k^2}{k^2 - 500} = \frac{k^2 - 500 + 500}{k^2 - 500} = 1 - \frac{500}{k^2 - 500},$$

and if $k^2 > 500$, that is, if $k \geq 23 = N_0$, the denominator increases with k. $\diamond$

Example 6.3.4. The real sequences defined by the formulas $a_k = (-1)^k k$ and $b_k = (-1)^k/k$ are *not* eventually non-decreasing. $\diamond$

Proposition 6.3.5. *If (a_k) is an eventually non-decreasing sequence, then*

(i) *(a_k) is bounded below.*

(ii) *(a_k) converges if and only if it is bounded above.*

Proof. By hypothesis, there exists an integer N_0 such that $a_k \leq a_{k'}$ whenever $N_0 \leq k \leq k'$. Particularly, $a_{N_0} \leq a_k$ if $N_0 \leq k$.

(i). The real number $m = \min(a_j)_{j=0}^{N_0}$ is a lower bound for (a_k): By construction, $m \leq a_k$ if $0 \leq k \leq N_0$. On the other hand, if $N_0 \leq k$, then $m \leq a_{N_0} \leq a_k$ as noted above.

(ii). Convergent implies bounded above: By Proposition 6.1.19, a convergent sequence is bounded, hence bounded above.

Bounded above implies convergent: Suppose there exists a real number M such that $a_k \leq M$ for all k. The set $A = \{a_k : N_0 \leq k\}$ is non-empty and bounded above by M, so $a_\infty := \sup A$ exists by the completeness axiom. It suffices to prove $(a_k) \to a_\infty$.

Assume $\varepsilon > 0$. By definition of a supremum, the number $a_\infty - \varepsilon < a_\infty$ is not an upper bound of A, so there exists an integer $N \geq N_0$ such that $a_\infty - \varepsilon < a_N$. If $k \geq N$, then $a_N \leq a_k \leq \sup A$, so

$$a_\infty - \varepsilon < a_N \leq a_k \leq a_\infty < a_\infty + \varepsilon.$$

That is, if $k \geq N$, then $|a_k - a_\infty| < \varepsilon$. Since ε was arbitrary, $(a_k) \to a_\infty$. $\square$

For practice, give definitions of *eventually non-increasing* and *eventually monotone* real sequences, and modify of the statements and proofs of Lemma 6.3.2 and Proposition 6.3.5 for eventually non-increasing sequences.

Example 6.3.6. Consider the sequence (a_k) in Example 6.1.4. As in the proof of Theorem 4.2.11, (a_k) is decreasing and bounded below, hence convergent. Further, the limit is $\sqrt{2}$, the unique positive real number whose square is 2. $\Diamond$

Proposition 6.3.7. *Assume (a_k) and (b_k) are convergent real sequences, with respective limits a_∞ and b_∞.*

(i) *If $0 \leq a_k$ for all but finitely many k, then $0 \leq a_\infty$.*

(ii) *If $a_k \leq b_k$ for all but finitely many k, then $a_\infty \leq b_\infty$.*

(iii) *If $c \leq a_k \leq d$ for all but finitely many k, then $c \leq a_\infty \leq d$.*

Proof. (i). We prove the contrapositive: If $a_\infty < 0$, then $a_k < 0$ for infinitely many k.

Put $\varepsilon = -a_\infty/2$, so $\varepsilon > 0$ by hypothesis. Since $(a_k) \to a_\infty$, there exists an N such that if $k \geq N$, then $|a_k - a_\infty| < \varepsilon$. But $|a_k - a_\infty| < \varepsilon$ if and only if $a_\infty/2 < a_k - a_\infty < -a_\infty/2$, and this implies $a_k < a_\infty/2 < 0$. That is, if $k \geq N$, then $a_k < 0$.

(ii). Define $c_k = b_k - a_k$. By hypothesis, $0 \leq c_k$ for all but finitely many k. By Proposition 6.2.1, (c_k) converges to $b_\infty - a_\infty$. Part (i) implies $0 \leq b_\infty - a_\infty$, or $a_\infty \leq b_\infty$.

(iii). This follows immediately from (ii). $\square$

Divergence to Infinity

The definition of sequential convergence makes no sense if $a_\infty = +\infty$ or $a_\infty = -\infty$. The objects $\infty := +\infty$ and $-\infty$ are not real numbers, so formal inequalities such as $|a_k - \infty| < \varepsilon$ have no meaning.

Nonetheless, it is useful to be able to study sequences that "approach" ∞ or $-\infty$. Such sequences diverge (do not have a real limit), but they still enjoy some special properties of convergent sequences.

Definition 6.3.8. Assume (a_k) is a real sequence. We say (a_k) *diverges to* ∞, denoted $(a_k) \to \infty$, if the following condition holds:

> For every real number M,
> > there exists an index N such that
> > > if $k \geq N$, then $a_k > M$.

We say (a_k) *diverges to* $-\infty$, denoted $(a_k) \to -\infty$, if:

For every real number M,
  there exists an index N such that
    if $k \geq N$, then $a_k < M$.

Remark 6.3.9. These conditions differ by one *character*. ◇

Example 6.3.10. The sequence $a_k = k$ diverges to ∞: By finitude, if M is an arbitrary real number, there exists a natural number N such that $N > M$. If $k \geq N$, then $a_k = k \geq N > M$. ◇

Example 6.3.11. If $x > 1$, then the sequence $a_k = x^k$ of Example 6.1.17 diverges to ∞. To prove this, note that $x = 1 + u$ for some positive u. By Proposition 3.3.5, $1 + ku \leq x^k = a_k$ for every natural number k. If $M \in \mathbf{R}$, there exists a natural number N such that $M < Nu$ by the accretion principle. If $k \geq N$, then

$$M < Nu < 1 + Nu \leq 1 + ku \leq x^k = a_k.$$

Since M was an arbitrary real number, $(x^k) \to \infty$. ◇

Remark 6.3.12. If (a_k) is a real sequence that is eventually non-decreasing, then either (a_k) is bounded above (hence convergent to a finite limit), or not bounded above (hence divergent to ∞).

Analogous remarks hold for a real sequence that is eventually non-increasing. Consequently, an eventually monotone sequence *always* has an extended real "limit." ◇

Proposition 6.3.13. *If $(a_k) \to \infty$ and $(b_k) \to b_\infty$ for some real number b_∞, then*

(i) $(a_k + b_k) \to \infty$.

(ii) *If $b_\infty > 0$, then $(a_k b_k) \to \infty$. If $b_\infty < 0$, then $(a_k b_k) \to -\infty$.*

(iii) *If $a_k \neq 0$ for all k, then $(b_k/a_k) \to 0$.*

Proof. (i). Since a convergent sequence is bounded (Proposition 6.1.19) and $(b_k) \to b_\infty$, there exists a positive real B such that $|b_k| \leq B$ for all k. Let M be arbitrary. Since $(a_k) \to \infty$, there exists an N such that if $k \geq N$, then $a_k > M + B$, which implies

$$a_k + b_k \geq a_k - |b_k| > (M + B) - B = M.$$

Since M was arbitrary, $(a_k + b_k) \to \infty$.

(ii). Assume $b_\infty > 0$. Taking $\varepsilon = b_\infty/2 > 0$, there exists an N_1 such that if $k \geq N_1$, then $|b_k - b_\infty| < \varepsilon = b_\infty/2$. Rearranging gives $b_\infty/2 < b_k$ if $k \geq N_1$.

Fix M arbitrarily. Since $(a_k) \to \infty$, there exists an index N greater than N_1 such that if $k \geq N$, then $a_k > 2M/b_\infty$. But this implies

$$a_k b_k > (2M/b_\infty) \cdot (b_\infty/2) = M.$$

Since M was arbitrary, $(a_k b_k) \to \infty$. To handle the case $b_\infty < 0$, multiply appropriately by -1 in the preceding proof.

(iii). Suppose $a_k \neq 0$ for all k, so the quotient sequence (b_k/a_k) is defined. As in (i), let B be a positive bound for $|b_k|$, and fix ε arbitrarily. Since $(a_k) \to \infty$, there is an N such that if $k \geq N$, then $a_k > B/\varepsilon$, which implies $b_k/a_k < B(\varepsilon/B) = \varepsilon$. Since ε was arbitrary, $(b_k/a_k) \to 0$.   □

Remark 6.3.14. Proposition 6.3.13 may be interpreted as assigning values to certain arithmetic expressions containing infinity: If $L > 0$ is real, then

$$\infty \pm L = \infty, \qquad \infty \cdot (\pm L) = \pm\infty, \qquad \pm L/\infty = 0.$$

Modifications of the preceding arguments, see Exercise 6.3.12, establish that

$$\infty + \infty = \infty, \qquad -\infty + (-\infty) = -\infty, \qquad \pm\infty \cdot \infty = \pm\infty.$$

However, ∞ and $-\infty$ are not real numbers, and the preceding "equations" must be understood as theorems about how particular sequences diverge, not as equalities in the sense of real numbers.

Moreover, the following expressions are *indeterminate*, loosely because the "value" depends on the approximating sequences:

$$\infty - \infty, \qquad 0 \cdot (\pm\infty), \qquad (\pm\infty)/(\pm\infty), \qquad 0/0.$$

Consequently, care is required when manipulating algebraic expressions involving ∞. For example, it is true in the "divergent limit" sense above that $\infty + 1 = \infty$, but not correct to subtract ∞, "deducing" that $1 = 0$.   ◇

Exercises for Section 6.3

Exercise 6.3.1. (★) Assume $(a_k)_{k=0}^\infty$ is a real sequence. Use mathematical induction to prove the following are equivalent: (i) $a_k \leq a_{k+1}$ for every natural number k. (ii) $a_k \leq a_{k+n}$ for all natural numbers k and n.

Exercise 6.3.2. (★) In each part, prove that the sequence with given kth term is eventually monotone.

(a) $a_k = 1/(2k - 13)$.   (b) $b_k = (8k - 3)/(2k - 13)$.

Exercise 6.3.3. Give a proof or counterexample: If $(a_k) \to a_\infty$ and $(b_k) \to b_\infty$ are convergent real sequences, then $\max(a_k, b_k) \to \max(a_\infty, b_\infty)$.

Exercise 6.3.4. (★) Construct a sequence **a** of positive real numbers such that $\mathbf{a} \to 0$ but **a** is not eventually monotone.

Exercise 6.3.5. True or false: If (a_k) is eventually positive and $\ell = \lim a_k$ exists, then $0 < \ell$.

Exercise 6.3.6. (★) (The squeeze theorem.) Suppose (a_k), (b_k), and (c_k) are real sequences, and assume there exists an N_0 such that if $k \geq N_0$, then $a_k \leq c_k \leq b_k$. Prove that if (a_k) and (b_k) converge to the same limit L, then $\lim_k c_k$ exists is equal to L. (Caution: Proposition 6.3.7 does not apply.)

Exercise 6.3.7. Assume (a_k) and (b_k) are real sequences. Prove that if (b_k) is bounded and $(a_k) \to 0$, then $(a_k b_k) \to 0$. (Caution: The hypotheses do not imply (b_k) converges.)

Exercise 6.3.8. Prove $(\sqrt{k})_{k=0}^\infty \to \infty$ as many ways as you can.

Exercise 6.3.9. (★) Assume (a_k) is a sequence of *positive* real numbers, and put $b_k = 1/a_k$. Prove that if $(a_k) \to 0$, then $(b_k) \to \infty$.

Exercise 6.3.10. Assume **a** is a non-increasing sequence of positive real numbers, and $\mathbf{a} \to 0$. Prove there exists a positive, non-increasing sequence **b** such that $\mathbf{b} \to 0$ but $(b_k/a_k) \to \infty$.

Exercise 6.3.11. (H) Assume $b > 0$, and define a sequence $(b_k)_{k=0}^\infty$ recursively by $b_0 = \sqrt{b}$, and $b_{k+1} = \sqrt{b \cdot b_k}$. Prove (b_k) converges, and find the limit.

Exercise 6.3.12. State and prove theorems asserting:

$$\infty + \infty = \infty, \qquad -\infty + (-\infty) = -\infty, \qquad \pm\infty \cdot \infty = \pm\infty.$$

6.4 Subsequences, Condensing Sequences

Suppose (a_k) is a real sequence. By selecting infinitely many terms in the same order, or "taking a subsequence," we can sometimes ensure "better" behavior, such as selecting a convergent subsequence from an arbitrary sequence. Separately, it is desirable to have a convergence criterion that refers only to the sequence itself, not to the limit. This section develops these tools.

Definition 6.4.1. An *index sequence* is a strictly increasing sequence ν (nu) of natural numbers; that is, $\nu(k) \in \mathbf{N}$ and $\nu(k) < \nu(k+1)$ for all k.

Lemma 6.4.2. *If ν is an index sequence, then $k \leq \nu(k)$ for all k. If the inequality is strict for some k_0, then the inequality is strict for all larger k.*

Proof. Immediate by mathematical induction.   □

Definition 6.4.3. Assume $\mathbf{a}$ is a real sequence and ν an index sequence. The sequence $\mathbf{b} = \mathbf{a} \circ \nu$ defined by $b_k = a_{\nu(k)}$ is called a *subsequence* of $\mathbf{a}$.

Remark 6.4.4. In words, a subsequence of (a_n) is a sequence obtained by selecting an infinite number of terms $a_{\nu(0)}, a_{\nu(1)}, \ldots, a_{\nu(k)}, \ldots,$ *in their original order*, namely, subject to $\nu(0) < \nu(1) < \nu(2) < \cdots < \nu(k) < \ldots.$ ◇

Example 6.4.5. The subsequence (a_{2k}), for which we take $\nu(k) = 2k$, consists of the *even terms* of (a_n). The subsequence (a_{2k+1}), taking $\nu(k) = 2k+1$, consists of the *odd terms* of (a_n).   ◇

Example 6.4.6. Assume $(a_n)_{n=0}^{\infty}$ is a real sequence. For each N in $\mathbf{N}$, the subsequence $(a_n)_{n=N}^{\infty} = (a_{N+k})_{k=0}^{\infty}$ is a *tail* of (a_n), obtained by discarding the head $(a_j)_{j=0}^{N-1}$. Here $\nu(k) = N + k$.   ◇

Remark 6.4.7. Convergence of a sequence is determined solely by convergence of an arbitrary tail; intuitively, prepending or omitting finitely many terms does not change the convergence or divergence of a sequence.   ◇

Remark 6.4.8. If some tail of a sequence has a property P, we say the original sequence is "eventually P." Terms such as "eventually non-decreasing" have already been introduced. Similarly, we might say a sequence is "eventually positive," "eventually less than 1 in absolute value," or "eventually constant."

Conditions such as "eventually bounded" or "eventually convergent" are syntactically well-formed, but convey no distinction: a tail is not bounded, unbounded, convergent, or divergent unless the original sequence already possesses the same property.   ◇

Lemma 6.4.9. *Assume (a_n) is a real sequence and $(a_n) \to a_\infty$. If $(b_k) = (a_{\nu(k)})$ is an arbitrary subsequence, then $(b_k) \to a_\infty$.*

Proof. Fix ε arbitrarily, and pick N such that if $n \geq N$, then $|a_n - a_\infty| < \varepsilon$. If $k \geq N$, then $\nu(k) \geq N$ by Lemma 6.4.2, so $|b_k - a_\infty| = |a_{\nu(k)} - a_\infty| < \varepsilon$.   □

Definition 6.4.10. Assume $\mathbf{a}$ is a real sequence. We say an index n is a *vista* of $\mathbf{a}$ if $m > n$ implies $a_m \leq a_n$.

Remark 6.4.11. Intuitively, a vista is a location n from which, standing at height a_n and looking to the right, we can see all the way to infinity.   ◇

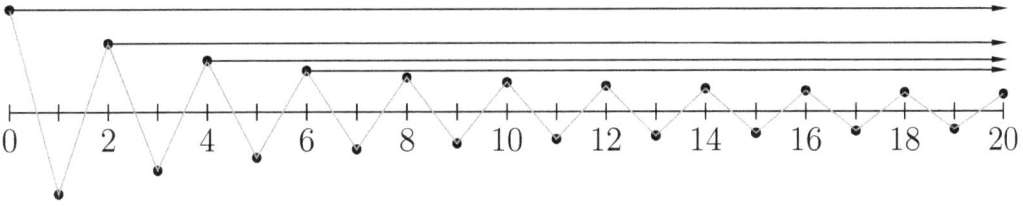

FIGURE 6.2
Vistas of the sequence in Example 6.4.12.

Example 6.4.12. If $a_n = (-1)^n 4/(4+n)$, then every even index is a vista, Figure 6.2. (Arrows indicate selected unobstructed lines of sight.) ◊

Example 6.4.13. A sequence $\mathbf{a}$ is non-increasing if and only if every natural number is a vista of $\mathbf{a}$.

If $\mathbf{a}$ is non-decreasing, then $\mathbf{a}$ has no vistas.

If $\mathbf{a}$ is unbounded above, then $\mathbf{a}$ has no vistas. ◊

Proposition 6.4.14. *Every real sequence has a monotone subsequence.*

Proof. We consider two cases: The sequence $\mathbf{a}$ has infinitely many vistas, or only finitely many. In each case, we construct a monotone subsequence recursively.

(Infinitely many vistas). Let $\nu(0)$ be a vista. Now assume inductively that we have constructed a finite increasing sequence $(\nu(k))_{k=0}^m$ of vistas. Since $\mathbf{a}$ has infinitely many vistas, there exists a vista $\nu(m+1)$ larger than $\nu(m)$. The subsequence $(a_{\nu(k)})$ is non-increasing: By definition of a vista, $a_{\nu(k+1)} \leq a_{\nu(k)}$ for all k.

(Finitely many vistas). There is an index $\nu(0)$ greater than every vista. Now assume inductively that we have constructed a finite increasing sequence $(\nu(k))_{k=0}^m$ such that if $0 \leq k < m$, then $a_{\nu(k)} < a_{\nu(k+1)}$. Since $\nu(m)$ is not a vista, there exists an index $\nu(m+1)$ greater than $\nu(m)$ such that $a_{\nu(m)} < a_{\nu(m+1)}$. The subsequence $(a_{\nu(k)})$ is strictly increasing by construction. □

Theorem 6.4.15 (Convergent subsequence theorem). *Every bounded real sequence has a convergent subsequence.*

Proof. If $\mathbf{a}$ is a bounded sequence, then by Proposition 6.4.14, there exists a monotone subsequence $(a_{\nu(k)})$. But a bounded, monotone sequence converges by Proposition 6.3.5. □

Condensing Sequences

Convergence of a sequence makes reference to a limit, which need not be a term of the sequence. The "condensing criterion" depends only on the sequence itself, and therefore provides a useful convergence-like condition when a limit is not known to exist.

Definition 6.4.16. A real sequence **a** is *condensing* if the following condition holds:

> For every ε,
> > there exists an index N such that
> > > if k and $k' \geq N$, then $|a_{k'} - a_k| < \varepsilon$.

Remark 6.4.17. In practice, we often write $k' = k + m$. The "condensing predicate" becomes "If $k \geq N$ and $m > 0$, then $|a_{k+m} - a_k| < \varepsilon$." ◇

Proposition 6.4.18. *If (a_k) is a real sequence, then (a_k) converges to some real number a_∞ if and only if (a_k) is condensing.*

Proof. (Convergent implies condensing). Assume $(a_k) \to a_\infty$, and fix ε arbitrarily. There exists an index N such that if $k \geq N$, then $|a_k - a_\infty| < \varepsilon/2$. If k and $k' \geq N$, the triangle inequality implies

$$|a_{k'} - a_k| \leq |a_{k'} - a_\infty| + |a_\infty - a_k| < \varepsilon/2 + \varepsilon/2 = \varepsilon.$$

Since ε was arbitrary, (a_k) is condensing.

(Condensing implies convergent). The proof proceeds as follows. We first prove that every condensing sequence is bounded. By Theorem 6.4.15, a condensing sequence has a convergent subsequence. Finally, we prove that a condensing sequence having a convergent subsequence is itself convergent to the same limit.

Taking $\varepsilon = 1$ in the condensing criterion, there is an index N such that if k and $k' \geq N$, then $|a_{k'} - a_k| < 1$. In particular, if $k \geq N$, then $|a_k - a_N| < 1$. Let $M = 1 + \max(\{|a_j|\}_{j=0}^{N})$. Just as in the proof of Proposition 6.1.19 (ii), it follows that $|a_k| \leq M$ for all k.

Since the condensing sequence (a_k) is bounded, Theorem 6.4.15 guarantees there is a subsequence $(a_{\nu(k)})$ converging to some real number a_∞. To complete the proof, it suffices to show $(a_k) \to a_\infty$.

Fix ε, and use the fact that $(a_{\nu(k)}) \to a_\infty$ to pick an index N_1 such that if $k \geq N_1$, then $|a_{\nu(k)} - a_\infty| < \varepsilon/2$. Now use the fact (a_k) is condensing to pick N greater than N_1 such that if k and $k' \geq N$, then $|a_k - a_{k'}| < \varepsilon/2$.

Since $\nu(N) \geq N$, we have $|a_{\nu(N)} - a_\infty| < \varepsilon/2$, and $|a_k - a_{\nu(N)}| < \varepsilon/2$ if $k \geq N$. By the triangle inequality, if $k \geq N$, then

$$|a_k - a_\infty| \leq |a_k - a_{\nu(N)}| + |a_{\nu(N)} - a_\infty| < \varepsilon/2 + \varepsilon/2 = \varepsilon.$$

Since ε was arbitrary, $(a_k) \to a_\infty$.   □

Remark 6.4.19. In Chapter 16, we use the term *complete* to refer to abstract settings in which every condensing sequence is convergent. Although this represents a formally distinct meaning of the term "complete" than in the axioms for the real number system, in the number line the general sense is logically equivalent to the completeness axiom. ◇

Limes Superior and Inferior

Assume (a_k) is an arbitrary real sequence. For each index n, consider the set $A_n = \{a_k : k \geq n\}$ of terms of the tail $(a_k)_{k=n}^{\infty}$, and define

$$\alpha_n = \inf A_n \leq \sup A_n = \beta_n.$$

The sets A_n are nested inward: $A_n \supseteq A_{n+1}$ for each n. By constriction, the sequence (α_n) is non-decreasing and the sequence (β_n) is non-increasing. Each sequence therefore has an extended real limit.

Definition 6.4.20. With the preceding notation,

$$\liminf a_k = \lim_{n} \inf_{k \geq n} a_k = \lim_{n} \alpha_n$$

is called the lim inf, or *limes inferior*, of (a_k), and

$$\limsup a_k = \lim_{n} \sup_{k \geq n} a_k = \lim_{n} \beta_n$$

is called the lim sup, or *limes superior*, of (a_k).

Example 6.4.21. If $a_k = (-1)^k$, then $\liminf a_k = -1$ and $\limsup a_k = 1$. If $b_k = k^{(-1)^k}$, then $\liminf a_k = 0$ and $\limsup a_k = \infty$. ◊

Proposition 6.4.22. *Assume (a_k) is a real sequence.*

(i) *There exists a subsequence of (a_k) converging to $\liminf a_k$, and there exists a subsequence of (a_k) converging to $\limsup a_k$.*

(ii) *If $(a_{\nu(k)})$ is a convergent subsequence of (a_k), then*

$$\liminf a_k \leq \lim a_{\nu(k)} \leq \limsup a_k.$$

(iii) *The sequence (a_k) converges if and only if $\liminf a_k = \limsup a_k \in \mathbf{R}$.*

Proof. See Exercise 6.4.14. □

Remark 6.4.23. In words, $\liminf a_k$ is the smallest subsequential limit of (a_k), and $\limsup a_k$ is the largest subsequential limit. ◇

Exercises for Section 6.4

Exercise 6.4.1.(★) Use Lemma 6.4.9 to prove the sequence $a_k = (-1)^k$ does not converge.

Exercise 6.4.2. Consider the real sequence defined by $a_k = k^{(-1)^k}$. Prove $\liminf a_k = 0$ and $\limsup a_k = \infty$. (This is claimed in Example 6.4.21.)

Exercise 6.4.3. Suppose (a_k) is unbounded above. Prove that (a_k) has no vistas.

Exercise 6.4.4. Assume **a** is a real sequence that is not bounded above. Without using Proposition 6.4.14, prove there exists a strictly increasing subsequence.

Exercise 6.4.5. (★) Assume **a** is a sequence of *positive* real numbers, and **a** → 0.

(a) Prove that for every natural number m, there exists a natural number $n > m$ such that $a_n < a_m$.

(b) Without Proposition 6.4.14, prove **a** has a strictly decreasing subsequence.

Exercise 6.4.6. (H) Assume **a** is a sequence that is not eventually constant.

(a) If **a** is non-increasing, prove **a** has a strictly decreasing subsequence.

(b) Prove **a** has a *strictly* monotone subsequence.

Exercise 6.4.7. (★) Assume α is an arbitrary real number.

(a) Prove there exists a rational sequence converging to α.

(b) Prove there exists a set of rational numbers whose supremum is α.

Exercise 6.4.8. Let $(a_k)_{k=0}^{\infty}$ be an enumeration of the rationals: For every rational number r, there exists a unique k such that $a_k = r$. Prove that if α is an arbitrary real number, there exists a subsequence $(a_{\nu(k)})$ converging to α. (This strengthens Exercise 6.4.7 (a) by obtaining the same conclusion after fixing an ordering of **Q**.)

Exercise 6.4.9. Assume $(a_k)_{k=0}^{\infty}$ is an integer sequence such that a_k positive if $k \geq 1$, and recursively define integer sequences $(p_n)_{n=-2}^{\infty}$ and $(q_n)_{n=-2}^{\infty}$ by

$$p_{-2} = 0, \qquad p_{-1} = 1, \qquad\qquad p_n = a_n p_{n-1} + p_{n-2},$$
$$q_{-2} = 1, \qquad q_{-1} = 0, \qquad\qquad q_n = a_n q_{n-1} + q_{n-2},$$

see Proposition 3.1.23.

(a) Prove that if $n \geq 0$, then

$$\frac{p_{n+1}}{q_{n+1}} - \frac{p_n}{q_n} = \frac{(-1)^n}{q_n q_{n+1}}.$$

Conclude that

$$\frac{p_n}{q_n} = [a_0, a_1, a_2, \ldots, a_n] = a_0 + \sum_{k=0}^{n-1} \frac{(-1)^k}{q_k q_{k+1}}.$$

(b) Prove the rational sequence (p_n/q_n) converges.

(c) If x is irrational, recursively define sequences $(a_k)_{k\in\mathbf{N}}$ and $(x_k)_{k\in\mathbf{N}}$ by
$x_0 = x$, $a_0 = \lfloor x_0 \rfloor$, $x_{k+1} = 1/(x_k - a_k)$, and $a_{k+1} = \lfloor x_{k+1} \rfloor$.

Prove a_k is a positive integer if $k \geq 1$, and the resulting sequence (p_n/q_n) converges to x.

(d) In the notation of (c), prove

$$\left| x - \frac{p_n}{q_n} \right| < \frac{1}{q_n q_{n+1}}, \qquad \text{for every natural number } n,$$

and if $n \geq 2$, then p_n/q_n is the point of $\mathbf{Q}_{q_n}$ closest to x.

Exercise 6.4.10. Express the following as continued fractions, see Exercise 6.4.9 (c), and calculate the first six rational approximations.

(a) $\sqrt{2}$.   (b) $\frac{1}{2}(1+\sqrt{5})$.   (c) $\sqrt{3}$.

Exercise 6.4.11. Write a general element $\mathbf{x}$ of the plane $\mathbf{R}^2$ in rectangular coordinates as (x_0, x_1). Define three functions on ordered pairs from $\mathbf{R}^2$:

$$d_1\left(\mathbf{x}, \mathbf{y}\right) = |y_0 - x_0| + |y_1 - x_1|;$$
$$d_2\left(\mathbf{x}, \mathbf{y}\right) = \sqrt{|y_0 - x_0|^2 + |y_1 - x_1|^2};$$
$$d_\infty\left(\mathbf{x}, \mathbf{y}\right) = \max\left(|y_0 - x_0|, |y_1 - x_1|\right).$$

Prove that $d_\infty\left(\mathbf{x}, \mathbf{y}\right) \leq d_2\left(\mathbf{x}, \mathbf{y}\right) \leq d_1\left(\mathbf{x}, \mathbf{y}\right) \leq 2d_\infty\left(\mathbf{x}, \mathbf{y}\right)$ for all $\mathbf{x}$ and $\mathbf{y}$.

Exercise 6.4.12. (H) If (x_k) and (y_k) are real sequences, we say (x_k, y_k) is the *plane sequence* with coordinates (x_k) and (y_k).

(a) Give formal ε-N definitions of *convergence* and *condensing* for a plane sequence.

(b) Your conditions in part (a) depend on a choice of distance function, or *metric*. To be precise, we should speak *a priori* of "convergence with respect to d_2" rather than simply "convergence." Happily, this is not necessary if we define "convergence" using the distance functions of Exercise 6.4.11: Prove that a plane sequence converges with respect to d_2 if and only if it converges with respect to d_∞, if and only if it converges with respect to d_1, if and only if both (x_k) and (y_k) converge as real sequences. Do analogous conclusions hold for condensing plane sequences?

(c) Prove that a plane sequence converges if and only if it is condensing.

Exercise 6.4.13. In this question, assume S and C are functions such that $C^2 + S^2 = 1$. Consider the plane sequence $\mathbf{x}_k = \big(C(k), S(k)\big)$, and assume (in the notation of Exercise 6.4.11) that $d_2(\mathbf{x}_k, \mathbf{x}_{k+1})$, the squared distance between $\mathbf{x}_k$ and $\mathbf{x}_{k+1}$, takes the same positive value for every integer k.

(a) Does $(\mathbf{x}_k)$ converge?

(b) Does each component of $(\mathbf{x}_k)$ have a convergent subsequence?

(c) Does $(\mathbf{x}_k)$ have a convergent subsequence?

Exercise 6.4.14. Prove Proposition 6.4.22.

7

Infinite Series

Some of the most useful sequences in mathematics arise by summing the terms of another sequence of numbers. On one hand, summing merely reformulates abstract definitions. On the other, summing provides a distinct and useful technical and psychological viewpoint.

7.1   Summability

Definition 7.1.1. Assume $(a_k)_{k=0}^{\infty}$ is a real sequence, and $(s_n)_{n=0}^{\infty}$ is the sequence of partial sums. We say (a_k) is *summable* if (s_n) converges to a real number s, and we write

$$\sum_{k=0}^{\infty} a_k = \lim_{n \to \infty} \sum_{k=0}^{n-1} a_k = \lim_{n \to \infty} s_n = s.$$

The expression on the left is called the *infinite series* with *terms a_k*, and is said to *converge* to s.

If (s_n) does not converge, we say the infinite series $\sum_k a_k$ *diverges*.

Remark 7.1.2. Think of a real sequence as an infinite list of credits (non-negative terms) and debits (negative terms). The partial sums are the "running totals" of the terms *taken in a specified order*. The sum of a series, if it exists, is the net value in the limit, when "all the terms have been added" in order.   ◇

Proposition 7.1.3. *Assume (a_k) and (b_k) are summable real sequences, and $c \in \mathbf{R}$. The sequences $(a_k + b_k)$ and (ca_k) are summable, and*

$$\sum_{k=0}^{\infty}(a_k + b_k) = \sum_{k=0}^{\infty} a_k + \sum_{k=0}^{\infty} b_k, \qquad \sum_{k=0}^{\infty}(ca_k) = c\sum_{k=0}^{\infty} a_k.$$

Proof. If (s_n) and (t_n) denote the respective sequences of partial sums of (a_k) and (b_k), then $(s_n + t_n)$ and (cs_n) are the respective partial sums of $(a_k + b_k)$ and (ca_k). The proposition follows immediately from Proposition 6.2.1.   □

DOI: 10.1201/9781003601357-7

Proposition 7.1.4. *If (a_k) is a real sequence, then for all natural numbers N and m, we have:*

$$s_{N+m} - s_N = \sum_{k=N}^{N+m-1} a_k, \quad \text{and therefore} \quad |s_{N+m} - s_N| \leq \sum_{k=N}^{N+m-1} |a_k|.$$

Proposition 7.1.5. *A real sequence (a_k) is summable if and only if the sequence of partial sums is condensing. In particular, if (a_k) is summable, then $(a_k) \to 0$.*

Proof. The first assertion is immediate from Proposition 6.4.18. For the second, fix ε arbitrarily and choose N such that if k and $k' \geq N$, then $|s_{k'} - s_k| < \varepsilon$. In particular, if $k \geq N$, then $|a_k| = |s_{k+1} - s_k| < \varepsilon$. This means $(a_k) \to 0$. $\square$

Definition 7.1.6. Assume a and r are real numbers. The infinite series

$$\sum_{k=0}^{\infty} ar^k = a + ar + ar^2 + ar^3 + \cdots$$

is called the *geometric series* with first term a and ratio r.

Theorem 7.1.7 (The geometric series formula). *If $a \neq 0$, the geometric series with first term a and ratio r converges if and only if $|r| < 1$, and the sum is*

$$\sum_{k=0}^{\infty} ar^k = \frac{a}{1 - r}.$$

Proof. Assume $a \neq 0$. If $|r| \geq 1$, then $|ar^k| = |a|\,|r|^k$ does not converge to 0, see Example 6.3.11, so the geometric series diverges.

Inversely, suppose $-1 < r < 1$. By Proposition 3.3.8, the partial sums are

$$s_n = \sum_{k=0}^{n-1} ar^k = a\frac{1 - r^n}{1 - r}.$$

By Example 6.1.17, $(r^n) \to 0$ since $-1 < r < 1$. Proposition 6.2.1 implies

$$\sum_{k=0}^{\infty} ar^k = \lim_{n \to \infty} \frac{a(1 - r^n)}{1 - r} = \frac{a(1 - \lim_n r^n)}{1 - r} = \frac{a}{1 - r}. \qquad \square$$

Example 7.1.8. The terms of the *harmonic series* decrease to 0,

$$\sum_{k=1}^{\infty} \frac{1}{k} = 1 + \frac{1}{2} + \frac{1}{3} + \cdots + \frac{1}{n} + \cdots.$$

Nonetheless, the harmonic series diverges, see also Exercise 4.3.10. To prove this, it suffices to show the partial sums, depicted in Figure 7.1, are unbounded.

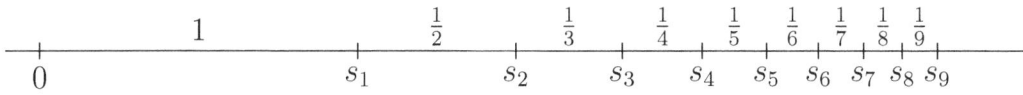

FIGURE 7.1
Partial sums of the harmonic series.

The first two terms add to $1 + (1/2)$. The next two terms, $1/3$ and $1/4$, are each no smaller than $1/4$, so their sum is greater than $2 \cdot 1/4 = 1/2$.

The next four terms, $1/5$, $1/6$, $1/7$, and $1/8$, are each at least $1/8$, so their sum is greater than $4 \cdot 1/8 = 1/2$.

Similarly, the next eight terms, $1/9$, $1/10$, ..., $1/16$, sum to at least $8 \cdot 1/16 = 1/2$, the sixteen terms after that sum to at least $1/2$, and so on *ad infinitum*. More formally, if m is a positive integer, then

$$\sum_{k=2^{m-1}+1}^{2^m} \frac{1}{k} \geq \sum_{k=2^{m-1}+1}^{2^m} \frac{1}{2^m} = \frac{2^{m-1}}{2^m} = \frac{1}{2}$$

since there are 2^{m-1} terms, each at least 2^{-m}. Consequently,

$$\sum_{k=1}^{2^n} \frac{1}{k} = 1 + \sum_{m=1}^{n} \left[\sum_{k=2^{m-1}+1}^{2^m} \frac{1}{k} \right] \geq 1 + \sum_{m=1}^{n} \frac{1}{2} = 1 + \frac{n}{2}.$$

The sequence of partial sums has an unbounded subsequence, hence is itself unbounded. ◇

Lemma 7.1.9. *Assume (a_k) and (b_k) are real sequences, with respective partial sums (s_n) and (t_n). If (b_k) is summable, and if there exists a natural number N_0 such that*

$$|s_{m'} - s_m| \leq |t_{m'} - t_m| \quad \text{whenever } m' > m \geq N_0,$$

then (a_k) is summable.

Proof. If (b_k) is summable, namely, if the sequence (t_n) of partial sums converges, then (t_n) is condensing: For every ε, there exists an N greater than N_0 such that if $m' > m \geq N$, then $|t_{m'} - t_m| < \varepsilon$. By the hypotheses of the lemma, if $m' > m \geq N$, then $|s_{m'} - s_m| < \varepsilon$ as well. This implies the sequence (s_n) is condensing, hence convergent by Proposition 6.4.18. □

Theorem 7.1.10 (Comparison). *Assume (a_k) and (b_k) are real sequences, and $|a_k| \leq b_k$ eventually. If (b_k) is summable, then (a_k) is summable. Contrapositively, if (a_k) is not summable, then (b_k) is not summable.*

Proof. Let (s_n) and (t_n) denote the sequences of partial sums of (a_k) and (b_k) respectively. By hypothesis, there exists an index N_0 such that $|a_k| \leq b_k$ for all k such that $k \geq N_0$. If $n \geq N_0$ and $m > 0$, then

$$|s_{n+m} - s_n| = \left| \sum_{k=n}^{n+m-1} a_k \right| \leq \sum_{k=n}^{n+m-1} |a_k| \leq \sum_{k=n}^{n+m-1} b_k = |t_{n+m} - t_n|.$$

The theorem follows from Lemma 7.1.9.   □

Exercises for Section 7.1

Exercise 7.1.1. (★) Determine whether the following converge. If so, find the sum.

(a) $\displaystyle\sum_{k=0}^{\infty} 4\left[-\frac{3}{5}\right]^k$.

(b) $\displaystyle\sum_{k=0}^{\infty} 0.01\left[\frac{7}{5}\right]^k$.

(c) $\displaystyle\sum_{k=0}^{\infty} 0 \cdot 1000^k$.

Exercise 7.1.2. (★) Write the repeating decimal $0.12345\overline{345}$ as a fraction. Suggestion: Multiply by 1000 and subtract.

Exercise 7.1.3. Prove that every repeating decimal represents a rational number.

Exercise 7.1.4. (H) Assume $(b_k)_{k \in \mathbb{N}}$ is a convergent sequence. Define the sequence (a_k) by $a_k = b_{k+1} - b_k$. Prove $(a_k)_{k=0}^{\infty}$ is summable, and find the sum. (The series $\sum_k a_k$ is said to be *telescoping*.)

Exercise 7.1.5. Each part refers to the series $\displaystyle\sum_{k=1}^{\infty} \frac{1}{k(k+1)}$.
    Calculate the first four partial sums, and evaluate the series with proof.

Exercise 7.1.6. (H) If a is a positive integer, evaluate the series $\displaystyle\sum_{k=1}^{\infty} \frac{a}{k(k+a)}$.

Exercise 7.1.7. Prove that if $1 < |x|$, then $-\displaystyle\sum_{k=1}^{\infty} \frac{1}{x^{k+1}} = \frac{1}{1-x}$.

Exercise 7.1.8. (★) Determine where each series converges, and find the sums as functions of x.

(a) $\displaystyle\sum_{k=0}^{\infty} x^{2k}$;

(b) $\displaystyle\sum_{k=0}^{\infty} (\tfrac{1}{2} + x^2)^k$;

(c) $\displaystyle\sum_{k=0}^{\infty} (1 - x^2)^k$.

Exercise 7.1.9. Assume (a_k) is a non-increasing positive sequence. Prove the *exponential sampling test*: $\sum_k a_k$ converges if and only if $\sum_k 2^k a_{2^k}$ converges. Suggestion: Modify the idea in Example 7.1.8.

Exercise 7.1.10. Assume p is a natural number. Use the exponential sampling test, Exercise 7.1.9, to prove $\sum_k k^{-p}$ converges if and only if $p > 1$.

Exercise 7.1.11. Assume $A = \{a_k\}_{k=1}^{\infty}$ is a countable set of real numbers. Prove that for every ε, there exists a countable collection of open intervals $\{O_k\}_{k=1}^{\infty}$ such that $A \subseteq \bigcup_k O_k$ and the sum of the lengths of the O_k is less than ε.

Exercise 7.1.12. Assume $(a_k)_{k=0}^{\infty}$ is a real sequence. For natural numbers m and n, define

$$t_m = \sum_{k=0}^{m-1} a_k, \qquad s_0 = 0, \quad s_n = \frac{1}{n} \sum_{m=0}^{n-1} t_m, \ n \geq 1.$$

If $(s_n) \to s$ for some real s, we say (a_k) is *mean summable* ("mean" in the sense of "average"), and we call s the *mean sum*.

(a) Prove that if (a_k) is summable, then (a_k) is mean summable, and the mean sum is the ordinary sum.

(b) Prove that if (a_k) is eventually non-negative, then (a_k) is mean summable if and only if (a_k) is summable.

(c) Show the alternating sequence $a_k = (-1)^k$ is mean summable, and find the mean sum.

Exercise 7.1.13. (H) Assume $(a_k)_{k=1}^{\infty}$ is a "digit sequence," namely, a sequence all of whose terms are integers between 0 and 9 inclusive.

(a) Prove that the series

$$\sum_{k=1}^{\infty} \frac{a_k}{10^k} = \frac{a_1}{10} + \frac{a_2}{100} + \frac{a_3}{1000} + \cdots = 0.a_1 a_2 a_3 \ldots$$

converges to a real limit between 0 and 1.

(b) Prove the digit sequences $(4, 9, 9, 9, \ldots)$ and $(5, 0, 0, \ldots)$ give the same sum in part (a).

(c) (H) Under what conditions do two distinct digit sequences define the same sum in part (a)?

Exercise 7.1.14. (A) Define $f : \mathscr{P}(\mathbf{Z}^+) \to \mathbf{R}$ by $f(A) = \sum_{k \in A} 2^{-k}$.

Find the image of f. Either prove f is injective or characterize distinct subsets on which f takes the same value.

Exercise 7.1.15. Assume $(a_k)_{k=1}^\infty$ is a binary sequence, all of whose terms are either 0 or 1, and $(b_k)_{k=1}^\infty$ a sequence all of whose terms are either 0 or 2.

(a) Prove that the series

$$\sum_{k=1}^\infty \frac{a_k}{2^k} = \frac{a_1}{2} + \frac{a_2}{2^2} + \frac{a_3}{2^3} + \cdots$$

converges to an element of the unit interval $[0,1]$. Conversely, show that every element of $[0,1]$ has a *binary representation* of this form.

(b) Prove that the series

$$x = \sum_{k=1}^\infty \frac{b_k}{3^k} = \frac{b_1}{3} + \frac{b_2}{3^2} + \frac{b_3}{3^3} + \cdots$$

converges to an element of the ternary set K. Conversely, show that every element of K can be expressed *uniquely* in this form.

Prove x is an endpoint of the set K_n if and only if (b_k) is eventually constant.

(c) Construct a surjection from K to $[0,1]$.

7.2 Absolute Summability

Definition 7.2.1. A real sequence (a_k) is *absolutely summable* if the sequence $(|a_k|)$ is summable.

Proposition 7.2.2. *If a real sequence (a_k) is absolutely summable, then it is summable, and*

$$\left| \sum_{k=0}^\infty a_k \right| \leq \sum_{k=0}^\infty |a_k|.$$

Proof. Let (s_n) and (t_n) denote the sequences of partial sums of (a_k) and $(|a_k|)$ respectively. By the triangle inequality,

$$|s_{n+m} - s_n| = \left| \sum_{k=n}^{n+m-1} a_k \right| \leq \sum_{k=n}^{n+m-1} |a_k| = |t_{n+m} - t_n|$$

for all natural numbers n and m. If (t_n) converges, then (s_n) converges by Lemma 7.1.9. Since

$$\left| \sum_{k=0}^{n-1} a_k \right| \leq \sum_{k=0}^{n-1} |a_k| \quad \text{if } n \geq 0$$

the inequality in the proposition holds by Proposition 6.3.7. $\qquad \square$

Remark 7.2.3. Alternatively, an infinite series $\sum_k a_k$ is *absolutely convergent* if $\sum_k |a_k|$ is convergent. ◇

Definition 7.2.4. If (a_k) is a real sequence, the real sequences (a_k^+) and (a_k^-) defined by

$$a_k^+ = \max(a_k, 0) = \frac{|a_k| + a_k}{2}, \qquad a_k^- = -\min(a_k, 0) = \frac{|a_k| - a_k}{2},$$

are called the sequence of *positive terms* of (a_k) and the sequence of *negative terms* of (a_k), respectively.

Example 7.2.5. If $a_k = (-1/2)^k$, then

$k =$	0	1	2	3	4	5 ...
$a_k =$	1	$-1/2$	$1/4$	$-1/8$	$1/16$	$-1/32$ ...
$|a_k| =$	1	$1/2$	$1/4$	$1/8$	$1/16$	$1/32$ ...
$a_k^+ =$	1	0	$1/4$	0	$1/16$	0 ...
$a_k^- =$	0	$1/2$	0	$1/8$	0	$1/32$ ...

◇

Remark 7.2.6. Each sequence $(a_k^\pm)$ is non-negative, hence summable if and only if its sequence of partial sums is bounded, Proposition 6.3.5 (ii). Further,

$$|a_k| = a_k^+ + a_k^-,$$
$$a_k = a_k^+ - a_k^-,$$

and $0 \leq a_k^\pm \leq |a_k|$ for all k. ◇

Proposition 7.2.7. *Assume (a_k) is a real sequence.*

(i) *(a_k) is absolutely summable if and only if (a_k^+) and (a_k^-) are summable.*

(ii) *If (a_k) is summable but not absolutely, (a_k^+) and (a_k^-) are non-summable.*

Proof. (i). If (a_k) is absolutely summable, namely, if $(|a_k|)$ is summable, then each sequence $(a_k^\pm)$ is summable by comparison with $(|a_k|)$.

Conversely, if (a_k^+) and (a_k^-) are both summable, then $(|a_k|) = (a_k^+ + a_k^-)$ is summable by Proposition 7.1.3.

(ii). By hypothesis, (a_k) is summable. Contrapositively, if *either* of $(a_k^\pm)$ is summable, then $(|a_k|)$ is as well by Proposition 7.1.3, since $|a_k| = 2a_k^\pm \mp a_k$.

Since $(|a_k|)$ is not summable, *both* of $(a_k^\pm)$ are non-summable. □

The Ratio Test

Theorem 7.2.8 (The ratio test). *Assume (a_k) is a real sequence. If the limiting ratio*

$$\rho = \lim_{k \to \infty} \left| \frac{a_{k+1}}{a_k} \right|$$

exists, and if $\rho < 1$, then (a_k) is absolutely summable.

Proof. Put $r = (1 + \rho)/2$, so that $\rho < r < 1$, and put $\varepsilon = r - \rho$, Figure 7.2. Since $|a_{k+1}/a_k| \to \rho$, there exists an index N such that if $k \geq N$, then

$$\left| \frac{a_{k+1}}{a_k} \right| < \rho + \varepsilon = r < 1.$$

Rearranging, $|a_{k+1}| < |a_k| r$ if $k \geq N$. By induction on m,

$$|a_{N+m}| < |a_N| r^m \quad \text{for all positive } m.$$

Consequently,

$$\sum_{k=N}^{\infty} |a_k| = \sum_{m=0}^{\infty} |a_{N+m}| \leq |a_N| \sum_{m=0}^{\infty} r^m.$$

This upper bound is a convergent geometric series, so $(|a_k|)$ is summable by comparison. $\square$

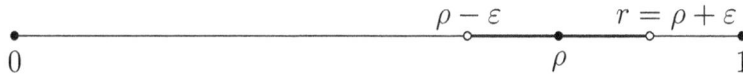

FIGURE 7.2
Bounding ratios in the ratio test.

Corollary 7.2.9. *Let r be a real number such that $|r| < 1$. For each positive integer m, the sequence $(k^m r^k)_{k \in \mathbf{N}}$ is absolutely summable. Consequently, $(k^m r^k)$ converges to 0, and in particular is bounded.*

Proof. Setting $a_k = k^m r^k$ and using Example 6.2.4, we have

$$\lim_{k \to \infty} \left| \frac{(k+1)^m r^{k+1}}{k^m r^k} \right| = \lim_{k \to \infty} \left(1 + (1/k) \right)^m |r| = |r| < 1,$$

so $(k^m r^k)$ is absolutely summable by the ratio test. The remaining assertions are immediate. $\square$

Products of Series

Definition 7.2.10. If $(a_j)_{j=0}^\infty$ and $(b_k)_{k=0}^\infty$ are summable sequences, their *product* is the infinite series

$$\sum_{k=0}^\infty \sum_{j=0}^k a_j b_{k-j} = a_0 b_0 + (a_0 b_1 + a_1 b_0) + (a_0 b^2 + a_1 b_1 + a_2 b_0) + \cdots.$$

Proposition 7.2.11. *If* $(a_j)_{j=0}^\infty$ *is absolutely summable with sum* A *and* $(b_k)_{k=0}^\infty$ *is absolutely summable with sum* B, *and if* (c_m) *is an arbitrary enumeration of the set* $\{a_j b_k : j \in \mathbf{N}, \ k \in \mathbf{N}\}$, *then* (c_m) *is absolutely summable, and*

$$\sum_{m=0}^\infty c_m = \left[\sum_{j=0}^\infty a_j\right]\left[\sum_{k=0}^\infty b_k\right].$$

Proof. For convenience, introduce

$$A_n = \sum_{j=0}^{n-1} a_j, \qquad B_n = \sum_{k=0}^{n-1} b_k, \qquad C_n = \sum_{m=0}^{n-1} c_m,$$

$$\mathbf{A}_n = \sum_{j=0}^{n-1} |a_j|, \qquad \mathbf{B}_n = \sum_{k=0}^{n-1} |b_k|.$$

Finally, put $A = \lim_n A_n$, $B = \lim_n B_n$, $\mathbf{A} = \lim_n \mathbf{A}_n$, and $\mathbf{B} = \lim_n \mathbf{B}_n$.

The sequence $A_n B_n$ converges to AB by Proposition 6.2.1 (ii). Similarly, $\mathbf{A}_n \mathbf{B}_n$ converges to $\mathbf{AB}$.

Fix ε arbitrarily, and choose N_1 such that if $n \geq N_1$, then

$$|AB - A_n B_n| < \varepsilon/2 \quad \text{and} \quad |\mathbf{AB} - \mathbf{A}_n \mathbf{B}_n| < \varepsilon/2.$$

Now choose N greater than N_1, so every product $a_j b_k$ such that $j, k < N_1$ is among the terms c_m such that $0 \leq m \leq N$. If $n \geq N$, then

$$|C_n - A_n B_n| \leq \sum_{j \text{ or } k > L} |a_j||b_k| = |\mathbf{AB} - \mathbf{A}_n \mathbf{B}_n| < \varepsilon/2.$$

Consequently, if $n \geq N$, then

$$|AB - C_n| \leq |AB - A_n B_n| + |A_n B_n - C_n| < \varepsilon. \qquad \square$$

Remark 7.2.12. The same conclusion holds if only one sequence is absolutely summable, though the estimates require more care. $\diamond$

Exercises for Section 7.2

Exercise 7.2.1. If (a_k) is absolutely summable, prove the sequence (a_k^2) is absolutely summable.

Exercise 7.2.2. (★) For which natural numbers p does

$$\sum_{k=0}^{\infty} \frac{(-1)^k}{(k+1)^p}$$

converge? Converge absolutely?

Exercise 7.2.3. For which real x does $\displaystyle\sum_{k=0}^{\infty} \frac{(x-5)^k}{(k+1)\cdot 3^k}$ converge absolutely?

Exercise 7.2.4. Recall that if $|x| < 1$, then

$$\frac{1}{1-x} = \sum_{k=0}^{\infty} x^k = 1 + x + x^2 + x^3 + x^4 + \cdots .$$

(a) Prove that $\displaystyle\frac{1}{(1-x)^2} = \sum_{k=0}^{\infty}(k+1)x^k$ if $|x| < 1$.

(b) Find a series representation of $\displaystyle\frac{1}{(1-x)^3}$ if $|x| < 1$.

Exercise 7.2.5. Suppose r and s are positive real numbers less than 1, and

$$\sum_{k=0}^{\infty} r^k = \sum_{k=0}^{\infty}(k+1)s^k.$$

Find r in terms of s. Hint: Use Exercise 7.2.4.

Exercise 7.2.6. (A) If $(a_i)_{i=0}^{\infty}$, $(b_j)_{j=0}^{\infty}$, and $(c_k)_{k=0}^{\infty}$ are real sequences such that

$$\left[\sum_{i=0}^{\infty} \frac{a_i x^i}{i!}\right]\left[\sum_{j=0}^{\infty} \frac{b_j x^j}{j!}\right] = \left[\sum_{k=0}^{\infty} \frac{c_k x^k}{k!}\right],$$

find a formula for c_k in terms of (a_i) and (b_j).

7.3   Alternating Series and Reordering

Alternating Series

Definition 7.3.1. If (a_k) is a sequence of positive terms, an infinite series $\pm\sum_k(-1)^k a_k$ is said to be *alternating*.

Theorem 7.3.2 (The alternating series test). *If (a_k) is a non-increasing positive real sequence, and if $(a_k) \to 0$, then the alternating series $\sum_k(-1)^k a_k$ converges, and*

$$|a_n - a_{n+1}| \leq \left|\sum_{k=n}^{\infty}(-1)^k a_k\right| \leq |a_n|.$$

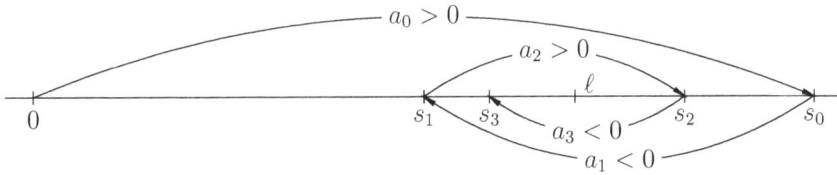

FIGURE 7.3
Partial sums of an alternating series.

Proof. Multiplying by -1 and re-indexing if necessary, we may assume without loss of generality that the series starts with a_0. The partial sums

$$s_n = \sum_{k=0}^{n-1}(-1)^k a_k$$

are shown qualitatively in Figure 7.3.

If n is an arbitrary odd index, then we have $s_{n+2} = s_n - a_n + a_{n+1} \leq s_n$ since $a_{n+1} \leq a_n$. With the same n, $s_{n+1} = s_{n-1} + a_{n-1} - a_n \geq s_{n-1}$. Writing $n = 2m + 1$ and combining these inequalities,

$$s_{2m} \leq s_{2m+2} \leq s_{2m+3} \leq s_{2m+1}$$

for every m. Consequently, the even partial sums form a non-decreasing sequence that is bounded above by every odd partial sum, and is therefore convergent to a real number $\ell^- = \sup_m s_{2m}$. Similarly, the odd partial sums form a non-increasing sequence that is bounded below by every even partial sum, and is therefore convergent to a real number $\ell^+ = \inf_m s_{2m+1}$. Since $(a_k) \to 0$ and

$$\ell^+ - \ell^- \leq s_{2m-1} - s_{2m} = a_{2m}$$

for all m, we have $\ell^+ = \ell^-$; that is, the sequence of partial sums converges to some real number ℓ. The bounds are Exercise 7.3.3. $\square$

Example 7.3.3. The alternating harmonic series

$$\sum_{k=0}^{\infty} \frac{(-1)^k}{k+1} = 1 - \tfrac{1}{2} + \tfrac{1}{3} - \tfrac{1}{4} + \tfrac{1}{5} - \tfrac{1}{6} + \tfrac{1}{7} - \tfrac{1}{8} + \tfrac{1}{9} + \cdots$$

converges to a real number between $s_1 = 1/2$ and $s_2 = 5/6$.   ◇

Reordering

When we add a finite list of real numbers, we may order the summands any way we like without changing the sum. The same turns out to be true for absolutely summable sequences, Theorem 7.3.8 (i). For summable sequences that are not absolutely summable, however, reordering the terms can change the sum to an arbitrary real number, or can result in a divergent series.

Definition 7.3.4. A *reordering* of $\mathbf{N}$ is a bijection $\nu : \mathbf{N} \to \mathbf{N}$.
    Assume $(a_k)_{k=0}^{\infty}$ is a sequence. A sequence $(b_k)_{k=0}^{\infty}$ is a *reordering* if there exists a reordering ν of $\mathbf{N}$ such that $b_k = a_{\nu(k)}$. In this situation, we say the series $\sum_k b_k$ is a *reordering* of $\sum_k a_k$.

Remark 7.3.5. If $m = \nu(k)$, it will be convenient to write $k = \nu^{-1}(m)$.   ◇

Remark 7.3.6. Intuitively, a reordering of a series is "the running total of the same summands, but in a different order."   ◇

Example 7.3.7. Consider the reordering of the alternating harmonic series in which each positive term is followed by two negative terms:

$$\left(1 - \tfrac{1}{2}\right) - \tfrac{1}{4} + \left(\tfrac{1}{3} - \tfrac{1}{6}\right) - \tfrac{1}{8} + \left(\tfrac{1}{5} - \tfrac{1}{10}\right) - \tfrac{1}{12} + \cdots .$$

The parentheses emphasize a subsequence of the partial sums:

$$\tfrac{1}{2} - \tfrac{1}{4} + \tfrac{1}{6} - \tfrac{1}{8} + \tfrac{1}{10} - \tfrac{1}{12} + \cdots ,$$

formally one-half the alternating harmonic series. *Reordering a convergent series can change the sum.*   ◇

Theorem 7.3.8 (Reordering). *Assume $(a_k)_{k=0}^{\infty}$ is a summable sequence.*

(i) *If (a_k) is absolutely summable, then every reordering is summable and has the same sum.*

(ii) *If (a_k) is summable but not absolutely, then for every real number S, there exists a reordering of (a_k) that sums to S.*

Proof. (i). Assume (a_j) is absolutely summable, and $(b_k) = (a_{\nu(k)})$ is a reordering. Fix ε arbitrarily, and use absolute summability to pick an index n such that

$$\sum_{j=n}^{\infty} |a_j| < \varepsilon.$$

Now pick $N > \max\{\nu^{-1}(j)\}_{j=0}^{n-1}$, so that if $k \geq N$, then $j = \nu(k) \geq n$. Every term of $(b_k)_{k=N}^{\infty}$ is a term of $(a_j)_{j=n}^{\infty}$, so

$$\sum_{k=N}^{\infty} |b_k| \leq \sum_{j=n}^{\infty} |a_j| < \varepsilon.$$

This proves (b_k) is absolutely summable. To evaluate the sum, note that

$$\left| \sum_{j=0}^{\infty} a_j - \sum_{k=0}^{N-1} b_k \right| \leq \left| \sum_{j=n}^{\infty} a_j \right| \leq \sum_{j=n}^{\infty} |a_j| < \varepsilon.$$

   (ii). Assume without loss of generality that $a_k \neq 0$ for all k. Since (a_k) is summable but not absolutely, the sequences $(a_k^{\pm})$ are separately non-summable: Each sums to ∞. Since (a_k) itself is summable, however, $(a_k^{\pm}) \to 0$.

   Suppose $S \geq 0$. Re-index each of $(a_k^{\pm})$ to remove all 0s from each sequence. Add terms of (a_k^+) sequentially until the partial sum exceeds S, then subtract terms of (a_k^-) sequentially until the partial sum is smaller than S. Perform this algorithm iteratively.

   If a_k^+ is a term that causes the partial sum s_n to exceed s, then $|s - s_n| \leq a_k^+$, and similarly if a_k^- is a term that causes the partial sum to be strictly smaller than s. Since $\lim_n \sup_{k \geq n} |a_k| = 0$, the partial sums are a condensing sequence converging to s.

   If $S < 0$, successively subtract positive terms of (a_k^-) until the partial sum is smaller than S, the proceed as above. $\square$

Exercises for Section 7.3

Exercise 7.3.1. ($\bigstar$) Give bounds on $\left| \sum_{k=n}^{\infty} \dfrac{(-1)^k}{k^2} \right|$ that decrease to 0.

Exercise 7.3.2. Each part refers to the series

$$\sum_{k=1}^{\infty} a_k = \sum_{k=1}^{\infty} \frac{(-1)^{k-1}}{\sqrt{k}}.$$

(a) Write out the sixth partial sum.

(b) Prove this series converges. Is the convergence absolute?

(c) Using the error bound for alternating series, how many terms suffice so that the partial sum approximates the sum to four decimal places, namely, with an error smaller than 0.5×10^{-4}?

(d) Does $\sum_k a_k^2$ converge?

Exercise 7.3.3. (★) In Theorem 7.3.2, establish the tail estimates

$$|a_n - a_{n+1}| \leq \left| \sum_{k=n}^{\infty} (-1)^k a_k \right| \leq |a_n|.$$

Exercise 7.3.4. (★) For each real x, define

$$C(x) = \sum_{k=0}^{\infty} \frac{(-1)^k x^{2k}}{(2k)!} = 1 - \frac{x^2}{2!} + \frac{x^4}{4!} - \frac{x^6}{6!} + \cdots.$$

(a) Prove the series converges absolutely for all real x.

(b) If C_n denotes the partial sum over $0 \leq k < n$, what is the smallest n that suffices to guarantee $|C(x) - C_n(x)| < 0.5 \times 10^{-6}$ for all x in $[-2, 2]$?

Exercise 7.3.5. For each real x, define

$$S(x) = \sum_{k=0}^{\infty} \frac{(-1)^k x^{2k+1}}{(2k+1)!} = x - \frac{x^3}{3!} + \frac{x^5}{5!} - \frac{x^7}{7!} + \cdots.$$

(a) Prove the series converges absolutely for all real x.

(b) If S_n denotes the partial sum over $0 \leq k < n$, what is the smallest n that suffices to guarantee $|S(x) - S_n(x)| < 0.5 \times 10^{-6}$ for all x in $[-1, 1]$?

(c) What is the smallest n that suffices to guarantee $|S(x) - S_n(x)| < 0.5 \times 10^{-6}$ for all x in $[-1/2, 1/2]$?

Exercise 7.3.6. Assume (a_k) is summable but not absolutely. Prove that some reordering of $\sum_k a_k$ diverges to ∞.

7.4   Power Series

A polynomial function in one variable depends on a finite sequence of coefficients, and is evaluated using only addition and multiplication. Power series are a generalization where we have a sequence of coefficients, and view evaluation as summing terms of successively higher degree.

Definition 7.4.1. Assume $(a_k)_{k=0}^{\infty}$ is a real sequence and x_0 is real. The infinite series

$$\sum_{k=0}^{\infty} a_k(x - x_0)^k = a_0 + a_1(x - x_0) + a_2(x - x_0)^2 + a_3(x - x_0)^3 + \cdots$$

is called the *power series* with *center* x_0 and *coefficients* (a_k). A power series is called a *germ* if it converges for some $x \neq x_0$.

Remark 7.4.2. The name *germ* is short for *germ of a real-analytic function*, and refers to "seed," not "pathogen." No other types of germ appear in this book, so the nickname suffices.

Let I denote the set of real x at which a power series converges. When $x = x_0$, every term (except possibly the 0th) vanishes, and the series converges; that is, $x_0 \in I$. As we will show in Proposition 7.4.9, if the power series converges at some real number x not equal to x_0, then the series converges *absolutely* at each point of the open ball with center x_0 and radius $|x - x_0|$; that is, I comprises either a single point or an interval centered at x_0 (possibly closed or half-open), called the *interval of convergence* of the power series. If I has positive length, then the germ defines a real-valued function f in I. $\diamond$

Example 7.4.3. By the geometric series formula, Theorem 7.1.7

$$\sum_{k=0}^{\infty} x^k = 1 + x + x^2 + x^3 + \cdots$$

converges if and only if $|x| < 1$. The interval of convergence is $I = (-1, 1)$, and

$$f(x) = \sum_{k=0}^{\infty} x^k = \frac{1}{1 - x}.$$

Replacing x by $-x$ gives the power series representation

$$\frac{1}{1 + x} = \frac{1}{1 - (-x)} = \sum_{k=0}^{\infty} (-x)^k = \sum_{k=0}^{\infty} (-1)^k x^k$$

on $(-1, 1)$. $\diamond$

Example 7.4.4. The algebraic identity

$$\frac{1}{1 - x^2} = \frac{1}{2}\left[\frac{1}{1 - x} + \frac{1}{1 + x}\right], \quad x \neq \pm 1,$$

may be checked by putting the right-hand side over a common denominator. If $-1 < x < 1$, however, this identity has a remarkable interpretation in terms of geometric series:

$$\frac{1}{1-x} = 1 + x + x^2 + x^3 + x^4 + x^5 + x^6 + x^7 + \cdots,$$

$$\frac{1}{1+x} = 1 - x + x^2 - x^3 + x^4 - x^5 + x^6 - x^7 + \cdots,$$

$$\frac{1}{1-x^2} = 1 \qquad + x^2 \qquad + x^4 \qquad + x^6 \qquad + \cdots;$$

the third line is half the term-by-term sum of the first two lines.

Generally, convergent power series can be added, subtracted, and multiplied by constants "just as if they were polynomials," because of Proposition 7.1.3. ◇

Example 7.4.5. Assume $c \neq 0$. By Example 7.4.3,

$$\frac{1}{c-x} = \frac{1}{c\left(1 - (x/c)\right)} = \frac{1}{c}\sum_{k=0}^{\infty}\left(\frac{x}{c}\right)^k = \sum_{k=0}^{\infty}\frac{x^k}{c^{k+1}}$$

provided $|x/c| < 1$, namely, on the interval $I = (-|c|, |c|)$. Similarly,

$$\frac{1}{c+x} = \frac{1}{c}\sum_{k=0}^{\infty}(-1)^k\left(\frac{x}{c}\right)^k = \sum_{k=0}^{\infty}\frac{(-1)^k}{c^{k+1}}x^k \quad \text{on } I = (-|c|, |c|). \qquad ◇$$

Definition 7.4.6. Assume I is a non-empty open set of real numbers. A function $f : I \to \mathbf{R}$ is *real-analytic* if for every x_0 in I, there is a positive R and a germ on $B_R(x_0)$ representing f, namely, there exist coefficients $(a_k)_{k=0}^{\infty}$ such that

$$f(x) = \sum_{k=0}^{\infty} a_k(x - x_0)^k \quad \text{if } |x - x_0| < R.$$

Example 7.4.7. The reciprocal function $f(x) = 1/x$ is real-analytic on $\mathbf{R}\setminus\{0\}$. Fix a non-zero x_0 arbitrarily. By Example 7.4.5 with $c = x_0$,

$$\frac{1}{x} = \frac{1}{x_0 + (x - x_0)} = \frac{1}{x_0}\sum_{k=0}^{\infty}(-1)^k\left(\frac{x - x_0}{x_0}\right)^k$$

provided $|(x - x_0)/x_0| < 1$, namely in the open ball $B_{|x_0|}(x_0)$. ◇

Remark 7.4.8. If

$$g(x) = \sum_{k=0}^{\infty} a_k(x - x_0)^k = a_0 + a_1(x - x_0) + a_2(x - x_0)^2 + \cdots$$

is a germ in $B_R(x_0)$, then $f(x) = g(x_0 + x)$ is a germ in $B_R(0) = (-R, R)$:

$$f(x) = \sum_{k=0}^{\infty} a_k x^k = a_0 + a_1 x + a_2 x^2 + a_3 x^3 + \cdots.$$

In proving theorems about germs, we may usually assume the center to be 0, gaining notational simplicity without loss of generality. ◇

Proposition 7.4.9. *Assume $(a_k)_{k=0}^{\infty}$ is a real sequence, x_0 a real number. If the series $\sum_k a_k R^k$ converges for some non-zero R, then the power series*

$$\sum_{k=0}^{\infty} a_k(x - x_0)^k$$

converges absolutely for every x such that $|x - x_0| < |R|$. That is, the interval of convergence I contains the open ball $B_{|R|}(x_0) = x_0 + (-|R|, |R|)$.

Proof. By hypothesis, the sequence $(a_k R^k)$ is summable, hence convergent to 0 by Proposition 7.1.5. Proposition 6.1.19 implies $(a_k R^k)$ is bounded; there exists a real number M such that

$$|a_k R^k| = |a_k| \cdot |R|^k \leq M \quad \text{for all } k.$$

By Remark 7.4.8, we may assume $x_0 = 0$. Let x be an arbitrary number such that $|x| < |R|$, and put $\rho = |x|/|R|$, so that $0 \leq \rho < 1$ and $|x| = |R| \cdot \rho$. We have

$$|a_k x^k| = |a_k| \cdot |x|^k = |a_k| \cdot |R|^k \cdot \rho^k \leq M\rho^k.$$

Since $M\rho^k$ is the general term of a convergent geometric series, $\left(|a_k x^k|\right)$ is summable by comparison. $\qquad\square$

Corollary 7.4.10. *Assume (a_k) is a real sequence and x_0 is real. There exists a unique non-negative extended real number R with the following properties: The power series $\sum_k a_k(x - x_0)^k$ converges absolutely for all real x such that $|x - x_0| < R$, and diverges for all x such that $R < |x - x_0|$.*

Proof. Consider the set of real numbers $J = \{x \text{ in } \mathbf{R} : \sum_k |a_k| \, |x|^k \text{ converges}\}$, and put $R = \sup J$. Since $0 \in J$, we have $0 \leq R$. Since $|-x| = |x|$ for all real x, the set J is symmetric about the origin.

Assume x is real and $|x| < R = \sup J$. By definition of a supremum, there exists an r in J such that $|x| < r$. By Proposition 7.4.9, $(-r, r) \subseteq J$. Since x in $(-R, R)$ is arbitrary, $(-R, R) \subseteq J$.

Conversely, if $R < |x| = r$, the power series $\sum_k a_k x^k$ diverges; if the series converged, Proposition 7.4.9 would imply that $(-r, r) \subseteq J$, contrary to the fact that $R = \sup J$. $\qquad\square$

Definition 7.4.11. The extended real number R in Corollary 7.4.10 is called the *radius* of the power series $\sum_k a_k(x - x_0)^k$.

Remark 7.4.12. If $R = 0$, the condition $|x - x_0| < R$ is empty, and the power series diverges if $x \neq x_0$. Similarly, if $R = \infty$, the condition $R < |x - x_0|$ is empty, and the series converges absolutely for all real x. The corollary gives no information about convergence if $|x - x_0| = R$. $\qquad\diamond$

Proposition 7.4.13. *A germ on $B_R(x_0)$ defines a real-analytic function f.*

Proof. At issue is whether $f(x)$ can be written as the sum of a convergent power series *centered at an arbitrary point* of $B_R(x_0)$. Without loss of generality, assume the germ is centered at 0,

$$f(x) = \sum_{k=0}^{\infty} a_k x^k \quad \text{if } |x| < R.$$

This frees up x_0 to stand for an arbitrary point of $(-R, R)$. If we write $x = (x - x_0) + x_0$, the binomial theorem implies

$$f(x) = \sum_{k=0}^{\infty} a_k \left((x - x_0) + x_0\right)^k = \sum_{k=0}^{\infty} a_k \sum_{j=0}^{k} \binom{k}{j} x_0^{k-j} (x - x_0)^j$$

$$= \sum_{j=0}^{\infty} b_j (x - x_0)^j, \quad \text{with} \quad b_j = \sum_{k=j}^{\infty} a_k \binom{k}{j} x_0^{k-j}.$$

This series has radius no smaller than $r = R - |x_0| > 0$, since if $|x - x_0| < r$, then $|x| \leq |x - x_0| + |x_0| < r + |x_0| = R$. $\qquad\square$

Theorem 7.4.14 (The ratio test). *Assume $(a_k)_{k=0}^{\infty}$ is a sequence of coefficients. If*

$$|x - x_0| \lim_{k\to\infty} \left| \frac{a_{k+1}}{a_k} \right| < 1,$$

namely, the limit exists and is smaller than 1, then the power series

$$\sum_{k=0}^{\infty} a_k (x - x_0)^k$$

converges absolutely at x.

Proof. For each real x, a power series is an ordinary numerical series, with terms $b_k = a_k (x - x_0)^k$. By the ratio test, if $|b_{k+1}/b_k| \to \rho$ and $\rho < 1$, the series converges absolutely. But by algebra,

$$\lim_{k\to\infty} \left| \frac{b_{k+1}}{b_k} \right| = \lim_{k\to\infty} \left| \frac{a_{k+1}(x - x_0)^{k+1}}{a_k(x - x_0)^k} \right| = |x - x_0| \lim_{k\to\infty} \left| \frac{a_{k+1}}{a_k} \right|. \qquad\square$$

Remark 7.4.15. The statement of Theorem 7.4.14 is convenient for many purposes, but the method of proof applies more generally to power series having some coefficients equal to 0. In general, let b_k denote the kth term (which is a function of x), and apply the ratio test to the series $\sum_k b_k$. $\qquad\diamond$

Example 7.4.16. The power series

$$\sum_{k=0}^{\infty}(-1)^k\frac{x^{2k+1}}{(2k+1)!} = x - \frac{x^3}{3!} + \frac{x^5}{5!} - \frac{x^7}{7!} + \cdots$$

converges for all real x, or $R = \infty$. Here, $b_k = (-1)^k x^{2k+1}/(2k+1)!$,

$$\lim_{k\to\infty}\left|\frac{b_{k+1}}{b_k}\right| = \lim_{k\to\infty}\left|\frac{(-1)^{k+1}x^{2k+3}}{(2k+3)!} \cdot \frac{(2k+1)!}{(-1)^k x^{2k+1}}\right|$$

$$= \lim_{k\to\infty}\left|\frac{x^2(2k+1)!}{(2k+3)!}\right| = \lim_{k\to\infty}\left|\frac{x^2}{(2k+3)(2k+2)}\right| = 0,$$

and this is less than 1 for all real x. $\diamondsuit$

Exercises for Section 7.4

Exercise 7.4.1. Find the set of real x for which the following converge:

(a) $\displaystyle\sum_{k=0}^{\infty}\frac{x^k}{k+1}$.   (b) $\displaystyle\sum_{k=0}^{\infty}k^2 x^k$.   (c) $\displaystyle\sum_{k=0}^{\infty}\frac{x^k}{k!}$.   (d) $\displaystyle\sum_{k=0}^{\infty}k!\,x^k$.

Exercise 7.4.2. Determine the interval of convergence of $\displaystyle\sum_{k=0}^{\infty}k^m x^k$, $m > 0$.

Exercise 7.4.3. If x is a real number, define

$$C(x) = \sum_{k=0}^{\infty}\frac{(-1)^k x^{2k}}{(2k)!} = 1 - \frac{x^2}{2!} + \frac{x^4}{4!} - \frac{x^6}{6!} + \frac{x^8}{8!} - \cdots,$$

$$S(x) = \sum_{k=0}^{\infty}\frac{(-1)^k x^{2k+1}}{(2k+1)!} = x - \frac{x^3}{3!} + \frac{x^5}{5!} - \frac{x^7}{7!} + \frac{x^9}{9!} - \cdots.$$

Prove that $C(x)$ and $S(x)$ converge absolutely for all real x.

Exercise 7.4.4. If n is a natural number, define

$$C_n(x) = \sum_{k=0}^{n-1}\frac{(-1)^k x^{2k}}{(2k)!}, \qquad\qquad C(x) = \sum_{k=0}^{\infty}\frac{(-1)^k x^{2k}}{(2k)!},$$

$$S_n(x) = \sum_{k=0}^{n-1}\frac{(-1)^k x^{2k+1}}{(2k+1)!}, \qquad\qquad S(x) = \sum_{k=0}^{\infty}\frac{(-1)^k x^{2k+1}}{(2k+1)!}.$$

(a) If x is real and $|x| < 2n$, prove that

$$|C(x) - C_n(x)| \le \frac{|x|^{2n}}{(2n)!}.$$

(b) Fix a positive r arbitrarily. Prove that for every positive ε, there exists an N such that if $n \geq N$, then $|C(x) - C_n(x)| < \varepsilon$ if $|x| \leq r$. State and prove analogous estimates for S_n and S.

Exercise 7.4.5. ($\bigstar$) The *Fibonacci sequence* (F_k) is defined recursively by

$$F_0 = F_1 = 1, \qquad F_{n+2} = F_n + F_{n+1}.$$

(a) Prove that $F_n \leq \phi^n := \left(\frac{1}{2}(1 + \sqrt{5})\right)^n$ for every natural number n.

(b) Write out the first seven terms of the power series

$$f(x) = \sum_{k=0}^{\infty} F_k x^k,$$

and show this series converges absolutely on some open interval containing 0.

(c) Prove that on the interval in part (b), $f(x) = 1/(1 - x - x^2)$.

Exercise 7.4.6. If x is a real number and n is a non-negative integer, define

$$E(x) = \sum_{k=0}^{\infty} \frac{x^k}{k!} = 1 + x + \frac{x^2}{2!} + \frac{x^3}{3!} + \frac{x^4}{4!} + \cdots,$$

$$E_n(x) = \sum_{k=0}^{n-1} \frac{x^k}{k!} = 1 + x + \frac{x^2}{2!} + \frac{x^3}{3!} + \cdots + \frac{x^{n-1}}{(n-1)!}.$$

(a) Prove $E(x)$ converges absolutely for all real x, and $E(x)E(y) = E(x+y)$ for all real x and y.

(b) Prove $|E(-1) - E_n(-1)| \leq 1/n!$.

(c) Prove $E(1) - E_n(1) = \displaystyle\sum_{k=0}^{\infty} \frac{1}{(n+k)!} \leq \frac{1}{(n-1) \cdot (n-1)!}$.

(d) What n suffices to ensure $E_n(1) = E(1) \pm 0.5 \times 10^{-6}$ (six-decimal approximation)?

Exercise 7.4.7. (H) With notation as in Exercise 7.4.6, prove:

(a) If q is a positive integer, then $q! E_q(1)$ is an integer.

(b) For every positive integer q, we have $E_q(1) < E(1) < E_q(1) + 1/(q \cdot q!)$.

(c) (H) $E(1)$ is irrational.

Exercise 7.4.8. (The root test.) Assume (a_k) is a real sequence and x_0 is a real number. Prove that the power series

$$\sum_{k=0}^{\infty} a_k (x - x_0)^k$$

converges absolutely if $\rho = \limsup |a_k|^{1/k} |x - x_0| < 1$ and diverges if $1 < \rho$.

8

Continuous Functions

Familiar objects do not teleport, do not change position "discontinuously" in an instant. If $f(t)$ represents the location of a particle at time t, intuition says we can ensure the change in $f(t)$ can be made as small as we like by looking at a sufficiently short interval of time. We are ready to formalize this intuition.

8.1 Continuity

Throughout this chapter, X denotes a non-empty set of real numbers, ε continues to denote an arbitrary positive real number, and δ (delta) denotes an arbitrary positive real number.

Definition 8.1.1. Assume $f : X \to \mathbf{R}$ is a function, and $\mathbf{x} = (x_k)$ a sequence in X, namely, a real sequence such that $x_k \in X$ for all k. We define the *image sequence* $\mathbf{y} = (y_k)$ by setting $y_k = f(x_k)$ for each k, namely, by applying f to the terms of (x_k).

Example 8.1.2. Suppose $f(x) = 1/x$ on the set where $x \neq 0$. For every positive integer k:

If $x_k = (k+1)/k$, then $y_k = k/(k+1)$.
If $x_k = 1/k$, then $y_k = k$.
If $x_k = (-1)^k/k$, then $y_k = (-1)^k k$. $\Diamond$

Example 8.1.3. The *signum* function sgn $: \mathbf{R} \to \mathbf{R}$ is defined by $\operatorname{sgn}(x) = 1$ if $x > 0$; $\operatorname{sgn}(x) = -1$ if $x < 0$, and $\operatorname{sgn}(0) = 0$.

If $x_k = 1/k$, then $y_k = 1$.
If $x_k = (-1)^k/k$, then $y_k = (-1)^k$. $\Diamond$

Definition 8.1.4. Assume $f : X \to \mathbf{R}$ is a function, and assume $x \in X$. We say f is *continuous at* x if the following condition holds:

> For *every* sequence (x_k) in X that converges to x, the image sequence $\big(f(x_k)\big)$ converges to $f(x)$.

If f is continuous at x for each x in X, we say f is *continuous on* X.

DOI: 10.1201/9781003601357-8

Remark 8.1.5. Symbolically, f is continuous at x if and only if

$$\lim_{k \to \infty} f(x_k) = f(x) = f\left(\lim_{k \to \infty} x_k\right)$$

whenever (x_k) is a sequence in X and $(x_k) \to x$.   ◊

Remark 8.1.6. If $f : X \to \mathbf{R}$ is continuous on X, and if A is a non-empty subset of X, the restriction $f|_A$ is continuous on A, since every sequence in A is a sequence in X.   ◊

The following observation is immediate, yet useful enough to state explicitly.

Lemma 8.1.7. *The identity function is continuous on* $\mathbf{R}$. *Every constant function is continuous on* $\mathbf{R}$.

Proposition 8.1.8. *Assume f and g are continuous functions on X. The functions $f \pm g$ and fg are continuous on X. If Z is the set of x in X such that $g(x) = 0$, then the quotient f/g is continuous on $X \setminus Z$.*

Proof. In effect, this restates Proposition 6.2.1: If x is an arbitrary element of X, and (x_k) is a sequence in X converging to x, then

$$\begin{aligned}
\lim_{k \to \infty} (f + g)(x_k) &= \lim_{k \to \infty} \big(f(x_k) + g(x_k)\big) && \text{definition of } f + g \\
&= \lim_{k \to \infty} f(x_k) + \lim_{k \to \infty} g(x_k) && \text{Proposition 6.2.1 (i)} \\
&= f(x) + g(x) && f, g \text{ continuous at } x \\
&= (f + g)(x).
\end{aligned}$$

The remaining assertions are proven entirely similarly.   □

Corollary 8.1.9. *Assume p and q are polynomials, $X = \{x \text{ in } \mathbf{R} : q(x) \neq 0\}$. The rational function $f = p/q : X \to \mathbf{R}$ is continuous on X. In particular, every polynomial function is continuous on* $\mathbf{R}$.

Proof. Intuitively, polynomials are obtained from the identity function and constant functions by adding and multiplying, so all polynomials are continuous. Exercise 8.1.3 requests a formal proof by induction.   □

Proposition 8.1.10. *If g and f are composable, f is continuous at x, and g is continuous at $y = f(x)$, then $g \circ f$ is continuous at x.*

Proof. Assume (x_k) is an arbitrary sequence in the domain of f converging to x, and let $(y_k) = \big(f(x_k)\big)$ be the image sequence. Since f is continuous at x, the image sequence converges to $y = f(x)$. Since g is continuous at y, the sequence $\big(g(y_k)\big) = \big((g \circ f)(x_k)\big)$ converges to $g(y) = (g \circ f)(x)$. Since (x_k) was arbitrary, $g \circ f$ is continuous at x.   □

Conceptualizing Continuous Functions

In calculus, continuous functions are normally depicted as "smooth" graphs with finitely many corners or cusps. "Zooming in" on most points causes the graph to "flatten out" into a line. As we will see in Chapter 10, this behavior corresponds to *differentiability.*

By contrast, a "typical" continuous function has a graph qualitatively similar to Figure 8.1, more like the tracing of an electrocardiogram, a seismograph, or stock price tracker.

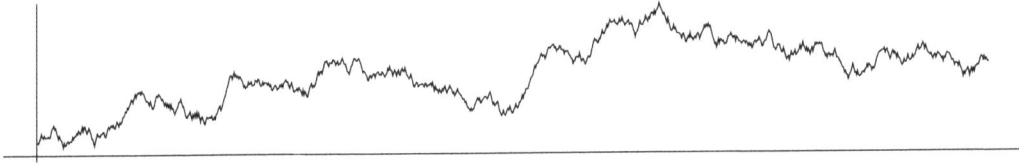

FIGURE 8.1
The graph of a "typical" continuous function.

Imagine the graph as a mountain range. Zooming in reveals cliffs and crags, then surfaces of rocks, which are "rough" at scales visible to the naked eye and do not appear more smooth when viewed under a microscope.

Mathematically, a "typical" continuous function possesses detail at arbitrarily small scales. Since the graph does not become more line-like upon zooming in on an arbitrary point, a typical continuous function is *nowhere differentiable.*

The full ramifications of these qualitative observations impose an intuition of continuous functions almost completely different from the picture presented in calculus. For instance, the graph of a typical continuous function is so rough that its length between *an arbitrary pair of points* is infinite. The only reason we can pretend to draw such a function is that a drawing instrument such as a sharp pencil, a laser printer, or chalk, does not delimit a mathematical point, but a small region. Our drawing instruments "cannot see" details smaller than the diameter of the tip, and they manage to "cover" the graph with a region of finite area, which can be "painted" in finite time.

Many theorems in this book apply to continuous functions. For visual simplicity, proofs are illustrated using "smooth" functions. To sharpen your understanding, it's a good idea to ponder how the same arguments succeed against a general continuous function.

Exercises for Section 8.1

Exercise 8.1.1. (★) Assume $f(x) = x$ if $x \leq 0$, and $1/x$ if $x > 0$.

(a) Sketch the graph of f. Suppose $x_k = -1/k$ if $k \geq 1$. Sketch this sequence in your graph. Does $f(x_k)$ converge to $f(0)$?

(b) Suppose $x_k = 1/k$ if $k \geq 1$. Sketch this sequence in your graph from (a). Does $f(x_k)$ converge to $f(0)$?

(c) Is f continuous at 0? Explain.

Exercise 8.1.2.(★) Suppose $f(x) = x/|x|$ if $x \neq 0$.

(a) Sketch the graph of f. Prove from the definition that f is continuous on its domain. Hint: Let ℓ be an arbitrary non-zero real number, and assume $(x_k) \to \ell$. By using a suitable choice of ε, prove $f(x_k)$ is eventually constant.

(b) Is there a way to define $f(0)$ to get a function continuous on $\mathbf{R}$? Hint: The sequence $x_k = (-1)^{k-1}/k$ converges to 0.

Exercise 8.1.3. Prove Corollary 8.1.9, using induction on the degree to handle polynomials.

Exercise 8.1.4.(★) Let $\chi_{\mathbf{Q}}$ be the indicator of the rationals,

$$\chi_{\mathbf{Q}}(x) = \begin{cases} 1 & \text{if } x \text{ is rational,} \\ 0 & \text{if } x \text{ is irrational.} \end{cases}$$

(a) Prove $\chi_{\mathbf{Q}}$ is discontinuous at every real number.

(b) Assume f is continuous. Prove that the product $f \cdot \chi_{\mathbf{Q}}$ is continuous at x if and only if $f(x) = 0$.

Exercise 8.1.5.(H) Assume f is the *denominator function*, Figure 8.2, defined by

$$f(x) = \begin{cases} 1/q & \text{if } x = p/q \text{ in lowest terms,} \\ 0 & \text{if } x \text{ is irrational.} \end{cases}$$

Prove f is discontinuous at every rational number and continuous at every irrational.

8.2   Limits and Continuity

For theoretical work, it is convenient to have an "adversarial" criterion for continuity that does not refer to arbitrary sequences. Recall that when $X \subseteq \mathbf{R}$, a *limit point* of X is a real number x_0 (possibly not an element of X) such that $B_\delta^\times(x_0) \cap X \neq \varnothing$ for every positive real number δ.

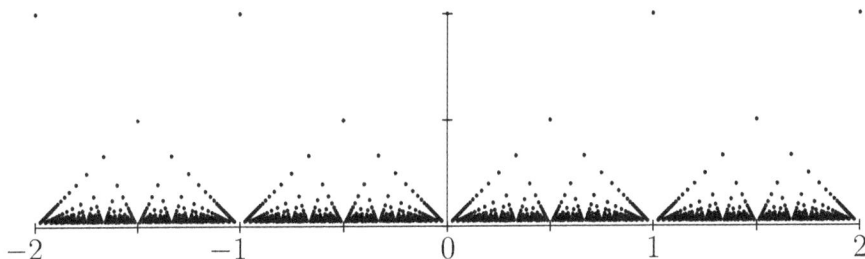

FIGURE 8.2
The graph of the denominator function.

Definition 8.2.1. Assume $f : X \to \mathbf{R}$ is a function, and x_0 is a limit point of X. We say the real number L is a *limit* of f at x_0 if the following condition holds:

> For every ε, there exists a δ such that if x is an element of $B_\delta^\times(x_0) \cap X$, namely $x \in X$ and $0 < |x - x_0| < \delta$, then $|f(x) - L| < \varepsilon$.

Remark 8.2.2. Just as with sequential limits, the preceding definition may be regarded as an adversarial game between Player ε, who issues a challenge, and Player δ, who attempts to respond, making the distance from $f(x)$ to L smaller than ε merely by constraining x to within distance δ of x_0 (but distinct from x_0). The number L is a limit of f at x_0 if and only if Player δ has a winning strategy against a perfect opponent, compare Figures 8.3a and 8.3b. ◇

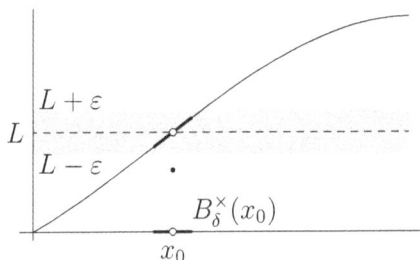

(a) The limit at x_0 equals L.

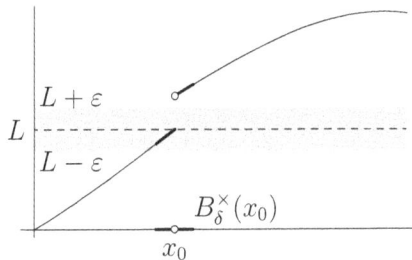

(b) The limit at x_0 does not exist.

FIGURE 8.3
A geometric interpretation of limits.

As with a sequential limit, a functional limit is unique if it exists.

Lemma 8.2.3. *If L and L' are limits of f at x_0, then $L = L'$.*

Proof. Let ε be arbitrary. Because L and L' are limits of f at x_0, there exists a δ such that if $x \in X$ and $0 < |x - x_0| < \delta$, then $|f(x) - L| < \varepsilon/2$ and $|f(x) - L'| < \varepsilon/2$. For any such x, the triangle inequality implies

$$|L - L'| \le |L - f(x)| + |f(x) - L'| < \varepsilon/2 + \varepsilon/2 = \varepsilon.$$

Since ε was arbitrary, $L = L'$. □

Remark 8.2.4. If f has limit L at x_0, Lemma 8.2.3 justifies our writing

$$L = \lim(f, x_0) \quad \text{or} \quad L = \lim_{x \to x_0} f(x)$$

and manipulating limits algebraically. ◇

Remark 8.2.5. If $L = \lim(f, x_0)$, we sometimes say $f(x)$ *approaches* L *as* x *approaches* x_0, or write "$f(x) \to L$ as $x \to x_0$." This "dynamical" language can be convenient, but is potentially misleading: The limit of f at x_0 either exists or it doesn't. There is no contingency on x approaching the point x_0, "getting closer and closer without reaching x_0." We do not have to "wait" (possibly forever) to find out whether or not the limit exists. ◇

Proposition 8.2.6. *Assume f is a real-valued function with domain X, and $x_\infty \in X$. The function f is continuous at x_∞ if and only if one of the following mutually exclusive conditions holds:*

(i) x_∞ *is an isolated point of X.*

(ii) x_∞ *is a limit point of X, and $\lim(f, x_\infty) = f(x_\infty)$.*

Proof. ((i) implies continuity). Suppose x_∞ is an isolated point of X. If (x_k) is a sequence in X that converges to x_∞, then (x_k) is eventually equal to x_∞, namely, there is an index N such that $x_k = x_\infty$ if $k \geq N$. Continuity of f at x_∞ follows immediately.

((ii) implies continuity). Suppose (ii) holds. Let ε be arbitrary. By (ii), there exists a positive δ such that if $x \in X$ and $0 < |x - x_\infty| < \delta$, then $|f(x) - f(x_\infty)| < \varepsilon$. Since $|f(x_\infty) - f(x_\infty)| = 0 < \varepsilon$ automatically, f satisfies the slightly stronger predicate:

If $x \in X$ and $|x - x_\infty| < \delta$, then $|f(x) - f(x_\infty)| < \varepsilon$.

Let (x_k) be an arbitrary sequence in X that converges to x_∞; using the δ of the preceding paragraph as a "challenge," there exists an index N such that if $k \geq N$, then $|x_k - x_\infty| < \delta$. But as just noted, this implies $|f(x_k) - f(x_\infty)| < \varepsilon$. Since ε was arbitrary, the image sequence $\big(f(x_k)\big)$ converges to $f(x_\infty)$; thus f is continuous at x_∞.

(Continuity implies (i) or (ii)). Contrapositively, we show that if both (i) and (ii) are false, then f is not continuous at x_∞.

Since (i) is false, x_∞ is a limit point of X. Since (ii) is false, Player ε has a winning strategy: There exists an ε such that no matter how δ is chosen, there is some x in X such that $|x - x_\infty| < \delta$ but $|f(x) - f(x_\infty)| \geq \varepsilon$.

Construct a sequence (x_k) as follows: For each positive integer k, pick x_k in X such that $|x_k - x_\infty| < 1/k = \delta$ but $|f(x_k) - f(x_\infty)| \geq \varepsilon$. The sequence (x_k) converges to x_∞, but the image sequence $\big(f(x_k)\big)$ does not converge to $f(x_\infty)$.

We have proven that if (i) and (ii) are false, then f is discontinuous at x_∞, which is the contrapositive of what was to be shown. □

Continuity of f at x_0 may be formulated as an adversarial game.

Corollary 8.2.7 (The ε-δ criterion). *If $f : X \to \mathbf{R}$ is a function and $x_0 \in X$, then f is continuous at x_0 if and only if:*

For every positive real ε,
  there exists a positive real δ such that
    if $x \in X$ and $|x - x_0| < \delta$, then $|f(x) - f(x_0)| < \varepsilon$.

In another direction, properties of a continuous function at one point "propagate" to some neighborhood. Two examples recur repeatedly:

Corollary 8.2.8. *Assume $f : X \to \mathbf{R}$ is continuous at some point x_0 in X.*

(i) *There exist positive real numbers M and δ such that $|f(x)| < M$ provided $x \in X$ and $|x - x_0| < \delta$.*

(ii) *If $f(x_0) \neq 0$, there is a δ such that if $x \in X$ and $|x - x_0| < \delta$, then $|f(x)| > |f(x_0)|/2$.*

Remark 8.2.9. Conclusion (i) is expressed by saying a continuous function is *locally bounded*. Conclusion (ii) says if a continuous function is non-zero at some point, then the function is *locally bounded away from zero*; not merely is $|f(x)|$ non-zero, there is a *positive lower bound*, $|f(x_0)|/2$. ◇

Proof. (i). Put $M = |f(x_0)| + 1$. By continuity of f at x_0 with $\varepsilon = 1$, there exists a δ such that if $x \in X$ and $|x - x_0| < \delta$, then $|f(x) - f(x_0)| < 1$. The triangle inequality implies

$$|f(x)| \leq |f(x_0)| + |f(x) - f(x_0)| < |f(x_0)| + 1 = M.$$

(ii). Take $\varepsilon = |f(x_0)|/2$, which is positive by hypothesis. Use continuity of f at x_0 to pick a positive δ such that if $x \in X$ and $|x - x_0| < \delta$, then $|f(x) - f(x_0)| < \varepsilon = |f(x_0)|/2$. By the reverse triangle inequality, if $x \in X$ and $|x - x_0| < \delta$, then

$$|f(x)| \geq |f(x_0)| - |f(x) - f(x_0)| > |f(x_0)|/2. \qquad □$$

Corollary 8.2.10. *Assume I is an interval of real numbers and $A \subseteq I$ a dense subset. If f and g are functions on I whose restrictions to A are equal, then $f = g$ on I.*

Proof. See Exercise 8.2.7. □

Proposition 8.2.11. *Assume X is an open interval of real numbers. If f is strictly monotone on X, then the inverse function $f^{-1} : f(X) \to X$ is continuous.*

Proof. See Exercise 8.2.8. □

One-Sided Limits

For every real x_0 and every δ, we have $B_\delta^\times(x_0) = (x_0 - \delta, x_0) \cup (x_0, x_0 + \delta)$. Using the interval components gives two useful "one-sided" criteria. When a property below corresponds to a property of ordinary limits, the proof is effectively identical, and is omitted.

Definition 8.2.12. Assume X is a non-empty set of real numbers, $f : X \to \mathbf{R}$ a function, and x_0 a real number. We say x_0 is *approachable from above* in X if $X \cap (x_0, x_0 + \delta)$ is non-empty for every positive δ.

Assume x is approachable from above in X. We say a real number L is a *limit from above* of f at x if for every sequence (x_k) in X such that $(x_k) \to x$ and $x_k > x$ for all k, we have $\left(f(x_k)\right) \to L$.

Proposition 8.2.13. *Assume $f : X \to \mathbf{R}$ is a function and x_0 a limit point of X. If x_0 is approachable from above in X, and if L and L' are both limits of f at x_0 from above, then $L = L'$.*

Remark 8.2.14. If the limit from above of f at x_0 exists, we denote it $f(x_0^+)$. Notations in other sources include $\lim(f, x_0^+)$ and $\lim_{x \to x_0^+} f(x)$. ◇

Proposition 8.2.15. *Assume $f : X \to \mathbf{R}$ is a function. If x_0 is approachable from above in X, then $L = f(x_0^+)$ if and only if:*

> *For every positive real ε,*
> *there exists a positive real δ such that*
> *if $x \in X \cap (x_0, x_0 + \delta)$, then $|f(x) - L| < \varepsilon$.*

Remark 8.2.16. Everything above regarding limits from above has a version for limits from below, and is left as an exercise for practice. If the limit from below of f at x_0 exists, we denote it $f(x_0^-)$. Notations in other sources include $\lim(f, x_0^-)$ and $\lim_{x \to x_0^-} f(x)$.

A limit from above or from below is called a *one-sided limit* of f at x_0. ◇

Proposition 8.2.17. *Assume X is a set of real numbers, $f : X \to \mathbf{R}$ a function, and x_0 a real number. If x_0 is approachable both from above and from below in X, then $\lim(f, x_0)$ exists and is equal to L if and only if both one-sided limits of f at x_0 exist and are equal to L.*

Definition 8.2.18. Assume f is defined in some open ball $B_r(x_0)$. If the one-sided limits $\lim(f, x_0^{\pm})$ exist and are distinct, we say f has a *jump* at x_0.

Example 8.2.19. The function $f(x) = x/|x|$ on $\mathbf{R} \setminus \{0\}$ is continuous everywhere in its domain. Every extension of f to $\mathbf{R}$ has a jump at 0. $\diamond$

Remark 8.2.20. If $\lim(f, x_0)$ exists and $\lim(f, x_0) \neq f(x_0)$, many authors say f has a *removable discontinuity* at x_0. Loosely, by redefining f at one point, the discontinuity can be removed. (Literally, the redefined function is not f.) $\diamond$

One-sided limits of monotone functions are analogous to limits of eventually-monotone sequences.

Proposition 8.2.21. *Assume* $f : X \to \mathbf{R}$ *is bounded and monotone.*

(i) *For every real* x_0, *if* x_0 *is approachable from above in* X, *then* $f(x_0^+)$ *exists.*

(ii) *For every real* x_0, *if* x_0 *is approachable from below in* X, *then* $f(x_0^-)$ *exists.*

Remark 8.2.22. The boundedness hypothesis may be dropped, with the understanding that the one-sided limits of f at the "extremities" of X may be infinite. $\diamond$

Remark 8.2.23. By Proposition 8.2.21, the only discontinuities of a monotone function on an open interval are jumps. $\diamond$

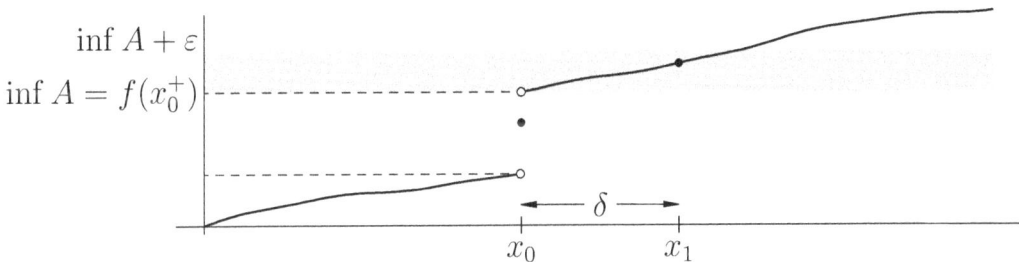

FIGURE 8.4
A monotone function has one-sided limits at each point.

Proof. (i). Assume x_0 is approachable from above in X. The non-empty set

$$A = \{f(x) : x \in X \text{ and } x_0 < x\}$$

is bounded below by $f(x_0)$. It suffices to show $f(x_0^+) = \inf A$.

Let ε be arbitrary. Since $\inf A + \varepsilon$ is not a lower bound of A, there exists an x_1 greater than x_0 in X such that $\inf A \leq f(x_1) < \inf A + \varepsilon$, Figure 8.4. Put $\delta = x_1 - x_0 > 0$.

If $x_0 < x < x_0 + \delta = x_1$, then $\inf A \leq f(x) \leq f(x_1) < \inf A + \varepsilon$ by monotonicity. Rearranging, $0 \leq f(x) - \inf A < \varepsilon$, so $|f(x) - \inf A| < \varepsilon$. Since ε was arbitrary, $f(x_0^+) = \inf A$.

The proof of (ii) is entirely similar, see Exercise 8.2.9. $\square$

Exercises for Section 8.2

Exercise 8.2.1. (★) Assume $f : X \to \mathbf{R}$ is continuous at some limit point c of X, and $g : X \to \mathbf{R}$ is equal to f everywhere in X except possibly at c.

Prove that $\lim(g, c)$ exists and is equal to $\lim(f, c)$. In words, *limits do not depend on function values at individual points.*

Exercise 8.2.2. (★) Show from the ε-δ condition that $f(x) = 3x - 5$ is continuous at c for every real c.

Exercise 8.2.3. (H) Show from the ε-δ condition that $f(x) = x^2$ is continuous at c for every real c.

Exercise 8.2.4. (H) Show from the ε-δ condition that $f(x) = 1/x$ is continuous at c for every non-zero real c.

Exercise 8.2.5. Recall that for each real number x, the *floor* $\lfloor x \rfloor$ of x is the largest integer n such that $n \le x$, and the *ceiling* $\lceil x \rceil$ of x is the smallest integer n such that $x \le n$.

(a) Sketch the graphs: $y = \lfloor x \rfloor$, $y = \lceil x \rceil$, $y = x - \lfloor x \rfloor$, $y = \lceil x \rceil - \lfloor x \rfloor$.

(b) Prove the function $f(x) = x - \lfloor x \rfloor$ on $\mathbf{R}$ has a jump at each integer.

(c) Let $g(x) = \lceil x \rceil - \lfloor x \rfloor$. At which points x_0 does $\lim(g, x_0)$ exist? At which points is g continuous?

Exercise 8.2.6. Prove that every continuous function of one variable can be written as a difference of continuous, *non-negative* functions.

Exercise 8.2.7. Prove Corollary 8.2.10.

Exercise 8.2.8. Prove Proposition 8.2.11.

Exercise 8.2.9. Prove Proposition 8.2.21 (ii).

Exercise 8.2.10. Assume f is the *denominator function*, see Exercise 8.1.5. Prove that $\lim(f, x_0) = 0$ for all real x_0.

Exercise 8.2.11. Assume X is non-empty. A subset O of X is *relatively open* (in X) if there exists an open subset G of $\mathbf{R}$ such that $O = X \cap G$. For example, $O = [0, 1)$ is relatively open in $X = [0, 2]$.

(a) Prove that $\varnothing$ and X are relatively open in X.

(b) Prove that if $\{O_i\}_{i \in \mathscr{I}}$ is an arbitrary collection of relatively open subsets of X, then $\bigcup_i O_i$ is relatively open in X.

(c) Prove that if $\{O_i\}_{i=0}^{n-1}$ is a finite collection of relatively open subsets of X, then $\bigcap_i O_i$ is relatively open in X.

(d) Suppose $f : X \to \mathbf{R}$ is a function. Prove that f is continuous on X if and only if: For every open set V of real numbers, the preimage $f^*(V)$ is relatively open in X.

Exercise 8.2.12. Assume X is a set of real numbers, $A \subseteq X$ a dense subset, and $f : X \to \mathbf{R}$ a continuous function. Prove $f(A)$ is dense in $f(X)$.

Exercise 8.2.13. Let $U : \mathbf{R} \to \mathbf{R}$ be the *unit step function*, defined by $U(x) = 0$ if $x < 0$, $U(x) = 1$ if $0 < x$, and $U(0) = 1/2$, namely $U = (1 + \mathrm{sgn})/2$.

Assume $(a_k)_{k=0}^{\infty}$ is a sequence enumerating the rationals (i.e., every rational number appears *exactly once* as an a_k), and define

$$f(x) = \sum_{k=0}^{\infty} \frac{U(x - a_k)}{2^{k+1}}.$$

(a) Prove the series $f(x)$ converges absolutely for every real x, the sum is between 0 and 1, and the function f is strictly increasing.

(b) Prove that f is continuous at every irrational number and discontinuous at every rational.

Exercise 8.2.14. (★) Assume $f : \mathbf{R} \to \mathbf{R}$ satisfies the "mean value property" $f\left(\frac{1}{2}(a+b)\right) = \frac{1}{2}\left(f(a) + f(b)\right)$ for all real a and b. What can you say about f? What if in addition f is continuous?

8.3 Continuity of Power Series

Analysis is founded on approximation and estimates. Convergent power series are one of the primary computational tools in this book. In this section we prove that real-analytic functions, those defined by convergent power series, are continuous. (This is something of a laughable understatement.) Before proving that, we introduce algebraic *little-o* and *big-O* notations for working with limits and approximations that are particularly well-suited to power series.

Little-o and Big-O Notation

Definition 8.3.1. Assume X is a real interval and x_0 is an interior point or endpoint of X, namely a point of the closure. If f is a function on X, we write $f \approx o(1)$ (read "f is little-o of 1") near x_0 if $\lim(f, x_0) = 0$.

Remark 8.3.2. If f is defined by a formula $f(x)$, we express the same idea by writing $f(x) \approx o(1)$ if $x \approx x_0$.   ◇

Remark 8.3.3. We use this notation in algebraic manipulation. For example, if g is defined on X, then $g \approx f + o(1)$ near x_0 means $g - f \approx o(1)$ near x_0. ◇

Example 8.3.4. If f is defined on an interval X and $x_0 \in X$, Corollary 8.2.7 asserts that $f \approx f(x_0) + o(1)$ near x_0 if and only if f is continuous at x_0. ◇

Example 8.3.5. Proposition 8.1.8 and the preceding example guarantee that approximations $f \approx f(x_0) + o(1)$ and $g \approx g(x_0) + o(1)$ near x_0 can be added, multiplied, and (if the denominator is non-zero) divided like equations:

$$f \pm g \approx f(x_0) \pm g(x_0) + o(1),$$
$$f \cdot g \approx f(x_0) \cdot g(x_0) + o(1),$$
$$f/g \approx f(x_0)/g(x_0) + o(1).$$

In particular, $o(1) \pm o(1) \approx o(1)$, $o(1) \cdot o(1) \approx o(1)$, and $M \cdot o(1) \approx o(1)$ for arbitrary real M.   ◇

Remark 8.3.6. The notational benefit of little-o quickly becomes clear; an "error term" $o(1)$ acts like a sponge, absorbing (and therefore hiding) "ignorable" cross terms. This magic works because we may "shrink δ inside $o(1)$," retaining an open interval on which the approximation of interest holds.   ◇

Definition 8.3.7. Assume g is non-zero on some punctured neighborhood of x_0. We write

$$f(x) \approx o\big(g(x)\big) \quad \text{if } x \approx x_0,$$

or "f is little-o of g near x_0," to mean $f(x)/g(x) \approx o(1)$ if $x \approx x_0$.

Remark 8.3.8. Intuitively, "$f \approx o(g)$ near x_0" means $|f(x)|$ is infinitesimally small compared to $|g(x)|$ provided x is sufficiently close to x_0.   ◇

Example 8.3.9. For every natural number k, we have $x^{k+1} \approx o(x^k)$ if $x \approx 0$, since $x^{k+1}/x^k = x \approx o(1)$ if $x \approx 0$.

   Inductively, if $m > 0$, then $x^{k+m} \approx o(x^k)$ if $x \approx 0$. If x_0 is an arbitrary real number and if $m > 0$, then $(x - x_0)^{k+m} \approx o(x - x_0)^k$ if $x \approx x_0$.   ◇

   Supplementary to little-o notation, which expresses relative vanishing, we have big-O notation, which expresses relative boundedness:

Definition 8.3.10. Assume f is a function on X. We write $f \approx O(1)$ near x_0, read "f is big-O of 1," if $|f|$ is locally bounded near x_0, namely, if there exist positive real numbers M and δ such that $|f(x)| \leq M$ provided $x \in X$ and $|x - x_0| < \delta$.

If g is non-vanishing in some punctured neighborhood of x_0, we write

$$f(x) \approx O\big(g(x)\big) \quad \text{if } x \approx x_0,$$

read "f is big-O of g," to mean $f/g \approx O(1)$ near x_0.

Example 8.3.11. If $f(x_0)$ is defined and $f(x) \approx f(x_0) + O(x - x_0)$ if $x \approx x_0$, then f is continuous at x_0: Indeed, there exist positive real numbers M and δ such that $|f(x) - f(x_0)| < M|x - x_0|$ provided $|x - x_0| < \delta$.

Generally, there is an infinite hierarchy of approximations. For each natural number m, $f(x) \approx O(x - x_0)^{m+1}$ implies $f(x) \approx o(x - x_0)^m$, which implies $f(x) \approx O(x - x_0)^m$. $\diamond$

Continuity of Power Series

Proposition 8.3.12. *If $\sum_k a_k z^k$ is a power series with positive radius R and x_0 is real, then the real-analytic function*

$$f(x) = \sum_{k=0}^{\infty} a_k (x - x_0)^k$$

is continuous on the interval $B_R(x_0) = x_0 + (-R, R)$.

Remark 8.3.13. As usual, replacing $x - x_0$ by x reduces to the case $x_0 = 0$.

The proof of Proposition 8.3.12 proceeds by showing that (i) for each r less than R, the partial sums of f "converge uniformly" to f on the interval $[-r, r]$, see Definition 8.3.14; and (ii) if a sequence of *continuous* functions converges uniformly to f, then f is continuous. For both clarity and subsequent use, we separate these steps. The proof of Proposition 8.3.12 given below establishes (ii) for the partial sums. The general case of (ii) is Exercise 8.3.5. $\diamond$

Definition 8.3.14. Assume X is a non-empty set of real numbers, $f : X \to \mathbf{R}$ is a function, and (f_n) is a sequence of functions on X. We say (f_n) *converges uniformly to f on X* if $\sup_{x \in X} |f_n(x) - f(x)| \to 0$.

Proposition 8.3.15. *Assume $f(x) = \sum_k a_k x^k$ is a convergent power series with positive radius R. For each natural number n, define*

$$f_n(x) = \sum_{k=0}^{n-1} a_k x^k = a_0 + a_1 x + a_2 x^2 + \cdots + a_{n-1} x^{n-1}.$$

If $0 < r < R$, then the sequence (f_n) converges uniformly to f on $[-r, r]$.

Proof. Fix positive ε and r such that $r < R$ arbitrarily. It suffices to show there exists an N such that if $n \geq N$, then $\sup_{|x| \leq r} |f(x) - f_n(x)| < \varepsilon$.

By Corollary 7.4.10, $\sum_k |a_k r^k|$ converges, so its tails converge to 0; specifically, there exists an N such that if $n \geq N$, then

$$\sum_{k=n}^{\infty} |a_k r^k| \leq \sum_{k=N}^{\infty} |a_k r^k| < \varepsilon/2.$$

But for all x in $[-r, r]$, namely if $|x| \leq r$, we have

$$|f(x) - f_n(x)| = \left| \sum_{k=0}^{\infty} a_k x^k - \sum_{k=0}^{n-1} a_k x^k \right|$$

$$= \left| \sum_{k=n}^{\infty} a_k x^k \right| \leq \sum_{k=n}^{\infty} |a_k x^k|$$

$$\leq \sum_{k=n}^{\infty} |a_k r^k| \leq \sum_{k=N}^{\infty} |a_k r^k| < \varepsilon/2.$$

Consequently, $\sup_{|x| \leq r} |f(x) - f_n(x)| \leq \varepsilon/2 < \varepsilon.$ $\qquad\qquad\square$

Proof of Proposition 8.3.12. Each partial sum f_n of the series for f is a polynomial, hence continuous on $(-R, R)$. If $x \in (-R, R)$, and ε is arbitrary, we will build a closed interval $[-r, r]$ contained in $(-R, R)$ on which $|f - f_N|$ is small, and having x as an interior point; and then a neighborhood $B_\delta(x)$ contained in $[-r, r]$ on which $|f_N(t) - f_N(x)|$ is small. Finally, we will use the triangle inequality to guarantee $|f(t) - f(x)|$ is small in $B_\delta(x)$, Figure 8.5.

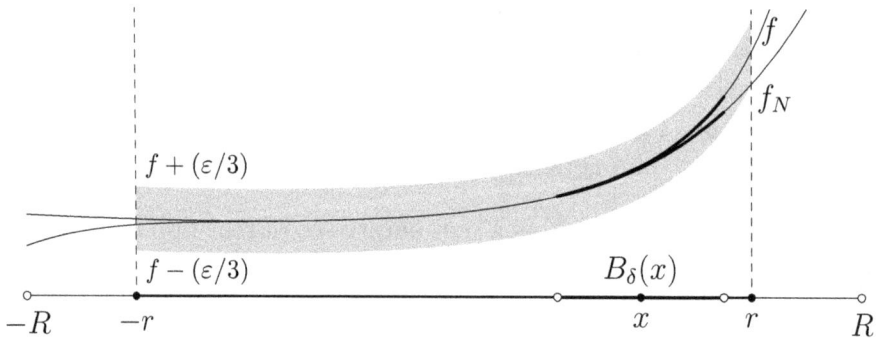

FIGURE 8.5
Approximating a power series by partial sums.

Fix x in $(-R, R)$ and ε arbitrarily. Put $r = \frac{1}{2}(|x| + R)$, so $|x| < r < R$, and use Proposition 8.3.15 to choose N such that $|f - f_N| < \varepsilon/3$ on $[-r, r]$.

Since f_N is continuous at x, there exists a δ such that if $|t - x| < \delta$, then $|f_N(t) - f_N(x)| < \varepsilon/3$. Without loss of generality, we may further assume $\delta < r - |x|$, so that $B_\delta(x) \subseteq [-r, r]$.

By the triangle inequality applied twice (Exercise 3.2.6), if $|t - x| < \delta$, then

$$|f(t) - f(x)| \leq \underbrace{|f(t) - f_N(t)|}_{<\varepsilon/3} + \underbrace{|f_N(t) - f_N(x)|}_{<\varepsilon/3} + \underbrace{|f_N(x) - f(x)|}_{<\varepsilon/3} < \varepsilon.$$

Since ε was arbitrary, f is continuous at x. Since x was an arbitrary point of $(-R, R)$, f is continuous on $(-R, R)$. $\qquad\square$

Corollary 8.3.16. *If $\sum_k a_k(x - x_0)^k$ converges with positive radius R, then for each non-negative integer n,*

$$\sum_{k=0}^{\infty} a_k(x - x_0)^k \approx \sum_{k=0}^{n-1} a_k(x - x_0)^k + O(x - x_0)^n \quad \text{if } x \approx x_0.$$

Proof. By re-indexing, the tail may be written

$$\sum_{k=n}^{\infty} a_k(x - x_0)^k = (x - x_0)^n \sum_{k=0}^{\infty} a_{n+k}(x - x_0)^k.$$

The convergent power series on the right is continuous by Proposition 8.3.12, hence is locally bounded near x_0. $\qquad\square$

Theorem 8.3.17 (The identity theorem for power series). *Assume*

$$f(x) = \sum_{k=0}^{\infty} a_k(x - x_0)^k$$

is a convergent power series with positive radius R. If $f(x) = 0$ for all x in some neighborhood of x_0, then $a_k = 0$ for all k.

Remark 8.3.18. Theorem 8.3.17 guarantees that if two convergent power series centered at x_0 agree on some neighborhood of x_0, they are identical (i.e., have the same coefficients). $\qquad\diamond$

Proof. As usual, assume $x_0 = 0$. We will establish the contrapositive: If some coefficient is non-zero, then $f(x)$ is not identically zero near x_0. Let n be the smallest index such that a_n is non-zero. We have

$$f(x) = a_n x^n + a_{n+1} x^{n+1} + a_{n+2} x^{n+2} + \cdots$$
$$= x^n \left(a_n + a_{n+1} x + a_{n+2} x^2 + \cdots \right) = x^n g(x).$$

The power series $g(x)$ is convergent with radius R. Proposition 8.3.12 implies g is continuous at 0, and $g(0) = a_n$ by direct evaluation. By Corollary 8.2.8, there is an open ball $B_\delta(0)$ on which $g(x) > |a_n|/2$. On the punctured ball $B_\delta^\times(0)$, we have $|f(x)| > |a_n/2| \, |x|^n > 0$. $\qquad\square$

An analytic function expanded in a power series centered at 0 is easily separated into its even and odd parts (Definition 5.2.6).

Proposition 8.3.19. *Assume $R > 0$, and $f(x) = \sum_\ell a_\ell x^\ell$ converges absolutely if $|x| < R$. In the interval $(-R, R)$, the even and odd parts of f are given by the convergent power series*

$$f_{\text{even}}(x) = \sum_{k=0}^{\infty} a_{2k} x^{2k} \qquad = a_0 + a_2 x^2 + a_4 x^4 + \cdots,$$

$$f_{\text{odd}}(x) = \sum_{k=0}^{\infty} a_{2k+1} x^{2k+1} = a_1 x + a_3 x^3 + a_5 x^5 + \cdots,$$

the sums of the even-degree terms and odd-degree terms, respectively.

Proof. Substituting $-x$ for x, we have

$$f(-x) = \sum_{\ell=0}^{\infty} a_\ell (-x)^\ell = \sum_{\ell=0}^{\infty} (-1)^\ell a_\ell x^\ell = a_0 - a_1 x + a_2 x^2 - a_3 x^3 + \cdots.$$

Adding this to the series for $f(x)$, the odd-degree terms cancel and the even-degree terms are doubled. The even part of f is therefore the sum of the even-degree terms of the power series.

Similarly, subtracting the series for $f(-x)$ from the series for $f(x)$, the even-degree terms cancel and the odd-degree terms are doubled. The odd part of f is therefore the sum of the odd-degree terms of the power series. $\qquad\square$

Exercises for Section 8.3

Exercise 8.3.1.($\bigstar$) Suppose $f(x) = x^2$ for all real x.

(a) Show that $f(x_0 + h) \approx x_0^2 + O(h)$ if $h \approx 0$.

(b) Show that $f(x_0 + h) \approx x_0^2 + 2x_0 h + O(h^2)$ if $h \approx 0$.

Exercise 8.3.2. Suppose $f(x) = x^3$ for all real x.

(a) Show that $f(x_0 + h) \approx x_0^3 + O(h)$ if $h \approx 0$.

(b) Show that $f(x_0 + h) \approx x_0^3 + 3x_0^2 h + O(h^2)$ if $h \approx 0$.

Exercise 8.3.3. Repeat Exercise 8.3.2 if n is an integer greater than 3 and $f(x) = x^n$. Hint: Use the binomial theorem.

Exercise 8.3.4. Suppose

$$\left.\begin{array}{l} f(x+h) = f(x) + Ah + o(h), \\ g(x+h) = g(x) + Bh + o(h) \end{array}\right\} \quad \text{if } h \approx 0.$$

(a) Find C so that $f + g \approx f(x) + g(x) + Ch + o(h)$.

(b) Write fg as an approximation in the same form.

(c) Show that if $x \neq 0$, then $\dfrac{1}{x+h} \approx \dfrac{1}{x} - \dfrac{h}{x^2} + O(h^2)$ if $h \approx 0$.

(d) Find $C = C(x, h)$ so that $\dfrac{1}{x+h} \approx \dfrac{1}{x} - \dfrac{h}{x^2} + Ch^2 + O(h^3)$ if $h \approx 0$.

Exercise 8.3.5. Assume $f : [a, b] \to \mathbf{R}$, and (f_k) is a sequence of continuous functions that converges uniformly to f on $[a, b]$. Prove f is continuous.

Exercise 8.3.6. Assume I is an open interval of real numbers, possibly unbounded, f is a function on I, and (f_k) is a sequence of continuous functions on I. Prove that if $(f_k) \to f$ uniformly on $[a, b]$ for every closed, bounded subinterval $[a, b] \subseteq I$, then f is continuous on I.

Exercise 8.3.7. If $f : \mathbf{R} \to \mathbf{R}$ is defined by a germ of infinite radius, prove f is continuous.

Exercise 8.3.8. This exercise describes a construction of the *ternary function* $f : [0, 1] \to [0, 1]$, which is continuous, non-decreasing, and maps the ternary set K onto $[0, 1]$.

Define $f_0(x) = x$ on $[0, 1]$. Inductively, for each component $[a, b]$ of K_n, if $f_n(a) = c$ and $f_n(b) = d$, then $f_{n+1}|_{[a,b]}$ is the piecewise-affine graph through the points (a, c), $\left(\frac{2a+b}{3}, \frac{c+d}{2}\right)$, $\left(\frac{a+2b}{3}, \frac{c+d}{2}\right)$, and (b, d). If $x \notin K_n$, define $f_{n+1}(x) = f_n(x)$. The left graph in Figure 8.6 shows the step from f_0 to f_1. The right graph shows the general effect on a component of K_n.

(a) Prove $f_n : [0, 1] \to [0, 1]$ is affine on each component of K_n, constant on each component of $[0, 1] \setminus K_n$, non-decreasing, continuous, and surjective.

(b) Prove $\sup\{|f_n(x) - f_{n+1}(x)| : 0 \leq x \leq 1\} \leq 2^{-(n+1)}$ for every n. (The optimal bound is $6^{-(n+1)}$, but requires a bit more work to establish. The weaker bound suffices here.) Figure 8.7 shows three consecutive terms.

Conclude that $\sup\{|f_n(x) - f_{n+k}(x)| : 0 \leq x \leq 1\} \leq 2^{-n}$ for every k, and $f(x) = \lim f_n(x)$ exists for all x in $[0, 1]$.

(c) Prove the function f in (b) is continuous (hint: Exercise 8.3.5), non-decreasing, and maps K onto $[0, 1]$.

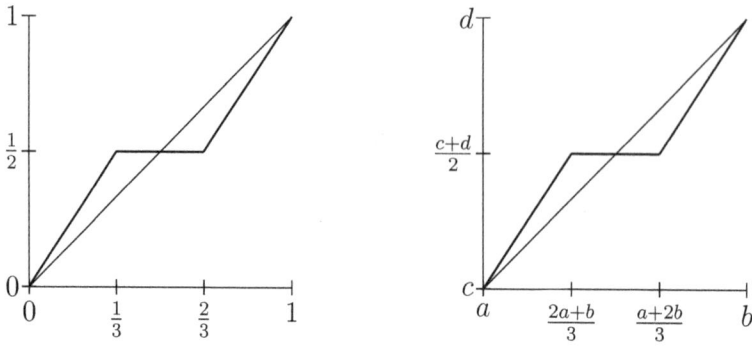

FIGURE 8.6
The iterative step in constructing the ternary function.

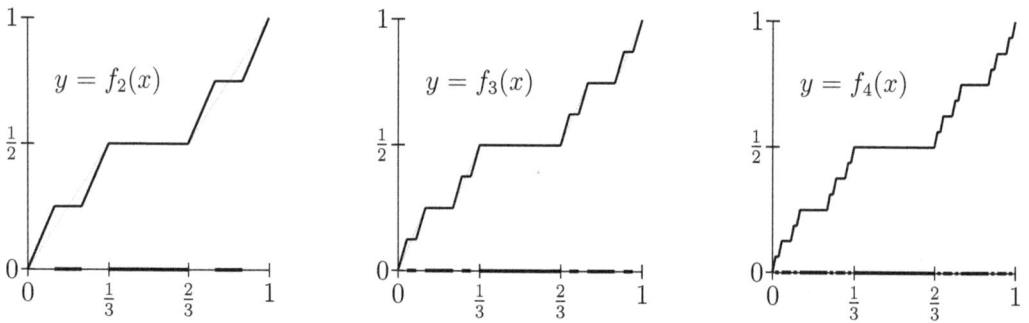

FIGURE 8.7
Successive approximations to the ternary function.

Exercise 8.3.9. This exercise constructs a continuous, nowhere-monotone function. Let $f_0(x) = x$ on $[0,1]$. Inductively, obtain f_{n+1} from f_n by performing the construction indicated in Figure 8.8 on each maximal affine segment of the graph:

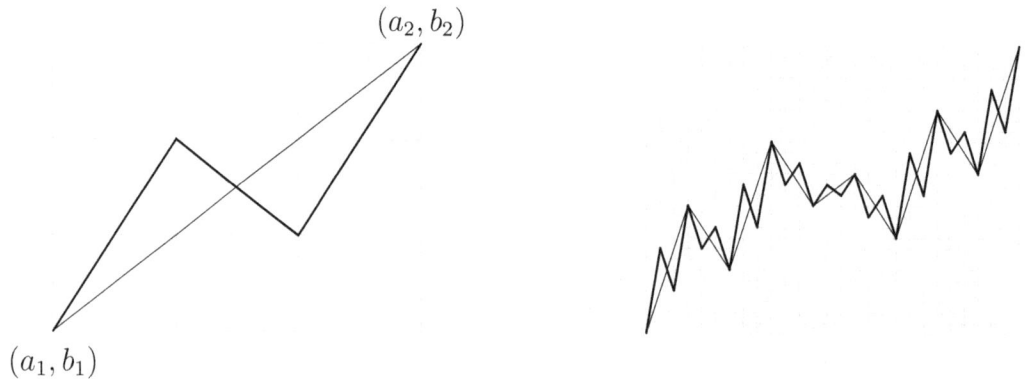

FIGURE 8.8
Constructing a nowhere-monotone function.

(a) Prove (f_n) converges to a continuous function f on $[0, 1]$. Exercise 8.3.5 may be helpful.

(b) If n and k are non-negative integers, and if $0 \le m \le 3^n$, the functions f_n and f_{n+k} (and therefore f_n and f) agree at every point of the form $a + m \cdot 3^{-n}(b - a)$.

(c) If $n \ge 0$ and $0 \le m < 3^n$, then the function f_{n+1} is not monotone on $[a + m \cdot 3^{-n}(b - a), a + (m + 1) \cdot 3^{-n}(b - a)]$. Conclude f is non-monotone on every interval.

8.4 The Intermediate Value Theorem

Theorem 8.4.1 (The intermediate value theorem). *Assume $f : [a, b] \to \mathbf{R}$ is a continuous function such that $f(a) \ne f(b)$. If y_0 is a real number between $f(a)$ and $f(b)$, then there exists an x_0 in (a, b) such that $f(x_0) = y_0$.*

Proof. Without loss of generality we may assume $y_0 = 0$ and $f(a) < 0 < f(b)$: For either choice of sign, the function g defined by $g(x) = \pm\big(f(x) - y_0\big)$ is continuous, and satisfies $g(x_0) = 0$ if and only if $f(x_0) = y_0$. For one choice of sign, $g(a) < 0 < g(b)$.

Let $A = \{x \text{ in } [a, b] : f(x) < 0\}$ be the set of points at which $f(x)$ is negative. Since $f(a) < 0$, we have $a \in A$; consequently, A is a non-empty set of real numbers that is bounded above (by b). Define $x_0 = \sup A$. It suffices to prove $x_0 \in (a, b)$ and $f(x_0) = 0$.

By Corollary 8.2.8, if $f(x) < 0$, there exists an open interval centered at x on which f is negative, and if $f(x) > 0$, there exists an open interval centered at x on which f is positive. In particular, there exists a positive δ_1 such that f is negative on $(a, a + \delta_1)$ (so $a < a + \delta_1 \le x_0$), and there exists a positive δ_2 such that f is positive on $(b - \delta_2, b)$ (so $x_0 \le b - \delta_2 < b$). This completes the proof that $a < x_0 < b$.

By trichotomy, exactly one of the following alternatives holds: $f(x_0) < 0$, $f(x_0) > 0$, or $f(x_0) = 0$. It suffices to prove the first two are false.

If $f(x_0)$ is negative, Corollary 8.2.8 implies there is a δ such that $f(x) < 0$ if $x_0 < x < x_0 + \delta$; this means x_0 is not an upper bound of A. Since $x_0 = \sup A$ *is* an upper bound of A, $f(x_0)$ is not negative.

If $f(x_0)$ is positive, Corollary 8.2.8 implies there is a δ such that $f(x) > 0$ if $x_0 - \delta < x < x_0$; this means $x_0 - \delta < x_0$ is an upper bound of A. Since $x_0 = \sup A$ is the *least* upper bound of A, $f(x_0)$ is not positive. $\square$

Remark 8.4.2. The same conclusion holds if $f(b) < 0 < f(a)$: Apply the theorem to the continuous function $g(x) = -f(x)$, which satisfies the condition $g(a) < 0 < g(b)$.   ◇

Remark 8.4.3. The intermediate value theorem guarantees that the image of a closed, bounded interval $[a, b]$ under a continuous function f is itself an interval. (In Section 8.5, see Theorem 8.5.1, we will prove the image is actually a closed, bounded interval.)   ◇

Example 8.4.4. The rational function $f(x) = 1/x$, defined if $x \neq 0$, is continuous on $\mathbf{R} \setminus \{0\}$. Moreover, $f(-1) = -1 < 0$ and $f(1) = 1 > 0$. However, $f(x) \neq 0$ for all x. This does not contradict the intermediate value theorem because the domain is not an interval.

In accord with the intermediate value theorem, the sign of f does not change on any interval contained in the domain of f.   ◇

Existence of Real Roots

Proposition 8.4.5. *If c is a positive real number and n a positive integer, there exists a unique positive real number x_0 such that $x_0^n = c$.*

Proof. Let $f : \mathbf{R} \to \mathbf{R}$ be the function $f(x) = x^n$. We have $f(0) = 0 < c$, while by Proposition 3.3.5,

$$f(1 + c) = (1 + c)^n \geq 1 + nc \geq 1 + c > c.$$

The intermediate value theorem applied to f on the interval $[0, 1+c]$ guarantees there exists a real number x_0 such that $0 < x_0 < 1 + c$ and $x_0^n = c$.

There cannot be more than one such number: By Lemma 5.3.16, f is strictly increasing, hence injective.   ☐

Definition 8.4.6. A number x satisfying $x^n = c$ is called an *nth root* of c. The unique positive nth root of a positive real number c is customarily denoted $\sqrt[n]{c}$ or $c^{1/n}$.

The number $\sqrt{c} = c^{1/2}$ is called the *square root* of c. The number $\sqrt[3]{c} = c^{1/3}$ is called the *cube root* of c.

Proposition 8.4.7. *If n is a positive integer, the nth root function, defined by $f(x) = x^{1/n}$ and viewed as a mapping $f : [0, \infty) \to [0, \infty)$, is bijective and continuous.*

Proof. The nth power function $g(x) = x^n$ is the inverse of f, so f is bijective.

Since g is strictly increasing, Proposition 8.2.11 implies $f = g^{-1}$ is continuous on $(0, \infty)$. To establish continuity at $x_0 = 0$, fix ε arbitrarily, put $\delta = \varepsilon^n$, and note that if $0 \leq x < \delta$, then $0 \leq x^{1/n} < \varepsilon$.   ☐

We can now extend the definition of exponentiation to rational exponents.

Definition 8.4.8. Assume x is a positive real. If m and n are integers and $n \neq 0$, we define
$$x^{m/n} = (x^{1/n})^m = (x^m)^{1/n},$$
the unique positive real number whose nth power is x^m.

Proposition 8.4.9. *Assume x and y are positive real numbers. If r and s are rational, then*

$$(xy)^r = x^r \cdot y^r, \qquad x^{r+s} = x^r \cdot x^s, \qquad x^{rs} = (x^r)^s.$$

Proof. Exercise 8.4.1. $\qquad\qquad\qquad\qquad\qquad\qquad\qquad\qquad\qquad\qquad$ $\square$

Remark 8.4.10. See Exercise 8.4.2 for the situation if $x \leq 0$. One point deserves emphasis: If $x < 0$, we insist $r = m/n$ is in lowest terms before we write x^r. For example, $1/3 = 2/6$ as rational numbers, but while $(-1)^{1/3} = -1$ unambiguously, $(-1)^{2/6}$ is multiply problematic: $[(-1)^2]^{1/6} = 1^{1/6} = 1$ is the wrong value, and $[(-1)^{1/6}]^2$ is not defined. $\qquad\qquad\qquad\qquad\qquad$ $\diamond$

Exercises for Section 8.4

Exercise 8.4.1. (H) Prove Proposition 8.4.9.

Exercise 8.4.2. If r is rational and positive, we define $0^r = 0$. If $x < 0$ and $r = m/n$ *in lowest terms, with n odd,* we define $x^r = (-1)^r |x|^r = (-1)^m |x|^r$. We leave $x^{m/n}$ *undefined* if n is even or if m and n have a common factor.

　　Does Proposition 8.4.9 hold for all non-zero x, y and all rational r and s with odd denominator? Explain why we do not define 0^r for negative r.

Exercise 8.4.3. ($\bigstar$) Assume $\rho > 0$. Prove that $\rho^{1/n} = \sqrt[n]{\rho} \to 1$ as $n \to \infty$. Suggestion: First assume $\rho \geq 1$.

Exercise 8.4.4. Use Exercise 3.3.3 to prove $n^{1/n} = \sqrt[n]{n} \to 1$ as $n \to \infty$.

Exercise 8.4.5. Assume y is an arbitrary real number. Determine how many real numbers x satisfy $2x/(1 + x^2) = y$. (The answer depends on y.)

Exercise 8.4.6. ($\bigstar$) Suppose $f : [-2, 2] \to \mathbf{R}$ is a continuous function satisfying $f(-2) = -5$, $f(-1) = -7$, $f(0) = 2$, $f(1) = -3$, and $f(2) = 0$. Based on this information, what is the smallest possible number of real solutions of $f(x) = 0$? Of $f(x) = -6$?

Exercise 8.4.7. (★) Prove $f(x) = x^5 + 3x + 1$ defines a bijection $f : \mathbf{R} \to \mathbf{R}$.

Exercise 8.4.8. Define $p : \mathbf{R} \to \mathbf{R}$ by $p(x) = x^5 - 5x$.

(a) Prove p is surjective.

(b) Assume p is strictly monotone on the intervals $(-\infty, -1]$, $[-1, 1]$, and $[1, \infty)$. If y is an arbitrary real number, determine how many real solutions x the equation $y = p(x)$ has. (The answer depends on y.)

Exercise 8.4.9. Assume

$$p(x) = a_0 + a_1 x + a_2 x^2 + \cdots + a_n x^n,$$
$$q(x) = b_0 + b_1 x + b_2 x^2 + \cdots + b_m x^m,$$

are polynomial functions, $a_n \neq 0$ and $b_m \neq 0$, and consider the rational function $f(x) = p(x)/q(x)$, defined on its natural domain. Prove:

(a) If $n < m$, then $f(x) \to 0$ as $|x| \to \infty$.

(b) If $n = m$, then $f(x) \to a_n/b_m$ as $|x| \to \infty$.

(c) If $n > m$, then $f(x) \to \mathrm{sgn}(a_n/b_m) \cdot \infty$ as $x \to \infty$.

Exercise 8.4.10. (★) Assume y_0 is a real number and n a positive integer.

(a) Prove that y_0 has at most two real nth roots. That is, the equation $x^n = y_0$ has at most two real solutions.

(b) Prove that if n is odd, then $f(x) = x^n$ is a bijection from $\mathbf{R}$ to $\mathbf{R}$.

Exercise 8.4.11. (H) Suppose $f : [a, b] \to [a, b]$ is a continuous mapping. Prove that f has a fixed point: There exists an x_0 in $[a, b]$ such that $f(x_0) = x_0$.

Exercise 8.4.12. If I is an interval and $f : I \to \mathbf{Q}$ a continuous function, prove f is constant.

Exercise 8.4.13. Assume I is a real interval and $f : I \to \mathbf{R}$ is continuous and injective. Prove f is strictly monotone.

Exercise 8.4.14. Assume I is a real interval and $f : I \to f(I)$ a bijection such that $f^{-1} = 1/f$, compare Exercise 5.3.6. Prove f is discontinuous.

8.5   The Extreme Value Theorem

Theorem 8.5.1 (The extreme value theorem). *If $f : [a, b] \to \mathbf{R}$ is a continuous function, then there exist points $x_{\min}$ and $x_{\max}$ in $[a, b]$ such that*

$$f(x_{\min}) \leq f(x) \leq f(x_{\max}) \quad \text{for all } x \text{ in } [a, b].$$

Proof. (Continuous implies bounded). Assume contrapositively that f is unbounded on $[a, b]$. We will prove there exists a point x_∞ at which f is discontinuous.

Since f is unbounded, for every natural number n, there exists a point x_n in $[a, b]$ such that $|f(x_n)| \geq n$. Choosing one such point for each natural number n gives a sequence (x_n) in $[a, b]$. The convergent subsequence theorem implies there exists a subsequence $(x_{\nu(k)})$ that converges to some real number x_∞. By construction, $a \leq x_{\nu(k)} \leq b$ for all k, so Proposition 6.3.7 implies $x_\infty \in [a, b]$. Further, for each positive integer k we have

$$|f(x_{\nu(k)})| \geq \nu(k) \geq k.$$

Particularly, the image sequence $y_k = f(x_{\nu(k)})$ is unbounded, and hence divergent. Since $(x_{\nu(k)}) \to x_\infty$ but the image sequence is not convergent, f is discontinuous at x_∞.

(Extrema are achieved). Since f is bounded, the image of f has a supremum β and an infimum α. It remains to prove there exist points $x_{\min}$ and $x_{\max}$ such that $f(x_{\min}) = \alpha$ and $f(x_{\max}) = \beta$.

If $k \geq 1$, then $\beta - (1/k) < \beta$, so $\beta - (1/k)$ is not an upper bound of the image of f: For each positive integer k, there exists a point x_k in $[a, b]$ such that $\beta - (1/k) < f(x_k) \leq \beta$. By the convergent subsequence theorem, (x_k) has a subsequence $(x_{\nu(k)})$ converging to some point $x_{\max}$ in $[a, b]$. Since

$$\beta - (1/k) \leq \beta - (1/\nu(k)) < f(x_{\nu(k)}) \leq \beta$$

for every k, continuity of f implies $f(x_{\max}) = \lim_k f(x_{\nu(k)}) = \beta$.

Achievement of the infimum α is proven entirely similarly. $\square$

Example 8.5.2. The function $f : [1, \infty) \to \mathbf{R}$ defined by $f(x) = 1/x$ is continuous throughout its domain and is bounded, but does not achieve a minimum. This does not contradict the extreme value theorem, because the domain of f is not a bounded interval. $\Diamond$

Example 8.5.3. The function $f : (-1, 1) \to \mathbf{R}$ defined by $f(x) = x/(1 - x^2)$ is continuous, but unbounded both above and below. This does not contradict the extreme value theorem because $(-1, 1)$ is not a *closed* interval. $\Diamond$

Limits at Infinity

Definition 8.5.4. Assume X is unbounded above, $f : X \to \mathbf{R}$ is a function, and L is a real number. If $\big(f(x_k)\big) \to L$ for every sequence (x_k) in X such that $(x_k) \to \infty$, we say L is a *limit of* f *at* ∞ and write $L = \lim(f, \infty) = \lim\limits_{x \to \infty} f(x)$.

Remark 8.5.5. If $L = \lim(f, \infty)$, we also write $f(x) \to L$ as $x \to \infty$. ◇

Remark 8.5.6. A limit at infinity is unique (if it exists). There is an analogous definition for the limit of f at $-\infty$ if the domain of f is unbounded below. ◇

Remark 8.5.7. If $L = \pm\infty$, the sequential condition in Definition 8.5.4 formally makes sense. We must remember, however, that "$f(x) \to \infty$," say, means "f diverges to ∞," and does *not* mean f has a limit at ∞. Particularly, it is best not to write "$\lim(f, \infty) = \infty$" and the like. ◇

Propositions 6.2.1 and 6.3.13 immediately imply corresponding properties for functional limits:

Proposition 8.5.8. *Assume X is unbounded above, and that f and g are functions on X.*

If $\lim(f, \infty)$ and $\lim(g, \infty)$ exist, then

(i) $\lim(f \pm g, \infty) = \lim(f, \infty) \pm \lim(g, \infty)$.

(ii) $\lim(fg, \infty) = \lim(f, \infty) \cdot \lim(g, \infty)$.

(iii) *If $\lim(g, \infty) \neq 0$, then $\lim(f/g, \infty) = \lim(f, \infty)/\lim(g, \infty)$.*

If $f \to \infty$ and $\lim(g, \infty) = L$ is finite, then

(iv) $f + g \to \infty$.

(v) *If $\pm L > 0$, then $fg \to \pm\infty$.*

(vi) *If $f(x) \neq 0$ for all x in X, then $\lim(g/f, \infty) = 0$.*

The analogous assertions hold at $-\infty$.

Example 8.5.9. If p is a *monic* polynomial, namely the degree is at least 1 and the leading coefficient is 1, then $p(x) \to \infty$ as $x \to \infty$, and $p(x) \to (-1)^n \infty$ as $x \to -\infty$. To prove this, write (if $x \neq 0$)

$$p(x) = x^n + a_{n-1}x^{n-1} + \cdots + a_1 x + a_0$$
$$= x^n(1 + a_{n-1}/x + \cdots + a_1/x^{n-1} + a_0/x^n).$$

Inside the parentheses, each summand except the first approaches 0 as $|x| \to \infty$, so the terms in parentheses approach 1. The assertions follow from Proposition 8.5.8 (v). ◇

Exercises for Section 8.5

Exercise 8.5.1.(★) Assume p is a polynomial of degree at most 3, and put $f(x) = p(x)/(1 + x^4)$. If $p(5) = 4.2$, prove f achieves an absolute maximum.

Exercise 8.5.2. Prove the function $p(x) = x^4 - 7x^3 + 5x^2 - \sqrt{3}x + 12$ achieves an absolute minimum on $\mathbf{R}$. Hint: First show that there exists a positive R so that if $|x| > R$, then $|p(x)| > 12$. Then use the extreme value theorem on $[-R, R]$ to finish the proof.

Exercise 8.5.3. Assume p is a non-constant polynomial of even degree whose top-degree coefficient is positive. Prove p achieves an absolute minimum on $\mathbf{R}$.

Exercise 8.5.4. Assume $p : \mathbf{R} \to \mathbf{R}$ is a polynomial function. Prove:

(a) If $\deg p$ is odd, then p is surjective.

(b) If $\deg p$ is even, then p has an absolute extremum.

Exercise 8.5.5. Fix n in $\mathbf{Z}^+$, and define $C = \{(x, y) \text{ in } \mathbf{R}^2 : x^n + y^n = 1\}$.

(a) Prove that the portion of C in the first quadrant is bounded.

(b) Under what condition on n is C bounded?

Exercise 8.5.6. Suppose $f : [a, b] \to \mathbf{R}$ is continuous. Prove that the image $f([a, b])$ is a closed, bounded interval.

Exercise 8.5.7.(★) Suppose that I is a half-open interval, and $f : I \to \mathbf{R}$ is continuous.

(a) Does it follow that f has *either* an absolute maximum or an absolute minimum in I?

(b) Does it follow that f is *either* bounded above or bounded below in I?

8.6   Discrete Dynamical Systems

Definition 8.6.1. Assume $f : X \to X$ is a function, so that f is composable with itself. We recursively define the (forward) *iterates* of f by

$$f^{[0]} = \iota, \qquad f^{[k+1]} = f \circ f^{[k]} \quad \text{if } k \geq 0.$$

That is, $f^{[1]} = f$, $f^{[2]} = f \circ f$, $f^{[3]} = f \circ f \circ f$, etc.
  The pair (X, f) is called a *discrete dynamical system*.

Example 8.6.2. If $f(x) = x^2 - 1$, then

$$f^{[2]}(x) = f(x)^2 - 1 = (x^2 - 1)^2 - 1 = x^4 - 2x^2,$$

$$f^{[3]}(x) = \left(f^{[2]}(x)\right)^2 - 1 = (x^4 - 2x^2)^2 - 1 = x^8 - 4x^6 + 4x^4 - 1,$$

$$f^{[4]}(x) = \left(f^{[3]}(x)\right)^2 - 1 = (x^8 - 4x^6 + 4x^4 - 1)^2 - 1,$$

and so forth.   ◇

Definition 8.6.3. If (X, f) is a discrete dynamical then for each "seed" or "initial value" x_0 in X, there is a *recursive sequence* (x_k) defined by

$$x_{k+1} = f(x_k) \quad \text{if } k \geq 0.$$

Figure 8.9 depicts the geometry: Starting at x_0 we alternately travel up or down to the graph of f, then left or right to the diagonal.

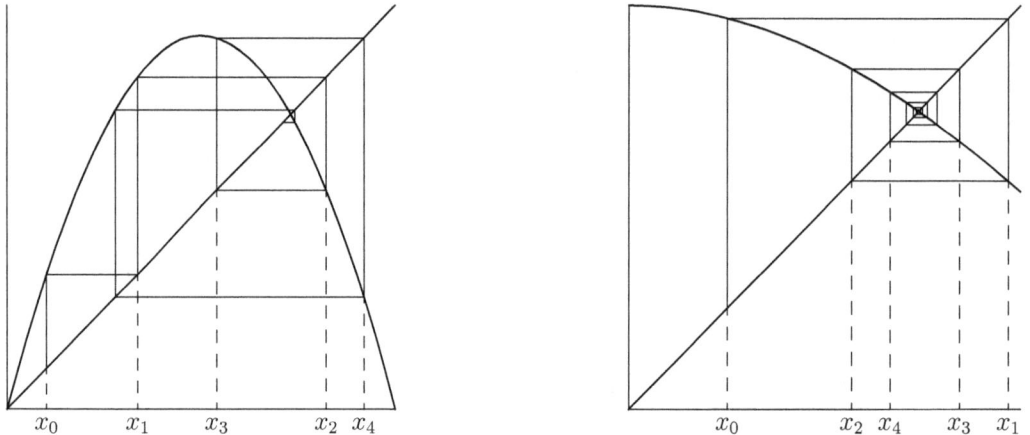

FIGURE 8.9
Iterating a function geometrically.

Proposition 8.6.4. *Assume X is non-empty, and $f : X \to X$ is continuous. If a recursive sequence converges to a point x_∞ in X, then $f(x_\infty) = x_\infty$.*

Proof. Since f is continuous,

$$f(x_\infty) = f\left(\lim_{k \to \infty} x_k\right) = \lim_{k \to \infty} f(x_k) = \lim_{k \to \infty} x_{k+1} = x_\infty. \qquad \square$$

Definition 8.6.5. Assume $f : X \to X$. A point x such that $f(x) = x$ is called a *fixed point* of f.

Remark 8.6.6. The proposition says that if a recursive sequence converges, the limit is a fixed point of f. This deceptively simple concept is connected to a variety of subtle mathematical problems.   ◇

Remark 8.6.7. Dynamical systems comprise an entire area of mathematics, whose surface this book barely scratches. Even "simple" mappings f can lead to "chaotic" behavior. Think of a kitchen mixer, whose blades move regularly in circles, but whose action quickly and thoroughly scrambles colloidal regions of space; or of plane fractals obtained by iterating a complex quadratic polynomial. ◇

Example 8.6.8 (Approximating square roots). Assume $c > 1$, and define the function $f : (0, \infty) \to (0, \infty)$ by

$$f(x) = \frac{1}{2}\left[x + \frac{c}{x}\right] = \frac{x^2 + c}{2x}.$$

Pick a seed x_0 satisfying $x_0^2 > c$, such as $x_0 = c$. If $k \geq 0$, algebra gives

$$x_{k+1}^2 - c = \frac{(x_k^2 + c)^2}{(2x_k)^2} - c$$

$$= \frac{(x_k^4 + 2cx_k^2 + c^2) - 4cx_k^2}{(2x_k)^2}$$

$$= \frac{x_k^4 - 2cx_k^2 + c^2}{(2x_k)^2} = \frac{(x_k^2 - c)^2}{(2x_k)^2},$$

which is positive by induction since $x_0^2 - c > 0$. Consequently,

$$x_k - x_{k+1} = x_k - \frac{1}{2}\left[x_k + \frac{c}{x_k}\right] = \frac{1}{2}\left[x_k - \frac{c}{x_k}\right] = \frac{x_k^2 - c}{2x_k} > 0.$$

We have shown that the recursive sequence (x_k) is positive (hence bounded below) and strictly decreasing, so (x_k) converges to a non-negative limit x_∞. By Proposition 8.6.4, we have

$$x_\infty = f(x_\infty) = \frac{x_\infty^2 + c}{2x_\infty}.$$

Rearranging, $x_\infty^2 = c$, so $x_\infty = \sqrt{c}$ since the limit is non-negative.

The convergence is remarkably rapid. For clarity, introduce $e_k := x_k^2 - c$. Since $\sqrt{c} < x_k$ for all k,

$$|x_k - \sqrt{c}| = x_k - \sqrt{c} < \frac{(x_k + \sqrt{c})(x_k - \sqrt{c})}{2\sqrt{c}} = \frac{x_k^2 - c}{2\sqrt{c}} = \frac{e_k}{2\sqrt{c}}.$$

As shown above,

$$e_{k+1} = x_{k+1}^2 - c = \left[\frac{x_k^2 - c}{2x_k}\right]^2 = \frac{e_k^2}{4x_k^2} < \frac{e_k^2}{4c}.$$

That is, the *number of decimals of accuracy more than doubles* with each step of the iteration. ◇

Exercises for Section 8.6

Exercise 8.6.1. If I is an interval, we say a function $T : I \to I$ is a *contraction* if there exists a positive real number λ (lambda) less than 1 such that if x, x' are elements of I, then $|T(x) - T(x')| \le \lambda|x - x'|$.

(a) Prove that a contraction has at most one fixed point.

(b) Assume x_0 is an element of I, and (x_k) is the recursive sequence with seed x_0. Prove that $|x_{k+1} - x_k| \le \lambda^k |x_1 - x_0|$ for every natural number k.

(c) Prove that for all non-negative k and all positive m,

$$|x_{k+m} - x_k| \le |x_1 - x_0| \frac{\lambda^k}{1 - \lambda}.$$

Conclude that (x_k) is condensing.

(d) If I is closed, prove a contraction on I has a fixed point.

Exercise 8.6.2. (★) Define $f : (1, \infty) \to (1, \infty)$ by $f(x) = x + (1/x)$.

(a) Prove that $|f(x) - f(x')| < |x - x'|$ for all positive real numbers x and x'. Is f a contraction on $(1, \infty)$?

(b) Assume (x_k) is the recursive sequence defined by f and having seed $x_0 = 2$. Is (x_k) condensing?

Exercise 8.6.3. Fix a positive real b, and define $f(x) = \sqrt{b + x}$ if $x \ge 0$.

(a) There is a unique positive fixed point of f; find it (in terms of b).

(b) For which b is f a contraction on $(0, \infty)$?
Suggestion: Multiply and divide by the conjugate expression.

(c) Assume x_0 is a positive seed for the recursive sequence (x_k) generated by f. Prove that (x_k) converges to the fixed point (whether or not f is a contraction).

(d) Evaluate $\sqrt{1 + \sqrt{1 + \sqrt{1 + \dots}}}$ and $\sqrt{2 + \sqrt{2 + \sqrt{2 + \dots}}}$.

Exercise 8.6.4. Assume $a > 0$, and define the function $f_a : [0, \infty) \to [0, \infty)$ by $f_a(x) = 1/(a + x)$.

(a) Prove $f_a^{[2]} = f_a \circ f_a$ is a contraction, and evaluate the continued fraction $[0, a, a, a, \dots]$, see Definition 3.1.18 and Exercise 6.4.9.

(b) Prove f_a is a contraction if and only if $a > 1$.

Exercise 8.6.5. Assume I is an interval and $f : I \to I$ a function. A fixed point x_∞ of f is *repelling* if there exists a positive δ and a λ greater than 1 such that if $|x - x_\infty| < \delta$, then $|f(x) - f(x_\infty)| \geq \lambda |x - x_\infty|$.

Suppose $x_0 \in I$. Prove that if the sequence (x_k) of iterates, $x_{k+1} = f(x_k)$, converges to x_∞, then $x_N = x_\infty$ for some N, and the sequence is eventually constant.

Exercise 8.6.6. (The logistic mapping.) If $0 \leq k \leq 4$, prove $f_k(x) = kx(1-x)$ maps $[0, 1]$ to itself. If $k > 1$, prove f_k has two fixed points: One repelling (Exercise 8.6.5) if $k > 1$ and both repelling if $k > 3$.

Exercise 8.6.7. Take $c = 3$ and $x_0 = 2$ in Example 8.6.8. Find the next three terms of the sequence. Using $1.7 < \sqrt{3} < 1.8$, find the smallest integer k such that $|x_k - \sqrt{3}| < 0.5 \times 10^{-8}$.

Exercise 8.6.8. Take $c = 5$ and $x_0 = 9/4$ in Example 8.6.8. Find the next two terms of the sequence. Using $2.2 < \sqrt{5} < 2.3$, what upper bound on $|x_2 - \sqrt{5}|$ does Example 8.6.8 guarantee?

9

Integration

Suppose f is a real-valued function on a closed, bounded interval $[a, b]$. In analysis, "integration" refers to a procedure that defines the signed area enclosed by the graph of f. There are multiple ways to define an integral, though all agree when f is continuous. Generally, the length of the definition correlates with the generality of functions that can be integrated. The definition here is simple, but suffices for the needs of this book.

9.1 Integrability

Recall that a *splitting* of the closed interval $I = [a, b]$ is a finite set $\Pi \subseteq [a, b]$ that contains both endpoints, Definition 4.1.15. Every splitting may be written uniquely as $\{t_i\}_{i=0}^{n}$ for some positive integer n, with $t_0 = a$, $t_n = b$, and $t_i < t_{i+1}$ if $0 \leq i < n$. The *ith piece* of Π, $I_i = [t_i, t_{i+1}]$, has length $\Delta t_i = t_{i+1} - t_i$.

Lower and Upper Sums

Definition 9.1.1. If Π and Π' are splittings of I, we say Π' is a *refinement* of Π if $\Pi \subseteq \Pi'$.

Remark 9.1.2. A refinement of Π is obtained by adding finitely many points to Π. If Π and Π' are splittings of I, their union $\Pi'' = \Pi \cup \Pi'$ is a "common refinement," namely, a refinement of each. ◇

Example 9.1.3. If n and n' are positive integers, an equal-length splitting $\Pi_{n'}$ is a refinement of Π_n if and only if n divides n'. ◇

Definition 9.1.4. Assume $f : [a, b] \to \mathbf{R}$ is a *bounded* function and Π is a splitting of $[a, b]$ with n pieces. For each i, put

$$m_i = \inf\{f(t) : t \in I_i\}, \qquad M_i = \sup\{f(t) : t \in I_i\}.$$

We define the *lower sum* and *upper sum* of f with respect to Π by

$$L(f, \Pi) = \sum_{i=0}^{n-1} m_i \, \Delta t_i, \qquad U(f, \Pi) = \sum_{i=0}^{n-1} M_i \, \Delta t_i.$$

DOI: 10.1201/9781003601357-9

Lemma 9.1.5. *Assume* $f : [a, b] \to \mathbf{R}$ *is a bounded function. If* Π *and* Π' *are splittings of* $[a, b]$ *and* Π' *is a refinement of* Π, *then*

$$L(f, \Pi) \leq L(f, \Pi') \leq U(f, \Pi') \leq U(f, \Pi).$$

Proof. Since every refinement of Π is obtained by appending finitely many points to Π, induction on the number of points in $\Pi' \setminus \Pi$ reduces the claim to the case $\Pi' = \Pi \cup \{z\}$ for some z.

Assume $z \in I_j$ for definiteness, and write $I_j = [t_j, z] \cup [z, t_{j+1}]$. As in Figure 9.1, let m'_j and m''_j denote the respective infima of f on the pieces $I'_j = [t_j, z]$ and $I''_j = [z, t_{j+1}]$. Since each piece is contained in I_j, we have $m_j \leq m'_j$ and $m_j \leq m''_j$ by constriction, Lemma 4.2.10. Consequently,

$$m_j \Delta t_j = m_j \left(\Delta t'_j + \Delta t''_j \right) \leq m'_j \Delta t'_j + m''_j \Delta t''_j.$$

Since the lower sums $L(f, \Pi)$ and $L(f, \Pi')$ have identical summands if $i \neq j$, we have $L(f, \Pi) \leq L(f, \Pi')$.

A completely analogous argument shows $U(f, \Pi') \leq U(f, \Pi)$. ☐

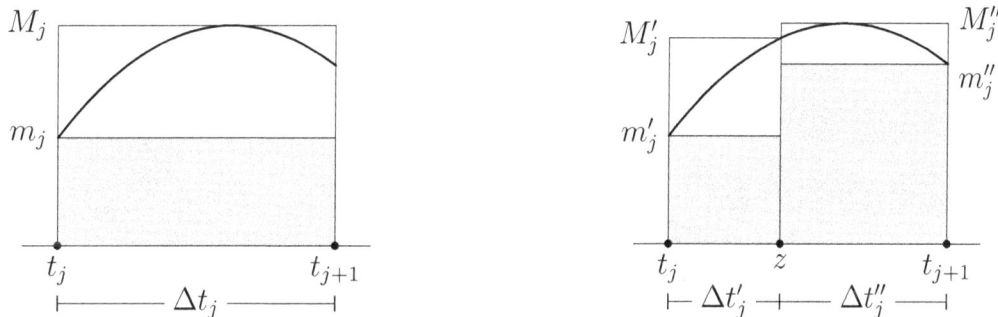

FIGURE 9.1
The effect of refinement on upper and lower sums.

Proposition 9.1.6. *Assume* $f : [a, b] \to \mathbf{R}$ *is a bounded function. If* Π *and* Π' *are arbitrary splittings of* $[a, b]$, *then* $L(f, \Pi) \leq U(f, \Pi')$.

Proof. The splitting $\Pi'' = \Pi \cup \Pi'$ is a refinement of both Π and Π'. By Lemma 9.1.5, $L(f, \Pi) \leq L(f, \Pi'') \leq U(f, \Pi'') \leq U(f, \Pi')$. ☐

Remark 9.1.7. Proposition 9.1.6 guarantees every lower sum of f is bounded above by every upper sum. Particularly, the set of lower sums of f,

$$\mathbf{L}(f, [a, b]) = \{L(f, \Pi) : \Pi \text{ a splitting of } [a, b]\},$$

is non-empty and bounded above, and the set of upper sums of f,

$$\mathbf{U}(f, [a, b]) = \{U(f, \Pi) : \Pi \text{ a splitting of } [a, b]\},$$

is non-empty and bounded below. ◇

Definition 9.1.8. Assume $f : [a, b] \to \mathbf{R}$ is bounded. The *lower integral* of f on $[a, b]$ is the real number $L(f, [a, b]) = \sup \mathbf{L}(f, [a, b])$. The *upper integral* of f on $[a, b]$ is the real number $U(f, [a, b]) = \inf \mathbf{U}(f, [a, b])$.

If $L(f, [a, b]) = U(f, [a, b])$, we say f is *integrable* on $[a, b]$, and call

$$L(f, [a, b]) = U(f, [a, b]) = \int_a^b f = \int_a^b f(t) \, dt$$

the *integral* of f over $[a, b]$.

Proposition 9.1.9. *A bounded function $f : [a, b] \to \mathbf{R}$ is integrable on $[a, b]$ if and only if the following condition holds:*

> *For every ε,*
>> *there exists a splitting Π of $[a, b]$ such that*
>>> $U(f, \Pi) - L(f, \Pi) < \varepsilon.$

Remark 9.1.10. Geometrically, we can make the shaded area in Figure 9.2 as small as we like for some splitting.   ◇

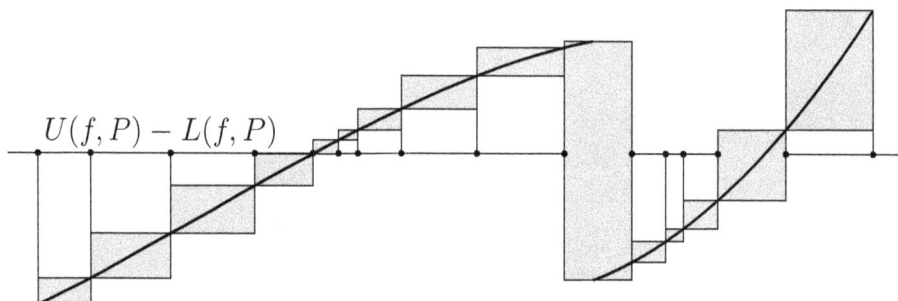

FIGURE 9.2
The difference between an upper and lower sum.

Proof. If f is integrable and $\varepsilon > 0$, then by definition of supremum and infimum there exist splittings Π' and Π'' such that

$$U(f, \Pi'') - \left[\int_a^b f \right] < \varepsilon/2, \qquad \left[\int_a^b f \right] - L(f, \Pi') < \varepsilon/2.$$

If $\Pi = \Pi' \cup \Pi''$, then $L(f, \Pi') \leq L(f, \Pi) \leq U(f, \Pi) \leq U(f, \Pi'')$, so

$$U(f, \Pi) - L(f, \Pi) \leq U(f, \Pi'') - L(f, \Pi') < \varepsilon.$$

Inversely, if f is *not* integrable and $\varepsilon = \inf \mathbf{U} - \sup \mathbf{L}$, then we have $U(f, \Pi) - L(f, \Pi) \geq \varepsilon > 0$ for every splitting Π.   □

Remark 9.1.11. Any particular upper or lower sum may be viewed as an approximation of f by a step function. ◇

We next compute several examples from the definition.

Example 9.1.12. If f is a constant function, say $f(t) = c$ for all t, then for every splitting of $[a, b]$ and for every piece of the splitting, $m_i = c = M_i$. Consequently, every lower sum and every upper sum is equal to $c(b - a)$, so

$$\int_a^b f = \int_a^b c\,dt = c(b - a).$$ ◇

Example 9.1.13. Assume $a < x_0 < b$, and $f = \chi_{\{x_0\}}$, namely, $f(t) = 1$ if $t = x_0$ and $f(t) = 0$ otherwise. We will prove the function f is integrable on $[a, b]$, and the integral is equal to 0.

Fix ε arbitrarily. We seek a splitting Π of the form

$$t_0 = a, \qquad t_1 = x_0 - \delta, \qquad t_2 = x_0 + \delta, \qquad t_3 = b$$

for which $U(f, \Pi) - L(f, \Pi) < \varepsilon$, Figure 9.3. It suffices to arrange that $a < x_0 - \delta$, $x_0 + \delta < b$, and $2\delta < \varepsilon$: Under these conditions, we have

i	m_i	M_i	Δt_i	$(M_i - m_i)\,\Delta t_i,$
0	0	0	$(x_0 - \delta) - a$	0
1	0	1	2δ	2δ
2	0	0	$b - (x_0 + \delta)$	0

Thus $U(f, \Pi) - L(f, \Pi) = 2\delta < \varepsilon$. Since ε was arbitrary, Proposition 9.1.9 implies f is integrable on $[a, b]$. Every lower sum is 0, so the integral of f is 0.

Modifications of this argument handle the possibilities $x_0 = a$ or $x_0 = b$. ◇

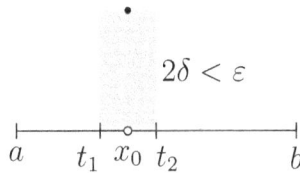

FIGURE 9.3
An upper sum for $\chi_{\{x_0\}}$ on $[a, b]$.

Example 9.1.14. If $f = \chi_{\mathbf{Q}}$ is the indicator of $\mathbf{Q}$, then f is *not* integrable on $[a, b]$. Indeed, let Π be an arbitrary splitting of $[a, b]$. In each piece, there exist both rational numbers and irrational numbers by density of the rationals and irrationals, so f takes both values 0 and 1 in *every* piece. Consequently, $m_i = 0$ and $M_i = 1$ for every i. Thus,

$$L(f, \Pi) = \sum_{i=0}^{n-1} 0\, \Delta t_i = 0, \qquad U(f, \Pi) = \sum_{i=0}^{n-1} 1\, \Delta t_i = (b - a),$$

independently of Π. This means $0 = L(f, [a, b]) < U(f, [a, b]) = b - a$. Proposition 9.1.9 implies f is not integrable on $[a, b]$. $\Diamond$

Proposition 9.1.15. *If $0 < a < b$ and k is a positive integer, then the monomial function $f(t) = t^k$ is integrable on $[a, b]$, and*

$$\int_a^b t^k \, dt = \frac{b^{k+1} - a^{k+1}}{k + 1}.$$

Proof. It is convenient to use a "geometric" splitting, for which the ratio of the lengths of consecutive intervals is constant, Figure 9.4.

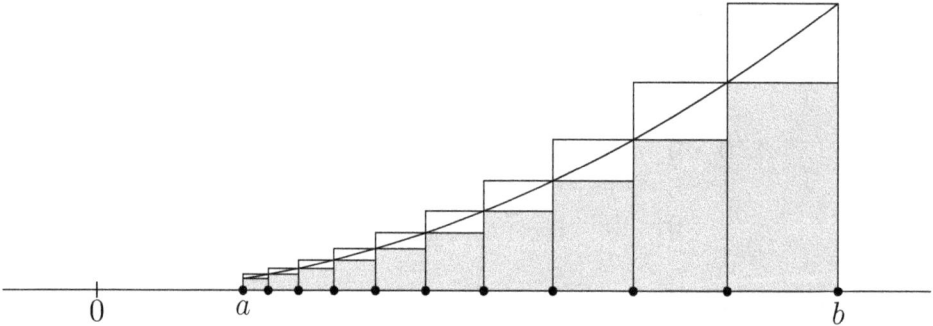

FIGURE 9.4
The lower and upper sums for a geometric splitting.

Assume n is a positive integer and put $\rho = (b/a)^{1/n}$, so $\rho > 1$ and $b = a\rho^n$. We'll use $\Pi = \{t_i\}_{i=0}^{n}$ with $t_i = a\rho^i$, for which $\Delta t_i = a\rho^{i+1} - a\rho^i = a(\rho - 1)\rho^i$ if $0 \leq i < n$. Since f is increasing on $[a, b]$, we have $m_i = f(t_i) = (a\rho^i)^k$ and $M_i = f(t_{i+1}) = (a\rho^{i+1})^k$. Because $M_i = \rho^k m_i$ for each i, the upper sum is ρ^k times the lower sum.

The general term of the lower sum is

$$f(t_i)\, \Delta t_i = (a\rho^i)^k\, a(\rho - 1)\rho^i = a^{k+1}(\rho - 1) \cdot (\rho^{k+1})^i.$$

The geometric sum formula $\sum_{i=0}^{n-1} r^i = \dfrac{r^n - 1}{r - 1}$ gives

$$L(f, \Pi) = \sum_{i=0}^{n-1} f(t_i) \, \Delta t_i = a^{k+1}(\rho - 1) \cdot \sum_{i=0}^{n-1} (\rho^{k+1})^i$$

$$= a^{k+1}(\rho - 1) \cdot \frac{(\rho^{k+1})^n - 1}{\rho^{k+1} - 1}$$

$$= a^{k+1}(\rho - 1) \cdot \frac{(\rho^n)^{k+1} - 1}{\rho^{k+1} - 1}$$

$$= a^{k+1}\left((\rho^n)^{k+1} - 1\right) \cdot \frac{\rho - 1}{\rho^{k+1} - 1}.$$

To simplify, recall that $\rho^n = b/a$, so

$$a^{k+1}\left((\rho^n)^{k+1} - 1\right) = a^{k+1}\left((b/a)^{k+1} - 1\right) = b^{k+1} - a^{k+1}.$$

Note further that

$$\frac{\rho^{k+1} - 1}{\rho - 1} = \sum_{i=0}^{k} \rho^i = 1 + \rho + \rho^2 + \cdots + \rho^k.$$

Call this sum $S(\rho)$ for brevity. By Exercise 8.4.3, $\rho = (b/a)^{1/n} \to 1$ as $n \to \infty$. Since S is a polynomial, $S(\rho) \to \lim(S, 1) = k + 1$. Consequently,

$$L(f, \Pi) = (b^{k+1} - a^{k+1}) \frac{\rho - 1}{\rho^{k+1} - 1} = \frac{b^{k+1} - a^{k+1}}{S(\rho)} \longrightarrow \frac{b^{k+1} - a^{k+1}}{k + 1}.$$

Since $L(f, \Pi) \le L(f, [a, b])$ for every Π, namely, the lower integral is no smaller than any particular lower sum,

$$\frac{b^{k+1} - a^{k+1}}{k + 1} = \lim_{n \to \infty} \frac{b^{k+1} - a^{k+1}}{S(\rho)} \le L(f, [a, b]).$$

As noted earlier, $U(f, \Pi) = \rho^k \cdot L(f, \Pi)$, so

$$U(f, [a, b]) \le \lim_{n \to \infty} U(f, \Pi) = \lim_{n \to \infty} \rho^k \cdot L(f, \Pi) = \frac{b^{k+1} - a^{k+1}}{k + 1}.$$

The lower and upper integrals therefore have the same value; this simultaneously proves that f is integrable on $[a, b]$, and evaluates the integral. $\qquad \square$

Exercises for Section 9.1

Exercise 9.1.1. (★) Assume $[a, b]$ is a real interval, and f and g are integrable on $[a, b]$. Prove that if $f \leq g$ on $[a, b]$, then $\int_a^b f \leq \int_a^b g$.

Exercise 9.1.2. Suppose $|f|$ is integrable on $[a, b]$. Is f necessarily integrable?

Exercise 9.1.3. (★) Assume $f(t) = t$ on $[0, b]$, and let $\Pi_n = \{ib/n\}_{i=0}^n$ be the equal-length splitting of $[0, b]$ with n pieces.

(a) Evaluate $L(f, \Pi_n)$ and $U(f, \Pi_n)$. Hint: Use Exercise 2.1.2 (a).

(b) Use (a) to prove $\sup L(f, \Pi) = \inf U(f, \Pi)$, and evaluate the integral $\int_0^b f$.

Exercise 9.1.4. Repeat Exercise 9.1.3 if $f(t) = t^2$. Hint: Use Exercise 2.1.1.

Exercise 9.1.5. (★) Define $f : [0, 1] \to \mathbf{R}$ by $f(t) = 1$ if $t = 1/k$ for some positive integer k, $f(t) = 0$ otherwise. Prove f is integrable on $[0, 1]$, and evaluate the integral.

Exercise 9.1.6. Prove the denominator function f of Exercise 8.1.5 is integrable on arbitrary intervals, and evaluate the integral.

9.2 Properties of the Integral

Calculating specific integrals from the definition is generally onerous. We *can*, however, deduce useful abstract properties of the integral that eventually lead us toward evaluating integrals easily.

Proposition 9.2.1. *Assume $a < b$, and f, g are bounded functions on $[a, b]$.*

(i) (*Linearity*) *If f and g are integrable on $[a, b]$, and if k is a real number, then the functions $f \pm g$ and kf are integrable on $[a, b]$, and*

$$\int_a^b (f + g) = \int_a^b f + \int_a^b g, \qquad \int_a^b (kf) = k \int_a^b f.$$

(ii) (*Monotonicity*) *If f and g are integrable on $[a, b]$, and if $f(t) \leq g(t)$ for all t in $[a, b]$, then*

$$\int_a^b f \leq \int_a^b g.$$

(iii) (*Triangle inequality*) *If f is integrable on* $[a, b]$, *then* $|f|$ *is integrable on* $[a, b]$, *and*

$$\left| \int_a^b f \right| \leq \int_a^b |f|.$$

(iv) (*Integral patching*) *If c is a real number and* $a < c < b$, *then f is integrable on* $[a, b]$ *if and only if f is integrable on the two intervals* $[a, c]$ *and* $[c, b]$, *and in this event*

$$\int_a^b f = \int_a^c f + \int_c^b f.$$

Proof. (Linearity). Let $\Pi = \{t_i\}_{i=0}^n$ be an arbitrary splitting of $[a, b]$. For each i such that $0 \leq i < n$, put $m_i^f = \inf\{f(t) : t \in I_i\}$, $m_i^g = \inf\{g(t) : t \in I_i\}$, and $m_i^{f+g} = \inf\{(f + g)(t) : t \in I_i\}$. Since

$$\{f(t) + g(t) : t \in I_i\} \subseteq \{f(s) + g(t) : s, t \in I_i\}$$

(a more restrictive predicate defines a smaller set), constriction implies

$$\begin{aligned} m_i^f + m_i^g &= \inf\{f(s) : s \in I_i\} + \inf\{g(t) : t \in I_i\} \\ &= \inf\{f(s) + g(t) : s, t \in I_i\} \\ &\leq \inf\{f(t) + g(t) : t \in I_i\} = m_i^{f+g}. \end{aligned}$$

(Loosely, we expect equality when the infima of f and g are achieved at the same point of I_i.) With analogous notation for the suprema, $M_i^{f+g} \leq M_i^f + M_i^g$. Adding up these inequalities,

$$L(f, \Pi) + L(g, \Pi) \leq L(f + g, \Pi) \leq U(f + g, \Pi) \leq U(f, \Pi) + U(g, \Pi)$$

for every splitting Π. Taking suprema over Π for the lower sums and infima over Π for the upper sums,

$$\int_a^b f + \int_a^b g \leq L(f + g, [a, b]) \leq U(f + g, [a, b]) \leq \int_a^b f + \int_a^b g.$$

This shows simultaneously that $f + g$ is integrable on $[a, b]$, and that the integral has the stated value.

Integrability of $f - g = f + (-1)g$ will follow once we prove kg is integrable for each real number k. To establish the latter claim, we consider two cases: $k \geq 0$, and $k = -1$.

Assume k is real. With notation as above, if Π is an arbitrary splitting of $[a, b]$ and $0 \leq i < n$, then

$$\{(kf)(t) : t \in I_i\} = k\{f(t) : t \in I_i\};$$

that is, the set of values of kf on I_i is k multiplied by the set of values of f (see Definition 4.1.11). If $k \geq 0$, we have

$$m_i^{kf} = \inf\{(kf)(t) : t \in I_i\} = k\inf\{f(t) : t \in I_i\} = km_i^f,$$
$$M_i^{kf} = \sup\{(kf)(t) : t \in I_i\} = k\sup\{f(t) : t \in I_i\} = kM_i^f$$

by Proposition 4.2.12 and Exercise 4.2.5. Forming lower and upper sums,

$$L(kf, \Pi) = \sum_{i=0}^{n-1} m_i^{kf} \Delta t_i = k\sum_{i=0}^{n-1} m_i^f \Delta t_i = kL(f, \Pi),$$
$$U(kf, \Pi) = \sum_{i=0}^{n-1} M_i^{kf} \Delta t_i = k\sum_{i=0}^{n-1} M_i^f \Delta t_i = kU(f, \Pi).$$

Since f is integrable on $[a, b]$,

$$
\begin{aligned}
L(kf, [a,b]) &= \sup_{\Pi} L(kf, \Pi) = \sup_{\Pi} kL(f, \Pi) \\
&= k\sup_{\Pi} L(f, \Pi) = k\inf_{\Pi} U(f, \Pi) \\
&= \inf_{\Pi} kU(f, \Pi) = \inf_{\Pi} U(kf, \Pi) = U(kf, [a,b]),
\end{aligned}
$$

proving that kf is integrable on $[a, b]$, and the value of the integral is k times the integral of f.

If instead $k = -1$, Proposition 4.2.12 and Exercise 4.2.5 imply

$$m_i^{-f} = \inf\{(-f)(t) : t \in I_i\} = -\sup\{f(t) : t \in I_i\} = -M_i^f,$$
$$M_i^{-f} = \sup\{(-f)(t) : t \in I_i\} = -\inf\{f(t) : t \in I_i\} = -m_i^f.$$

Forming lower and upper sums gives

$$L(-f, \Pi) = \sum_{i=0}^{n-1} m_i^{-f} \Delta t_i = -\sum_{i=0}^{n-1} M_i^f \Delta t_i = -U(f, \Pi),$$
$$U(-f, \Pi) = \sum_{i=0}^{n-1} M_i^{-f} \Delta t_i = -\sum_{i=0}^{n-1} m_i^f \Delta t_i = -L(f, \Pi).$$

Taking the supremum of the lower sums and the infimum of the upper sums,

$$L(-f, [a,b]) = -U(f, [a,b]) = -L(f, [a,b]) = U(-f, [a,b]),$$

so $-f$ is integrable on $[a, b]$, and the integral of $-f$ is minus the integral of f.

(Monotonicity). This was proven in Exercise 9.1.1.

(Triangle inequality). Assume f is integrable on $[a, b]$. Fix ε arbitrarily and choose a splitting Π of $[a, b]$ such that $U(f, \Pi) - L(f, \Pi) < \varepsilon$.

By the reverse triangle inequality,

$$|f(x)| - |f(y)| \leq |f(x) - f(y)| \quad \text{for all } x, y \text{ in } [a,b].$$

In particular, letting I_i be the ith piece of Π and taking the supremum over x in I_i, then taking the infimum over y in I_i, we have

$$M_i^{|f|} - m_i^{|f|} \leq |M_i^f - m_i^f| = M_i^f - m_i^f.$$

Multiplying by Δt_i and summing over i,

$$U(|f|, \Pi) - L(|f|, \Pi) \leq U(f, \Pi) - L(f, \Pi) < \varepsilon.$$

Since ε was arbitrary, $|f|$ is integrable on $[a,b]$.

Finally, $-|f(t)| \leq f(t) \leq |f(t)|$ for all t in $[a,b]$. Monotonicity gives

$$-\int_a^b |f| \leq \int_a^b f \leq \int_a^b |f|, \quad \text{or} \quad \left| \int_a^b f \right| \leq \int_a^b |f|.$$

(Integral patching). Assume $f : [a,b] \to \mathbf{R}$ is a function, $a < c < b$, and fix ε arbitrarily.

Suppose f is integrable on $[a,b]$. Choose a splitting Π of $[a,b]$ such that

$$\sum_{i=0}^{n-1} (M_i^f - m_i^f) \Delta t_i = U(f, \Pi) - L(f, \Pi) < \varepsilon.$$

Replacing Π with $\Pi \cup \{c\}$ if necessary, we may assume $c \in \Pi$ while preserving the preceding inequality.

The set $\Pi' = [a,c] \cap \Pi$ is a splitting of $[a,c]$. Since every piece of Π' is a piece of Π, and every summand in $U(f, \Pi) - L(f, \Pi)$ is non-negative, $U(f, \Pi') - L(f, \Pi') < \varepsilon$. By Proposition 9.1.9, f is integrable on $[a,c]$. A similar argument proves f is integrable on $[c,b]$.

Conversely, suppose f is integrable on $[a,c]$ and on $[c,b]$. Choose splittings Π' and Π'' of $[a,c]$ and $[c,b]$ such that

$$U(f, \Pi') - L(f, \Pi') < \varepsilon/2, \qquad U(f, \Pi'') - L(f, \Pi'') < \varepsilon/2.$$

The union $\Pi = \Pi' \cup \Pi''$ is a splitting of $[a,b]$ for which

$$U(f, \Pi) - L(f, \Pi) = \Big(U(f, \Pi') - L(f, \Pi') \Big) + \Big(U(f, \Pi'') - L(f, \Pi'') \Big) < \varepsilon.$$

Since ε was arbitrary, f is integrable on $[a,b]$.

In either situation, $L(f, \Pi) = L(f, \Pi') + L(f, \Pi'')$, which proves

$$\int_a^b f = \int_a^c f + \int_c^b f.$$

This completes the proof of Proposition 9.2.1. $\qquad \square$

Corollary 9.2.2. *Assume f and g are functions on some interval $[a, b]$, and $f(t) = g(t)$ except at finitely many points. If f is integrable on $[a, b]$, then g is integrable on $[a, b]$, and the integral of g is equal to the integral of f.*

Proof. Consider the function $h = g - f$, which by hypothesis is non-zero at only finitely many points. That is, there exist points $\{x_j\}_{j=0}^{m-1}$ in $[a, b]$ and real numbers $\{k_j\}_{j=0}^{m-1}$ such that

$$h = \sum_{j=0}^{m-1} k_j \, \chi_{\{x_j\}}$$

By Example 9.1.13, each function $\chi_{\{x_j\}}$ is integrable and has integral equal to 0. By linearity of the integral, h itself is integrable, and

$$\int_a^b h = \sum_{j=0}^{m-1} k_j \int_a^b \chi_{\{x_j\}} = 0.$$

Since $g = f + h$, linearity implies g is integrable, and

$$\int_a^b g = \int_a^b (f + h) = \int_a^b f + \int_a^b h = \int_a^b f. \qquad \square$$

Corollary 9.2.3. *If $f : [a, b] \to \mathbf{R}$ is integrable, and if $m \leq f(t) \leq M$ for all but finitely many t in $[a, b]$, then*

$$m(b - a) \leq \int_a^b f \leq M(b - a).$$

Proof. This follows immediately from monotonicity and Corollary 9.2.2. $\quad \square$

When we work with integrals, "closed intervals" $[a, b]$ such that $b < a$ naturally arise. The following convention turns out to be particularly useful.

Definition 9.2.4. If I is a real interval and $f : I \to \mathbf{R}$ is integrable, we define

$$\int_b^a f = - \int_a^b f \quad \text{for all } a \text{ and } b \text{ in } I.$$

Particularly, $\displaystyle\int_a^a f = 0$ for all a in I.

Corollary 9.2.5 (The cocycle property). *Assume I is an interval of real numbers, and $f : I \to \mathbf{R}$ is integrable. For all a, b, c in I,*

$$\int_a^b f + \int_b^c f + \int_c^a f = 0.$$

Proof. First assume $a \leq c \leq b$. By integral patching and our conventions on swapping limits of integration,

$$\int_a^b f = \int_a^c f + \int_c^b f = -\left[\int_b^c f + \int_c^a f\right],$$

so the cocycle property holds in this case. But the stated equality is invariant under permutation of the symbols a, b, and c, and is therefore true regardless of the ordering of the numbers a, b, and c. □

Theorem 9.2.6 (The triangle inequality for integrals). *If a and b are real and f is integrable on the interval with endpoints a and b, then*

$$\left|\int_a^b f\right| \leq \left|\int_a^b |f|\right|.$$

Proof. If $a < b$, the conclusion is Proposition 9.2.1 (iii). Since each side is unchanged by swapping a and b, the conclusion holds if $b \leq a$ as well. □

Remark 9.2.7. When we study functions defined by integration, there is real convenience in a triangle inequality that applies even then the limits of integration are "out of order." ◇

A Transformation Theorem

For immediate use and for illustration, we prove an "affine change of variables" result. Below, τ (tau) connotes transformation and μ (mu) multiplication.

Proposition 9.2.8. *Assume $I = [a, b]$ is a domain of integration, and f is integrable on I.*

(i) *For every real number c,*

$$\int_{a+c}^{b+c} f(t - c)\, dt = \int_a^b f(t)\, dt.$$

(ii) *If μ is real and non-zero, then*

$$\mu \int_{a/\mu}^{b/\mu} f(\mu t)\, dt = \int_a^b f(t)\, dt.$$

Proof. (i). Define $\tau : I + c \to I$ by $\tau(t) = t - c$. For each splitting $\Pi = \{t_i\}_{i=0}^n$ of $[a + c, b + c]$, there is a unique splitting $\Pi' = \{t_i'\}_{i=0}^n$ of $[a, b]$ defined by $t_i' = \tau(t_i) = t_i - c$.

Since $(f \circ \tau)(I_i) = f\big(\tau(I_i)\big) = f(I_i')$ as sets, we have

$$m_i'^f = m_i^{f \circ \tau}, \qquad\qquad M_i'^f = M_i^{f \circ \tau}.$$

Further, $\Delta t_i' = (t_i - c) - (t_{i-1} - c) = \Delta t_i$. Consequently,

$$L(f, \Pi') = \sum_{i=0}^{n-1} m_i'^f \Delta t_i' = \sum_{i=0}^{n-1} m_i^{f \circ \tau} \Delta t_i = L(f \circ \tau, \Pi),$$

and similarly $U(f, \Pi') = U(f \circ \tau, \Pi)$. The theorem follows at once by taking the supremum of the lower sums and the infimum of the upper sums.

(ii). Define $\tau : I/\mu = [a/\mu, b/\mu] \to I$ by $\tau(t) = \mu t$. The preceding proof goes through with straightforward modifications, notably that $\Delta t_i' = \mu \, \Delta t_i$.   □

Corollary 9.2.9. *Assume $a > 0$ and f is integrable on $[-a, a]$.*

(i) *If f is even, namely, if $f(-t) = f(t)$ for all t, then*

$$\int_{-a}^{a} f(t) \, dt = 2 \int_0^a f(t) \, dt.$$

(ii) *If f is odd, namely, if $f(-t) = -f(t)$ for all t, then*

$$\int_{-a}^{a} f(t) \, dt = 0.$$

Proof. By part (ii) of Proposition 9.2.8, with $\mu = -1$,

$$\int_{-a}^{0} f(t) \, dt = - \int_a^0 f(-t) \, dt = \int_0^a f(-t) \, dt.$$

Breaking up the integral at 0,

$$\int_{-a}^{a} f(t) \, dt = \int_{-a}^{0} f(t) \, dt + \int_0^a f(t) \, dt$$
$$= \int_0^a f(-t) \, dt + \int_0^a f(t) \, dt = \int_0^a \big(f(-t) + f(t) \big) \, dt.$$

Both parts follow immediately.   □

Exercises for Section 9.2

Exercise 9.2.1. (★) Assume k is a natural number. Prove

$$\int_a^b t^k \, dt = \frac{b^{k+1} - a^{k+1}}{k + 1} \qquad \text{for all real } a \text{ and } b.$$

Hint: First establish integrability on $[0, b]$, using an equal-length splitting Π and noting that $U(f, \Pi) - L(f, \Pi)$ is telescoping.

Exercise 9.2.2. Prove the conclusion of Proposition 9.2.1 (ii) under the weaker hypothesis that $f(t) \leq g(t)$ with at most finitely many exceptions.

Exercise 9.2.3. Assume f is a function on $[0, 1]$, and f is integrable on $[\delta, 1]$ for every δ in $(0, 1)$. Give a proof or counterexample to each of the following:

(a) If $\lim_{\delta \to 0} \int_\delta^1 f(t) \, dt$ exists, then f is integrable on $[0, 1]$.

(b) If $\lim(f, 0^+)$ exists, then f is integrable on $[0, 1]$.

Exercise 9.2.4. (★) Assume $f = \chi_{\mathbf{Q}}$ is the indicator of $\mathbf{Q}$.

(a) Prove there exists a sequence (f_k) of integrable functions such that for all real x, we have $f_k(x) \to f(x)$.

(b) Prove that for every ε, there exists a sequence $(I_k)_{k=0}^\infty$ of closed, bounded intervals of total length ε whose union contains $\mathbf{Q}$.

(c) What numerical value is sensible to assign to the integral of f over $[a, b]$?

Exercise 9.2.5. (H) Suppose (f_k) is a sequence of integrable functions on $[0, 1]$, and that for every ε, there exists an N such that if $N \leq k < k'$, then $\sup\{|f_{k'}(x) - f_k(x)| : x \in [0, 1]\} < \varepsilon$. Prove the sequence converges pointwise, namely, $f(x) = \lim f_k(x)$ exists for each x in $[0, 1]$; the limit function is integrable; and the integral of the limit is the limit of the integrals:

$$\lim_{k \to \infty} \int_0^1 f_k = \int_0^1 f = \int_0^1 \lim_{k \to \infty} f_k.$$

9.3   Criteria for Integrability

In this section we prove monotone functions are integrable, continuous functions are integrable, and products of integrable functions are integrable.

Proposition 9.3.1. *If $f : [a, b] \to \mathbf{R}$ is monotone, then f is integrable on $[a, b]$.*

Proof. Replacing f by $-f$ if necessary, we may assume f is non-decreasing. Particularly, $f(a) \leq f(t) \leq f(b)$ for all t in $[a, b]$, so f is bounded.

Fix ε arbitrarily, and use reciprocal finitude to choose a positive integer n such that $\big(f(b) - f(a)\big)(b - a)/n < \varepsilon$. Let Π_n be the equal-length splitting with n pieces. On the ith piece $I_i = [t_i, t_{i+1}]$, we have $m_i = f(t_i)$, $M_i = f(t_{i+1})$, and $\Delta t_i = (b - a)/n$ independently of i. Writing Δt instead of Δt_i,

$$U(f, \Pi_n) - L(f, \Pi_n) = \sum_{i=0}^{n-1} (M_i - m_i) \Delta t_i = \left[\sum_{i=0}^{n-1} \big(f(t_{i+1}) - f(t_i)\big)\right] \Delta t.$$

The sum in square brackets is telescoping, so

$$U(f, \Pi_n) - L(f, \Pi_n) = \big(f(t_n) - f(t_0)\big)\Delta t = \big(f(b) - f(a)\big)\frac{b-a}{n} < \varepsilon.$$

Since ε was arbitrary, f is integrable on $[a, b]$. $\qquad\square$

Integrability of Continuous Functions

Definition 9.3.2. A function $f : [a, b] \to \mathbf{R}$ is *uniformly continuous* on $[a, b]$ if: For every positive ε, there exists a positive δ such that if t and $t' \in [a, b]$ and $|t - t'| \le \delta$, then $|f(t) - f(t')| < \varepsilon$.

Remark 9.3.3. Comparing with Corollary 8.2.7, for every ε, there is a *single* δ that "wins" the ε-δ game for Player δ at *each point of* $[a, b]$. $\qquad\diamond$

Proposition 9.3.4. *Every continuous function $f : [a, b] \to \mathbf{R}$ whose domain is a closed, bounded interval of real numbers is uniformly continuous.*

Proof. We proceed contrapositively. If f is not uniformly continuous on $[a, b]$, then Player ε can force a win; that is, there exists an ε such that for every δ, there are points t and t' in $[a, b]$ such that $|t - t'| \le \delta$ but $|f(t) - f(t')| \ge \varepsilon$. We will prove that in this situation, f is discontinuous at some point of $[a, b]$.

Fix ε as in the preceding paragraph. For each positive integer n, choose points t_n and t'_n in $[a, b]$ such that $|t_n - t'_n| \le 1/n$, but $|f(t_n) - f(t'_n)| \ge \varepsilon$. Since $[a, b]$ is a closed, bounded interval, Theorem 6.4.15 implies there exists a subsequence $(t_{\nu(k)})$ of (t_n) convergent to some point t_∞ in $[a, b]$. Since

$$|t'_{\nu(k)} - t_\infty| \le |t'_{\nu(k)} - t_{\nu(k)}| + |t_{\nu(k)} - t_\infty| \le 1/k + |t_{\nu(k)} - t_\infty|,$$

the sequence $(t'_{\nu(k)})$ also converges to t_∞. But by construction, the image sequences $\big(f(t_{\nu(k)})\big)$ and $\big(f(t'_{\nu(k)})\big)$ do not have the same limit (even if both limits exist), because $|f(t_{\nu(k)}) - f(t'_{\nu(k)})| \ge \varepsilon$ for all k. By definition, f is discontinuous at t_∞. $\qquad\square$

Proposition 9.3.5. *A continuous function $f : [a, b] \to \mathbf{R}$ is integrable.*

Proof. Fix ε arbitrarily. Since f is uniformly continuous by Proposition 9.3.4, there exists a δ such that if t and t' are elements of $[a, b]$ such that $|t - t'| \le \delta$, then

$$|f(t) - f(t')| < \varepsilon/(b - a).$$

By reciprocal finitude, there exists a positive integer n such that $(b - a)/n < \delta$. Let $\Pi = \Pi_n$ be the equal-length splitting of $[a, b]$ into n pieces.

If $0 \le i < n$, and if $\{t, t'\} \subseteq I_i$, then $|t - t'| \le (b - a)/n < \delta$, which implies $|f(t) - f(t')| < \varepsilon/(b - a)$. Since f is continuous, f achieves a maximum value and a minimum value in I_i by the extreme value theorem. Consequently, $M_i^f - m_i^f < \varepsilon/(b-a)$. Summing over i,

$$U(f, \Pi) - L(f, \Pi) = \sum_{i=0}^{n-1} (M_i^f - m_i^f)\, \Delta t_i < \frac{\varepsilon}{b-a} \sum_{i=0}^{n-1} \Delta t_i = \varepsilon.$$

Since ε was arbitrary, f is integrable by Proposition 9.1.9. $\square$

Integrals of Products

Proposition 9.3.6. *If f and g are integrable functions on $[a, b]$, then the product fg is integrable on $[a, b]$.*

Remark 9.3.7. There is no general formula for the integral of fg in terms of the integrals of f and g. $\diamond$

Proof. By the "polarization identity"

$$fg = \tfrac{1}{2}\big((f+g)^2 - f^2 - g^2\big),$$

it suffices to prove that if f is integrable, then f^2 is integrable.

Fix ε. Since f is integrable, there exists a positive real number M such that $|f(t)| \leq M$ for all t in $[a, b]$, and there exists a splitting Π of $[a, b]$ such that

$$U(f, \Pi) - L(f, \Pi) < \varepsilon/(2M).$$

Assume $\{t, t'\} \subseteq I_i$ for some i. By the difference of squares identity,

$$f(t)^2 - f(t')^2 = \big(f(t) + f(t')\big) \cdot \big(f(t) - f(t')\big) \leq 2M \cdot (M_i^f - m_i^f).$$

Taking the supremum over t and then the infimum over t',

$$M_i^{f^2} - m_i^{f^2} \leq 2M \cdot (M_i^f - m_i^f) \quad \text{for each } i.$$

Multiplying by Δt_i and summing over i,

$$U(f^2, \Pi) - L(f^2, \Pi) \leq 2M \cdot \big(U(f, \Pi) - L(f, \Pi)\big) < \varepsilon.$$

Since ε was arbitrary, f^2 is integrable on $[a, b]$. $\square$

Numerical Methods

In practice, infima and suprema of f on I_i may be difficult to calculate. "Sampled sums" give a convenient way of approximating an integral numerically.

Definition 9.3.8. Assume $f : [a, b] \to \mathbf{R}$, and $\Pi = \{t_i\}_{i=0}^{n}$ is a splitting of $[a, b]$. A set $t^* = \{t_i^*\}_{i=0}^{n-1}$ such that $t_i \leq t_i^* \leq t_{i+1}$ for each i is called a set of *sample points from* Π. The corresponding *sampled sum from* Π is the expression

$$S(f, \Pi, t^*) = \sum_{i=0}^{n-1} f(t_i^*) \, \Delta t_i.$$

If $t_i^* = t_i$ for each i, $\text{LEFT}(f, \Pi) := S(f, \Pi, t^*)$ is called the *left-hand sum*.
If $t_i^* = t_{i+1}$ for each i, $\text{RIGHT}(f, \Pi) := S(f, \Pi, t^*)$ is the *right-hand sum*.
If $t_i^* = \frac{1}{2}(t_i + t_{i+1})$ for each i, $\text{MID}(f, \Pi) := S(f, \Pi, t^*)$ is the *midpoint sum*.
The average of the left and right sums is the *trapezoid sum* $\text{TRAP}(f, \Pi)$.

Averages

Definition 9.3.9. If f is integrable on $[a, b]$, we define the *average* of f on $[a, b]$ to be the real number

$$\overline{f} = \frac{1}{b-a} \int_a^b f = \frac{1}{b-a} \int_a^b f(t) \, dt.$$

Remark 9.3.10. Exercise 9.3.11 shows that if f is continuous, the average value is, in a precise sense, the limit of the average of sampled values.   ◇

Exercises for Section 9.3

Exercise 9.3.1. (★) Assume (a, b) is a bounded real interval and $f : (a, b) \to \mathbf{R}$ is continuous and bounded. Prove f is integrable on (a, b) in the sense that if we define $f(a)$ and $f(b)$ arbitrarily, then f is integrable on $[a, b]$ and the integral does not depend on the endpoint values.

Exercise 9.3.2. (★) Assume f is continuous and non-negative on $[a, b]$ but not identically 0. Prove the integral is strictly positive.

Exercise 9.3.3. (H) Assume $f : (0, \infty) \to \mathbf{R}$ is continuous, and that for all positive a and b, we have

$$\int_1^a f(t) \, dt = \int_b^{ab} f(t) \, dt.$$

Prove $f(t) = f(1)/t$ for all positive t, and give a geometric interpretation.

Exercise 9.3.4. Assume f is bounded on $[a, b]$ and Π is an arbitrary splitting. Prove $L(f, \Pi) \leq S(f, \Pi, t^*) \leq U(f, \Pi)$ for every sampled sum of f from Π.

Exercise 9.3.5. In each part, f is a function on $[a, b]$ and Π is a splitting.

(a) Prove that if f is non-decreasing on $[a, b]$, then

$$\text{LEFT}(f, \Pi) \leq \int_a^b f \leq \text{RIGHT}(f, \Pi).$$

(b) Prove that $\text{TRAP}(f, \Pi)$ is the sum of the areas of the trapezoids bounded by the secant lines joining the points $\big(t_i, f(t_i)\big)$.

(c) Prove that $\text{MID}(f, \Pi)$ is the sum of the areas of the trapezoids bounded by arbitrary non-vertical lines passing through the points $\big(\bar{t}_i, f(\bar{t}_i)\big)$.

Exercise 9.3.6. (★) Assume $f(t) = t^2$ on $[a, b]$, and $\Pi = \{a, b\}$ is the splitting with one piece. Calculate the left, right, trapezoid, and midpoint sums, and illustrate each sum with a sketch. Prove that the *parabolic sum*,

$$\text{PARA}(f, \Pi) := \tfrac{1}{3}\big(\text{TRAP}(f, \Pi) + 2\,\text{MID}(f, \Pi)\big)$$

gives the exact value of the integral.

Exercise 9.3.7. Assume $f(t) = t^3$ on $[a, b]$, and $\Pi = \{a, b\}$ is the splitting with one piece. Calculate the left, right, trapezoid, and midpoint sums. By how much does the parabolic sum $\tfrac{1}{3}\big(\text{TRAP}(f, \Pi) + 2\,\text{MID}(f, \Pi)\big)$ differ from the exact value of the integral?

Exercise 9.3.8. (A) Assume a and b are real numbers and $a < b$.

(a) Calculate the average value of $f(t) = t$ over $[a, b]$.

(b) Prove the average value of $f(t) = t^2$ over $[a, b]$ is $(a^2 + ab + b^2)/3$.

(c) Assuming $k > 0$, calculate the average value of $f(t) = t^k$ over $[a, b]$.

Exercise 9.3.9. Suppose f is integrable on $[a, b]$, and let $\overline{f}$ be the average value of f on $[a, b]$.

(a) Prove that the integral of $f - \overline{f}$ over $[a, b]$ is equal to 0.

(b) (The *mean value theorem for integrals*) Prove that if f is continuous on $[a, b]$, there exists a number c in (a, b) such that $f(c) = \overline{f}$.

(c) If $f(t) = t^2$ on $[0, 1]$, find the value of c in $(0, 1)$ that satisfies the mean value theorem for integrals, and carefully sketch f and its average value.

Exercise 9.3.10. Assume f is integrable on every closed subinterval of $B_r(x_0)$.

(a) Prove that if $0 < |h| < r$, the average value of f on the closed interval with endpoints x_0 and $x_0 + h$ is

$$\frac{1}{h} \int_{x_0}^{x_0+h} f.$$

(b) Assume f is continuous at x_0. Prove

$$\lim_{h \to 0} \frac{1}{h} \int_{x_0}^{x_0+h} f(t)\, dt = f(x_0).$$

(c) Show by example that the result of part (b) may be either true or false if f is discontinuous at x_0.

Exercise 9.3.11. This exercise justifies the definition of the average value of a *continuous* function f on $[a, b]$.

(a) Fix ε arbitrarily. By Proposition 9.3.4, there exists a δ such that if t and t' are points in $[a, b]$ and $|t - t'| < \delta$, then $|f(t) - f(t')| < \varepsilon/[2(b - a)]$. Prove that if Π is a splitting such that $\Delta t_i < \delta$ for all i, then

$$\left| S(f, \Pi, t^*) - \int_a^b f \right| < \varepsilon \quad \text{for every sampled sum of } f \text{ from } \Pi.$$

(b) Assume n is a positive integer, and $\Delta t = (b - a)/n$. Prove

$$\lim_{n \to \infty} \frac{1}{n} \sum_{i=0}^{n-1} f(a + i\, \Delta t) = \frac{1}{b - a} \int_a^b f(t)\, dt.$$

Exercise 9.3.12. Assume ρ is a continuous, non-negative (but not identically zero) function on an interval $[a, b]$. The *ρ-weighted average* of an integrable function f is defined to be

$$\overline{f}_\rho = \left[\int_a^b f\rho \right] \Big/ \left[\int_a^b \rho \right].$$

State and prove a mean value theorem for weighted averages.

Exercise 9.3.13. Fix natural numbers k and n. Assume $0 < a < b$, and $\rho(t) = t^k$. Calculate the ρ-weighted average of $f(t) = t^n$ on $[a, b]$.

Exercise 9.3.14. Assume $[a, b]$ is a closed, bounded interval of real numbers, and $f : [a, b] \to \mathbf{R}$ is bounded. Let D denote the set of x in $[a, b]$ such that f is discontinuous at x, and assume that for every ε, there exist finitely many open intervals of total length at most ε whose union contains D. Prove f is integrable on $[a, b]$.

Exercise 9.3.15. (H) For x real define $f(x) = (-1)^k$ if $k < x < k + 1$ for some positive integer k, and $f(x) = 0$ otherwise. Prove f is integrable over every closed, bounded interval $[a, b]$.

9.4   Definite Integrals

Using integrals to define functions genuinely enlarges our class of functions. For example, the integral of the reciprocal function is not rational.

Definition 9.4.1. Assume $f : [a, b] \to \mathbf{R}$ is integrable. For each x in $[a, b]$, the function f is integrable on $[a, x]$. The function $F : [a, b] \to \mathbf{R}$ defined by

$$F(x) = \int_a^x f = \int_a^x f(t)\, dt$$

is called the *definite integral* of f from a.

Lemma 9.4.2. *The signum function* $\mathrm{sgn} : \mathbf{R} \to \mathbf{R}$, *defined by* $\mathrm{sgn}(t) = t/|t|$ *if* $t \neq 0$, *and* $\mathrm{sgn}(0) = 0$, *is integrable on every closed, bounded interval* $[a, b]$, *and*

$$\int_0^x \mathrm{sgn}(t)\, dt = |x| \quad \text{for all real } x.$$

Proof. See Exercise 9.4.1. ☐

Remark 9.4.3. If f is integrable on some interval containing a, x_1, and x_2, and F is the definite integral of f, the cocycle property gives

$$F(x_2) - F(x_1) = \int_a^{x_2} f - \int_a^{x_1} f = \int_{x_1}^{x_2} f. \qquad \diamond$$

Proposition 9.4.4. *Assume f is integrable on $[a, b]$, and define F on $[a, b]$ by*

$$F(x) = \int_a^x f.$$

(i) *If $|f(t)| \leq M$ for all but finitely many t, then*

$$|F(x_2) - F(x_1)| \leq M|x_2 - x_1| \quad \text{for all } x_1, x_2 \text{ in } [a, b].$$

In particular, F is uniformly continuous on $[a, b]$.

(ii) *If f is non-negative, then F is non-decreasing.*

(iii) *If f is continuous, non-negative, and non-zero somewhere in every open interval, then F is strictly increasing on $[a, b]$.*

Proof. (i). We may as well assume $|f(t)| \leq M$ for all t, since by Corollary 9.2.2 we may change the value of f at finitely many points without changing the integral. If $a \leq x_1, x_2 \leq b$, the triangle inequality (Theorem 9.2.6) implies

$$|F(x_2) - F(x_1)| = \left| \int_{x_1}^{x_2} f \right| \leq \left| \int_{x_1}^{x_2} |f| \right| \leq \left| \int_{x_1}^{x_2} M \right| = M|x_2 - x_1|.$$

To prove F is uniformly continuous, fix ε arbitrarily, and put $\delta = \varepsilon/(M+1)$. (Adding 1 accommodates $M = 0$.) If x_1 and x_2 are arbitrary points of $[a, b]$, then $|x_2 - x_1| < \delta$ implies $|F(x_2) - F(x_1)| \leq M|x_2 - x_1| \leq M\delta < \varepsilon$.

(ii). Suppose f is non-negative. If $a \leq x_1 < x_2 \leq b$, then

$$0 = \int_{x_1}^{x_2} 0 \, dt \leq \int_{x_1}^{x_2} f(t) \, dt = F(x_2) - F(x_1), \quad \text{or } F(x_1) \leq F(x_2).$$

(iii). Assume $a \leq x_1 < x_2 \leq b$. By hypothesis, there exists a point t_0 such that $x_1 < t_0 < x_2$ and $0 < f(t_0)$. By Exercise 9.3.2,

$$0 < \int_{x_1}^{x_2} f(t) \, dt = F(x_2) - F(x_1).$$

Since x_1 and x_2 were arbitrary, F is strictly increasing. $\qquad \square$

Proposition 9.4.5. *If k is a natural number and x_1, x_2 are real, then*

$$\int_{x_1}^{x_2} t^k \, dt = \frac{x_2^{k+1} - x_1^{k+1}}{k+1}.$$

Proof. Although this restates Exercise 9.2.1, we give another proof to illustrate our new tools. Since each side is continuous in x_1, taking limits as $x_1 \to 0$ gives

$$\int_0^{x_2} t^k \, dt = \frac{x_2^{k+1}}{k+1} \quad \text{for all non-negative } x_2.$$

If $w \leq 0$, namely $0 \leq -w$, then taking $\mu = -1$ in Proposition 9.2.8 gives

$$\int_0^w t^k \, dt = -\int_0^{-w} (-t)^k \, dt = (-1)^{k+1} \int_0^{-w} t^k \, dt = \frac{w^{k+1}}{k+1}.$$

We have established that $F(w)$, the integral of t^k from 0 to w, is equal to $w^{k+1}/(k+1)$ for all real w. If x_1, x_2 are arbitrary, then

$$\int_{x_1}^{x_2} t^k \, dt = F(x_2) - F(x_1) = \frac{x_2^{k+1} - x_1^{k+1}}{k+1}. \qquad \square$$

Example 9.4.6. Assume p is a polynomial, say

$$p(x) = \sum_{k=0}^{n} a_k x^k = a_0 + a_1 x + a_2 x^2 + \cdots + a_n x^n.$$

Since p is continuous on $\mathbf{R}$, p is integrable on every interval $[0, x]$, and

$$\int_0^x p(t) \, dt = \int_0^x \sum_{k=0}^{n} a_k t^k \, dt = \sum_{k=0}^{n} a_k \int_0^x t^k \, dt = \sum_{k=0}^{n} \frac{a_k}{k+1} x^{k+1}.$$

If we call this new polynomial P, then

$$\int_a^b p(t) \, dt = \int_0^b p(t) \, dt - \int_0^a p(t) \, dt = P(b) - P(a). \qquad \diamondsuit$$

Remark 9.4.7. To evaluate a definite integral from the definition for even a single x, we must compute the supremum of the set of lower sums of f as Π ranges over the set of splittings of $[a, x]$. In general this is somewhere between laborious and genuinely difficult.

The preceding example shows that for a polynomial function, the definite integral (laborious to evaluate) is equal to a particular polynomial (trivial to evaluate) that can be found by inspection. That these functions are one and the same is a substantial and non-trivial piece of information. $\diamond$

Integration and O Notation

Proposition 9.4.8. *Assume f is integrable on some interval containing x_0, and that for some natural number k, $f(x) \approx O(x - x_0)^k$ if $x \approx x_0$. If*

$$F(x) = \int_{x_0}^{x} f(t)\, dt$$

is the definite integral of f from x_0, then $F(x) \approx O(x - x_0)^{k+1}$ if $x \approx x_0$.

Proof. By hypothesis, there exist positive real numbers M and δ such that $|f(t)| \leq M|t - x_0|^k$ if $|t - x_0| < \delta$. By Theorem 9.2.6, if $|x - x_0| < \delta$, then

$$|F(x)| \leq \left| \int_{x_0}^{x} |f(t)|\, dt \right| \leq M \left| \int_{x_0}^{x} |t - x_0|^k\, dt \right| = \frac{M}{k+1} |x - x_0|^{k+1}.$$

Particularly, $F(x) \approx O(x - x_0)^{k+1}$ if $x \approx x_0$. $\square$

Remark 9.4.9. Qualitatively, integrating from x_0 increases by 1 the degree of approximation near x_0. $\diamond$

Integration of Power Series

A convergent power series can be integrated *term by term*, in exactly the same way (formally) as a polynomial, compare Example 9.4.6.

Proposition 9.4.10. *Assume f is defined by a germ of radius R:*

$$f(x) = \sum_{k=0}^{\infty} a_k (x - x_0)^k$$
$$= a_0 + a_1(x - x_0) + a_2(x - x_0)^2 + \cdots .$$

For every x such that $|x - x_0| < R$, f is integrable on $[x_0, x]$, and

$$\int_{x_0}^{x} f(t)\, dt = \sum_{k=0}^{\infty} \frac{a_k}{k+1} (x - x_0)^{k+1}$$
$$= a_0(x - x_0) + \frac{a_1}{2}(x - x_0)^2 + \frac{a_2}{3}(x - x_0)^3 + \cdots .$$

Proof. As usual, assume $x_0 = 0$ without loss of generality. Proposition 8.3.12 implies f is continuous on $(-R, R)$, and therefore integrable on every closed, bounded subinterval by Proposition 9.3.5.

For each natural number n, introduce the polynomials

$$f_n(t) = \sum_{k=0}^{n-1} a_k t^k, \qquad F_n(x) = \int_0^x f_n(t)\, dt = \sum_{k=0}^{n-1} \frac{a_k}{k+1} x^{k+1},$$

and define

$$F(x) = \int_0^x f(t)\, dt, \qquad G(x) = \sum_{k=0}^{\infty} \frac{a_k}{k+1} x^{k+1}.$$

The power series $G(x)$ converges absolutely for every x such that $|x| < R$, by comparison with the series $x f(x) = \sum_k a_k x^{k+1}$.

To prove $F(x) = G(x)$ if $|x| < R$, it suffices to prove that if $0 < r < R$, and if ε is arbitrary, then $|F - G| < \varepsilon$ on $[-r, r]$. By Proposition 8.3.15, there exists an index N such that

$$\sup_{|x| \leq r} |f(x) - f_N(x)| < \varepsilon/(2r).$$

If $x \in [-r, r]$, the triangle inequality for integrals, Theorem 9.2.6, guarantees

$$|F(x) - F_N(x)| \leq \left| \int_0^x |f(t) - f_N(t)|\, dt \right| < |x|\, \varepsilon/(2r) \leq \varepsilon/2.$$

Furthermore,

$$|G(x) - F_N(x)| \leq \sum_{k=N}^{\infty} \left| \frac{a_k}{k+1} \right| |x|^{k+1} \leq |r| \sum_{k=N}^{\infty} |a_k|\, |r|^k < \varepsilon/2.$$

That is, $|F_N - G| < \varepsilon/2$ on $[-r, r]$. Adding these estimates,

$$|F - G| \leq |F - F_N| + |F_N - G| < \varepsilon \quad \text{on } [-r, r].$$

Since ε was arbitrary, $F \equiv G$ on $[-r, r]$. Since r was an arbitrary positive real number less than R, $|x| < R$ implies $F(x) = G(x)$. $\qquad \square$

The Natural Logarithm

One of the most important functions in analysis may be defined as a definite integral.

Definition 9.4.11. The *natural logarithm* function $\log : (0, \infty) \to \mathbf{R}$ is defined by

$$\log x = \int_1^x \frac{dt}{t}.$$

Proposition 9.4.12. *For all positive real numbers x and y,*

$$\log xy = \log x + \log y, \qquad\qquad \log(1/x) = -\log x.$$

The natural logarithm is strictly increasing and surjective. There exists a unique real number e such that $\log e = 1$, and $2 < e < 3$.

Proof. Exercise 9.4.5.   □

Corollary 9.4.13. *If $x_0 > 0$, then*

$$\log x - \log x_0 = \sum_{k=0}^{\infty} \frac{(-1)^k}{x_0^{k+1}} \frac{(x - x_0)^{k+1}}{k+1} \qquad on\ (0, 2x_0).$$

Proof. See Exercise 9.4.7.   □

Exercises for Section 9.4

Exercise 9.4.1.(★) Prove Lemma 9.4.2.

Exercise 9.4.2. Prove that $\int_0^x |t|\,dt = \dfrac{x|x|}{2}$ for all real x.

Exercise 9.4.3. Assume a, b, c are real. Evaluate $\int_a^b (c+t)^2\,dt$ in two ways.

Exercise 9.4.4. Assume f is integrable on $[a, b]$. Prove there exists an x such that $a \le x \le b$ and

$$\int_a^x f(t)\,dt = \int_x^b f(t)\,dt.$$

Exercise 9.4.5. This exercise establishes the properties of log in Proposition 9.4.12. Throughout, x and y denote positive real numbers.

(a) Prove that $\log(xy) = \log x + \log y$. Hints: Split the integral over $[1, xy]$ into $[1, x] \cup [x, xy]$, then use Proposition 9.2.8 with $\mu = x$.

(b) Prove that $\log(1/x) = -\log x$.

(c) Prove log is strictly increasing, and $\log 2 < 1 < \log 3$. Show there exists a unique real number e such that $\log e = 1$. Hint: To prove $1 < \log 3$, first show $1 - (t/4) \le 1/t$ for all positive t, with equality if and only if $t = 2$.

(d) Use parts (a) and (b) to prove $\log(e^n) = n$ for every integer n. Conclude that $\log : (0, \infty) \to \mathbf{R}$ is surjective.

Exercise 9.4.6. (H) This exercise continues Exercise 9.4.5, expressing exponentiation with rational exponents in terms of log.

(a) Prove $\log(x^p) = p \log x$ for all positive real x, and for all integers p.

(b) If $r = p/q$ is rational, prove $\log(x^r) = r \log x$ for all positive real x.

Exercise 9.4.7. (H) Prove Corollary 9.4.13.

Exercise 9.4.8. Define

$$f(t) = \sum_{k=0}^{\infty} \frac{(-1)^k t^{2k}}{2^k \, k!}, \qquad\qquad F(x) = \int_0^x f(t) \, dt.$$

Verify the series for f has infinite radius, expand F as a power series, and
determine how many terms suffice to approximate $F(x)$ on $[-\sqrt{2}, \sqrt{2}]$ with
error at most 0.5×10^{-5}.

10

Differentiation

For functions describing natural phenomena, incrementing the input by a small amount typically results in a nearly proportional increment in the output. Symbolically, if f is a function and x_0 is a point of the domain, then for small Δx (Delta x) we have $f(x_0 + \Delta x) - f(x_0) \approx m(x_0)\,\Delta x$ for some function m. Functions with this property are the subject of this chapter.

10.1 Differentiability

Definition 10.1.1. Assume X is a non-empty set of real numbers, $f : X \to \mathbf{R}$ a function, and x_0 an interior point of X; that is, there exists a δ such that $B_\delta(x_0) \subseteq X$. If $0 < |h| < \delta$ and $x = x_0 + h$, define the *difference quotient*

$$\Delta f(x_0, h) = \frac{f(x_0 + h) - f(x_0)}{h} = \frac{f(x) - f(x_0)}{x - x_0}.$$

If $\Delta f(x_0, h)$ has a limit as $h \to 0$, we say f is *differentiable at x_0*, and call

$$f'(x_0) = \lim_{h \to 0} \Delta f(x_0, h) = \lim_{h \to 0} \frac{f(x_0 + h) - f(x_0)}{h} = \lim_{x \to x_0} \frac{f(x) - f(x_0)}{x - x_0}$$

the *derivative* of f at x_0.

If X is open and f is differentiable at x_0 for each x_0 in X, we say f is *differentiable on X*.

Remark 10.1.2. The difference quotient $\Delta f(x_0, h)$ represents the "average rate of change" of f on the interval between x_0 and $x = x_0 + h$, namely, the change in f on this interval divided by the length of the interval.

To say f is differentiable at x_0 means these rates of change approach a finite limit as $h \to 0$, in which case the derivative represents, by definition, the "instantaneous rate of change" of f at x_0. ◇

Example 10.1.3. If f is a constant function on some open interval X, then f is differentiable on X. Indeed, $f'(x_0) = 0$ for all x_0 since every difference quotient of f is zero. ◇

DOI: 10.1201/9781003601357-10

Example 10.1.4. The signum function sgn is differentiable on $\mathbf{R} \setminus \{0\}$, and $\text{sgn}'(x) = 0$ for all x in $\mathbf{R} \setminus \{0\}$, but sgn is not constant. $\Diamond$

Example 10.1.5. The identity function ι is differentiable on an arbitrary open interval. Since every difference quotient is equal to 1, we have $\iota'(x_0) = 1$ for all x_0. $\Diamond$

Example 10.1.6. The quadratic polynomial $f(x) = 1 + 2x + x^2$ is differentiable on $\mathbf{R}$, and $f'(x_0) = 2 + 2x_0$. To prove this, we calculate

$$\Delta f(x_0, h) = \frac{f(x) - f(x_0)}{x - x_0} = \frac{(1 + 2x + x^2) - (1 + 2x_0 + x_0^2)}{x - x_0}$$
$$= \frac{2(x - x_0) + (x^2 - x_0^2)}{x - x_0} = 2 + (x + x_0),$$

since $x \neq x_0$. Fixing x_0 and taking the limit as $x \to x_0$, we have

$$f'(x_0) = \lim_{x \to x_0} \frac{f(x) - f(x_0)}{x - x_0} = \lim_{x \to x_0} 2 + (x + x_0) = 2 + 2x_0. \qquad \Diamond$$

Example 10.1.7. The square root function $f(x) = \sqrt{x}$ is differentiable on $(0, \infty)$, and $f'(x_0) = 1/(2\sqrt{x_0})$: If $x_0 > 0$ and $|h| < x_0$, then

$$\Delta f(x_0, h) = \frac{\sqrt{x_0 + h} - \sqrt{x_0}}{h} = \frac{\sqrt{x_0 + h} - \sqrt{x_0}}{h} \cdot \frac{\sqrt{x_0 + h} + \sqrt{x_0}}{\sqrt{x_0 + h} + \sqrt{x_0}}$$
$$= \frac{(x_0 + h) - x_0}{h(\sqrt{x_0 + h} + \sqrt{x_0})} = \frac{1}{\sqrt{x_0 + h} + \sqrt{x_0}}.$$

Since the square root function is continuous,

$$f'(x_0) = \lim_{h \to 0} \Delta f(x_0, h) = \lim_{h \to 0} \frac{1}{\sqrt{x_0 + h} + \sqrt{x_0}} = \frac{1}{2\sqrt{x_0}}.$$

The square root function is not differentiable at 0. First, 0 is not an interior point of the domain $X = [0, \infty)$. Even if we restrict to $h \to 0^+$, however, the difference quotient $\Delta f(0, h) = \sqrt{h}/h = 1/\sqrt{h}$ diverges to ∞. $\Diamond$

Local Extrema

Definition 10.1.8. Assume $f : X \to \mathbf{R}$ is a function and x_0 is a limit point of X. We say x_0 is a *local minimum* of f in X if there exists a δ such that $f(x_0) \leq f(x)$ for all x in $X \cap B_\delta(x_0)$, namely, for all x in X such that $|x - x_0| < \delta$. The real number $f(x_0)$ is called a *local minimum value* of f in X.

If instead $x \in X \cap B_\delta^\times(x_0)$ implies $f(x_0) < f(x)$, we say x_0 is a *strict local minimum* of f.

Analogously, we define *local maximum* and *strict local maximum* points of X. A *local extremum* of f in X is a point that is either a local minimum or local maximum.

Example 10.1.9. On the set $X = [0, 1]$, the function $f(x) = x$ has a local minimum at 0 and a local maximum at 1.

On the open interval $(0, 1)$, the function $f(x) = x$ is bounded, but has no local extrema. ◇

Proposition 10.1.10. *If $f : [a, b] \to \mathbf{R}$ is a continuous function, and x_0 is a local extremum of f in X, then x_0 is precisely one of the following:*

(i) *An endpoint of $[a, b]$.*

(ii) *An interior point at which $f'(x_0)$ does not exist.*

(iii) *An interior point at which $f'(x_0) = 0$.*

Proof. Suppose x_0 is a local extremum. If x_0 is an endpoint, or if $f'(x_0)$ does not exist, there is nothing to prove. Assume, therefore, that $x_0 \in (a, b)$, and

$$f'(x_0) = \lim_{x \to x_0} \frac{f(x) - f(x_0)}{x - x_0}$$

exists. We may assume without loss of generality that x_0 is a local minimum; if not, replace f by $-f$. By definition of a local minimum, $f(x_0) \leq f(x)$ if $x \approx x_0$. That is, the numerator of the difference quotient is non-negative.

Letting $x \to x_0$ from below, that is, assuming $x - x_0 < 0$, we have $f'(x_0) \leq 0$. Letting $x \to x_0$ from above, that is, assuming $x - x_0 > 0$, we have $f'(x_0) \geq 0$. Consequently, $f'(x_0) = 0$. □

Approximate Linearity

Proposition 10.1.11. *Assume m is a real number. A function f is differentiable at an interior point x_0, with $m = f'(x_0)$, if and only if*

$$f(x_0 + h) \approx f(x_0) + mh + o(h) \quad \text{if } h \approx 0.$$

Proof. By definition, f is differentiable at x_0 and $m = f'(x_0)$ if and only if

$$0 = \lim_{h \to 0} \frac{f(x_0 + h) - f(x_0)}{h} - m = \lim_{h \to 0} \frac{f(x_0 + h) - f(x_0) - mh}{h}.$$

Again by definition, $f(x_0 + h) - f(x_0) - mh \approx o(h)$ if $h \approx 0$. □

Definition 10.1.12. If f is differentiable at x_0 with derivative $f'(x_0)$, the polynomial $P_{x_0}^1 f(x) = f(x_0) + f'(x_0)(x - x_0)$ is called the *first-degree germ* of f at x_0. The graph $y = P_{x_0}^1 f(x)$ is the *tangent line* to $y = f(x)$ at x_0.

Remark 10.1.13. The first-degree germ $P_{x_0}^1 f$ is the *unique* polynomial p of degree at most 1 satisfying $f(x) \approx p(x) + o(x - x_0)$ if $x \approx x_0$.

Geometrically, $f'(x_0)$ exists if and only if zooming in on the graph at x_0 causes the graph to look more and more like a line with slope $f'(x_0)$. ◇

Proposition 10.1.14. *If f is differentiable at x_0, then f is continuous at x_0.*

Proof. If f is differentiable at x_0, Proposition 10.1.11 implies

$$f(x) \approx f(x_0) + f'(x_0)(x - x_0) + o(x - x_0)$$
$$\approx f(x_0) + o(1) \quad \text{if } x \approx x_0. \qquad \square$$

Example 10.1.15. The converse of Proposition 10.1.14 is false. The absolute value function, $f(x) = |x|$, is continuous everywhere, but not differentiable at 0:

$$\Delta f(0, h) = \frac{f(h) - f(0)}{h - 0} = \frac{|h|}{h} = \operatorname{sgn} h,$$

which has no limit at 0. (Geometrically, zooming in at the origin preserves the graph, and the graph is not a line.) As noted in Chapter 8, a "typical" continuous function is *nowhere* differentiable: Zooming in does not stabilize the graph at all, much less cause the graph to look like a line.   ◇

Remark 10.1.16. A functional relation $y = f(x)$ determines the "dependent variable" y in terms of the "independent variable" x, and a difference quotient may be written $\Delta y / \Delta x = \Delta y(x_0, \Delta x)/\Delta x$. If f is differentiable at x, the limit of $\Delta y / \Delta x$ as $\Delta x \to 0$ may be written in *differential notation*, $dy/dx = dy/dx(x_0)$ rather than $f'(x_0)$, and viewed as a formal ratio of "infinitesimal increments" dy and dx. Literally, reciprocal finitude of the reals, Corollary 4.3.7, asserts no such real numbers exist. Nonetheless, properties we will establish in Section 10.2 may be viewed as granting license to manipulate ratios of infinitesimals as if they were ordinary (real) fractions, see Exercise 10.2.9. The rules for symbolically representing differential expressions and manipulating them to obtain correct results is self-descriptively called *differential calculus*.   ◇

Derivatives as Functions

If f is differentiable at each point of its domain, we may view the derivative f' as a function, and ask about properties of f', such as continuity.

Definition 10.1.17. Assume X is a non-empty open set in $\mathbf{R}$ and $f : X \to \mathbf{R}$ is a function. If f is differentiable on X, the function $f' : X \to \mathbf{R}$ is the *(first)* *derivative* of f, and we say f is a *primitive* of f'. (Thus, "primitive" is a noun in real analysis, not an adjective.) If in addition f' is a continuous function on X, we say f is *of class $\mathscr{C}^1$ on X.*

Remark 10.1.18. In an inductive spirit, we may ask whether $f^{(1)} := f'$ is itself differentiable. If so, we may form the *second derivative* $f^{(2)} := (f^{(1)})' = f''$, and ask the same questions recursively. Doing so builds a tower of increasingly stringent conditions a function may or may not satisfy.   ◇

Definition 10.1.19. Assume X is a non-empty open set of real numbers, n a positive integer, and $f : X \to \mathbf{R}$ an n-times differentiable function with nth derivative function $f^{(n)}$. If $f^{(n)}$ is continuous on X, we say f is *of class* $\mathscr{C}^n$ on X.

If $f^{(n)}$ is differentiable on X, we inductively define $f^{(n+1)} = (f^{(n)})'$, and say f is $(n + 1)$ *times differentiable on* X

Most functions in this book are n times differentiable for *every* $n \geq 1$.

Definition 10.1.20. Assume X is a non-empty open set of real numbers, and $f : X \to \mathbf{R}$ a function. If f is n times differentiable for every positive integer n, we say f is *smooth*, or *of class* $\mathscr{C}^\infty$, on X.

Remark 10.1.21. The nth derivative of a smooth function is automatically continuous by Proposition 10.1.14.

A function of class $\mathscr{C}^n$ is sometimes said to be *n times continuously differentiable*, and a smooth function is sometimes said to be *infinitely differentiable*. This book does not use either term. ◇

Exercises for Section 10.1

Exercise 10.1.1. (H) If $f(x) = \sqrt{5}x^2 - 4\sqrt[3]{7}x + 10^{40}$, calculate $f'(x)$ from the definition.

Exercise 10.1.2. (★) Analyze the difference quotient at 0 if $f_{1/3}(x) = x^{1/3}$.

Exercise 10.1.3. Analyze the difference quotient at 0 if $f_{2/3}(x) = x^{2/3}$.

Exercise 10.1.4. Analyze the difference quotient at 0 if $f_{4/3}(x) = x^{4/3}$.

Exercise 10.1.5. Assume $f(x) = \sqrt[3]{x}$ if $x > 0$. Calculate f' from the definition.

Exercise 10.1.6. (★) Assume $f : \mathbf{R} \to \mathbf{R}$ is differentiable and ℓ-periodic. Prove f' is ℓ-periodic. Does f necessarily have a periodic primitive?

Exercise 10.1.7. Suppose f is differentiable in a neighborhood of x_0, and $f'(x_0) > 0$. Prove there is a δ such that if $0 < h < \delta$ then $f(x_0 - h) < f(x_0)$ and $f(x_0) < f(x_0 + h)$.

Exercise 10.1.8. (★) Suppose we are solving an equation $f(x) = g(x)$ for x. May we differentiate both sides to obtain $f'(x) = g'(x)$? Explain carefully.

Exercise 10.1.9. (★) In each part, assume $f : \mathbf{R} \to \mathbf{R}$ is a function.

(a) If $|f(x)| \le x^2$ for all real x, prove f is differentiable at 0 and $f'(0) = 0$.

(b) Construct a function that is differentiable at 0 and discontinuous at x for all non-zero x.

Exercise 10.1.10. For each positive integer N, there is a function $f : \mathbf{R} \to \mathbf{R}$ that is discontinuous at x for every non-zero x, but is $o(|x|^{1+(1/N)})$ if $x \approx 0$, so $f'(0) = 0$.

Exercise 10.1.11. ($\bigstar$) Assume f is defined in a neighborhood of x_0, and put $y_0 = f(x_0)$. Find the equation of the graph $y = f(x)$ after zooming in at (x_0, y_0) with factor c. Prove f is differentiable at x_0 if and only if the limit as $c \to \infty$ of the zoomed-in graph is a line.

10.2 Differentiation Rules

Theorem 10.2.1 (Linearity of the derivative). *Assume f and g are differentiable at x and c is a real number. The functions $f \pm g$ and cf are differentiable at x, and*

$$(f \pm g)'(x) = f'(x) \pm g'(x), \qquad\qquad (cf)'(x) = c\,f'(x).$$

Proof. By hypothesis, there exist real numbers $f'(x)$ and $g'(x)$ such that

$$\left.\begin{aligned} f(x+h) &\approx f(x) + h\,f'(x) + o(h) \\ g(x+h) &\approx g(x) + h\,g'(x) + o(h) \end{aligned}\right\} \quad \text{if } h \approx 0.$$

Adding,

$$(f+g)(x+h) \approx (f+g)(x) + h\left(f'(x) + g'(x)\right) + o(h).$$

By Proposition 10.1.11, this simultaneously proves $f + g$ is differentiable at x and shows the derivative is $f'(x) + g'(x)$.

The proofs for $f - g$ and cf are entirely similar. $\qquad\qquad\square$

Theorem 10.2.2 (The product and quotient rules). *If f and g are differentiable at x, then fg is differentiable at x, and*

$$(fg)'(x) = f'(x)g(x) + f(x)g'(x).$$

If $g(x) \ne 0$, then f/g is differentiable at x, and

$$\left(\frac{f}{g}\right)'(x) = \frac{f'(x)g(x) - f(x)g'(x)}{g(x)^2}.$$

Proof. By hypothesis,

$$\left.\begin{aligned} f(x+h) &\approx f(x) + h\,f'(x) + o(h) \\ g(x+h) &\approx g(x) + h\,g'(x) + o(h) \end{aligned}\right\} \quad \text{if } h \approx 0.$$

Multiplying these approximations gives

$$\begin{aligned} (fg)(x+h) &= \big[f(x) + h\,f'(x) + o(h)\big]\big[g(x) + h\,g'(x) + o(h)\big] \\ &= (fg)(x) + h\big[f'(x)g(x) + f(x)g'(x)\big] + o(h). \end{aligned}$$

By Proposition 10.1.11, this establishes the product rule.

For quotients, algebra and the preceding substitutions give

$$\begin{aligned} \Delta(f/g)(x,h) &= \frac{1}{h}\left[\frac{f(x+h)}{g(x+h)} - \frac{f(x)}{g(x)}\right] \\ &= \frac{1}{h}\left[\frac{f(x+h)g(x) - f(x)g(x+h)}{g(x+h)g(x)}\right] \\ &= \frac{1}{h}\left[\frac{h\big(f'(x)g(x) - f(x)g'(x)\big) + o(h)}{g(x+h)g(x)}\right] \\ &= \frac{f'(x)g(x) - f(x)g'(x)}{g(x+h)g(x)} + o(1). \end{aligned}$$

Since g is differentiable at x, $g(x+h) \approx g(x) + O(h)$ if $h \approx 0$. $\qquad\square$

Corollary 10.2.3. *Assume n is a natural number.*

(i) $f(x) = x^n$ *is differentiable on* $\mathbf{R}$, *and* $f'(x) = nx^{n-1}$.

(ii) $g(x) = x^{-n}$ *is differentiable on* $\mathbf{R} \setminus \{0\}$, *and* $g'(x) = -nx^{-n-1}$.

Proof. (i). Although the function $f(x) = x^0 = 1$ is constant by Definition 3.3.1, the formula $f'(x) = nx^{n-1} = 0 \cdot x^{-1}$ is an edge case: We interpret it as meaning $f'(x) = 0$ for all real x, including by continuity at $x = 0$.

Properly, therefore, our induction starts with $n = 1$, for which $f(x) = x$ is the identity function ι, and $f'(x) = 1 = x^0 = nx^{n-1}$ for all real x.

Assume inductively for some positive integer n that the derivative of $f(x) = x^n$ is $f'(x) = nx^{n-1}$ for all real x. Taking $g(x) = x$ and using the product rule, the derivative of $(fg)(x) = x^{n+1}$ is

$$(fg)'(x) = f'(x)g(x) + f(x)g'(x) = (nx^{n-1}) \cdot x + x^n \cdot 1 = (n+1)x^n.$$

(ii). Assume $n \geq 1$ and $g(x) = x^{-n} = 1/x^n = 1/f(x)$. By the quotient rule,

$$g'(x) = \left(\frac{1}{f}\right)'(x) = -\frac{f'(x)}{f(x)^2} = -\frac{nx^{n-1}}{(x^n)^2} = -nx^{-n-1}. \qquad\square$$

Corollary 10.2.4. *A polynomial function*

$$p(x) = \sum_{k=0}^{n} a_k x^k = a_0 + a_1 x + a_2 x^2 + \cdots + a_n x^n$$

is differentiable on $\mathbf{R}$, *and*

$$p'(x) = \sum_{k=1}^{n} k a_k x^{k-1} = a_1 + 2a_2 x + 3a_3 x^2 + \cdots + n a_n x^{n-1}.$$

Since p' *is itself a polynomial,* p *is smooth.*

Example 10.2.5. Every rational function is smooth in its natural domain: If p and q are polynomials and $f(x) = p(x)/q(x)$, then f is differentiable for all real x where $q(x) \neq 0$, and f' is a rational function with the same natural domain as f. For example, using differential notation for convenience,

$$\frac{d}{dx} \frac{1}{1+x+x^2} = -\frac{1+2x}{(1+x+x^2)^2},$$

$$\frac{d}{dx} \frac{x}{1+x^2} = \frac{(1+x^2) - x(2x)}{(1+x^2)^2} = \frac{1-x^2}{(1+x^2)^2}. \qquad \Diamond$$

Theorem 10.2.6 (The chain rule). *Assume* g *and* f *are composable functions. If* f *is differentiable at* x *and* g *is differentiable at* $y = f(x)$, *then* $g \circ f$ *is differentiable at* x, *and*

$$(g \circ f)'(x) = g'\big(f(x)\big) \cdot f'(x) = g'(y) \cdot f'(x).$$

Proof. By hypothesis,

$$f(x+h) \approx f(x) + h\, f'(x) + o(h) \quad \text{if } h \approx 0,$$
$$g(y+k) \approx g(y) + k\, g'(y) + o(k) \quad \text{if } k \approx 0.$$

If we write $y = f(x)$ and $y + k = f(x+h)$, then

$$k = f(x+h) - f(x) \approx h\, f'(x) + o(h).$$

Thus $k \approx O(h)$ if $h \approx 0$, so $o(k) \approx o(h)$. Consequently,

$$\begin{aligned}
(g \circ f)(x+h) - (g \circ f)(x) &= g(y+k) - g(y)\\
&\approx k\, g'(y) + o(k)\\
&\approx \big(h\, f'(x) + o(h)\big) g'(y) + o(h)\\
&\approx h\, g'\big(f(x)\big) \cdot f'(x) + o(h).
\end{aligned}$$

By Proposition 10.1.11, $(g \circ f)'(x) = g'\big(f(x)\big) \cdot f'(x)$. $\qquad \square$

Example 10.2.7. If n is an integer, then for all real x,

$$\frac{d}{dx}(1 + x + x^2)^n = n(1 + x + x^2)^{n-1}(1 + 2x),$$

$$\frac{d}{dx}\left[\frac{x}{1 + x^2}\right]^n = n\left[\frac{x}{1 + x^2}\right]^{n-1}\frac{1 - x^2}{(1 + x^2)^2}. \qquad \Diamond$$

Remark 10.2.8. The preceding functions would be all but impossible to differentiate without the chain rule; the only recourse would be to multiply out, differentiate term by term, and attempt to factor the result. $\qquad \diamond$

Example 10.2.9. The absolute value function $a(x) = |x|$ is differentiable except at 0, and $a'(x) = x/|x|$. If f is differentiable, then $|f|$ is differentiable except possibly where $f = 0$, and $|f|'(x) = f(x)f'(x)/|f(x)|$. $\qquad \Diamond$

Differentiability of Inverse Functions

Theorem 10.2.10 (The inverse function theorem in one variable). *Assume I is an open interval in $\mathbf{R}$, and $f : I \to \mathbf{R}$ a continuous, strictly monotone function. If f is differentiable at some x_0 in I, then the inverse $g = f^{-1}$ is differentiable at $y_0 = f(x_0)$ if and only if $f'(x_0) \neq 0$, and in this event*

$$g'(y_0) = \frac{1}{f'(x_0)}.$$

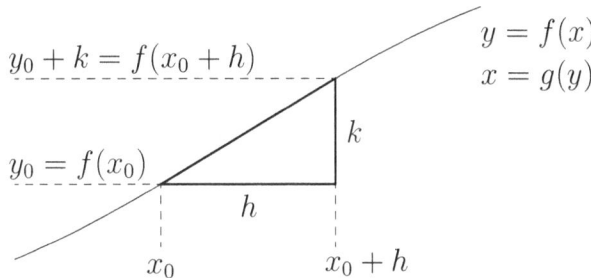

FIGURE 10.1
The difference quotient of an inverse function.

Proof. For every real number h such that $x_0 + h = x \in I$, we may write $y = f(x) = f(x_0 + h)$ and $k = y - y_0 = f(x_0 + h) - f(x_0)$, see Figure 10.1. Since f is injective, k is uniquely determined by h, and since f and f^{-1} are continuous (see Proposition 8.2.11), $o(h) \approx o(k)$ if $k \approx 0$. By algebra, or reading off Figure 10.1, $\Delta(f^{-1})(y_0, k) = h/k = 1/\Delta f(x_0, h)$. If $f'(x_0) \neq 0$, then

$$(f^{-1})'(y_0) = \lim_{k \to 0} \Delta(f^{-1})(y_0, k) = \lim_{h \to 0} \frac{1}{\Delta f(x_0, h)} = \frac{1}{f'(x_0)}.$$

Conversely, $x = f^{-1}\big(f(x)\big)$ for all x in I. If f^{-1} is differentiable at $y_0 = f(x_0)$, the chain rule implies $1 = (f^{-1})'(y_0) \cdot f'(x_0)$, and therefore $f'(x_0) \neq 0$.   □

Corollary 10.2.11. *If r is rational and $f : (0, \infty) \to \mathbf{R}$ is defined by $f(x) = x^r$, then f is differentiable, and $f'(x) = rx^{r-1}$.*

Proof. See Exercise 10.2.5.   □

Exercises for Section 10.2

Exercise 10.2.1. (★) Find *all* the derivatives of the following functions:

(a) $s(x) = x - (x^3/3!)$.

(b) $c(x) = 1 - (x^2/2!) + (x^4/4!)$.

(c) $e(x) = 1 + x + (x^2/2!) + (x^3/3!)$.

Exercise 10.2.2. (★) Assume n is a natural number and $f(x) = x^n$. Find all the derivatives of f, expressing them in terms of factorials and powers of x. In particular, show that $f^{(n)}(x) = n!$, and that $f^{(k)}(x) = 0$ if $k > n$.

Exercise 10.2.3. If $n \geq 2$ and $f(x) = |x|^n$, find a formula for $f'(x)$.

Exercise 10.2.4. Define $f : (0, \infty) \to \mathbf{R}$ by $f(x) = 1/\sqrt{x}$.

(a) Prove f is differentiable and calculate $f'(x)$ from the definition.

(b) Prove f is differentiable and calculate $f'(x)$ using Theorem 10.2.10.

Exercise 10.2.5. (★) Prove Corollary 10.2.11.

Exercise 10.2.6. Define $f : (-1, \infty) \to \mathbf{R}$ by $f(x) = (1 + x)^{1/2}$. Find all the derivatives of f. Express $f^{(k)}(0)$ in terms of factorials.

Exercise 10.2.7. In each part, m and n denote natural numbers. Use algebraic identities, and the product, quotient, and chain rules to calculate the derivatives of the indicated functions. Re-use answers to earlier parts where possible. In your answers, it should be possible to tell by inspection where the derivative is zero or undefined.

(a) $f(t) = (1 + t)^n (1 - t)^n$, $g(t) = (1 + t^2)^n (1 - t^2)^n$.

(b) $f(t) = (1 + t)^m (1 - t)^n$, $g(t) = (1 + t^2)^m (1 - t^2)^n$.

(c) $f(t) = \dfrac{1-t}{1+t}$, $\qquad g(t) = \dfrac{1-t^2}{1+t^2}$, $\qquad h(t) = \left[\dfrac{1-t^2}{1+t^2}\right]^n$.

Exercise 10.2.8.($\bigstar$) In each part, n denotes a natural number. By reverse-engineering the chain rule if possible, or other means if not, find primitives of the indicated functions. Note that if r is rational and $r \neq -1$, then $\Phi(u) = \frac{1}{r+1}u^{r+1}$ is a primitive of $\phi(u) = \Phi'(u) = u^r$.

(a) $f(t) = (1+t)^n$, $\qquad g(t) = t(1+t^2)^n$, $\qquad h(t) = (1+t^2)^n$.

(b) $f(t) = (1+t)^{n/2}$, $\qquad g(t) = 2t(1+t^2)^{n/2}$, $\qquad h(t) = t/(2+t^2)^{n/2}$, $n \neq 2$.

Exercise 10.2.9.($\bigstar$) Match each differential formula with the corresponding proposition or theorem from this chapter, and specify the appropriate functional relationships.

(a) $\dfrac{d(cy+z)}{dx} = c\dfrac{dy}{dx} + \dfrac{dz}{dx}$. $\qquad\qquad$ (b) $\dfrac{d(uv)}{dx} = u\dfrac{dv}{dx} + v\dfrac{du}{dx}$.

(c) $\dfrac{dz}{dx} = \dfrac{dz}{dy}\dfrac{dy}{dx}$. $\qquad\qquad\qquad$ (d) $\dfrac{dx}{dy} = \dfrac{1}{dy/dx}$.

Exercise 10.2.10. Assume f is a function of class $\mathscr{C}^2$; that is, f'' exists and is continuous. Assume $f(x_0) > 0$, and define $g(x) = 1/f(x)$ provided $f(x) \neq 0$.

Find a formula for g'' in a neighborhood of x_0. Your formula should depend only on the values of f and its first two derivatives at x_0.

If $f''(x_0) > 0$ what (if anything) is guaranteed about the sign of $g''(x_0)$? What if $f''(x_0) < 0$?

Exercise 10.2.11. Assume f is of class $\mathscr{C}^2$, and $f'(x_0) > 0$, so f is invertible near x_0, and set $y_0 = f(x_0)$.

If g is the branch of f^{-1} satisfying $g(y_0) = x_0$, prove g is twice differentiable at y_0, and find an expression for g'' near y_0.

If $f''(x_0) > 0$ what (if anything) is guaranteed about the sign of $g''(y_0)$? What if $f''(x_0) < 0$?

Exercise 10.2.12. Suppose $f : \mathbf{R} \to \mathbf{R}$ is differentiable and satisfies $f' = 1-f^2$. Prove f is smooth, and calculate f'' and f''' as functions of f.

Exercise 10.2.13. Prove that a differentiable function $f : \mathbf{R} \to \mathbf{R}$ satisfies

$$f'\left(\frac{s+t}{2}\right) = \frac{f(t)-f(s)}{t-s} \quad \text{for all real } s, t$$

if and only if f is a polynomial of degree at most 2.

Exercise 10.2.14. (Unexpected tangents.) Assume f is a differentiable function, ℓ a non-vertical line through an arbitrary point $\big(x_0, f(x_0)\big)$, and ε arbitrary. This three-part exercise constructs a differentiable function g that differs from f everywhere by at most ε, and has ℓ as a tangent line at x_0.

(a) Define $\phi(x) = x/(1 + x^2)$. Calculate ϕ', and determine the absolute maximum and minimum values of ϕ. Let m be an arbitrary real number, and define $\phi_m(x) = \phi(mx)$. Show that the line $y = mx$ is tangent to the graph of ϕ_m at $x = 0$.

(b) If n is a positive integer, define $\psi_n(x) = (1/n)\phi_m(nx)$. Prove that the line $y = mx$ is tangent to the graph of ψ_n at the origin, and for sufficiently large n, $|\psi_n(x)| < \varepsilon$ for all real x.

(c) Assume $f : \mathbf{R} \to \mathbf{R}$ is differentiable, and ℓ is a non-vertical line through an arbitrary point $\big(x_0, f(x_0)\big) = (x_0, y_0)$ on the graph of f.

   Use a function ψ_n from part (b) to prove that for every ε, there exists a differentiable function g, defined on the same set as f, satisfying the conditions: (i) $g(x_0) = f(x_0)$, (ii) The line ℓ is tangent to the graph of g at (x_0, y_0), and (iii) $|g(x) - f(x)| < \varepsilon$ for all x in the domain of f.

In words, by perturbing f by an arbitrarily small amount in the vertical direction, we can make the graph "unexpectedly" tangent to an arbitrary line. Strictly speaking, we have no visual basis for saying any particular line is mathematically tangent to any particular graph.

10.3   The Mean Value Theorem

Theorem 10.3.1 (The mean value theorem). *If $f : [a, b] \to \mathbf{R}$ is a continuous function that is differentiable on (a, b), then there exists an x_0 in (a, b) such that*

$$f'(x_0) = \frac{f(b) - f(a)}{b - a}.$$

Remark 10.3.2. Analytically, the derivative at *some* point of (a, b) is equal to the secant slope over $[a, b]$. In terms of position and speed: If on a car trip you cover 60 miles in a certain one-hour period of time, then at some instant during that hour your speed must have been exactly 60 miles per hour.

   The true power of the mean value theorem, however, arises because if x_1 and x_2 are *arbitrary* real numbers such that $a \leq x_1 < x_2 \leq b$, we can apply the mean value theorem to f on the interval $[x_1, x_2]$. That is, when traveling from Point A to Point B, your average speed over *every* time interval is equal to your instantaneous speed at *some instant* during that interval. ◇

Proof. By subtracting the affine interpolation $f_{a,b}$, we may as well assume our function vanishes at a and b. Specifically, define $g : [a, b] \to \mathbf{R}$ by

$$g(x) = f(x) - f_{a,b}(x) = f(x) - \left[f(a) + \frac{f(b) - f(a)}{b - a} (x - a) \right].$$

The function g is continuous on $[a, b]$, differentiable on (a, b) with derivative

$$g'(x) = f'(x) - \frac{f(b) - f(a)}{b - a} \quad \text{for all } x \text{ in } (a, b),$$

and $g(a) = 0 = g(b)$. It suffices to prove $g'(x_0) = 0$ for some x_0 in (a, b).

By the extreme value theorem, there exist points $x_{\min}$ and $x_{\max}$ in $[a, b]$ such that

$$g(x_{\min}) \leq g(x) \leq g(x_{\max}) \quad \text{for all } x \text{ in } [a, b].$$

Suppose at least one of $x_{\min}$ and $x_{\max}$ is in (a, b), and call it x_0. By Proposition 10.1.10, $g'(x_0) = 0$ and the proof is complete.

Otherwise, each of our points $x_{\min}$ and $x_{\max}$ is an endpoint of $[a, b]$. Since the endpoint values $g(a) = 0 = g(b)$ are equal, the *extreme* values are equal; that is, $g(x) = 0$ for all x. Consequently, $g'(x_0) = 0$ for *every* point x_0 in (a, b). □

Theorem 10.3.3 (The identity theorem). *Assume f and g are differentiable functions on some interval I. If $f'(x) = g'(x)$ for every interior point x of I, then there exists a real number C such that $f(x) = g(x) + C$ for all x in I.*

Proof. The function $h = f - g$ is differentiable, and by hypothesis the derivative $h'(x) = f'(x) - g'(x)$ is 0 for all x in the interior of I. It suffices to prove h is constant. Since the hypothesis consists of infinitely many conditions, no finite number of which imply the conclusion, we prove the contrapositive.

If h is non-constant, there exist numbers a and b in I such that $a < b$ and $h(a) \neq h(b)$. By the mean value theorem applied to h on $[a, b]$, there exists an x_0 in (a, b) such that

$$h'(x_0) = \frac{h(b) - h(a)}{b - a} \neq 0. \qquad \square$$

Example 10.3.4. The floor function on $\mathbf{R} \setminus \mathbf{Z}$ has derivative identically 0, but is non-constant, equal to n on the open interval $(n, n + 1)$ for each integer n. This type of example shows the importance of assuming the domain is an *interval* in the identity theorem. ◇

Definition 10.3.5. Assume that X is a non-empty set of real numbers and $f : X \to \mathbf{R}$ is a function. We say f has *bounded stretch* on X if there exists a real M such that $|f(x_2) - f(x_1)| \leq M|x_2 - x_1|$ for all x_1 and x_2 in X. The infimum of all such M is called the *stretch* of f (on X).

We say f has *locally bounded stretch* on X if for every x_0 in X, there exists a positive r such that f has bounded stretch on $B_r(x_0)$.

Remark 10.3.6. Assume $\{x_1, x_2\} \subseteq X$. Since $|f(x_2) - f(x_1)| \leq M|x_2 - x_1|$ is automatic if $x_1 = x_2$, f has bounded stretch on X if and only if there is a real M such that

$$\left| \frac{f(x_2) - f(x_1)}{x_2 - x_1} \right| \leq M \quad \text{if } x_1 < x_2.$$

If f is differentiable and $|f'| \leq M$ on some interval $I \subseteq X$, then f has bounded stretch on I: By the mean value theorem, the quotient on the left is $|f'(x_0|$ for some x_0 in (x_1, x_2). Every $\mathscr{C}^1$ function has locally bounded stretch.   ◇

Monotonicity

Together with the intermediate value theorem (Theorem 8.4.1), derivatives provide a powerful computational tool for showing a function is injective and/or surjective on a real interval.

Theorem 10.3.7 (The monotonicity theorem). *If $f : [a, b] \to \mathbf{R}$ is continuous, differentiable on (a, b), and if $f'(z) > 0$ for all z in (a, b), then f is strictly increasing in $[a, b]$, and f is a bijection to the closed interval $[f(a), f(b)]$.*

Proof. Assume x and y are arbitrary numbers such that $a \leq x < y \leq b$. By the mean value theorem applied to f on $[x, y]$, there is an x_0 in (x, y) such that

$$\frac{f(y) - f(x)}{y - x} = f'(x_0).$$

By hypothesis, $f'(x_0) > 0$, and since $x < y$, it follows that $f(x) < f(y)$. Since x and y were arbitrary, f is strictly increasing on $[a, b]$.

In particular, f is injective, and the image of f is contained in $[f(a), f(b)]$: If $a \leq x \leq b$, then $f(a) \leq f(x) \leq f(b)$. Conversely, if $f(a) < y < f(b)$, the intermediate value theorem guarantees there exists an x in (a, b) such that $y = f(x)$; that is, $[f(a), f(b)]$ is contained in the image of f.   □

Remark 10.3.8. An entirely analogous argument shows that if f is continuous on $[a, b]$ and $f'(x) < 0$ for all x in (a, b), then f is decreasing on $[a, b]$, and the image of f is $[f(b), f(a)]$.   ◇

Example 10.3.9. The polynomial function $f(x) = x^3 - 3x$ is differentiable on $\mathbf{R}$, and $f'(x) = 3x^2 - 3 = 3(x - 1)(x + 1)$. Since $f'(x) > 0$ if $x < -1$, f is

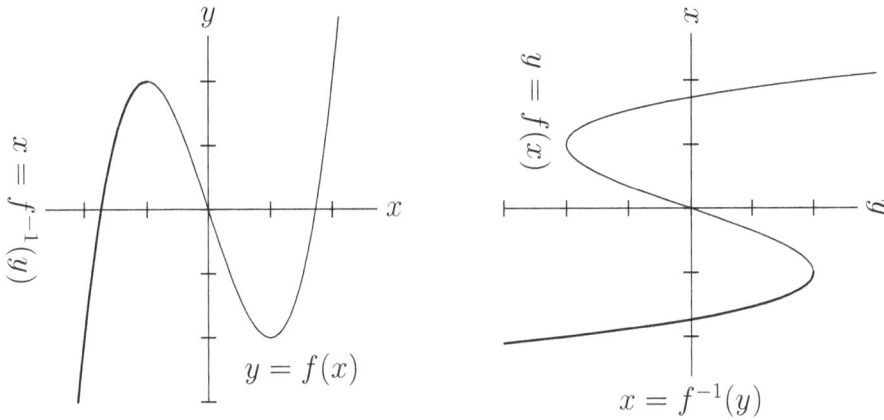

FIGURE 10.2
Branches of f^{-1} for $f(x) = x^3 - 3x$.

strictly increasing on the *closed* interval $(-\infty, -1]$. Further, the image of f on this interval is the interval $(-\infty, 2]$, since $f(x) \to -\infty$ as $x \to -\infty$ and $f(-1) = 2$, Figure 10.2.

Since $f'(x) < 0$ if $-1 < x < 1$, f is strictly decreasing on the closed interval $[-1, 1]$, and f maps this interval to $[-2, 2]$. Similarly, f is strictly increasing on $[1, \infty)$, and f maps this interval to $[-2, \infty)$.

It follows from this analysis that $f : \mathbf{R} \to \mathbf{R}$ is surjective; the equation $y = x^3 - 3x$ has at least one solution x for every real y. In fact, we have shown more. If $|y| < 2$, the equation $y = x^3 - 3x$ has precisely three solutions: one of them less than -1, one between -1 and 1, and one greater than 1. If $y = \pm 2$, there are precisely two solutions (one being ∓ 1). If $|y| > 2$, there is exactly one solution of $y = x^3 - 3x$.

This analysis also demonstrates the existence of a unique continuous function $g : (-\infty, 2] \to (-\infty, -1]$ satisfying $g(x^3 - 3x) = (g \circ f)(x) = x$ if $x \le -1$ and $g(y)^3 - 3g(y) = (f \circ g)(y) = y$ if $y \le 2$. The function g, a *branch of f^{-1}*, is differentiable on the open interval $(-\infty, 2)$, and

$$g'(y) = \frac{1}{f'(x)} = \frac{1}{3(x^2 - 1)} = \frac{1}{3(g(y)^2 - 1)}.$$

Similarly, there is a continuous branch of f^{-1} from $[-2, 2]$ to $[-1, 1]$, and a continuous branch of f^{-1} from $[-2, \infty)$ to $[1, \infty)$. $\diamond$

Exercises for Section 10.3

Exercise 10.3.1. ($\bigstar$) Assume n is a positive integer, b a positive real number, and $f(x) = x^n$. Prove there is a unique c in $(0, b)$ satisfying the mean value theorem for f on $[0, b]$.

Exercise 10.3.2. (A) In each part, define $f(x) = (x^2 - 1)^2$.

(a) Use the techniques of Example 10.3.9 to find branches of f^{-1}.

(b) Solve the equation $y = f(x)$ for x in terms of y. Match each formula you find with a branch of f^{-1} found in (b).

Exercise 10.3.3. In each part, define $f(x) = x/(1 + x^2)$.

(a) Use the techniques of Example 10.3.9 to find branches of f^{-1}.

(b) Solve the equation $y = f(x)$ for x in terms of y. Match each formula you find with a branch of f^{-1} found in (a).

Exercise 10.3.4. (A) Suppose I is an open set of real numbers and f is a differentiable function on I. Proof or counterexample:

(a) If $f'(x) = 0$ for all x in I, then f is constant on I.

(b) If $f'(x) > 0$ for all x in I, then f is strictly increasing on I.

Exercise 10.3.5. (★) Assume f is defined in some neighborhood of x_0.

(a) Assuming f is differentiable at x_0, evaluate

$$\lim_{h \to 0} \frac{f(x_0 + h) - f(x_0 - h)}{2h}.$$

(b) If the limit in part (a) exists at x_0, does it follow that f is differentiable at x_0? Continuous?

Exercise 10.3.6. If f is of class $\mathscr{C}^2$ in a neighborhood x_0, prove

$$f''(x_0) = \lim_{h \to 0^+} \frac{f(x_0 + h) + f(x_0 - h) - 2f(x_0)}{h^2}.$$

Exercise 10.3.7. (H) Assume $f, g : \mathbf{R} \to \mathbf{R}$ are differentiable functions satisfying $f' = f$ and $g' = g$. If g is non-vanishing, prove that $f(x) = f(0)g(x)/g(0)$.

Exercise 10.3.8. (H) This exercise gives a generalization of the mean value theorem. Assume f and g are continuous on some interval $[a, b]$ and differentiable on (a, b). Prove there exists a point x_0 in (a, b) such that

$$f'(x_0)\big(g(b) - g(a)\big) = g'(x_0)\big(f(b) - f(a)\big).$$

Exercise 10.3.9. (H) This exercise outlines the *zooming rule*, a computational procedure for evaluating certain indeterminate limits of the form $0/0$.

(a) Suppose f and g are differentiable in $(c, c + r)$ for some r, g is non-vanishing, and $\lim(f, c^+) = \lim(g, c^+) = 0$. Prove that if $\lim(f'/g', c^+) = \ell$, then $\lim(f/g, c^+) = \ell$.

In words, if the (one-sided or two-sided) limit of a quotient is formally $0/0$, try differentiating the numerator and denominator and re-evaluating. If the limit is ℓ, the original limit is also ℓ.

(b) Suppose f and g are differentiable and g is non-vanishing on some interval (R, ∞), and $\lim(f, \infty) = 0$, $\lim(g, \infty) = 0$. Prove that if $\lim(f'/g', \infty) = \ell$, then $\lim(f/g, \infty) = \ell$.

Exercise 10.3.10. In each part, assume r is a rational number and c a positive real. Use Exercise 10.3.9 to evaluate the stated limit.

(a) $\displaystyle\lim_{x \to c} \frac{x^r - c^r}{x - c}$.

(b) $\displaystyle\lim_{x \to c} \frac{(x^2 + c^2)^r - (2c^2)^r}{x - c}$.

10.4 Applications

The mean value theorem connects properties of f on an interval, such as monotonicity, to the derivative f'. This section gives additional applications.

The Intermediate Value Property

Definition 10.4.1. Assume I is an interval of real numbers. A function $f : I \to \mathbf{R}$ satisfies the *intermediate value property* (on I) if the following holds: For all a and b in I, and for every real number m between $f(a)$ and $f(b)$, there exists a real c in $[a, b]$ such that $f(c) = m$.

Remark 10.4.2. This can be phrased succinctly at the level of sets: If we put $A = \min\big(f(a), f(b)\big)$ and $B = \max\big(f(a), f(b)\big)$, then $[A, B] \subseteq f\big([a, b]\big)$. ◇

Proposition 10.4.3. *If the function f is differentiable on some open interval I, then its derivative f' satisfies the intermediate value property on I.*

Proof. See Exercise 10.4.5 □

Remark 10.4.4. Exercises 10.4.8 and 10.4.9 show that a derivative need not be continuous. Examples are harder to visualize than might be expected. ◇

Smooth Patching

Suppose f is given "piecewise" by formulas on abutting intervals. Theorem 10.4.5 guarantees the patched function is differentiable *if* it is continuous ("the graphs have the same height where they meet") and "the one-sided slopes agree." These hypotheses are not necessary, see Exercises 10.4.8 and 10.4.9.

Theorem 10.4.5 (Smooth patching). *Assume f is defined on some open ball $B_r(x_0)$, differentiable on the punctured ball $B_r^\times(x_0)$, and the one-sided limits $f'(x_0^-)$ and $f'(x_0^+)$ exist. Under these hypotheses, f is differentiable at x_0 if and only if f is continuous at x_0 and $f'(x_0^-) = f'(x_0^+)$.*

Proof. See Exercise 10.4.6.   □

Example 10.4.6. Assume $f(x) = ax^2 + bx + c$ if $x < 1$ and $f(x) = x^3$ if $1 \le x$. For which a, b and c is f differentiable?

Both formulas define smooth functions on **R**. If $x < 1$, Corollary 10.2.4 gives $f'(x) = 2ax + b$, while if $1 < x$ (strict inequality) we have $f'(x) = 3x^2$.

The one-sided limits of f and f' may be calculated by evaluation. Particularly, f is continuous at 1 if and only if

$$a + b + c = f(1^-) = f(1^+) = 1,$$

and "f has equal slopes" at 1 if and only if

$$2a + b = f'(1^-) = f'(1^+) = 3.$$

By smooth patching, f is differentiable on **R** if and only if f is differentiable at 1, if and only if $a + b + c = 1$ and $2a + b = 3$.

The second gives $b = 3 - 2a$; substituting in the first gives $a + (3 - 2a) + c = 1$, or $c = a - 2$. For each real a, the resulting function, which is given by $f(x) = ax^2 + (3 - 2a)x + (a - 2)$ if $x < 1$, is differentiable.   ◊

Convexity

Recall Definition 5.1.30: A function f defined on an interval is convex if f lies below its secants. We start with useful alternative characterizations.

Proposition 10.4.7. *Assume I is an interval of real numbers and $f : I \to \mathbf{R}$ is a function. If x, y, and z denote arbitrary points of I, then the following are equivalent:*

(i) *f is convex on I:* $f(z) \le f_{x,y}(z) = f(x) + \dfrac{f(y) - f(x)}{y - x}(z - x)$ *if $x < z < y$.*

(ii) $f\big((1 - s)x + sy\big) \le (1 - s)f(x) + sf(y)$ *if $x < y$ and $0 < s < 1$.*

(iii) $\dfrac{f(z) - f(x)}{z - x} \leq \dfrac{f(y) - f(x)}{y - x} \leq \dfrac{f(y) - f(z)}{y - z}$ *if* $x < z < y$.

Proof. Since each z in (x, y) may be written $x + s(y - x) = (1 - s)x + sy$ for a unique s in $(0, 1)$, convexity is equivalent to (ii). Operationally, applying a convex function to a convex linear combination "distributes upward."

If $[x, y] \subseteq I$ and $0 < s < 1$, then (ii) may be rearranged to

$$f\big((1 - s)x + sy\big) - f(x) \leq s\big(f(y) - f(x)\big) \leq f(y) - f\big(sx + (1 - s)y\big).$$

Taking $s = (z - x)/(y - x)$ in the first and $s = (y - z)/(y - x)$ in the second shows (ii) is equivalent to (iii). Operationally, if $[x, y]$ is split into two pieces at z, the secant slope over $[x, z]$ is no larger, and the secant slope over $[z, y]$ no smaller, than the secant slope over $[x, y]$. $\square$

Proposition 10.4.8. *Assume I is an open interval. If $f : I \to \mathbf{R}$ is convex, then f has locally bounded stretch in I. Particularly, $f(x) - f(x_0) \approx O(x - x_0)$ for each x_0.*

Proof. See Exercise 10.4.12. $\square$

Remark 10.4.9. A convex function can be discontinuous at endpoints of its domain. For example, $f(x) = 0$ if $0 \leq x < 1$ and $f(1) = 1$ is convex. $\diamond$

Lemma 10.4.10. *If f is continuous on $[a, b]$, vanishes at the endpoints, is twice-differentiable on (a, b), and if $f''(x) \geq 0$ for all x in (a, b), then $f(z) \leq 0$ for all z in (a, b).*

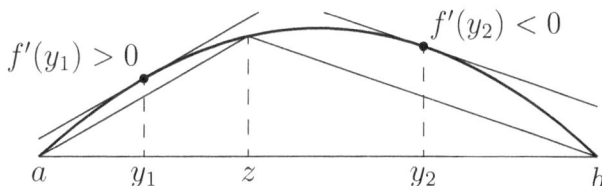

FIGURE 10.3
Determining the sign of f'' from the value of f.

Proof. Assume contrapositively that there exists a z in (a, b) such that $f(z) > 0$. It suffices to prove $f''(x) < 0$ for some x in (a, b). By the mean value theorem applied to f on $[a, z]$, there exists a y_1 in (a, z) such that

$$f'(y_1) = \frac{f(z) - f(a)}{z - a} = \frac{f(z)}{z - a} > 0.$$

Similarly, there is a point y_2 in (z, b) such that $f'(y_2) < 0$, Figure 10.3.

Applying the mean value theorem to f' on $[y_1, y_2]$, there is an x in (y_1, y_2) such that

$$f''(x) = \frac{f'(y_2) - f'(y_1)}{y_2 - y_1} < 0. \qquad \square$$

Proposition 10.4.11. *If f is twice differentiable on $[a, b]$, and $f''(z) \geq 0$ for all z in (a, b), then f is convex on $[a, b]$.*

Proof. Assume $a \leq x < y \leq b$, and consider the function

$$g(z) = f(z) - f_{x,y}(z) = f(z) - \left[f(x) + \frac{f(y) - f(x)}{y - x}(z - x) \right].$$

By construction, $g(x) = g(y) = 0$ and $g'' = f'' \geq 0$. Lemma 10.4.10 implies $g(z) \leq 0$ on $[x, y]$. Since x and y were arbitrary, f is convex on $[a, b]$. $\qquad \square$

Corollary 10.4.12. *If f is continuous on $[a, b]$, twice-differentiable, $f'' \geq 0$ on (a, b), and if $f'' = 0$ only at isolated points in (a, b), then f is strictly convex on $[a, b]$.*

Proof. Assume $a \leq x < y \leq b$. In the notation above, it suffices to prove $g(z) < 0$ (strict inequality) if $x < z < y$. By Lemma 10.4.10, $g \leq 0$ on (x, y). However, g cannot be identically zero, because $g'' = 0$ only at isolated points. Thus, $g(z_0) < 0$ for some z_0 in (x, y).

Since g is convex by Proposition 10.4.11 and $g(x) = 0$,

$$g(z) \leq g(x) + \frac{g(z_0) - g(x)}{z_0 - x}(z - x) = \frac{g(z_0)}{z_0 - x}(z - x) < 0$$

if $x < z \leq z_0$. A similar argument shows $g(z) < 0$ if $z_0 < z < y$. $\qquad \square$

Definition 10.4.13. Assume f is continuous in some neighborhood of a real number x_0. We say the point $\left(x_0, f(x_0) \right)$ on the graph $y = f(x)$ is an *inflection point* if "the concavity of f changes at x_0," namely, if there exists an open ball $B_r(x_0)$ such that f is strictly convex on $(x_0 - r, x_0)$ and strictly concave on $(x_0, x_0 + r)$, or else is strictly concave on $(x_0 - r, x_0)$ and strictly convex on $(x_0, x_0 + r)$.

Example 10.4.14. The function $f(x) = x^{1/3}$ is continuous on $\mathbf{R}$, though not differentiable at 0, and $f''(x) = (-2/9)x^{-5/3}$ changes sign at $x = 0$. Corollary 10.4.12 implies f is convex on $[-1, 0]$ and concave on $[0, 1]$, so the graph of f has an inflection point at the origin. $\qquad \Diamond$

Example 10.4.15. The function $f(x) = x^4$ is smooth, and $f''(x) = 12x^2 \geq 0$ for all real x, with equality only if $x = 0$. By Corollary 10.4.12, f is strictly convex on $\mathbf{R}$. $\qquad \Diamond$

Remark 10.4.16. In calculus, convexity tends to get conflated with positivity of the second derivative. In fact, a convex function can fail to be twice-differentiable at *every* point of its domain, see Exercise 10.4.11. $\qquad \Diamond$

Exercises for Section 10.4

Exercise 10.4.1. Find all values of a_1, a_2, b_1, b_2 such that the function

$$f(x) = \begin{cases} a_1 + (a_2/x) & x < -1, \\ 4 - x^2 & -1 \le x \le 1, \\ b_1 + (b_2/x^2) & 1 < x \end{cases}$$

is differentiable on $\mathbf{R}$.

Exercise 10.4.2. (★) Define $f : \mathbf{R} \to \mathbf{R}$ by $f(x) = (x^2 - 1)^2$. Find the maximal intervals of monotonicity and convexity of f, zeros of f, and any local extrema and inflection points.

Exercise 10.4.3. Define $f(x) = x/(1 + x^2)$. Find the maximal intervals of monotonicity and convexity of f, zeros of f, and any local extrema and inflection points. Use this to sketch the graph $y = f(x)$.

Exercise 10.4.4. Suppose $f(x) = \dfrac{2x}{1 - x^2}$ if $x \ne \pm 1$.

(a) Calculate $f'(x)$ and $f''(x)$. Find the intervals of monotonicity and convexity of f, the zeros of f, and any local extrema and inflection points. Use this information to sketch the graph $y = f(x)$.

(b) Use techniques of Example 10.3.9 to find corresponding branches of f^{-1}.

(c) Solve the equation $y = f(x)$ for x in terms of y. Match each formula you find with a branch of f^{-1} found in (a).

Exercise 10.4.5. Prove Proposition 10.4.3. Hint: In the notation of the theorem, put $g(x) = f(x) - mx$ on the interval $[a, b]$ and prove g takes its absolute minimum in (a, b).

Exercise 10.4.6. (H) Prove Theorem 10.4.5, smooth patching. Ideas: If (i) holds and (ii) does not, use Proposition 10.4.3 to prove f is not differentiable at x_0.

Inversely, if (i) and (ii) are both true, apply the mean value theorem to intervals contained in $B_r(x_0)$ and having x_0 as an endpoint.

Exercise 10.4.7. (A) Prove there exists a non-constant, 1-periodic function ψ (psi) of class $\mathscr{C}^1$ on $\mathbf{R}$.

Exercise 10.4.8. Let $\psi : \mathbf{R} \to \mathbf{R}$ be a non-constant, 1-periodic function of class $\mathscr{C}^1$, so $\psi(x + 1) = \psi(x)$ for all real x. (See Exercise 10.4.7.) Define $f(x) = x^2 \psi(1/x)$ if $x \ne 0$, and $f(0) = 0$.

(a) Prove that ψ is bounded.

(b) For $x \neq 0$, calculate $f'(x)$ in terms of $\psi(x)$ and $\psi'(x)$. Prove f' is locally bounded near 0.

(c) Prove $f'(0)$ exists, but f' is discontinuous at 0.

Exercise 10.4.9. With notation of Exercise 10.4.8, define $f(x) = x^2\psi(1/x^2)$ if $x \neq 0$, and $f(0) = 0$. Prove that f is differentiable on $\mathbf{R}$, but f' is unbounded in every neighborhood of 0.

Exercise 10.4.10. Assume I is an interval, $f : I \to \mathbf{R}$ a non-decreasing (hence integrable) function, $a \in I$, and F the definite integral of f from a. Prove F is convex, and f is strictly increasing if and only if F is strictly convex. Caution: f need not be continuous on any interval.

Exercise 10.4.11. Let f be the function constructed in Exercise 8.2.13, and define

$$F(x) = \int_0^x f(t)\, dt.$$

Prove that F is strictly increasing and strictly convex, but is differentiable at x if and only if x is irrational, and therefore twice-differentiable nowhere.

Exercise 10.4.12. (H) Assume I is a real interval and $f : I \to \mathbf{R}$ convex.

(a) If x_0 is an interior point of I, prove $f(x) - f(x_0) \approx O(x - x_0)$ near x_0. Explicitly, prove there exists an M and a positive r such that $B_r(x_0) \subseteq I$ and $|f(x) - f(x_0)| \leq M(x - x_0)$ if $|x - x_0| \leq r$.

(b) Prove Proposition 10.4.8.

Exercise 10.4.13. ($\bigstar$) In this exercise we'll compare several types of continuity on an interval $[a, b]$, and examine the adversarial games corresponding to three of them: ordinary continuity, uniform continuity, and bounded stretch.

(a) Write out formal definitions of "f is continuous on $[a, b]$"; "f is uniformly continuous on $[a, b]$"; "f has bounded stretch on $[a, b]$." Treat this either as a review exercise or a short research assignment with the book's index.

(b) Prove that bounded stretch implies uniform continuity, but not conversely. (If you have done Exercise 10.4.9, give a *differentiable* function that is uniformly continuous but does not have bounded stretch.)

(c) If f is continuous on $[a, b]$, then for each x_0 in $[a, b]$ and each positive ε, Player δ has a "largest winning response" $\delta(x_0, \varepsilon)$ against Player ε. Write $\delta(x_0, \varepsilon)$ as a supremum, and prove it is non-decreasing in ε.

(d) What condition on $\delta(x_0, \varepsilon)$ is equivalent to uniform continuity? What condition is equivalent to bounded stretch?

(e) Prove that if f has bounded stretch in $[a, b]$, then $f(x) \approx f(x_0) + O(x - x_0)$ for every x_0. Is the converse true?

(f) Assume r is a rational number in $(0, 1)$. We'll say a function $f : [a, b] \to \mathbf{R}$ is *r-continuous* on $[a, b]$ if there exists a positive real number M such that $|f(x_2) - f(x_1)| \leq M|x_2 - x_1|^r$ for all x_1 and x_2 in $[a, b]$. (Once we define powers with irrational exponent we can assume r is real.) Assume $0 < s < r < 1$. Prove that bounded stretch implies r-continuous, which implies s-continuous, which implies uniformly continuous.

(g) In the notation of part (f), find all functions that are "1.0001-continuous" on $[a, b]$.

11

The Fundamental Theorems of Calculus

At first glance, integrals and derivatives are not closely related. On closer inspection, they are nearly inverses as operators on functions. In Chapter 9 we calculated the integral of $f(x) = x^n$ over an interval $[a, b]$. Despite fairly substantial effort, the end result was simple: Find a *primitive*, a function satisfying $F' = f$, such as $F(x) = x^{n+1}/(n + 1)$. The integral of $F' = f$ turned out to be $F(b) - F(a) = (b^{n+1} - a^{n+1})/(n+1)$. This chapter establishes sweeping generalizations.

11.1 Integrals and Derivatives

There are two fundamental theorems of calculus. One describes differentiating a definite integral, the other integrating a derivative.

Theorem 11.1.1 (Derivative of an integral). *Assume f is integrable on $[a, b]$, and F is the definite integral*

$$F(x) = \int_a^x f, \quad \text{defined for } x \text{ in } [a, b].$$

If f is continuous at a point x_0 in (a, b), then F is differentiable at x_0, and $F'(x_0) = f(x_0)$. In differential notation,

$$\frac{d}{dx} \int_a^x f(t)\, dt = f(x) \quad \text{if } f \text{ is continuous at } x.$$

Proof. If $x_0 + h$ lies in (a, b), then

$$\Delta F(x_0, h) = \frac{F(x_0 + h) - F(x_0)}{h}$$
$$= \frac{1}{h}\left[\int_a^{x_0+h} f - \int_a^{x_0} f\right] = \frac{1}{h}\int_{x_0}^{x_0+h} f,$$

see Figure 11.1. Viewing $f(x_0)$ as a constant function, we deduce

$$\Delta F(x_0, h) - f(x_0) = \frac{1}{h}\left[\int_{x_0}^{x_0+h} f\right] - f(x_0) = \frac{1}{h}\int_{x_0}^{x_0+h} \big(f - f(x_0)\big).$$

DOI: 10.1201/9781003601357-11

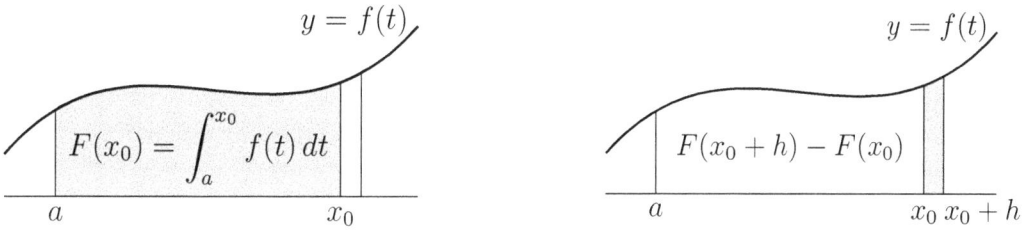

FIGURE 11.1
The increment of a definite integral.

Since f is continuous at x_0, $f(x) - f(x_0) \approx o(1)$ if $x \approx x_0$, so

$$\Delta F(x_0, h) - f(x_0) \approx \frac{1}{h} \int_{x_0}^{x_0+h} o(1) \approx o(1) \quad \text{if } h \approx 0.$$

Letting $h \to 0$ shows $F'(x_0) = f(x_0)$. $\qquad\qquad\square$

Theorem 11.1.2 (Integral of a derivative). *Assume F is of class $\mathscr{C}^1$ on some open interval I. If $a \in I$, then*

$$\int_a^x F'(t)\, dt = F(x) - F(a) \quad \text{for all } x \text{ in } I.$$

Proof. Consider the definite integral

$$G(x) = \int_a^x F'(t)\, dt.$$

By Theorem 11.1.1, G is differentiable in I, and $G'(x) = F'(x)$ for all x in I. By the identity theorem, there exists a real number C such that $G(x) = F(x) + C$.

To evaluate C, set $x = a$: We deduce $0 = G(a) = F(a) + C$, so $C = -F(a)$, or $G(x) = F(x) - F(a)$. $\qquad\qquad\square$

Remark 11.1.3. The hypotheses in Theorem 11.1.2 can be weakened. For example, if F is differentiable and $f = F'$ is integrable, the same conclusion holds, see Exercise 11.1.14. $\qquad\qquad\diamond$

Change of Variables

Proposition 9.2.8 has a useful, far-reaching generalization.

Theorem 11.1.4 (Change of variables). *If $a < b$, τ is a function of class $\mathscr{C}^1$ on $[a, b]$, and f is a continuous function whose domain contains the image of τ, then*

$$\int_{\tau(a)}^{\tau(b)} f = \int_a^b (f \circ \tau) \cdot \tau'.$$

Remark 11.1.5. In differential notation, put $t = \tau(s)$, so that $dt = \tau'(s)\,ds$. The conclusion of the theorem reads

$$\int_{\tau(a)}^{\tau(b)} f(t)\,dt = \int_a^b f\big(\tau(s)\big) \cdot \tau'(s)\,ds.$$

This type of formal symbolic manipulation for correctly adding up infinitely many infinitesimal increments is self-descriptively called *integral calculus.*   ◇

Proof. Consider the functions

$$G(x) = \int_{\tau(a)}^x f, \qquad\qquad F(s) = (G \circ \tau)(s) = \int_{\tau(a)}^{\tau(s)} f,$$

which are of class $\mathscr{C}^1$. By the chain rule and Theorem 11.1.1,

$$F' = (G' \circ \tau) \cdot \tau' = (f \circ \tau) \cdot \tau'.$$

By Theorem 11.1.2,

$$\int_{\tau(a)}^{\tau(b)} f = F(b) - F(a) = \int_a^b F' = \int_a^b (f \circ \tau) \cdot \tau'. \qquad\qquad \square$$

Differentiation of Power Series

Proposition 11.1.6. *Suppose*

$$f(x) = \sum_{k=0}^{\infty} a_k(x - x_0)^k$$

$$= a_0 + a_1(x - x_0) + a_2(x - x_0)^2 + a_3(x - x_0)^3 + \cdots$$

is a germ on $B_R(x_0)$. The real-analytic function f is differentiable on $B_R(x_0)$, and

$$f'(x) = \sum_{k=1}^{\infty} k a_k(x - x_0)^{k-1} = \sum_{k=0}^{\infty} (k+1)a_{k+1}(x - x_0)^k$$

$$= a_1 + 2a_2(x - x_0) + 3a_3(x - x_0)^2 + 4a_4(x - x_0)^3 + \cdots.$$

Proof. As usual, assume $x_0 = 0$. Throughout the proof, let

$$f_n(x) = \sum_{k=0}^{n} a_k x^k, \qquad\qquad \text{the polynomial approximators of } f(x);$$

$$g(x) = \sum_{k=1}^{\infty} k a_k x^{k-1}, \qquad\qquad \text{the termwise derived series;}$$

$$f_n'(x) = \sum_{k=1}^{n} k a_k x^{k-1}, \qquad\qquad \text{the polynomial approximators of } g(x).$$

Our first task is to prove that the series $g(x)$ converges absolutely for all x in $(-R, R)$. Write $r = \frac{1}{2}(|x| + R)$ and $|x| = r\rho$, so that $|x| < |r| < R$ and $\rho < 1$. The terms of $g(x)$ are therefore bounded in absolute value by

$$|ka_k x^{k-1}| \le |a_k r^{k-1}| \cdot |k\rho^{k-1}|.$$

By Corollary 7.2.9, $(k\rho^{k-1}) \to 0$, so the second factor is bounded. The first factor is, aside from a "missing" factor of r, the general term of the series for $f(r)$, which converges absolutely by hypothesis. The derived series $g(x)$ therefore converges absolutely on $(-R, R)$, and consequently may be integrated term by term:

$$\int_0^x g(t)\, dt = \int_0^x \lim_{n \to \infty} f_n'(t)\, dt$$
$$= \lim_{n \to \infty} \int_0^x f_n'(t)\, dt \qquad \text{Proposition 9.4.10}$$
$$= \lim_{n \to \infty} f_n(x) - f_n(0) \qquad \text{Theorem 11.1.2}$$
$$= f(x) - f(0).$$

By Theorem 11.1.1,

$$f'(x) = \frac{d}{dx} \int_0^x g(t)\, dt = g(x). \qquad \square$$

Remark 11.1.7. Computationally, a convergent power series can be differentiated term by term in its interval of convergence, just as if it were a polynomial.

Proposition 11.1.6 guarantees the termwise derived series has the same radius as the original power series. This has an important "bootstrapping" consequence: Since f' is real-analytic on $(-R, R)$, the *second* derivative f'' can be represented by a convergent power series on $(-R, R)$, obtained by differentiating the series for f' term by term, and so forth. That is, a real-analytic function is *smooth*. ◇

Exercises for Section 11.1

Exercise 11.1.1. (★) Let U be the unit step function, $U(t) = 0$ if $t < 0$, $U(t) = 1$ if $0 < t$, and $U(0) = 1/2$. Calculate $F(x) = \int_0^x U$ and sketch the graph of F.

Exercise 11.1.2. Define $F : \mathbf{R} \to \mathbf{R}$ by $F(x) = \int_0^x \frac{(1+t)}{(2+2t+t^2)^3}\, dt$.

Evaluate $F(x)$ as an algebraic formula, and find $F'(x)$ in two ways.

Exercise 11.1.3.(A) Suppose $f : \mathbf{R} \to \mathbf{R}$ is continuous.

(a) Define $G : \mathbf{R} \to \mathbf{R}$ by

$$G(x) = \int_0^{x^2} f(t)\, dt.$$

Prove that G is differentiable, and find $G'(x)$.

(b) Define $H : \mathbf{R} \to \mathbf{R}$ by

$$H(x) = \int_x^{x^2} f(t)\, dt.$$

Show that H is differentiable, and find $H'(x)$.

(c) Assume ϕ and ψ are differentiable on (α, β), and define

$$\Phi(x) = \int_{\psi(x)}^{\phi(x)} f(t)\, dt \quad \text{for } x \text{ in } (\alpha, \beta).$$

Show that Φ is differentiable, and find $\Phi'(x)$ in terms of f, ϕ, and ψ.

Exercise 11.1.4. Consider the functions F, $G : \mathbf{R} \to \mathbf{R}$ defined by

$$F(x) = \int_0^{x^2} \frac{t\, dt}{\sqrt[3]{1+t^3}}, \qquad G(x) = \int_0^x \frac{t^2\, dt}{\sqrt[3]{1+t^6}}.$$

Calculate F' and G'. Determine which (if either) is larger: $F(1/2)$ or $G(1/2)$.

Exercise 11.1.5.($\bigstar$) Assume I is an open interval and $f : I \to \mathbf{R}$ continuous. Prove that for every c in $\mathbf{R}$ and every x_0 in I, there exists a unique $\mathscr{C}^1$ function $F : I \to \mathbf{R}$ satisfying the "initial-value problem"

$$F' = f, \qquad f(x_0) = c.$$

Exercise 11.1.6.($\bigstar$) Does there exist an integrable function $f : [-1,1] \to \mathbf{R}$ such that

$$\int_{-1}^x f(t)\, dt = \sqrt{1 - x^2} \quad \text{for all } x \text{ in } [-1,1]?$$

Exercise 11.1.7. Assume $A : (-1,1) \to A(-1,1)$ is the $\mathscr{C}^1$ function defined by

$$A'(x) = \frac{1}{\sqrt{1 - x^2}}, \qquad A(0) = 0.$$

(a) Prove A is invertible. If $S = A^{-1}$, prove $S' = \sqrt{1 - S^2}$.

(b) Prove S is $\mathscr{C}^2$, and find S'' in terms of S.

Exercise 11.1.8. (★) Assume $f : \mathbf{R} \to \mathbf{R}$ is continuous. This exercise constructs a function F such that $F'' = f$.

(a) Find the derivatives of $g(x) = \int_0^x t f(t)\, dt$ and $h(x) = \int_0^x x f(t)\, dt$.

(b) Prove $F(x) = \int_0^x (x - t) f(t)\, dt = \int_0^x \left[\int_0^s f(t)\, dt \right] ds$ satisfies $F'' = f$.

Exercise 11.1.9. (H) Assume I is an open interval and $f : I \to \mathbf{R}$ is continuous. Prove that for every c_0 and c_1 in $\mathbf{R}$ and every x_0 in I, there exists a unique $\mathscr{C}^2$ function $F : I \to \mathbf{R}$ satisfying the initial-value problem

$$ F'' = f, \qquad f(x_0) = c, \quad f'(x_0) = c_1. $$

Exercise 11.1.10. (H) Assume I is an open interval, x_0 a point of I, and $f : I \to \mathbf{R}$ a continuous function. Prove that the initial-value problem

$$ y'' = f(y), \qquad y(x_0) = y_0, \quad y'(x_0) = y_0' $$

has a solution in some neighborhood of x_0 if $y_0' \neq 0$.

Exercise 11.1.11. (H) If n is an integer and $n \geq 2$, define $f_n(t) = t^n (1 - t)^n$ if $0 \leq x \leq 1$ and 0 otherwise, and put

$$ c_n = \int_0^1 f_n(t)\, dt, \qquad\qquad F_n(x) = \frac{1}{c_n} \int_0^x f_n(t)\, dt. $$

(a) Prove that F_n is of class $\mathscr{C}^n$, non-decreasing, and that $F_n(x) = 0$ if $x \leq 0$, $F_n(x) = 1$ if $x \geq 1$.

(b) Assume $[c, d] \subseteq (a, b)$. Use part (a) to show there exists a non-negative function f of class $\mathscr{C}^n$ defined on $\mathbf{R}$ such that $f(x) = 1$ if $c \leq x \leq d$; $f(x) = 0$ if $x \leq a$ or $b \leq x$; and $0 \leq f(x) \leq 1$ for all real x.

Exercise 11.1.12. Assume $f : \mathbf{R} \to \mathbf{R}$ is continuous and ℓ-periodic. Prove:

(a) For every real number a, $\int_a^{a+\ell} f(t)\, dt = \int_0^\ell f(t)\, dt$.

(b) There exists an ℓ-periodic function F such that $F' = f$ if and only if

$$ \int_0^\ell f(t)\, dt = 0. $$

(c) The function $F(x) = \int_0^x f(t)\, dt - \frac{x}{\ell} \int_0^\ell f(t)\, dt$ is ℓ-periodic.

Exercise 11.1.13. (H) Assume F is of class $\mathscr{C}^1$ on some interval I. Prove there exist *non-decreasing* $\mathscr{C}^1$-functions F_+ and F_- on I such that $F = F_+ - F_-$.

Exercise 11.1.14. Assume F is a differentiable function on $[a,b]$, and the derivative $f = F'$ is integrable (though not necessarily continuous). Assume $\Pi = \{t_i\}_{i=0}^n$ is an arbitrary splitting of $[a,b]$. Prove there exists a set t^* of sample points from Π such that $F(t_{i+1}) - F(t_i) = f(t_i^*)\,\Delta t_i$. Conclude

$$\int_a^b f(t)\,dt = F(b) - F(a).$$

11.2   Approximation by Germs

Theorem 11.2.1 (Integration by parts). *If u and v are of class $\mathscr{C}^1$ on $[a,b]$, then*

$$\int_a^b uv' = uv\Big|_a^b - \int_a^b vu'.$$

Proof. By the product rule for derivatives, $(uv)' = u'v + uv'$, or

$$uv' = (uv)' - vu'.$$

The proposition follows from Theorem 11.1.2 by integrating over $[a,b]$. $\square$

Definition 11.2.2. If f is n times differentiable in some neighborhood of x_0, the *nth-degree germ* of f at x_0 is the polynomial

$$P_{x_0}^n f(x) = \sum_{k=0}^n \frac{f^{(k)}(x_0)}{k!}(x - x_0)^k$$

$$= f(x_0) + f'(x_0)(x - x_0) + \frac{f''(x_0)}{2!}(x - x_0)^2 + \cdots + \frac{f^{(n)}(x_0)}{n!}(x - x_0)^n.$$

The *nth-degree remainder* is $R_{x_0}^n f(x) = f(x) - P_{x_0}^n f(x)$.

Remark 11.2.3. If f is a germ (convergent power series) at x_0, each partial sum is an nth-degree germ at x_0 for some n, and each tail is a remainder. $\diamond$

Theorem 11.2.4 (The remainder theorem). *If n is a natural number and f is of class $\mathscr{C}^{n+1}$ in some open interval X containing x_0, then for each x in X:*

(i) (*Integral form*)

$$R_{x_0}^n f(x) = \frac{1}{n!}\int_{x_0}^x f^{(n+1)}(t)(x - t)^n\,dt.$$

(ii) (*Sampled form*) *There is a* z_{n+1} *between* x_0 *and* x *such that*

$$R_{x_0}^n f(x) = \frac{f^{(n+1)}(z_{n+1})}{(n+1)!}(x-x_0)^{n+1}.$$

Proof. If $n = 0$, Theorem 11.1.2 gives the integral form of the remainder:

$$R_{x_0}^0 f(x) = f(x) - P_{x_0}^0 f(x) = f(x) - f(x_0) = \int_{x_0}^x f'(t)\,dt.$$

Assume inductively that

$$R_{x_0}^k f(x) = \frac{1}{k!}\int_{x_0}^x f^{(k+1)}(t)(x-t)^k\,dt$$

for some natural number k. Integrating by parts with

$$u(t) = f^{(k+1)}(t), \qquad\qquad v(t) = -\frac{1}{(k+1)!}(x-t)^{k+1},$$

$$u'(t) = f^{(k+2)}(t), \qquad\qquad v'(t) = \frac{1}{k!}(x-t)^k,$$

gives

$$R_{x_0}^k f(t) = -\frac{f^{(k+1)}(t)}{(k+1)!}(x-t)^{k+1}\Big|_{t=x_0}^{t=x} + \frac{1}{(k+1)!}\int_{x_0}^x f^{(k+2)}(t)(x-t)^{k+1}\,dt$$

$$= \frac{f^{(k+1)}(x_0)}{(k+1)!}(x-x_0)^{k+1} + \frac{1}{(k+1)!}\int_{x_0}^x f^{(k+2)}(t)(x-t)^{k+1}\,dt,$$

or

$$f(x) = P_{x_0}^k f(x) + R_{x_0}^k f(x)$$

$$= P_{x_0}^{k+1} f(x) + \frac{1}{(k+1)!}\int_{x_0}^x f^{(k+2)}(t)(x-t)^{k+1}\,dt,$$

the asserted form of $R_{x_0}^{k+1} f(x)$. This establishes the inductive step.

To establish the sampled form of the remainder, note that $f^{(n+1)}$ is continuous on the interval between x_0 and x, so $f^{(n+1)}$ achieves an absolute minimum value m and an absolute maximum value M in this interval. By monotonicity of the integral,

$$\frac{1}{n!}\int_{x_0}^x m(x-t)^n\,dt \le \frac{1}{n!}\int_{x_0}^x f^{(n+1)}(t)(x-t)^n\,dt \le \frac{1}{n!}\int_{x_0}^x M(x-t)^n\,dt,$$

or

$$m\frac{(x-x_0)^{n+1}}{(n+1)!} \le R_{x_0}^n f(x) \le M\frac{(x-x_0)^{n+1}}{(n+1)!}.$$

Again since $f^{(n+1)}$ is continuous, the intermediate value theorem guarantees there exists a z_{n+1} between x_0 and x such that

$$R_{x_0}^n f(x) = \frac{f^{(n+1)}(z_{n+1})}{(n+1)!}(x-x_0)^{n+1}. \qquad\qquad \square$$

Remark 11.2.5. Philosophically, the remainder theorem says that on an interval about a point x_0, a function of class $\mathscr{C}^{n+1}$ "behaves like a polynomial of degree at most n up to order $(n+1)$." Particularly, if f is smooth, then for every n,

$$f(x) \approx P_{x_0}^n f(x) + O(x - x_0)^{n+1} \quad \text{if } x \approx x_0,$$

with the constant in O bounded by the maximum of $|f^{(n+1)}|$ on X. When $x \approx x_0$, a larger degree corresponds to a better approximation.   ◇

Applications

The remainder theorem has a multitude of applications. Two are introduced here, and explored in the exercises.

In calculus, a function f is often defined to be "convex" if the graph of f "lies above each tangent line," namely if $P_{x_0}^1 f(x) \leq f(x)$ for all x_0 and x. The secant criterion for convexity in Definition 5.1.30, which is standard in real analysis, makes no assumption of differentiability, much less continuity of the second derivative, and is therefore both simpler and more general than the common calculus definition.

Proposition 11.2.6. *If f'' is continuous and $f''(x_0) > 0$, then on some open ball $B_r(x_0)$, the graph of f lies above each of its tangent lines.*

Proof. See Exercise 11.2.6.   □

The second application is error bounds on numerical methods of integration, see also the discussion of sampled sums in Chapter 9.

Definition 11.2.7. Assume $f : [a, b] \to \mathbf{R}$ is a function and $\Pi = \{x_i\}_{i=0}^n$ is a splitting of $[a, b]$.

The *trapezoid sum* for Π is the average of the left- and right-hand sums:

$$\text{TRAP}(f, \Pi) = \sum_{i=0}^{n-1} \tfrac{1}{2}\left[f(x_i) + f(x_{i+1})\right] \Delta x_i.$$

The *midpoint sum* for Π is the sampled sum with $x_i^* = \overline{x}_i = \tfrac{1}{2}(x_i + x_{i+1})$:

$$\text{MID}(f, \Pi) = S(f, \Pi, x^*) = \sum_{i=0}^{n-1} f(\overline{x}_i) \, \Delta x_i.$$

The *parabolic sum* is the weighted average

$$\text{PARA}(f, \Pi) = \tfrac{1}{3}\left[\text{TRAP}(f, \Pi) + 2\,\text{MID}(f, \Pi)\right]$$

$$= \sum_{i=0}^{n-1} \tfrac{1}{6}\left[f(x_i) + 4f(\overline{x}_i) + f(x_{i+1})\right] \Delta x_i.$$

Remark 11.2.8. Exercise 9.3.5 gives useful geometric interpretations of the trapezoid and midpoint sums. The parabolic sum with one piece is the area enclosed by the quadratic graph through the endpoints and midpoint, see also Exercise 9.3.6.

Error bounds for equal-length splittings are developed in Exercises 11.2.9, 11.2.10, and 11.2.11, assuming f is sufficiently smooth. Particularly, if f is of class $\mathscr{C}^4$, the error for the parabolic sum is no larger than $O(1/n^4)$, with constant jointly proportional to $K_4 := \max |f^{(4)}|$ and $(b-a)^5$. ◇

Exercises for Section 11.2

Exercise 11.2.1. By Definition 9.4.11, the natural logarithm log is the unique function satisfying $\log 1 = 0$ and $\log' t = 1/t$ for all positive t.

(a) If r is rational and $r \neq -1$, use integration by parts to evaluate

$$f(x) = \int_e^x t^r \log t \, dt, \quad x \text{ positive.}$$

(b) Use substitution to evaluate the integral in part (a) if $r = -1$.

Exercise 11.2.2. Use Definition 11.2.2 to expand $f(x) = \dfrac{1}{1-x}$ about $x_0 = 0$.

Exercise 11.2.3. Assume n is a natural number, x_0 is real, I is an open interval containing x_0 (possibly all of $\mathbf{R}$), and f is a real-valued function of class $\mathscr{C}^n$ on I.

(a) Assume q^n is a polynomial, $\deg q^n \leq n$, and $f(x) \approx q^n(x) + o(x - x_0)^n$ if $x \approx x_0$. Prove q^n is the nth-degree germ of f at x_0.

(b) If f is a polynomial of degree n, then the nth-degree germ of f at x_0 is f itself, expanded in powers of $(x - x_0)$.

(c) Use Definition 11.2.2 to expand the polynomial $f(x) = ax^2 + bx + c$ in powers of $(x - x_0)$. Use algebra to verify that the result is equal to f.

(d) Similarly, expand $f(x) = x^n$ in powers of $(x - x_0)$. (The result is consistent with the binomial theorem.)

Exercise 11.2.4. (★) Assume f is of class $\mathscr{C}^n$ on an interval I symmetric about 0, and let p^n denote the nth-degree germ of f at 0. Find the nth-degree germs of the even part and odd part of f.

Exercise 11.2.5. ($\bigstar$) Suppose f is smooth on $\mathbf{R}$, and p^n is the nth-degree germ of f at 0. If $g(x) = f(x^2)$, prove $q^{2n}(x) := p^n(x^2)$ is the $(2n+1)$th-degree germ of g at 0.

Exercise 11.2.6. Prove Proposition 11.2.6.

Exercise 11.2.7. (H) Assume f is of class $\mathscr{C}^{n+1}$ in some open ball $B_r(x_0)$, and $f^{(n+1)}(x_0) \neq 0$.

If $|x - x_0| < r$, the number z_n in the sampled remainder can be written uniquely as a convex linear combination $z_n = x_0 + t_n(x - x_0)$ for some real number t_n in $(0, 1)$. Prove $t_n = 1/(n+1) + o(1)$.

Exercise 11.2.8. Assume f is of class $\mathscr{C}^2$ on $[a, b]$, and let Π be an arbitrary splitting. Prove that if $f'' \geq 0$ on $[a, b]$, then

$$\text{MID}(f, \Pi) \leq \int_a^b f \leq \text{TRAP}(f, \Pi),$$

and illustrate with a sketch.

Exercise 11.2.9. (H) (Midpoint sum error bound). Assume f is of class $\mathscr{C}^2$ on $[a, b]$, and that $|f''| \leq K_2$ on $[a, b]$. Prove that for every positive integer n, if Π is the equal-length splitting of $[a, b]$ with n pieces, then

$$\left| \int_a^b f - \text{MID}(f, \Pi) \right| \leq \frac{K_2(b - a)^3}{24n^2}.$$

Exercise 11.2.10. (H) (Trapezoid sum error bound). Assume f is of class $\mathscr{C}^2$ on $[a, b]$, and that $|f''| \leq K_2$ on $[a, b]$. Prove that for every positive integer n, if Π is the equal-length splitting of $[a, b]$ with n pieces, then

$$\left| \int_a^b f - \text{TRAP}(f, \Pi) \right| \leq \frac{K_2(b - a)^3}{12n^2}.$$

Exercise 11.2.11. (H) (Parabolic sum error bound). Assume f is of class $\mathscr{C}^4$ on $[a, b]$, and that $|f^{(4)}| \leq K_4$ on $[a, b]$. Prove that for every positive integer n, if Π is the equal-length splitting of $[a, b]$ with n pieces, then

$$\left| \int_a^b f - \text{PARA}(f, \Pi) \right| \leq \frac{K_4(b - a)^5}{2880n^4}.$$

Exercise 11.2.12. ($\bigstar$) We are asked to tabulate values of

$$F(x) = \int_0^x \sqrt{1 + t^4}\, dt, \quad 0 \leq x \leq 1$$

correct to four decimal places, an error of at most 0.5×10^{-4}. Determine how many pieces suffice for the midpoint sum, according to the error bound of Exercise 11.2.9.

Exercise 11.2.13. (H) Repeat the analysis of Exercise 11.2.12 for the parabolic sum, using the error bound in Exercise 11.2.11.

Exercise 11.2.14. Assume f is a smooth, real-valued function on some open interval I, and that for every x_0 in I, there exist positive real numbers M and R such that

$$\left| \frac{f^{(n+1)}(z) R^{n+1}}{(n+1)!} \right| \le M \quad \text{for all } z \text{ such that } |z - x_0| < R.$$

Prove f is real-analytic on I: The remainder in Theorem 11.2.4 converges to 0 as $n \to \infty$.

Exercise 11.2.15. (★) Use Exercises 10.2.6 and 11.2.14 to prove the function $f(x) = (1+x)^{1/2}$ is real-analytic on $(-1, \infty)$.

Exercise 11.2.16. Assume p is a rational number that is not an integer. Prove that the function $f(x) = (1+x)^p$ is real-analytic on $(-1, \infty)$.

Exercise 11.2.17. Use differentiation and algebra to manipulate the geometric series with first term 1 and ratio x on $(-1, 1)$, giving closed-form expressions for the power series:

(a) $\displaystyle\sum_{k=1}^{\infty} k x^k = x + 2x^2 + 3x^3 + \cdots.$ (b) $\displaystyle\sum_{k=1}^{\infty} k^2 x^k = x + 4x^2 + 9x^3 + \cdots.$

Exercise 11.2.18. Assume f is an increasing function of class $\mathscr{C}^1$ on some interval $[a, b]$. Use the substitution $y = f(x)$ and integration by parts to prove that

$$\int_{f(a)}^{f(b)} f^{-1}(y) \, dy = b f(b) - a f(a) - \int_a^b f(x) \, dx,$$

and illustrate with a sketch.

Exercise 11.2.19. Assume p and q are positive rational numbers satisfying $1/p + 1/q = 1$. (The arguments here apply to powers with irrational exponent, but we have not yet defined these, so must restrict the statement for now.)

(a) Prove the graphs $y = x^{p-1}$ and $x = y^{q-1}$ in the first quadrant are identical.

(b) Show that if a and b are positive, then

$$ab \le \frac{a^p}{p} + \frac{b^q}{q}.$$

Suggestion: Make a sketch using part (a), and interpret each term of the inequality as an area.

11.3 Improper Integrals

Definition 11.3.1. An integral $\int_a^b f$ is called a *basic improper integral* if one of the following holds:

(i) Exactly one endpoint a or b is infinite, and f is bounded.

(ii) Both endpoints are finite and f is unbounded, but only in a neighborhood of one endpoint.

An integral is *improper* if the interval of integration can be divided into finitely many subintervals, on each of which the integral is a basic improper integral.

Example 11.3.2. If $p > 0$ is real, the following are basic improper integrals:

$$\int_0^1 \frac{dx}{x^p}, \qquad\qquad \int_{-3}^0 \frac{dx}{|x|^p}, \qquad\qquad \int_1^\infty \frac{dx}{x^p}.$$

The following are improper and are split into basic improper pieces:

$$\int_0^\infty \frac{dx}{x^p} = \int_0^1 \frac{dx}{x^p} + \int_1^\infty \frac{dx}{x^p}, \qquad \int_{-1}^\infty \frac{dx}{|x|^p} = \int_{-1}^0 \frac{dx}{|x|^p} + \int_0^1 \frac{dx}{|x|^p} + \int_1^\infty \frac{dx}{|x|^p}. \;\Diamond$$

Definition 11.3.3. A basic improper integral $\int_a^\infty f$ is said to *converge* to L if

$$L = \lim_{b\to\infty} \int_a^b f.$$

Basic improper integrals $\int_{-\infty}^b f$ are handled similarly.

A basic improper integral $\int_a^b f$ with f unbounded near b is said to *converge* to L if
$$L = \lim_{x\to b^-} \int_a^x f.$$

An analogous definition is made for $\int_a^b$ if f is unbounded near a.

A basic improper integral that does not converge is said to *diverge*.

An improper integral is said to *converge to L* if the integral can be split into finitely many basic sub-integrals, each of which converges, and the sum of the sub-integrals is L. Otherwise, the improper integral is said to *diverge*.

Proposition 11.3.4. *If $p > 0$, define $f_p(x) = 1/x^p$ if $x > 0$, and $f_p(0) = 0$.*

(i) *The improper integral $\int_0^1 f_p$ converges to $\dfrac{1}{1-p}$ if and only if $p < 1$.*

(ii) *The improper integral $\int_1^\infty f_p$ converges to $\dfrac{1}{p-1}$ if and only if $p > 1$.*

Both integrals diverge otherwise.

Remark 11.3.5. We have not yet defined exponentiation with irrational exponents, so strictly speaking the proof below applies only to rational exponents p. Thanks to Proposition 12.1.6 in Chapter 12, the proof below goes through without modification for arbitrary positive real p. ◇

Proof. First suppose $p = 1$. The definite integral of $1/x$ from 1 is the natural logarithm, see Definition 9.4.11. In Chapter 9, we showed there exists a real number e such that $2 < e < 3$ and

$$\int_1^{e^n} \frac{dx}{x} = n = \int_{e^{-n}}^1 \frac{dx}{x}.$$

In particular, the basic improper integrals

$$\int_0^1 \frac{dx}{x} \quad \text{and} \quad \int_1^\infty \frac{dx}{x}$$

both diverge.

If $p \neq 1$, the function $F_p(x) = x^{1-p}/(1-p)$ is a primitive of f_p on the set of positive real numbers.

(i). If $0 < p < 1$, then $1 - p > 0$, so

$$\lim_{a \to 0^+} \int_a^1 \frac{dx}{x^p} = \lim_{a \to 0^+} \frac{x^{1-p}}{1-p}\Big|_{x=a}^{x=1} = \lim_{a \to 0^+} \frac{1 - a^{1-p}}{1-p} = \frac{1}{1-p}.$$

If $p > 1$ instead, then $1 - p < 0$, and $a^{1-p} \to \infty$ as $a \to 0^+$, so the integral diverges.

(ii). If $1 < p$, then $1 - p < 0$, so

$$\lim_{b \to \infty} \int_1^b \frac{dx}{x^p} = \lim_{b \to \infty} \frac{x^{1-p}}{1-p}\Big|_{x=1}^{x=b} = \lim_{b \to \infty} \frac{b^{1-p} - 1}{1-p} = \frac{1}{p-1}.$$

If $0 < p < 1$ instead, then $1 - p > 0$, and $b^{1-p} \to \infty$ as $b \to \infty$, so the integral diverges. □

Definition 11.3.6. An improper integral $\int_a^b f$ *converges absolutely* if the improper integral $\int_a^b |f|$ converges.

Remark 11.3.7. To avoid stating results with four (or more) closely related parts, we focus for the rest of this section on basic improper integrals over an unbounded interval $[a, \infty)$. Analogous results are true for other types of improper integral. ◇

Proposition 11.3.8. *Assume a is a real number and f, g are integrable on every interval $[a, b]$ such that $a < b$.*

(i) *If f is non-negative, then*

$$\int_a^\infty f \quad converges\ if\ and\ only\ if \quad F(x) = \int_a^x f \quad is\ bounded.$$

(ii) *If $f^+ = \max(f, 0)$ and $f^- = -\min(f, 0)$, then*

$$\int_a^\infty f \quad converges\ absolutely\ if\ and\ only\ if \quad \int_a^x f^\pm \quad both\ converge,$$

and this implies $\int_a^\infty f$ converges.

(iii) *If $|g| \leq |f|$ except at finitely many points, and if $\int_a^\infty f$ converges absolutely, then $\int_a^\infty g$ converges absolutely, and*

$$\left| \int_a^\infty g \right| \leq \int_a^\infty |f|.$$

Proof. These assertions follow immediately from the corresponding facts about ordinary (proper) integrals, infinite sequences, and series. Indeed, if (b_k) is a real sequence diverging to ∞, consider the sequence

$$B_k = \int_a^{b_k} f$$

and note that the improper integral of f converges to L if and only if every such sequence (B_k) converges to L. $\qquad\square$

Proposition 11.3.9. *If f is integrable on $[a, b]$ for all b greater than a, and if $\tau : [\alpha, \beta) \to [a, \infty)$ is of class $\mathscr{C}^1$ and satisfies $\tau(\alpha) = a$ and $\tau(x) \to \infty$ as $x \to \beta^-$, then*

$$\int_\alpha^\beta f\big(\tau(s)\big) \cdot \tau'(s)\, ds = \int_a^\infty f(t)\, dt,$$

in the sense that the right-hand integral converges if and only if the left-hand integral converges, and in this event the two have the same value.

Proof. By the change of variables theorem for ordinary integrals,

$$\int_\alpha^x f\big(\tau(s)\big) \cdot \tau'(s)\, ds = \int_a^{\tau(x)} f(t)\, dt$$

if $\alpha \leq x < \beta$. The proposition follows by taking $x \to \beta^-$. $\qquad\square$

Example 11.3.10. If f is continuous and bounded on $[1, \infty)$, then

$$\int_0^1 \frac{f(1/t)}{t^2}\, dt = -\int_1^0 \frac{f(1/t)}{t^2}\, dt = \int_1^\infty f(x)\, dx,$$

in the sense that both integrals converge or diverge, and if both converge, they have the same value. $\qquad\Diamond$

Remark 11.3.11. In Chapters 12 and 13 we will meet new functions for which "improper" change of variables is more interesting a tool than it might appear given the scant examples we currently have. $\qquad\Diamond$

The Integral Test for Series

Proposition 11.3.12. *Assume f is a non-increasing, positive, real-valued function, defined for all non-negative x. For every positive integer n,*

$$\sum_{k=1}^{n} f(k) \le \int_{0}^{n} f(x)\,dx \le \sum_{k=0}^{n-1} f(k).$$

In particular, the real sequence $(a_k)_{k=0}^{\infty}$ defined by $a_k = f(k)$ is summable if and only if f is improperly integrable.

Proof. Let k be an arbitrary natural number. Because f is non-increasing,

$$0 < f(k+1) \le f(x) \le f(k) \quad \text{if } k \le x \le k+1.$$

Integrating over $[k, k+1]$,

$$f(k+1) \le \int_{k}^{k+1} f(x)\,dx \le f(k).$$

Summing over k and patching integrals gives, see Figure 11.2,

$$\sum_{k=1}^{n} f(k) = \sum_{k=0}^{n-1} f(k+1) \le \int_{0}^{n} f(x)\,dx \le \sum_{k=0}^{n-1} f(k).$$

All three are non-decreasing in n, and the two sums differ by $f(0) - f(n)$, which is bounded, so all three converge or diverge together. $\qquad\square$

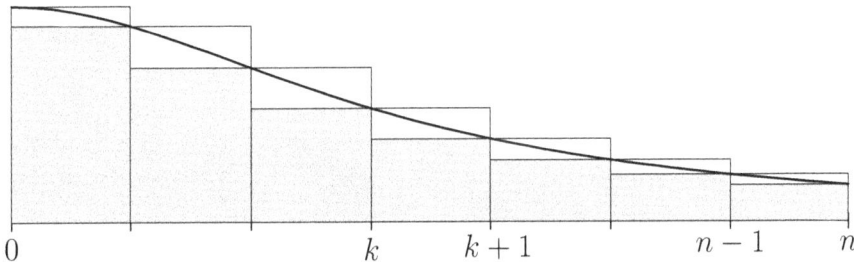

FIGURE 11.2
Bounding an integral by partial sums of a series.

Corollary 11.3.13. *If f is non-increasing, positive, and improperly integrable on $[0, \infty)$, then for every positive integer n,*

$$\sum_{k=0}^{n-1} f(k) + \int_{n}^{\infty} f(x)\,dx \le \sum_{k=0}^{\infty} f(k) \le \sum_{k=0}^{n-1} f(k) + \int_{n-1}^{\infty} f(x)\,dx.$$

Proof. See Exercise 11.3.6. $\qquad\square$

Exercises for Section 11.3

Exercise 11.3.1. (★) Write each improper integral as a sum of basic improper integrals if necessary, and determine convergence. (Do not attempt to evaluate.)

(a) $\displaystyle\int_{-\infty}^{\infty} \frac{dx}{1+x^2}$.   (b) $\displaystyle\int_0^1 \frac{dx}{\sqrt{1-x^2}}$.   (c) $\displaystyle\int_0^{\infty} \frac{x^3 \, dx}{x^4 + 1}$.   (d) $\displaystyle\int_{-1}^1 \frac{dx}{x^2}$.

Exercise 11.3.2. Assume m is a non-negative integer, r is a positive real number, and $f : [0, \infty) \to \mathbf{R}$ is integrable on $[0, b]$ for every positive b. If

$$I_m(r) := \int_0^{\infty} \frac{f(rt)}{t^m} \, dt$$

converges, express $I_m(r)$ in terms of $I_m(1)$. Particularly, prove I_1 is constant.

Exercise 11.3.3. Assume p is a positive rational. Determine whether each improper integral converges. If so, evaluate.

(a) $\displaystyle\int_e^{\infty} \frac{dx}{(\log x)^p x}$.   (b) $\displaystyle\int_{e^e}^{\infty} \frac{dx}{[\log(\log x)]^p (\log x) x}$.

Exercise 11.3.4. (★) Evaluate

$$\int_0^x \frac{dt}{1 - t^2}.$$

For which real x does the integral converge?

Exercise 11.3.5. (A) Assume $-1 < a$, and m is a positive integer.

(a) Prove the improper integral

$$I_m(a) := \int_1^{\infty} \frac{dt}{t^m(t+a)}$$

converges, $I_m(a)$ decreases to 0 as $m \to \infty$, and evaluate.

Suggestion: Use partial fractions and write $\dfrac{t^m - (-a)^m}{t - (-a)}$ as a geometric sum.

(b) Use (a) to give a formula for $\log(1 + a)$ if $-1 < a$, and evaluate the alternating harmonic series

$$\sum_{k=1}^{\infty} \frac{(-1)^{k-1}}{k} = 1 - \frac{1}{2} + \frac{1}{3} - \frac{1}{4} + \cdots.$$

Exercise 11.3.6.

(a) Prove Corollary 11.3.13.

(b) In the same notation, put $I_n = \int_{n-1}^{\infty} f(x)\, dx$. If

$$\sum_{k=0}^{\infty} f(k) = \sum_{k=0}^{n-1} f(k) + \tfrac{1}{2}(I_n + I_{n+1}) + E_n,$$

prove $|E_n| \le \tfrac{1}{2}(I_n - I_{n+1}) = \dfrac{1}{2}\int_{n-1}^{n} f(x)\, dx.$

Exercise 11.3.7. The sums of the infinite series

$$\sum_{k=0}^{\infty} \frac{1}{(k+1)^2}, \qquad \sum_{k=0}^{\infty} \frac{1}{(k+1)^3}, \qquad \sum_{k=0}^{\infty} \frac{1}{(k+1)^5}$$

are to be estimated to within $\varepsilon = 0.5 \times 10^{-6}$. With notation as in Exercise 11.3.6, for the estimates S_n given, determine the smallest n such that $E_n < \varepsilon$.

(a) $S_n = \sum_{k=0}^{n-1} f(k)$ and $E_n = I_n$. (This is a standard calculus estimate.)

(b) $S_n = \sum_{k=0}^{n-1} f(k) + \tfrac{1}{2}(I_n + I_{n+1})$ and $E_n = \tfrac{1}{2}(I_n - I_{n+1}).$

12

Exponential Functions

In this chapter we use our collection of tools—integrals, derivatives, power series, the fundamental theorems, and the mean value theorem—to define and study exponential functions. The roots of the story go back to Definition 9.4.11, see also Figure 12.1, which introduced the natural logarithm $\log : (0, \infty) \to \mathbf{R}$ by

$$\log x = \int_1^x \frac{1}{t}\, dt.$$

Proposition 9.4.12 guarantees the "multiplication-to-addition" identities

$$\log xy = \log x + \log y, \qquad \log(1/x) = -\log x \quad \text{for all positive } x,\, y.$$

There exists a unique real number e satisfying $\log e = 1$, for which $2 < e < 3$. Since log is continuous and $\log(e^n) = n$ for every integer n, the logarithm is surjective by the intermediate value theorem.

By Theorem 11.1.1, log is differentiable, $\log' x = 1/x > 0$, and therefore is strictly increasing by Theorem 10.3.7. Although the tangent lines to the graph $y = \log x$ become arbitrarily close to horizontal, log is unbounded above.

12.1 The Natural Exponential Function

Definition 12.1.1. The *natural exponential function* $\exp : \mathbf{R} \to (0, \infty)$ is the inverse of log.

Remark 12.1.2. If y is real and $x > 0$, then $x = \exp y$ if and only if $y = \log x$, Figure 12.2. We need not be fussy about the codomain; if convenient we view $\exp : \mathbf{R} \to \mathbf{R}$ as real-valued. In any case, $\exp y > 0$ for all real y. ◇

The Differential Characterization of exp

Proposition 12.1.3. *For all real x, $\exp' x = \exp x$. Conversely, if $f : \mathbf{R} \to \mathbf{R}$ is a differentiable function such that $f' = f$, then $f(x) = f(0) \exp x$ for all x.*

Proof. For all real x, we have $x = \log(\exp x)$. Since log is differentiable and has non-vanishing derivative, its inverse function exp is differentiable, and by

DOI: 10.1201/9781003601357-12

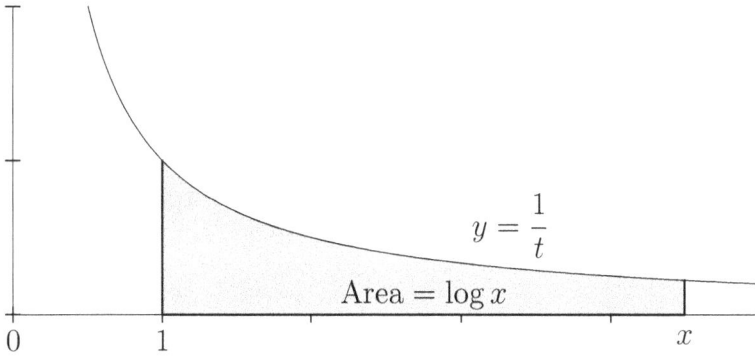

FIGURE 12.1
The natural logarithm as an area.

the inverse function theorem, Theorem 10.2.10,

$$\exp' x = \frac{1}{\log'(\exp x)} = \frac{1}{1/\exp x} = \exp x.$$

Conversely, assume $f : \mathbf{R} \to \mathbf{R}$ is differentiable and $f' = f$. Since $\exp x > 0$ for all real x, the quotient $g(x) := f(x)/\exp x$ is differentiable. Since $\exp' = \exp$, the quotient rule gives

$$g'(x) = \frac{(\exp x)f'(x) - f(x)(\exp x)}{(\exp x)^2} = \frac{f'(x) - f(x)}{\exp x} = 0 \quad \text{for all } x.$$

By the identity theorem, g is a constant function: $g(x) = g(0) = f(0)$ for all real x. That is, $f(x) = f(0) \exp x$ for all x. $\qquad\qquad\square$

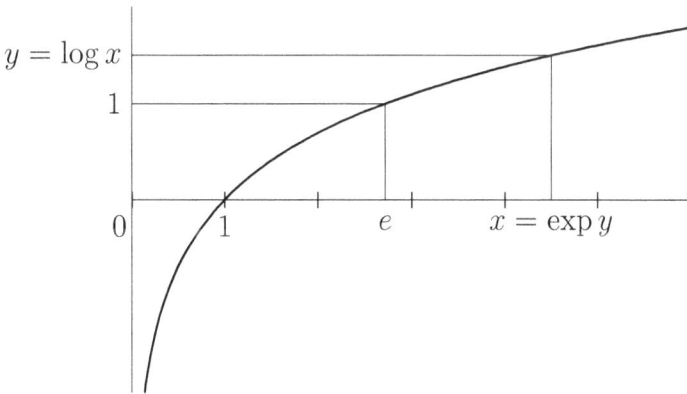

FIGURE 12.2
The natural exponential function is the inverse of log.

Real Exponentiation

Proposition 12.1.4. *If $a > 0$ and $r = p/q$ is rational, then*

$$\log a^r = r \log a, \quad \text{namely,} \quad a^r = \exp(r \log a).$$

Proof. Mathematical induction together with $\log xy = \log x + \log y$ shows that $\log b^m = m \log b$ for every positive real number b and every integer m.

Set $b = a^r = a^{p/q}$, so that $b^q = a^p$. Taking logarithms and using the preceding paragraph, $q \log b = p \log a$, or

$$\log a^r = \log b = (p/q) \log a = r \log a. \qquad \square$$

Proposition 12.1.4 suggests the definition of exponentiation with arbitrary real exponent.

Definition 12.1.5. *If $a > 0$ and x is real, we define*

$$a^x = \exp(x \log a).$$

In particular, $e^x = \exp x$ for all real x.

Proposition 12.1.6. *If r is real and $f : (0, \infty) \to \mathbf{R}$ is defined by $f(x) = x^r$, then f is differentiable, and $f'(x) = rx^{r-1}$.*

Proof. By definition, $f(x) = \exp(r \log x)$. The chain rule gives

$$f'(x) = \exp'(r \log x)\, r \log' x = x^r (r/x) = rx^{r-1}. \qquad \square$$

Proposition 12.1.7. *Assume $b > 0$. The function $\exp_b : \mathbf{R} \to \mathbf{R}$ defined by $\exp_b x = b^x = \exp(x \log b)$ is differentiable, and*

$$\exp_b' x = (\log b) \exp_b x \quad \text{for all real } x.$$

Conversely, if k is a real number and f is a differentiable function satisfying $f' = kf$, then $f(x) = f(0) \exp(kx)$ for all x.

Proof. If k is real and $f(x) = \exp(kx)$, Proposition 12.1.3 and the chain rule give $f'(x) = kf(x)$. The first assertion follows by taking $k = \log b$.

The converse is analogous to Proposition 12.1.3, see Exercise 12.1.10. $\quad\square$

Corollary 12.1.8. $e^{x+y} = e^x e^y$ *and* $e^{xy} = (e^y)^x$ *for all real x and y.*

Proof. Fix a real number y arbitrarily, and define $f(x) = e^{x+y}$. By the chain rule, f is differentiable, and $f' = f$. Since $f(0) = e^y$, Proposition 12.1.3 implies $e^{x+y} = e^x e^y$ for all real x.

To prove the second assertion, consider $g(x) = (e^y)^x$. By the first part of Proposition 12.1.7 with $b = e^y$,

$$g'(x) = \log(e^y)(e^y)^x = yg(x).$$

By the second part of Proposition 12.1.7 $g(x) = e^{xy}$, since $g(0) = 1$.

As a fringe benefit, $(e^y)^x = (e^x)^y$, since each is equal to e^{xy}. $\quad\square$

Remark 12.1.9. The identity $\exp(x + y) = \exp x \exp y$ may also be proven algebraically. Set $u = \exp x$ and $v = \exp y$. Applying exp to both sides of the identity $x + y = \log u + \log v = \log uv$ gives

$$\exp(x + y) = \exp(\log uv) = uv = \exp x \exp y. \qquad \diamond$$

Corollary 12.1.10. *If $b > 0$, then $b^{x+y} = b^x b^y$ and $b^{xy} = (b^x)^y$ for all real x and y.*

Proof. Exercise 12.1.11. $\qquad\square$

Remark 12.1.11. Although we define $0^0 = 1$, if L is real and $0 \leq L < 1$, there exist functions f and g such that $\lim(f, 0) = 0$ and $\lim(g, 0) = 0$ but $\lim(f^g, 0) = L$, see Exercise 12.1.6. $\qquad \diamond$

Definition 12.1.12. If $b > 0$, the inverse of $\exp_b$, denoted $\log_b$, is called the *base-b logarithm.*

Proposition 12.1.13. *For all positive b and x, we have $\log_b x = (\log x)/(\log b)$.*

Proof. The equation $y = \log_b x$ means $x = b^y = \exp(y \log b)$. Taking the natural logarithm, $y = (\log x)/(\log b)$. $\qquad\square$

Exercises for Section 12.1

Exercise 12.1.1. (★) Assume a, b, and c are positive real numbers. Is it more sensible to agree that a^{b^c} is equal to $(a^b)^c$ or to $a^{(b^c)}$, or does it matter?

Exercise 12.1.2. If a and b are positive real numbers, then $b^{\log a} = a^{\log b}$.

Exercise 12.1.3. (A) One sometimes reads whimsical claims along the lines of, "If a hundred monkeys type characters at random, they will eventually type the complete works of Shakespeare." Is this true? Discuss.

Exercise 12.1.4. (★) Assume u is differentiable. Find the derivative of $\exp \circ u$, and the derivative of $\log \circ |u|$ at points where u is non-zero.

Exercise 12.1.5. (A) Find the even and odd parts of $f(t) = t/(1 - e^t)$. As needed, write each in a form that is manifestly symmetric.

Exercise 12.1.6. Assume L is a real number such that $0 \leq L < 1$. Find a pair of positive functions f and g such that $\lim(f, 0) = \lim(g, 0)$ but $\lim(f^g, 0) = L$, see Remark 12.1.11.

Exercise 12.1.7. (H) Define $f : \mathbf{R} \to \mathbf{R}$ by $f(x) = (x^2 - 1)e^{-x}$.

(a) Sketch the graph of f, using information about the first two derivatives to determine intervals on which f is monotone, convex, and concave.

(b) Assume y is real. Determine, with justification, how many solutions the equation $f(x) = y$ has. (The answer depends on y.)

Exercise 12.1.8. For x real and non-negative, define

$$F(x) = \int_0^{x^2} \frac{e^{t^3}}{\sqrt{t^3 + 1}}\, dt, \qquad G(x) = \int_0^{x^3} \frac{e^{t^2}}{\sqrt{t^2 + 1}}\, dt.$$

Which is larger, $F(1/2)$ or $G(1/2)$?

Exercise 12.1.9. Assume n is a positive integer, and define $f_n(x) = x^n e^{-x}$ on $[0, \infty)$. Prove f_n has a unique maximum, and find it with justification.

Exercise 12.1.10. Complete the proof of Proposition 12.1.7: If k is a real number and f is a differentiable function satisfying $f' = kf$, then we have $f(x) = f(0) \exp(kx)$ for all real x.

Exercise 12.1.11. Prove Corollary 12.1.10: If $b > 0$, then $b^{x+y} = b^x b^y$ and $b^{xy} = (b^x)^y$ for all real x, y.

Exercise 12.1.12. (★) This question relates base-10 logarithms with the number of digits in decimal notation.

(a) If n and N are positive integers, prove that

$$n \le \log_{10} N < n + 1 \quad \text{if and only if} \quad 10^n \le N < 10^{n+1},$$

if and only if N is an integer having $n + 1$ digits. In words, the integer part of the base 10 logarithm of N is one less than the number of digits of N.

(b) Which is larger, $2^{2^{2^{2^{2^2}}}}$ or $10^{10^{10^{100}}}$? About how many digits does each have?

Exercise 12.1.13. Define $f : \mathbf{R} \to \mathbf{R}$ by $f(x) = \exp(-1/x^2)$ if $x > 0$ and $f(x) = 0$ if $x \le 0$. Prove that $f^{(k)}(0)$ exists and is equal to zero for every natural number k. Conclude that f is smooth, but not real-analytic in any neighborhood of 0. Suggestion: First use induction on the degree to prove that for every polynomial p,

$$\lim_{x \to 0} p(1/x)\, f(x) = 0.$$

Then show inductively that every derivative of f is of this form.

Exercise 12.1.14. This exercise constructs *smooth bump functions*. Assume $a < c < d < b$, and let f be the smooth function of Exercise 12.1.13.

(a) Prove that $g(x) = f(x)f(1-x)$ is smooth, strictly positive in $(0,1)$, and identically 0 off $[0,1]$, and that

$$h(x) = \int_0^x g(t)\,dt \bigg/ \int_0^1 g(t)\,dt$$

is smooth, non-decreasing, identically 0 if $x \leq 0$, identically 1 if $1 \leq x$, and strictly increasing on $[0,1]$.

(b) Prove that

$$\phi(x) = h\left(\frac{x-a}{c-a}\right) h\left(\frac{b-x}{b-d}\right)$$

is smooth, and satisfies $0 \leq \phi(x) \leq 1$ for all x, $\phi(x) = 1$ if and only if $x \in [c,d]$, and $\phi(x) = 0$ if and only if $x \notin (a,b)$.

Exercise 12.1.15. (H) (The convexity bound.) If f is integrable on $[a,b]$ and if $p > 1$, define the *p-norm* of f to be the non-negative real number

$$\|f\|_p = \left[\frac{1}{b-a}\int_a^b |f|^p\right]^{1/p}.$$

This exercise outlines a proof that if p, q are positive real numbers such that $1/p + 1/q = 1$, and if f and g are integrable on $[a,b]$, then

$$\frac{1}{b-a}\int_a^b |fg| \leq \|f\|_p \|g\|_q = \left[\frac{1}{b-a}\int_a^b |f|^p\right]^{1/p}\left[\frac{1}{b-a}\int_a^b |g|^q\right]^{1/q}. \quad (\ddagger)$$

(a) Assume $0 < \alpha < 1$. Prove $t^\alpha \leq \alpha t + (1-\alpha)$ for all non-negative t.

(b) Let $\beta = 1 - \alpha$, so that $0 < \beta < 1$. Prove

$$u^\alpha v^\beta \leq \alpha u + \beta v \quad \text{for all positive } u, v.$$

Suggestion: Set $t = u/v$ in part (a).

(c) Assume $p > 1$, and put $q = p/(p-1)$, so $1/p + 1/q = 1$. Prove

$$AB \leq A^p/p + B^q/q \quad \text{for all non-negative } A, B.$$

(d) (H) If $1/p + 1/q = 1$ and $(a_k)_{k=0}^{n-1}$ and $(b_k)_{k=0}^{n-1}$ are finite real sequences, prove that

$$\sum_{k=0}^{n-1} |a_k b_k| \leq \left[\sum_{k=0}^{n-1} |a_k|^p\right]^{1/p}\left[\sum_{k=0}^{n-1} |b_k|^q\right]^{1/q}.$$

(e) (H) Prove the inequality $(\ddagger)$.

12.2   Representations of exp

In this section we establish two famous representations of e^x: As a limit of "discrete compounding" and as a power series. Many authors take one of these as the definition.

Proposition 12.2.1. *For all real x,*

$$e^x = \lim_{n\to\infty} \left[1 + \frac{x}{n}\right]^n.$$

Proof. The theorem is immediate if $x = 0$. Otherwise, if $0 < h < 1/|x|$ then $1 + xh > 0$. Since $\log 1 = 0$, we may write

$$\frac{1}{h}\log(1 + xh) = \frac{\log(1 + xh) - \log 1}{h} = x\frac{\log(1 + xh) - \log 1}{xh}.$$

As $h \to 0$, this approaches $x \log' 1 = x$. Writing $h = 1/n$ and using continuity of exp,

$$\lim_{n\to\infty} \left[1 + \frac{x}{n}\right]^n = \lim_{h\to 0^+} (1 + xh)^{1/h}$$

$$= \lim_{h\to 0^+} \exp\left(\frac{1}{h}\log(1 + xh)\right)$$

$$= \exp\left(\lim_{h\to 0^+} \frac{1}{h}\log(1 + xh)\right) = e^x. \qquad \square$$

Remark 12.2.2. Proposition 12.2.1 characterizes the natural exponential function as a limit of geometric growth. If, for example, x is the annual interest rate on a savings account, and there are n compoundings per year, then the multiplier on the right gives the factor by which the savings increase over one year. As the number of compoundings per year grows without bound, the balance does not become infinite in a finite time. Instead, if \$1 is allowed to accrue interest with continuous compounding, then in the time it would take the savings to double without compounding, the balance increases to \$2.72 (rounded to the nearest penny).   ◇

Proposition 12.2.3. $e^x = \displaystyle\sum_{k=0}^{\infty} \frac{x^k}{k!}$ *for all real x.*

Proof. Let $f(x)$ denote the sum of the power series. The coefficients are $a_k = 1/k!$, and

$$\lim_{k\to\infty} \left|\frac{a_{k+1}x^{k+1}}{a_k x^k}\right| = \lim_{k\to\infty} \frac{k!\,|x|}{(k+1)!} = \lim_{k\to\infty} \frac{|x|}{k+1} = 0.$$

Since this is less than 1 independently of x, the series converges absolutely for all x. By Proposition 11.1.6, the series may be differentiated term by term:

$$f'(x) = \sum_{k=1}^{\infty} \frac{1}{k!} kx^{k-1} = \sum_{k=0}^{\infty} \frac{x^k}{k!} = f(x).$$

Since $f(0) = 1$, Proposition 12.1.3 implies $f(x) = e^x$ for all x.   □

Remark 12.2.4. The exponential series could have been found using Definition 11.2.2, since $f^{(n)}(0) = \exp 0 = 1$ for all n. The series can be easily remembered: The constant term is $1 = e^0$, and each successive term is the integral from 0 of the preceding term.   ◇

Corollary 12.2.5. *For every real M,* $\displaystyle\lim_{x\to\infty} \frac{x^M}{\exp x} = 0$.

For every positive ε, $\displaystyle\lim_{x\to\infty} \frac{\log x}{x^\varepsilon} = 0$.

Proof. For the first, use finitude to pick an integer $N > |M|$. By Proposition 12.2.3, $x^{N+1}/(N+1)! < \exp x$ for all non-negative x, so

$$\lim_{x\to\infty} \left| \frac{x^M}{\exp x} \right| \leq \lim_{x\to\infty} (N+1)! \left| \frac{x^N}{x^{N+1}} \right| = \lim_{x\to\infty} \frac{(N+1)!}{x} = 0.$$

For the second, substitute $x = \exp(t/\varepsilon)$, so

$$\lim_{x\to\infty} \frac{\log x}{x^\varepsilon} = \lim_{t\to\infty} \frac{t/\varepsilon}{\exp t} = 0.$$ □

Corollary 12.2.6. $\displaystyle e = \sum_{k=0}^{\infty} \frac{1}{k!} = 1 + \frac{1}{1!} + \frac{1}{2!} + \frac{1}{3!} + \frac{1}{4!} + \cdots$ *is irrational.*

Proof. (This was also established in Exercise 7.4.7.) The series representation follows by setting $x = 1$ in the exponential power series. We will prove that if e is rational, then there exist integers N and M such that $N < M < N + 1$. Contrapositively, since such integers do not exist, e is irrational.

If ℓ and m are natural numbers, Exercise 3.2.13 gives

$$\frac{1}{(\ell+m)!} \leq \frac{1}{(\ell+1)^m \, \ell!},$$

with strict inequality if $m > 1$. By the geometric series formula,

$$\sum_{m=1}^{\infty} \frac{1}{(\ell+1)^m} = \frac{1/(\ell+1)}{1 - 1/(\ell+1)} = \frac{1}{(\ell+1)-1} = \frac{1}{\ell},$$

and so

$$\sum_{k=\ell+1}^{\infty} \frac{1}{k!} = \sum_{m=1}^{\infty} \frac{1}{(\ell+m)!} < \sum_{m=1}^{\infty} \frac{1}{(\ell+1)^m \ell!} = \frac{1}{\ell \cdot \ell!}.$$

Since every term in the series for e is positive, we have

$$\sum_{k=0}^{\ell} \frac{1}{k!} < e = \sum_{k=0}^{\infty} \frac{1}{k!} = \sum_{k=0}^{\ell} \frac{1}{k!} + \sum_{k=\ell+1}^{\infty} \frac{1}{k!} < \sum_{k=0}^{\ell} \frac{1}{k!} + \frac{1}{\ell \cdot \ell!}$$

for every positive integer ℓ. Multiplying through by $\ell!$,

$$\sum_{k=0}^{\ell} \frac{\ell!}{k!} < \ell! \, e < \sum_{k=0}^{\ell} \frac{\ell!}{k!} + \frac{1}{\ell}.$$

Each term $\ell!/k!$ of the sum is an integer since $0 \le k \le \ell$, so the sum represents some integer $N(\ell)$ depending on ℓ.

If e is rational, then there exist positive integers p and q such that $e = p/q$. But then $N(q) = N$ and $q! \, e = p(q-1)! = M$ are integers, and the preceding inequality implies $N < M < N + (1/q) < N + 1$. Since no such integers M and N exist, e is not rational. □

Remark 12.2.7. The preceding proof shows that for every positive integer ℓ,

$$e = \left[\sum_{k=0}^{\ell} \frac{1}{k!}\right] + \frac{1}{2\ell \cdot \ell!} + E_\ell, \quad |E_\ell| < \frac{1}{2\ell \cdot \ell!}.$$

For example, taking $\ell = 10$ gives a rational estimate for e that is accurate to within $\varepsilon = 1/(20 \cdot 10!) = 1/72,576,000 < 1.4 \times 10^{-8}$:

$$e \approx \frac{197,282,021}{72,576,000} \approx 2.71828\,1815.$$

(To fifteen decimals, the value of e is $2.71828\,18284\,59045\ldots$.) ◇

Exercises for Section 12.2

Exercise 12.2.1. (★) Evaluate the following limits:

(a) $\lim_{x\to 0^+} x \log x.$ (b) $\lim_{x\to 0^+} x^x.$ (c) $\lim_{x\to\infty} x^{1/x}.$

Exercise 12.2.2. Define $f(x) = x^x$ if $x > 0$. Calculate f', and find $f'(0^+)$.
Prove f has a unique minimum, and find the location and minimum value.

Exercise 12.2.3. Evaluate the following limits:

(a) $\lim\limits_{x \to 0} \log(1 + x)/x.$ (b) $\lim\limits_{x \to 0}(1 + x)^{1/x}.$ (c) $\lim\limits_{x \to \infty}(1 + x)^{1/x}.$

Exercise 12.2.4. Prove that $\lim\limits_{x \to \infty} \dfrac{(\log x)^n}{x} = 0$ for every positive integer n.

Exercise 12.2.5.($\bigstar$) Determine which of the following converge:

(a) $\sum\limits_{n=1}^{\infty} \dfrac{\log n}{n^{3/2}}.$ (b) $\sum\limits_{n=1}^{\infty}(-1)^n \dfrac{\log n}{n}.$ (c) $\sum\limits_{n=2}^{\infty} \dfrac{1}{n(\log n)^2}.$

Exercise 12.2.6. Define functions $F, G : (e, \infty) \to \mathbf{R}$ by

$$F(x) = \int_{e^e}^{x} \frac{dt}{t \log t[\log(\log t)]}, \qquad G(x) = \int_{e^e}^{x} \frac{dt}{t \log t[\log(\log t)]^{1.0001}},$$

compare Exercise 11.3.3, and write $f = F'$ and $g = G'$.

(a) Show $\sup F = \infty$; calculate $\sup G$.

(b) For what x is $F(x) = 100$? For what x' is $G(x') = 100$? Which number is larger? Use a calculator to estimate how many digits x has.

(c) Prove $f(x)/g(x)$ is increasing and unbounded. Use a calculator to estimate f/g at 10^{100}. For what x is $f(x)/g(x) = 100$, namely, at which point is "F increasing 100 times faster than G?" Use a calculator to estimate how many digits x has in decimal notation.

Exercise 12.2.7. Using the estimate of Remark 12.2.7, what is the smallest multiple of a hundred terms guaranteeing an accuracy $\varepsilon < 0.5 \times 10^{-1000}$?

Exercise 12.2.8. Assume n is a positive integer. Prove $\lim\limits_{x \to \infty} x^n e^{-x} = 0$ in as many ways as you can, and prove $\lim\limits_{x \to 0^+} x^{-n} e^{-1/x} = 0$.

Exercise 12.2.9. If $b \geq 1$, define a non-decreasing sequence $\big(x_k(b)\big)_{k=0}^{\infty}$ by

$$x_0(b) = 1, \qquad x_{k+1}(b) = b^{x_k(b)} \quad \text{if } k \geq 0.$$

Thus $x_1(b) = b$, $x_2(b) = b^b$, $x_3(b) = b^{b^b}$, and so forth. If $\big(x_k(b)\big)$ converges, denote the limit by $x_\infty(b)$.

(a) Prove x_∞ is increasing. For which positive b is $x_\infty(b)$ defined?

(b) Two people are arguing. One says that if $b = 2^{1/2}$, then $x_\infty(b) = 2$, since $(2^{1/2})^2 = 2$; the other says $x_\infty(b) = 4$ because $(2^{1/2})^4 = 4$. Who, if either, is correct, and why?

Exercise 12.2.10. Fix a real number b greater than 1. Prove that $\exp_b$ is not an algebraic function: For every positive integer N, there do not exist polynomial functions $(p_k)_{k=0}^{N}$ such that

$$0 = \sum_{k=0}^{N} p_k(x)y^k \quad \text{if and only if} \quad y = b^x.$$

12.3   Hyperbolic Functions

Definition 12.3.1. The *hyperbolic cosine* and *hyperbolic sine* functions, Figure 12.3, are defined to be the even and odd parts of exp:

$$\cosh x = \frac{e^x + e^{-x}}{2}, \qquad\qquad \sinh x = \frac{e^x - e^{-x}}{2}.$$

The *hyperbolic tangent* and *hyperbolic secant* functions, shown in Figure 12.4, are defined by

$$\tanh x = \frac{\sinh x}{\cosh x} = \frac{e^x - e^{-x}}{e^x + e^{-x}}, \qquad\qquad \operatorname{sech} x = \frac{1}{\cosh x}.$$

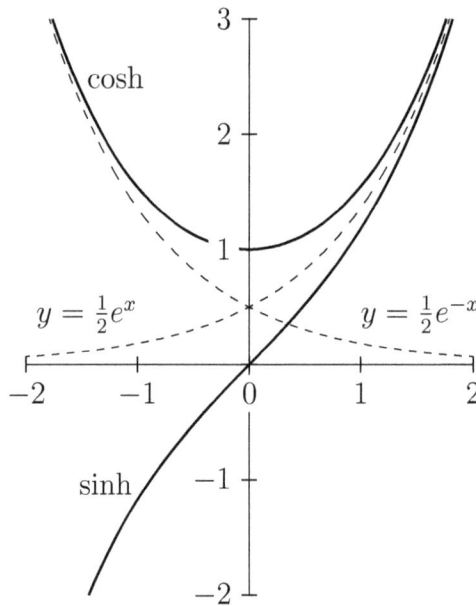

FIGURE 12.3
Hyperbolic cosine and sine.

Remark 12.3.2. The functions "cosh" and "sech" are pronounced phonetically; sinh is pronounced "cinch," and tanh rhymes with "ranch." As the notation suggests, these functions have a close kinship with the circular functions of ordinary trigonometry.   ◇

Proposition 12.3.3. *The functions* cosh, sinh, tanh, *and* sech *are defined for all real numbers, and satisfy the identities* $\cosh^2 - \sinh^2 = 1$,

$$\cosh' = \sinh, \qquad\qquad \sinh' = \cosh, \qquad\qquad \tanh' = \operatorname{sech}^2.$$

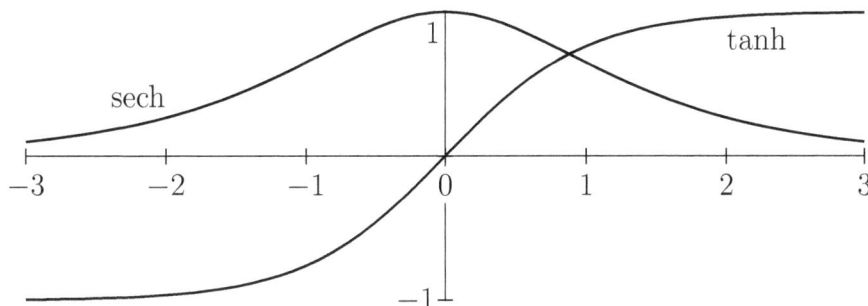

FIGURE 12.4
Hyperbolic tangent and secant.

cosh *maps* $[0, \infty)$ *bijectively to* $[1, \infty)$. sinh *maps* $\mathbf{R}$ *bijectively to* $\mathbf{R}$. tanh *maps* $\mathbf{R}$ *bijectively to* $(-1, 1)$.

   Every point (x, y) *on the right nappe of the hyperbola* $x^2 - y^2 = 1$ *is of the form* $(\cosh t, \sinh t)$ *for a unique real* t.

Proof. See Exercise 12.3.1 □

   Proposition 8.3.19 and the power series for exp immediately give series expansions for cosh and sinh.

Proposition 12.3.4. *For all real* x,

$$\cosh x = \sum_{k=0}^{\infty} \frac{x^{2k}}{(2k)!} = 1 + \frac{x^2}{2!} + \frac{x^4}{4!} + \frac{x^6}{6!} + \cdots,$$

$$\sinh x = \sum_{k=0}^{\infty} \frac{x^{2k+1}}{(2k+1)!} = x + \frac{x^3}{3!} + \frac{x^5}{5!} + \frac{x^7}{7!} + \cdots.$$

Proposition 12.3.5. *For all real* x *and* y,

$$\cosh(x + y) = \cosh x \cosh y + \sinh x \sinh y,$$
$$\sinh(x + y) = \cosh x \sinh y + \sinh x \cosh y,$$
$$\tanh(x + y) = \frac{\tanh x + \tanh y}{1 + \tanh x \tanh y}.$$

Particularly, for all real x,

$$\cosh(2x) = \cosh^2 x + \sinh^2 x, \qquad \sinh(2x) = 2 \cosh x \sinh x.$$

Inverse Hyperbolic Functions

Proposition 12.3.6. *The inverse hyperbolic functions are as follows:*

$$\cosh^{-1} x = \log\left(x \pm \sqrt{x^2 - 1}\right), \quad x \geq 1;$$
$$\sinh^{-1} x = \log\left(x + \sqrt{x^2 + 1}\right), \quad x \text{ real};$$
$$\tanh^{-1} x = \frac{1}{2} \log\left(\frac{1+x}{1-x}\right), \quad -1 < x < 1.$$

Their derivatives are

$$(\cosh^{-1})'(x) = \frac{\pm 1}{\sqrt{x^2 - 1}}, \quad x \geq 1;$$

$$(\sinh^{-1})'(x) = \frac{1}{\sqrt{x^2 + 1}}, \quad x \ \textit{real};$$

$$(\tanh^{-1})'(x) = \frac{1}{1 - x^2}, \quad -1 < x < 1.$$

Remark 12.3.7. Because cosh is even, the two branches of $\cosh^{-1}$ must be negatives of each other. By the difference of squares identity, $x \geq 1$ implies

$$x - \sqrt{x^2 - 1} = \frac{1}{x + \sqrt{x^2 - 1}} > 0,$$

so indeed, $\log\left(x \pm \sqrt{x^2 - 1}\right) = \pm \log\left(x + \sqrt{x^2 - 1}\right)$. Similar arguments show $\sinh^{-1}$ and $\tanh^{-1}$ are odd.   ◇

Proof. If $x \geq 1$, then $x = \cosh y = (e^y + e^{-y})/2$ if and only if

$$(e^y)^2 - (2x)e^y + 1 = 0.$$

This is a quadratic in e^y. By the quadratic formula, $e^y = x \pm \sqrt{x^2 - 1}$, or

$$y = \cosh^{-1} x = \log\left(x \pm \sqrt{x^2 - 1}\right).$$

Similarly, $x = \sinh y = (e^y - e^{-y})/2$ if and only if

$$(e^y)^2 - (2x)e^y - 1 = 0.$$

The quadratic formula gives $e^y = x + \sqrt{x^2 + 1}$, or

$$y = \sinh^{-1} x = \log\left(x + \sqrt{x^2 + 1}\right).$$

There is no sign ambiguity because only this choice leads to a real-valued function when x is real.

Finally,

$$x = \tanh y = \frac{e^y - e^{-y}}{e^y + e^{-y}} = \frac{e^{2y} - 1}{e^{2y} + 1}$$

if and only if $xe^{2y} + x = e^{2y} - 1$, or $e^{2y} = (1 + x)/(1 - x)$, or

$$y = \tanh^{-1} x = \frac{1}{2} \log\left(\frac{1 + x}{1 - x}\right).$$

The derivatives are found by direct calculation, see Exercise 12.3.4.   □

Exercises for Section 12.3

Exercise 12.3.1. Prove Proposition 12.3.3.

Exercise 12.3.2. Define $f : \mathbf{R} \to \mathbf{R}^2$ by $f(t) = (\operatorname{sech} t, \tanh t)$. Prove the image is contained in the unit circle $\{(x, y) : x^2 + y^2 = 1\}$. Describe the image exactly.

Exercise 12.3.3. (★) Establish the given identities where each makes sense:

(a) $\cosh(\sinh^{-1} t) = \sqrt{t^2 + 1}$. (b) $\sinh(\cosh^{-1} t) = \pm\sqrt{t^2 - 1}$.

Exercise 12.3.4. Finish the proof of Proposition 12.3.6 by calculating the derivatives of $\cosh^{-1}$, $\sinh^{-1}$, and $\tanh^{-1}$.

Exercise 12.3.5. (★) Use the substitution $u = \sinh t$ to show

$$\int_0^x \sqrt{u^2 + 1}\, du = \tfrac{1}{2}\left(x\sqrt{x^2 + 1} + \log(x + \sqrt{x^2 + 1})\right).$$

Exercise 12.3.6. Evaluate $\int_1^x \sqrt{u^2 - 1}\, du$ if $x \geq 1$.

Exercise 12.3.7. (★) Assuming $0 \leq t$, calculate the shaded area in Figure 12.5 as a function of t.

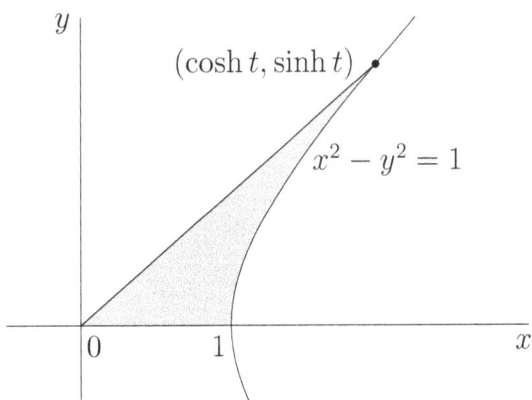

FIGURE 12.5
The hyperbolic sector for $[0, t]$.

Exercise 12.3.8. Evaluate $\int_{-\infty}^{\infty} \operatorname{sech}^2 x \, dx$.

Exercise 12.3.9. (H) If x and y are in $(-1, 1)$, define

$$x \oplus y = \frac{x+y}{1+xy}.$$

Prove this is a binary operation on $(-1, 1)$, and determine whether $\oplus$ is associative, commutative, has an identity element, and if so, which elements have inverses.

Exercise 12.3.10. Each part refers to the function $f(t) = (e^t - 1)/t$, extended by continuity so $f(0) = 1$, and its reciprocal $g(t) = 1/f(t)$.

(a) Define the hyperbolic cotangent $\coth = \cosh/\sinh$. Prove $g(t) + \frac{t}{2} = \frac{t}{2} \coth \frac{t}{2}$.

(b) Write f and g as power series

$$f(t) = \sum_{i=0}^{\infty} a_i \frac{t^i}{i!}, \qquad\qquad g(t) = \sum_{j=0}^{\infty} b_j \frac{t^j}{j!}.$$

Find a formula for the a_i, and prove $b_k = -\dfrac{1}{k+1} \sum_{j=0}^{k-1} \binom{k+1}{j} b_j$.

(c) Use the (b_k) to express $t \coth t$ as a power series.

12.4 Factorials

An Exponential Estimate

Factorials are closely connected with the natural exponential function.

Theorem 12.4.1 (Factorial growth rate). *For every positive integer* n,

$$e^{7/8} \left(\frac{n}{e}\right)^n \sqrt{n} < n! < e \left(\frac{n}{e}\right)^n \sqrt{n}.$$

Proof. Integrating by parts with $u = \log x$, we have

$$\int_1^n \log x \, dx = x \log x - x \Big|_{x=1}^{n} = n \log n - n + 1 = \log\left(\frac{n^n}{e^n} \cdot e\right).$$

The natural logarithm function is concave, so for every splitting Π of $[1, n]$, we have $\mathrm{TRAP}(\log, \Pi) < \int_1^n \log$, see Figure 12.6. Particularly, if Π is the

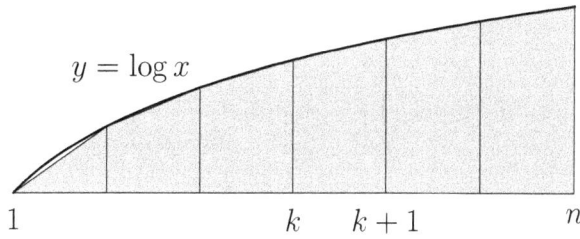

FIGURE 12.6
Trapezoid approximation of the integral.

equal-length splitting of $[1, n]$ with $n - 1$ pieces of length 1, the inequality
$\text{TRAP}(\log, \Pi) < \int_1^n \log$ becomes

$$\frac{1}{2}(\log 1 + \log 2) + \frac{1}{2}(\log 2 + \log 3) + \cdots + \frac{1}{2}\Big(\log(n - 1) + \log n\Big)$$

$$= -\frac{1}{2}\log n + \sum_{k=1}^{n} \log k = \log\left(\frac{n!}{\sqrt{n}}\right) < \log\left(\frac{n^n}{e^n} \cdot e\right).$$

Exponentiating and rearranging gives the asserted upper bound for $n!$.

On the other hand, $[1, n] = [1, \frac{3}{2}] \cup [\frac{3}{2}, n - \frac{1}{2}] \cup [n - \frac{1}{2}, n]$, see Figure 12.7. The (unshaded) "end piece" integrals satisfy

$$\int_1^{3/2} \log x \, dx < \int_1^{3/2} (x - 1) \, dx = 1/8, \qquad \int_{n-(1/2)}^{n} \log x \, dx < (1/2) \log n.$$

For the middle interval, let Π be the equal-length splitting with $(n - 2)$ pieces of length 1. Concavity of log implies $\int_{3/2}^{n-(1/2)} \log < \text{MID}(\log, \Pi)$, compare Exercise 11.2.8. The midpoints from Π are integers k such that $2 \leq k \leq n - 1$, so $\text{MID}(\log, \Pi) = \sum_k \log k$.

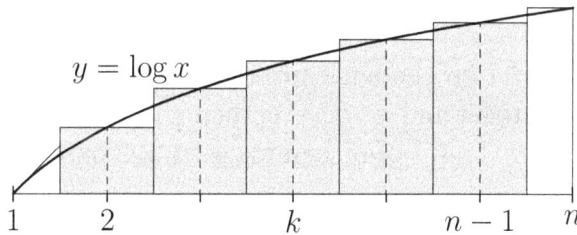

FIGURE 12.7
Modified midpoint approximation of the integral.

Summing these estimates gives

$$\log\left(\frac{n^n}{e^n}\cdot e\right) = \int_1^n \log < \left[\int_1^{3/2}\log + \int_{n-(1/2)}^n \log\right] + \mathrm{MID}(\log, \Pi)$$

$$= (1/8) + (1/2)\log n + \sum_{k=2}^{n-1}\log k$$

$$= (1/8) - (1/2)\log n + \sum_{k=1}^n \log k = \log\left(e^{1/8}\cdot \frac{n!}{\sqrt{n}}\right).$$

Rearranging gives the asserted lower bound for $n!$.   □

The Gamma Function

The Gamma function is a "smooth interpolation" of factorials for positive real numbers. The development here follows an exercise in Rudin, [25].

Lemma 12.4.2. *For every positive real number x, $t^x/e^t \to 0$ as $t \to \infty$.*

Proof. The series representation for e^t guarantees that if n is an integer greater than x and if $t > 0$, then $e^t > t^{n+1}/(n+1)!$. Consequently, if $t > 0$, then $t^x/e^t \le t^n/e^t < (n+1)!/t$, and the upper bound goes to 0 as $t \to \infty$.   □

Proposition 12.4.3. *For every real $x > 0$, the improper integral*

$$\int_0^\infty t^{x-1}e^{-t}\,dt$$

converges.

Proof. Since $e^{-t} < 1$ if $t > 0$,

$$\int_0^1 t^{x-1}e^{-t}\,dt < \int_0^1 t^{x-1}\,dt = \frac{1}{x},$$

so the improper integral on the left converges.

If n is a positive integer and $n \ge x - 1$, then $e^t > t^{n+2}/(n+2)!$ for positive t by the same idea as in the proof of Lemma 12.4.2. Thus

$$\int_1^\infty t^{x-1}e^{-t}\,dt \le \int_1^\infty t^n e^{-t}\,dt \le (n+2)!\int_1^\infty t^{-2}\,dt = (n+2)!,$$

so the improper integral on the left converges.   □

Definition 12.4.4. *The Gamma function* $\Gamma : (0, \infty) \to \mathbf{R}$ *is defined by*

$$\Gamma(x) = \int_0^\infty t^{x-1}e^{-t}\,dt.$$

The main result of this section is the following characterization of Γ, due to Harald Bohr and Johannes Mollerup.

Proposition 12.4.5. *The function Γ satisfies the following properties:*

(i) $\Gamma(x+1) = x\Gamma(x)$ *for all x.*

(ii) $\Gamma(n+1) = n!$ *for every positive integer n.*

(iii) $\log \Gamma$ *is convex.*

Conversely, if $f : (0, \infty) \to \mathbf{R}$ satisfies these properties, then $f = \Gamma$.

Proof. (i). If $b > 0$, integrating by parts with $u = t^x$ and $dv = e^{-t}\,dt$ gives

$$\int_0^b t^x e^{-t}\,dt = -t^x e^{-t}\Big|_{t=0}^{t=b} + x\int_0^b t^{x-1}e^{-t}\,dt$$
$$= -b^x e^{-b} + x\int_0^b t^{x-1}e^{-t}\,dt.$$

By Lemma 12.4.2, $-b^x e^{-b} \to 0$ as $b \to \infty$, so

$$\Gamma(x+1) = \int_0^\infty t^x e^{-t}\,dt = x\int_0^\infty t^{x-1}e^{-t}\,dt = x\Gamma(x).$$

(ii). If $n = 0$, we have

$$\Gamma(n+1) = \int_0^\infty e^{-t}\,dt = \lim_{b\to\infty} -e^{-t}\Big|_{t=0}^{t=b} = \lim_{b\to\infty}(1 - e^{-b}) = 1 = 0!.$$

By (i) and mathematical induction, $\Gamma(n+1) = n!$ if $n \geq 0$.

(iii). If $1/p + 1/q = 1$, the integrand of $\Gamma\big((x/p) + (y/q)\big)$ may be written

$$t^{(x/p)+(y/q)-1}e^{-t} = t^{(x/p)+(y/q)-(1/p)-(1/q)}e^{-t[(1/p)+(1/q)]}$$
$$= (t^{x-1}e^{-t})^{1/p}(t^{y-1}e^{-t})^{1/q}.$$

Since $t^{x-1}e^{-t} > 0$ on $(0, \infty)$, the convexity bound, Exercise 12.1.15, implies

$$\Gamma\big((x/p) + (y/q)\big) \leq \left[\int_0^\infty t^{x-1}e^{-t}\,dt\right]^{1/p}\left[\int_0^\infty t^{y-1}e^{-t}\,dt\right]^{1/q} = \Gamma(x)^{1/p}\Gamma(y)^{1/q}.$$

Taking logarithms,

$$\log \Gamma\big((x/p) + (y/q)\big) \leq (1/p)\log \Gamma(x) + (1/q)\log \Gamma(y),$$

which is the desired convexity statement.

Conversely, assume $f : (0, \infty) \to \mathbf{R}$ satisfies properties (i), (ii), and (iii), and set $\varphi = \log f$. Property (i) says $\varphi(x + 1) = \varphi(x) + \log x$ if $x > 0$; that is, the difference quotient of φ on $[y, y + 1]$ is $\log y$. By induction on n,

$$\varphi(x + n + 1) = \varphi(x) + \log \prod_{j=0}^{n} (x + j). \qquad (*)$$

Property (ii) says $\varphi(n + 1) = \log(n!)$ for every positive integer n and (iii) says φ is convex.

Fix a positive integer n. If $0 < x < 1$ arbitrarily, consider the four points $n < n + 1 < n + 1 + x < n + 2$. Since φ is convex, the difference quotients of φ on the intervals $[n, n + 1]$, $[n + 1, n + 1 + x]$, and $[n + 1, n + 2]$ are in non-decreasing order by Proposition 10.4.7 (iii), namely

$$\log n \leq \frac{\varphi(n + 1 + x) - \varphi(n + 1)}{x} \leq \log(n + 1).$$

Multiplying by x and substituting $(*)$,

$$x \log n \leq \varphi(x) + \left[\log \prod_{j=0}^{n} (x + j) \right] - \log(n!) \leq x \log(n + 1),$$

or

$$0 \leq \varphi(x) - \log\left[\frac{n! \, n^x}{\prod_{j=0}^{n} (x + j)} \right] \leq x \log\left(1 + \frac{1}{n} \right).$$

As $n \to \infty$, the upper bound goes to 0, so

$$\varphi(x) = \lim_{n \to \infty} \log\left[\frac{n! \, n^x}{\prod_{j=0}^{n} (x + j)} \right] \quad \text{if } 0 < x < 1.$$

By property (i), φ satisfies this equation for all positive x.

We have shown that if f satisfies properties (i)–(iii), then $\log f$ is the limit on the right. Since Γ satisfies (i)–(iii), the limit on the right must be equal to $\log \Gamma(x)$. $\qquad \square$

Remark 12.4.6. In particular, we obtain the representation

$$\Gamma(x) = \lim_{n \to \infty} \frac{n! \, n^x}{\prod_{j=0}^{n} (x + j)} \quad \text{if } 0 < x,$$

analogous to the characterization of exp in Proposition 12.2.1. $\qquad \diamond$

Corollary 12.4.7. *If $a > 0$, then $\int_{0}^{\infty} t^{x-1} e^{-t/a} \, dt = a^x \cdot \Gamma(x)$.*

Proof. Making the substitution $t = au$,

$$\int_{0}^{\infty} t^{x-1} e^{-t/a} \, dt = a^{x-1} \cdot a \int_{0}^{\infty} u^{x-1} e^{-u} \, du = a^x \cdot \Gamma(x). \qquad \square$$

Definition 12.4.8. The β *function* is defined by

$$\beta(x, y) = \int_0^1 t^{x-1}(1-t)^{y-1}\, dt \quad x, y \text{ positive.}$$

Proposition 12.4.9. $\beta(x, y) = \dfrac{\Gamma(x)\Gamma(y)}{\Gamma(x+y)}$ *for all positive x, y.*

Proof. See Exercise 12.4.9. $\qquad\square$

Definition 12.4.10. Define the ζ *function* (zeta) on $(1, \infty)$ by

$$\zeta(s) = \sum_{m=1}^{\infty} \frac{1}{m^s} = 1 + \frac{1}{2^s} + \frac{1}{3^s} + \cdots.$$

Proposition 12.4.11. *If $s > 1$, then* $\zeta(s) = \dfrac{1}{\Gamma(s)} \displaystyle\int_0^{\infty} \dfrac{x^{s-1}}{e^x - 1}\, dx.$

Proof. See Exercise 12.4.10. $\qquad\square$

We end with a widely cited result proven in the late 19th century.

Theorem 12.4.12 (*e* is transcendental)**.** *There exists no non-zero polynomial p with integer coefficients such that $p(e) = 0$.*

Proof. See Exercise 12.4.11. $\qquad\square$

Exercises for Section 12.4

Exercise 12.4.1.($\bigstar$) Without calculation, explain why:

(a) $\displaystyle\sum_{n=0}^{\infty} \frac{x^n}{n!\, e^x} = 1$ for all real x.   (b) $\displaystyle\int_0^{\infty} \frac{x^n\, dx}{n!\, e^x} = 1$ for all n in $\mathbf{N}$.

Exercise 12.4.2. For every natural number n, $\displaystyle\int_0^1 (\log x)^n\, dx = (-1)^n n!$.

Exercise 12.4.3. If m, n are natural numbers, evaluate $\displaystyle\int_0^1 x^m (\log x)^n\, dx$.

Exercise 12.4.4.($\bigstar$) Prove that $\displaystyle\int_0^{\infty} e^{-t^2}\, dt$ converges. (Do not evaluate.)

Exercise 12.4.5. Granted that $\displaystyle\int_{-\infty}^{\infty} e^{-x^2}\, dx = \sqrt{\pi}$ (the number π is defined in Chapter 13, see also Corollary 13.2.12), evaluate the following in terms of π:

(a) $\int_0^\infty \frac{e^{-x}}{\sqrt{x}}\, dx = \Gamma(1/2).$

(c) $\int_0^\infty x^n e^{-x^2/a^2}\, dx.$

(b) $\int_{-\infty}^\infty e^{-x^2/a^2}\, dx,\ a > 0.$

(d) $\int_{-\infty}^\infty x^n e^{-x^2/a^2}\, dx,\ a > 0.$

Exercise 12.4.6. (★) Find lower and upper bounds for $\binom{2n}{n} = (2n)!/(n!)^2$ that do not contain factorials.

Exercise 12.4.7. Determine which of the following converge:

(a) $\displaystyle\sum_{n=1}^\infty \frac{(n+1)^n}{n^{n+1}}.$

(b) $\displaystyle\sum_{n=1}^\infty \frac{n!}{n^n}.$

(c) $\displaystyle\sum_{n=1}^\infty \frac{(n!)^2}{(2n)!}.$

Exercise 12.4.8. This question asks you to mimic the estimates in the proof of Theorem 12.4.1 for the function $f(x) = 1/x$ on the interval $[1, n]$, with n a positive integer. (The limit in (c) is often denoted γ (gamma).)

(a) Use secant lines on the intervals $[k, k+1]$ to prove

$$\log n < \frac{1}{2} + \left[\sum_{k=2}^{n-1} \frac{1}{k}\right] + \frac{1}{2n}.$$

(b) Use tangent lines at $k = 2, 3, \ldots, n-1$, and suitably handle the intervals at the ends, to prove

$$\log 3/2 + \left[\sum_{k=2}^{n-1} \frac{1}{k}\right] + \frac{1}{2n} < \log n.$$

(c) Prove that $\left[\displaystyle\sum_{k=1}^n \frac{1}{k}\right] - \log n$ is increasing and bounded above.

Exercise 12.4.9. (H) Prove Proposition 12.4.9.

Exercise 12.4.10. (H) Prove Proposition 12.4.11, including that the integral converges.

Exercise 12.4.11. This exercise outlines a proof of Theorem 12.4.12, that e is *transcendental*, not the root of a polynomial equation with integer coefficients. The argument here is adapted from Lecture VII. of Klein, [18]. This exercise is unusual, filling in details in a provided sketch rather than requesting a proof from scratch. The mathematical importance of the result justifies its inclusion, but the proof here hinges on a particular sequence of polynomials with integer coefficients and all terms of large degree:

$$q_m(x) = x^m \left[\prod_{j=1}^n (x - j)\right]^{m+1} = x^m \left[(x-1)(x-2)\cdots(x-n)\right]^{m+1}.$$

The strategy is contrapositive: If there exist integers $(a_k)_{k=0}^n$ such that $a_0 \neq 0$ and

$$0 = \sum_{k=0}^n a_k e^k = a_0 + a_1 e + \cdots + a_n e^n, \qquad (*)$$

then there exist a real number r such that $|r| < 1$ and a non-zero integer N such that $0 = r + N$. Since no such r and N exist, there exist no integers (a_k) as above.

To prove r and N exist if $(*)$ is satisfied, multiply $(*)$ by $\frac{1}{m!} q_m(x) e^{-x} \, dx$ and integrate from 0 to ∞, splitting the kth integral at k:

$$0 = \sum_{k=0}^n \frac{a_k e^k}{m!} \int_0^k q_m(x) e^{-x} \, dx + \sum_{k=0}^n \frac{a_k e^k}{m!} \int_k^\infty q_m(x) e^{-x} \, dx$$

$$= \underbrace{\sum_{k=1}^n \frac{a_k e^k}{m!} \int_0^k q_m(x) e^{-x} \, dx}_{=r_m} + \underbrace{\sum_{k=0}^n \frac{a_k}{m!} \int_0^\infty q_m(x+k) e^{-x} \, dx}_{=N_m}.$$

Establish the following two claims and complete the proof.

(i) There is an M such that $\left| (x-1)(x-2) \cdots (x-n) \right| \leq M$ if $0 \leq x \leq n$, and

$$|r_m| := \left| \sum_{k=1}^n \frac{a_k e^k}{m!} \int_0^k q_m(x) e^{-x} \, dx \right| \leq \sum_{k=1}^n \frac{|a_k| e^k (kM)^{m+1}}{(m+1)!} \to 0.$$

(ii) If

$$N_{k,m} := \frac{e^k}{m!} \int_k^\infty q_m(x) e^{-x} \, dx = \frac{1}{m!} \int_0^\infty q_m(x+k) e^{-x} \, dx$$

for all k such that $0 \leq k \leq n$, then $N_{0,m} = a_0(-n!)^{m+1} + (m+1)(\text{integer})$, while if $1 \leq k \leq n$, then $N_{k,m} = (m+1)(\text{integer})$. If m is sufficiently large, the sum N_m is a non-zero integer. Hints: For non-negative integer m, we have $\int_0^\infty x^m e^{-x} \, dx = m!$. The q_m have integer coefficients and all terms have large degree.

13

Circular Functions

Trigonometric functions are usually introduced via geometry, either as ratios of sides in a right triangle, or in terms of points on the unit circle. Our analytic development loosely follows Ahlfors, [1]. One advantage over a geometric approach is that the definitions, based on the axioms of $\mathbf{R}$, fit neatly into the framework established so far. Fringe benefits include illustration of solving differential equations by power series, the ability to express and prove geometric theorems analytically, and easy generalization to complex variables.

13.1 Cosine and Sine

Proposition 13.1.1. *There exist unique functions C and $S : \mathbf{R} \to \mathbf{R}$ of class $\mathscr{C}^2$ satisfying*

$$
\begin{aligned}
C'' + C = 0, && C(0) = 1, && C'(0) = 0; \\
S'' + S = 0, && S(0) = 0, && S'(0) = 1.
\end{aligned}
$$

Definition 13.1.2. The function C in Proposition 13.1.1 is called the *cosine* function, $\cos : \mathbf{R} \to \mathbf{R}$. The function S is called the *sine* function $\sin : \mathbf{R} \to \mathbf{R}$.

Proof of Proposition 13.1.1. (Existence). We use an *Ansatz*, or "guess." Suppose provisionally that there exists a real-analytic solution y of the second-order differential equation $y'' + y = 0$ in some neighborhood of 0. Write

$$
y(x) = \sum_{k=0}^{\infty} a_k x^k = a_0 + a_1 x + a_2 x^2 + \cdots ,
$$

and use term by term differentiation and index-shifting to deduce

$$
y'(x) = \sum_{k=0}^{\infty} (k+1)a_{k+1} x^k,
$$

$$
y''(x) = \sum_{k=0}^{\infty} (k+2)(k+1)a_{k+2} x^k.
$$

DOI: 10.1201/9781003601357-13

By assumption, y satisfies

$$0 = y'' + y = \sum_{k=0}^{\infty}(k+2)(k+1)a_{k+2}x^k + \sum_{k=0}^{\infty}a_k x^k$$
$$= \sum_{k=0}^{\infty}\Big[(k+2)(k+1)a_{k+2} + a_k\Big]x^k.$$

The coefficients of this series must all be zero by the identity theorem for power series (Theorem 8.3.17), so we get a recursion relation

$$a_{k+2} = -\frac{a_k}{(k+2)(k+1)}, \quad k \geq 0.$$

The initial conditions $y(0) = a_0$ and $y'(0) = a_1$ determine the first two coefficients. The first two coefficients, together with the recursion relation, determine all the remaining coefficients:

$$a_{2k} = \frac{(-1)^k a_0}{(2k)!}, \qquad a_{2k+1} = \frac{(-1)^k a_1}{(2k+1)!}.$$

The respective initial conditions for C ($a_0 = 1$ and $a_1 = 0$) and S ($a_0 = 0$ and $a_1 = 1$) give

$$C(x) = \sum_{k=0}^{\infty}(-1)^k \frac{x^{2k}}{(2k)!} = 1 - \frac{x^2}{2!} + \frac{x^4}{4!} - \frac{x^6}{6!} + \cdots,$$
$$S(x) = \sum_{k=0}^{\infty}(-1)^k \frac{x^{2k+1}}{(2k+1)!} = x - \frac{x^3}{3!} + \frac{x^5}{5!} - \frac{x^7}{7!} + \cdots.$$

To determine the interval of convergence for each series, apply the ratio test. For $C(x)$, we have

$$\lim_{k\to\infty}\left|\frac{x^{2k+2}/(2k+2)!}{x^{2k}/(2k)!}\right| = \lim_{k\to\infty}\frac{|x|^2\,(2k)!}{(2k+2)!} = \lim_{k\to\infty}\frac{|x|^2}{(2k+2)(2k+1)},$$

which is 0 for all real x; that is, the series for $C(x)$ converges absolutely for every real x. The calculation for $S(x)$ is similar.

In summary, there exist real-analytic functions C and S satisfying the conditions of Proposition 13.1.1.

(Uniqueness). The existence proof implicitly showed that the conditions of Proposition 13.1.1 uniquely define *real-analytic* functions C and S. However, there could conceivably exist non-analytic solutions.

Lemma 13.1.3. *Assume* $y : \mathbf{R} \to \mathbf{R}$ *is a twice-differentiable function satisfying* $y'' + y = 0$ *on* $\mathbf{R}$. *If* $y(0) = y'(0) = 0$, *then* $y(x) = 0$ *for all* x.

Proof. If $y'' + y = 0$ for all real x, then

$$\left((y')^2 + y^2\right)' = 2y'y'' + 2yy' = 2y' \cdot (y'' + y) = 0.$$

By the identity theorem (Theorem 10.3.3), the function $(y')^2 + y^2$ is constant:

$$y'(x)^2 + y(x)^2 = y'(0)^2 + y(0)^2 = 0 \quad \text{for all real } x.$$

In particular, $y(x) = 0$ for all real x.   □

We now complete the proof of uniqueness. Assume a and b are arbitrary real numbers, and f is any twice-differentiable function satisfying $f'' + f = 0$, $f(0) = a$ and $f'(0) = b$. By linearity of the derivative, the function

$$y(x) = f(x) - \left(aC(x) + bS(x)\right)$$

satisfies $y'' + y = 0$, $y(0) = 0$, and $y'(0) = 0$. By Lemma 13.1.3, $y(x) = 0$ for all real x, so $f(x) = aC(x) + bS(x)$. That is, the differential equation $f'' + f = 0$ and the conditions $f(0) = a$ and $f'(0) = b$ uniquely determine f. This completes the proof of Proposition 13.1.1.   □

For later use, we record the final step of the proof:

Proposition 13.1.4. *If a and b are real numbers and f is a twice-differentiable function satisfying $f'' + f = 0$, $f(0) = a$ and $f'(0) = b$, then*

$$f(x) = a\cos x + b\sin x \quad \text{for all real } x.$$

Proposition 13.1.5. *The sine and cosine functions satisfy the following identities for all real x, y:*

(i) $\cos(-x) = \cos x$, $\quad \sin(-x) = -\sin x$.

(ii) $\sin'(x) = \cos(x)$, $\quad \cos'(x) = -\sin(x)$.

(iii) $\sin^2 x + \cos^2 x = 1$.

(iv) $\cos(x + y) = \cos x \cos y - \sin x \sin y$,
    $\sin(x + y) = \sin y \cos x + \sin x \cos y$.

   In particular, $\sin(2x) = 2\sin x \cos x$ *and* $\cos(2x) = \cos^2 x - \sin^2 x$.

Proof. (i). It is apparent that sin is an odd function from its power series; however, a direct proof (in the spirit of the proposition) can be given using Proposition 13.1.4: The function $f(x) = \sin(-x)$ satisfies $f'' + f = 0$, $f(0) = 0$, and (by the chain rule) $f'(0) = -1$, so $f = -\sin$. Evenness of cos is similar.

(ii). If $f = \sin'$, then $f'' + f = 0$ by differentiating $\sin'' + \sin = 0$. But $f(0) = \sin' 0 = 1$ by definition of sin, and $f'(0) = \sin''(0) = -\sin 0 = 0$. By Proposition 13.1.4, $\sin' = \cos$. A similar argument shows $\cos' = -\sin$.

(iii). Consider the function $f = \sin^2 + \cos^2$. Part (ii) gives

$$f' = 2\sin\sin' + 2\cos\cos' = 2\sin\cos + 2\cos(-\sin) = 0,$$

so f is constant. Thus $f(x) = f(0) = \sin^2 0 + \cos^2 0 = 1$ for all x.

(iv). To prove the addition formulas, fix a real number y and consider the function $f(x) = \sin(x + y)$. The chain rule gives $f'' + f = 0$, and the derivative formula for sin implies $f'(x) = \cos(x + y)$ for all x. Substituting $x = 0$, we find $f(0) = \sin y$ and $f'(0) = \cos y$. Proposition 13.1.4 with $a = \sin y$ and $b = \cos y$ gives

$$\sin(x + y) = f(x) = \sin y \cos x + \sin x \cos y$$

The addition formula for cos is similar. ◻

Several auxiliary formulas are listed here for reference. These need not be memorized, but their technique of proof is worth remembering.

Corollary 13.1.6. *For all real x and y,*

$$\cos x \cos y = \tfrac{1}{2}\big(\cos(x - y) + \cos(x + y)\big),$$
$$\sin x \sin y = \tfrac{1}{2}\big(\cos(x - y) - \cos(x + y)\big),$$
$$\sin x \cos y = \tfrac{1}{2}\big(\sin(x + y) + \sin(x - y)\big);$$
$$\cos x + \cos y = 2\cos\big(\tfrac{1}{2}(x + y)\big)\cos\big(\tfrac{1}{2}(x - y)\big),$$
$$\cos x - \cos y = 2\sin\big(\tfrac{1}{2}(x + y)\big)\sin\big(\tfrac{1}{2}(x - y)\big),$$
$$\sin x + \sin y = 2\sin\big(\tfrac{1}{2}(x + y)\big)\cos\big(\tfrac{1}{2}(x - y)\big).$$

Proof. From the addition theorem for cos and the fact that sin is odd,

$$\cos(x - y) = \cos x \cos y + \sin x \sin y,$$
$$\cos(x + y) = \cos x \cos y - \sin x \sin y.$$

Adding and subtracting these (and dividing by 2) gives the first two formulas. The third is derived similarly from the addition theorem for sin.

To derive the fourth, write the first as

$$\cos(u + v) + \cos(u - v) = 2\cos u \cos v.$$

Put $u = \tfrac{1}{2}(x + y)$ and $v = \tfrac{1}{2}(x - y)$, noting that $x = u + v$ and $y = u - v$. The fifth and sixth formulas are similar. ◻

Exercises for Section 13.1

Exercise 13.1.1. (★) Use termwise differentiation to give an alternative proof that $\sin' = \cos$ and $\cos' = -\sin$.

Exercise 13.1.2. Use the trigonometric power series to evaluate:

(a) $\displaystyle\lim_{x \to 0} \frac{1 - \cos x}{x^2}$.

(b) $\displaystyle\lim_{x \to 0} \int_0^{x^2} \frac{1 - \cos t}{x^6}\, dt$.

Exercise 13.1.3. Show that the power series

$$f(x) = \sum_{k=0}^{\infty} (-1)^k \frac{x^k}{(2k)!} = 1 - \frac{x}{2!} + \frac{x^2}{4!} - \frac{x^3}{6!} + \cdots$$

has infinite radius, and $f(x^2) = \cos x$ and $f(-x^2) = \cosh x$ if $x \ge 0$.

Exercise 13.1.4. Using the sampled form of the remainder, how large could

$$\left| \cos x - \sum_{k=0}^{n-1} (-1)^k \frac{x^{2k}}{(2k)!} \right|$$

be if $|x| \le \pi/2 < 1.6$? How many terms are sufficient to estimate $\cos x$ on this interval to within $\varepsilon < 10^{-6}$? $\varepsilon < 10^{-8}$?

13.2 Periodicity

Proposition 13.2.1. *There exists a unique positive real number τ such that $\cos$ is positive on $(-\tau, \tau)$, and $\cos \tau = 0$. Further,*

$$\sqrt{2} < \tau < \sqrt{6 - \sqrt{12}}.$$

Proof. To prove there exists a smallest positive root τ of the cosine function, we establish the estimates

$$x - \frac{x^3}{3!} \le \sin x \le x \quad \text{for all non-negative real } x,$$

$$1 - \frac{x^2}{2!} \le \cos x \le 1 - \frac{x^2}{2!} + \frac{x^4}{4!} \quad \text{for all real } x.$$

The identity $\sin^2 t + \cos^2 t = 1$ implies $-1 \le \cos t \le 1$ for every real t. Integrating from 0 to x gives, for every non-negative x,

$$-x \le \int_0^x \cos t\, dt = \sin t \Big|_{t=0}^{x} = \sin x \le x.$$

Since x was arbitrary, we may replace x with t, obtaining $-t \leq \sin t \leq t$ for non-negative t. Integrating *this* from 0 to x gives

$$-\frac{x^2}{2} \leq \int_0^x \sin t \, dt = 1 - \cos x \leq \frac{x^2}{2}.$$

Again replacing x with t and rearranging, we find $1 - (t^2/2) \leq \cos t \leq 1$ for non-negative t. Another integration gives

$$x - \frac{x^3}{6} \leq \sin x \leq x,$$

and a fourth gives $x^2/2 - x^4/24 \leq 1 - \cos x \leq x^2/2$, or

$$1 - \frac{x^2}{2} \leq \cos x \leq 1 - \frac{x^2}{2} + \frac{x^4}{24}$$

for all non-negative x. This inequality is true if $x < 0$ as well because each term is even in x. Figure 13.1 depicts these inequalities geometrically.

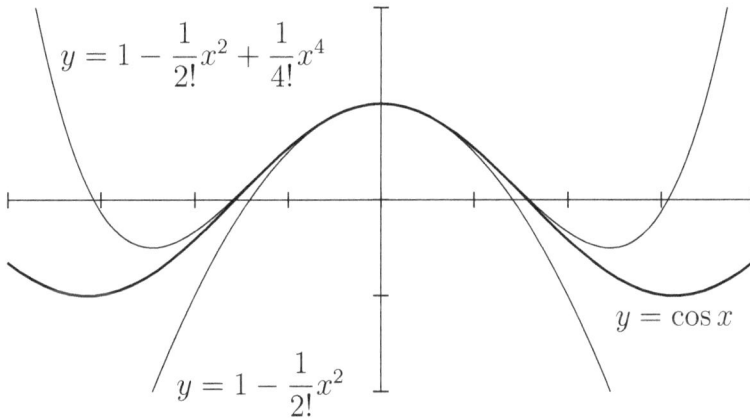

FIGURE 13.1
Upper and lower bounds on the cosine function.

Put $p^2(x) = 1 - (x^2/2)$ and $p^4(x) = 1 - (x^2/2) + (x^4/24)$. We have $p^2(x) > 0$ if $0 < x < \sqrt{2}$, and by the quadratic formula, $p^4(x) > 0$ if $0 < x < \sqrt{6 - \sqrt{12}}$. The preceding inequality for cos implies

$$\cos(\sqrt{6 - \sqrt{12}}) \leq 0 \leq \cos \sqrt{2}.$$

By the intermediate value theorem, cos has a zero between $\sqrt{2} \approx 1.41421$ and $\sqrt{6 - \sqrt{12}} \approx 1.59245$, Figure 13.2. This is the unique zero in $[0, \sqrt{6}]$, because

$$\cos' x = -\sin x \leq -x + \frac{x^3}{6} < 0 \quad \text{if } 0 < x < \sqrt{6}.$$

Evenness of cos implies $0 < \cos x$ if $-\tau < x < \tau$. $\qquad \square$

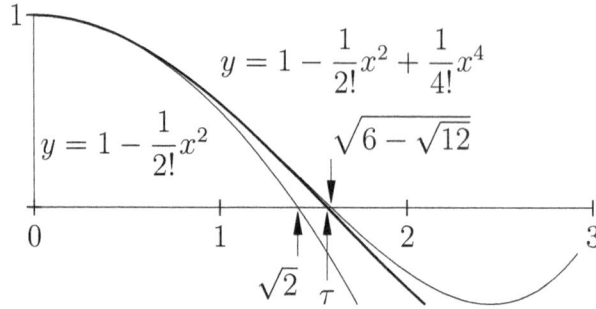

FIGURE 13.2
The smallest positive zero of the cosine function.

Definition 13.2.2. $\pi := 2\tau$.

Remark 13.2.3. The number π is the familiar ratio of the circumference of a circle to its diameter, see Remark 13.2.14. Proposition 13.2.1 guarantees

$$2.82842 \leq \pi \leq 3.1849. \qquad \diamond$$

Proposition 13.2.4. *The functions* cos *and* sin *are* 2π-*periodic:*

$$\cos(x + 4\tau) = \cos x, \qquad \sin(x + 4\tau) = \sin x, \quad \textit{for all real } x.$$

Proof. Since $\cos(\pi/2) = 0$, we have $\sin(\pi/2) = \pm 1$. The proof of Proposition 13.2.1 guarantees $0 \leq x - x^3/6 \leq \sin x$ if $0 \leq x \leq \sqrt{6}$, so we have $\sin(\pi/2) = 1$. The addition formulas for sin and cos give

$$\sin(x + \pi/2) = \sin x \cos(\pi/2) + \cos x \sin(\pi/2) = \quad \cos x,$$
$$\cos(x + \pi/2) = \cos x \cos(\pi/2) - \sin x \sin(\pi/2) = -\sin x.$$

Bootstrapping, we have $\sin(x + \pi) = \cos(x + \pi/2) = -\sin x$, and therefore $\sin(x + 2\pi) = -\sin(x + \pi) = \sin x$ for all real x. The cosine function is 2π-periodic as well since $\cos x = \sin(x + \pi/2)$ for all x. $\qquad \square$

Remark 13.2.5. The proof shows sin and cos are π-anti-periodic:

$$\sin(x + \pi) = -\sin x, \qquad \cos(x + \pi) = -\cos x \quad \text{for all real } x.$$

Consequently, $\sin x = 0$ if and only if $x = k\pi$ for some integer k, while $\cos x = 0$ if and only if $x = (k + \frac{1}{2})\pi$ for some integer k. $\qquad \diamond$

Proposition 13.2.6. *For all real* x, $\cos x = \sin(\pi/2 - x)$. *Further,*

$$\sin(\pi/4) = \cos(\pi/4) = \sqrt{2}/2,$$
$$\sin(\pi/6) = \cos(\pi/3) = 1/2,$$
$$\cos(\pi/6) = \sin(\pi/3) = \sqrt{3}/2.$$

Proof. The identity $\cos x = \sin(\pi/2 - x)$ follows from the addition formula for sin and evenness of cos. Taking $x = \pi/4$ shows that $\cos(\pi/4) = \sin(\pi/4)$; since each is positive and the sum of their squares is 1, each is equal to $\sqrt{1/2} = \sqrt{2}/2$.

Repeated use of the addition formulas gives, for all real x,

$$\cos 3x = \cos(2x + x) = \cos(2x)\cos x - \sin(2x)\sin x$$
$$= (\cos^2 x - \sin^2 x)\cos x - (2\sin x \cos x)\sin x$$
$$= \cos x(\cos^2 x - 3\sin^2 x) = \cos x(1 - 4\sin^2 x).$$

If $x = \pi/6$, then $\cos 3x = \cos(\pi/2) = 0$. Since $\cos x > 0$, we have $1 = 4\sin^2 x$. But $\sin x > 0$ since $0 < x < \pi/2$. We conclude that $\sin(\pi/6) = 1/2$, and therefore that $\cos(\pi/3) = \sin(\pi/6) = 1/2$.

If $0 \leq x \leq \pi/2$, then $\cos x = \sqrt{1 - \sin^2 x}$ since both cos and sin are non-negative on this interval. The preceding conclusions immediately imply $\cos(\pi/6) = \sin(\pi/3) = \sqrt{3}/2$. $\qquad\square$

Remark 13.2.7. The series of sin at 0 contains only terms of odd degree. For each n, consequently, the $(2n + 1)$th-degree germ at 0 is the $(2n + 2)$th-degree germ, and the remainder is $O(x^{2n+3})$. Exercise 13.2.4 guarantees the odd-degree germs alternately under- and over-estimate, so the "first omitted term" bounds the remainder. Figure 13.3 depicts the germs, with shades of gray depicting regions bounded by successive $(2n + 1)$th-degree germs, darker with increasing degree. Table 13.1 shows the rate of convergence on the half-period $[-\frac{\pi}{2}, \frac{\pi}{2}]$, which in principle suffices to determine the function everywhere. Similar remarks are true for cos.

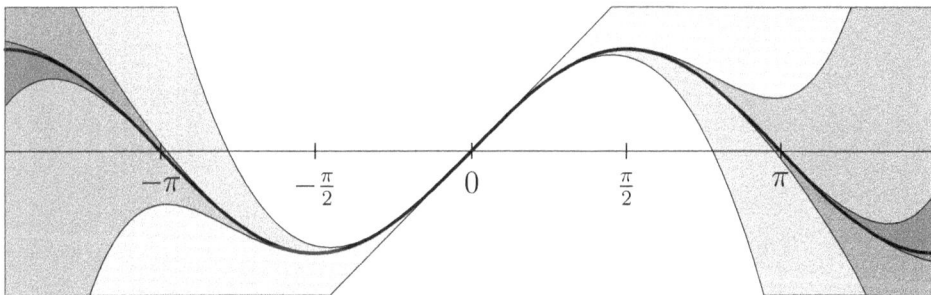

FIGURE 13.3
Polynomial germs of sin at 0.

Proposition 13.2.8. *Assume n is a positive integer. For all real numbers a and b,*

$$\int_a^b \cos^n x \, dx = \frac{1}{n}\cos^{n-1} x \sin x \Big|_{x=a}^{x=b} + \frac{n-1}{n}\int_a^b \cos^{n-2} x \, dx,$$

TABLE 13.1

Approximating $\sin x$ on $[-\frac{\pi}{2}, \frac{\pi}{2}]$ by nth-degree germs.

Degree $(2n+2)$	$P_0^{2n+2} \sin(x)$	Error Bound
2	x	0.646
4	$x - \dfrac{x^3}{3!}$	0.08
6	$x - \dfrac{x^3}{3!} + \dfrac{x^5}{5!}$	0.0047
8	$x - \dfrac{x^3}{3!} + \dfrac{x^5}{5!} - \dfrac{x^7}{7!}$	0.000161
10	$x - \dfrac{x^3}{3!} + \dfrac{x^5}{5!} - \dfrac{x^7}{7!} + \dfrac{x^9}{9!}$	0.0000036

$\diamond$

$$\int_a^b \sin^n x \, dx = -\frac{1}{n} \sin^{n-1} x \cos x \Big|_{x=a}^{x=b} + \frac{n-1}{n} \int_a^b \sin^{n-2} x \, dx,$$

Proof. See Exercise 13.2.5. $\square$

Corollary 13.2.9. *For every natural number n, if*

$$I(n) = \int_0^{\pi/2} \cos^n x \, dx = \int_0^{\pi/2} \sin^n x \, dx,$$

then

$$I(2k) = \frac{(2k-1)}{(2k)} \cdot \frac{(2k-3)}{(2k-2)} \cdots \frac{5}{6} \cdot \frac{3}{4} \cdot \frac{1}{2} \cdot \frac{\pi}{2} = \frac{(2k)!}{(2^k \cdot k!)^2} \cdot \frac{\pi}{2},$$

$$I(2k+1) = \frac{(2k)}{(2k+1)} \cdot \frac{(2k-2)}{(2k-1)} \cdots \frac{6}{7} \cdot \frac{4}{5} \cdot \frac{2}{3} = \frac{(2^k \cdot k!)^2}{(2k+1)!}.$$

Proof. See Exercise 13.2.6. $\square$

Proposition 13.2.10. $2 \displaystyle\int_0^{\pi/2} (\sin t)^{2x-1}(\cos t)^{2y-1} \, dt = \dfrac{\Gamma(x)\Gamma(y)}{\Gamma(x+y)}.$

Proof. Substitute $t = \sin^2 t$ in the definition of the β function and use Proposition 12.4.9. $\square$

Remark 13.2.11. Taking $x = 1/2$ or $y = 1/2$ expresses the integrals of powers of sin and cos in terms of Γ, compare Corollary 13.2.9. $\diamond$

Corollary 13.2.12. $\displaystyle\int_{-\infty}^{+\infty} e^{-u^2} \, du = \sqrt{\pi}.$

Proof. See Exercise 13.2.7. $\square$

The Unit Circle

Proposition 13.2.13. *Assume x and y are real numbers satisfying $x^2 + y^2 = 1$. There exists a unique real number θ (theta) such that $0 \le \theta < 2\pi$, $x = \cos\theta$, and $y = \sin\theta$.*

Proof. Since cos is continuous and strictly decreasing on $[0, \pi]$, and $\cos 0 = 1$, $\cos\pi = -1$, Theorem 10.3.7 implies cos maps $[0, \pi]$ bijectively to $[-1, 1]$. A similar argument shows cos maps $(\pi, 2\pi)$ bijectively to $(-1, 1)$.

If a point (x, y) on the circle $x^2 + y^2 = 1$ lies in the closed upper half-plane, namely, if $0 \le y \le 1$, then $y = \sqrt{1 - x^2}$. By the preceding paragraph there exists a unique θ in $[0, \pi]$ such that $x = \cos\theta$. Since $\sin\theta \ge 0$ if $0 \le \theta \le \pi$, the point $(\cos\theta, \sin\theta)$ is equal to (x, y).

If instead (x, y) lies in the open lower half-plane, namely, $-1 \le y < 0$, then $y = -\sqrt{1 - x^2}$ and $-1 < x < 1$. There is a unique θ in $(\pi, 2\pi)$ such that $x = \cos\theta$. Since $\sin\theta < 0$ if $\pi < \theta < 2\pi$, the point $(\cos\theta, \sin\theta)$ is equal to (x, y).

In summary, the mapping $\theta \mapsto (\cos\theta, \sin\theta)$ is a bijection from the half-open interval $[0, 2\pi) = [0, \pi] \cup (\pi, 2\pi)$ to the unit circle. $\qquad\square$

Remark 13.2.14. If a particle has position $(\cos\theta, \sin\theta)$ for θ in $[0, 2\pi]$, the preceding argument shows that the particle traces the unit circle once. The particle's speed at time θ is $\sqrt{(\cos'\theta)^2 + (\sin'\theta)^2} = 1$, so the distance traveled, a.k.a., the circumference of the unit circle, is 2π. $\qquad\diamond$

Definition 13.2.15. If (x, y) is a point of the plane, we say a real ordered pair (r, θ) is a set of *polar coordinates* for (x, y) if

$$x = r\cos\theta,$$
$$y = r\sin\theta.$$

Corollary 13.2.16. *If (x, y) is a point of the plane, there exists a set of polar coordinates for (x, y). If $(x, y) \ne (0, 0)$, we may assume $r > 0$. Subject to this condition, $r = \sqrt{x^2 + y^2}$ and θ is unique up to an added integer multiple of 2π.*

Proof. For all real θ, $(0, \theta)$ is a set of polar coordinates for $(0, 0)$. For the remainder of the proof, assume $(x, y) \ne (0, 0)$.

(Existence). Set $r = \sqrt{x^2 + y^2}$. By Proposition 13.2.13 applied to the numbers $u = x/r$ and $v = y/r$, which satisfy $u^2 + v^2 = 1$, there exists a real θ in $[0, 2\pi)$ such that $u = x/r = \cos\theta$ and $v = y/r = \sin\theta$.

(Uniqueness). If r is real, then $(x/r)^2 + (y/r)^2 = (x^2 + y^2)/r^2 = 1$ if and only if $r^2 = x^2 + y^2$, if and only if $r = \pm\sqrt{x^2 + y^2}$. Only one branch is positive. Now suppose θ is real and

$$u = \frac{x}{\sqrt{x^2 + y^2}} = \cos\theta, \qquad\qquad v = \frac{y}{\sqrt{x^2 + y^2}} = \sin\theta.$$

The real number $\theta/(2\pi)$ is written uniquely as an integer $k = \lfloor \theta/(2\pi) \rfloor$ plus a real number x' such that $0 \leq x' < 1$, Corollary 4.3.2. Writing $\theta_0 = 2\pi x'$, we have $0 \leq \theta_0 < 2\pi$ and $\theta = 2\pi k + \theta_0$. Since cos and sin are 2π-periodic, we have

$$u = \cos \theta = \cos \theta_0, \qquad\qquad v = \sin \theta = \sin \theta_0.$$

By Proposition 13.2.13, θ_0 is the only real number in $[0, 2\pi)$ satisfying the preceding equations; that is, θ is unique up to an added integer multiple of 2π. $\qquad\qquad\qquad\qquad\qquad\qquad\qquad\qquad\qquad\qquad\qquad\qquad\quad$ $\square$

Exercises for Section 13.2

Exercise 13.2.1. ($\bigstar$) Evaluate $\sin(\pi/8)$ and $\cos(\pi/8)$ exactly. (Your answers should involve only square roots and rational numbers.) Hint: Use the double-angle formula for cos to develop half-angle formulas for cos and sin.

Exercise 13.2.2. Evaluate $\sin(\pi/12)$ and $\cos(\pi/12)$ exactly. (Your answers should involve only square roots and rational numbers.)

Exercise 13.2.3. (H) Evaluate the improper integral

$$\int_0^\pi \log \sin x \, dx.$$

Exercise 13.2.4. By continuing the estimates in the proof of Proposition 13.2.1, prove that for every natural number n and all non-negative real x,

$$\sum_{k=0}^{2n+1} (-1)^k \frac{x^{2k}}{(2k)!} \leq \cos x \leq \sum_{k=0}^{2n} (-1)^k \frac{x^{2k}}{(2k)!},$$

$$\sum_{k=0}^{2n+1} (-1)^k \frac{x^{2k+1}}{(2k+1)!} \leq \sin x \leq \sum_{k=0}^{2n} (-1)^k \frac{x^{2k+1}}{(2k+1)!}.$$

In words, successive germs of cos and sin alternately under-estimate and over-estimate the function values on the positive reals. (The fact that the coefficients alternate in sign is not enough to deduce this result, but *is* a good way of remembering it.)

Exercise 13.2.5. Prove Proposition 13.2.8.

Exercise 13.2.6. Prove Corollary 13.2.9.

Exercise 13.2.7. (H) Prove Corollary 13.2.12.

Exercise 13.2.8. (H) Assume F is differentiable in some neighborhood of 0, and that $f = F'$ is differentiable at 0. Must there exist a punctured neighborhood of 0 on which f is continuous?

Exercise 13.2.9. (H) Assume α is irrational, and define $c_\alpha(t) = \cos t + \cos(\alpha t)$ and $s_\alpha(t) = \sin t + \sin(\alpha t)$. Prove these functions are not periodic, and determine whether each has an absolute maximum or an absolute minimum.

13.3 Auxiliary Functions

Definition 13.3.1. On the set of real numbers x such that $\cos x \neq 0$, define the *secant* function $\sec x = 1/\cos x$.

Remark 13.3.2. The secant is even, π-anti-periodic, and $|\sec x| \geq 1$ for all x in the domain, Figure 13.4. (Tick marks are integers.) ◇

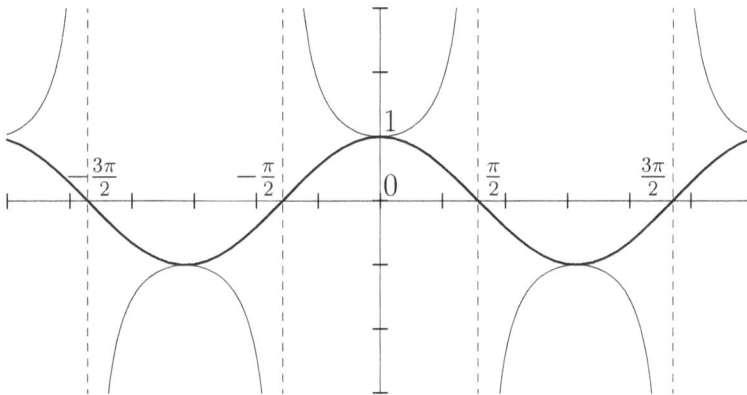

FIGURE 13.4
The cosine and secant functions.

Definition 13.3.3. On the set of real numbers x such that $\sin x \neq 0$, define the *cosecant* function $\csc x = 1/\sin x$.

Remark 13.3.4. The cosecant function is odd, π-anti-periodic, and satisfies the identity $\csc(x + \frac{\pi}{2}) = \sec x$, Figure 13.5. (Tick marks are integers, dashed lines are zeros of sin.) ◇

Definition 13.3.5. On the set of real numbers x such that $\cos x \neq 0$, define the *tangent* function, $\tan x = \sin x / \cos x$.

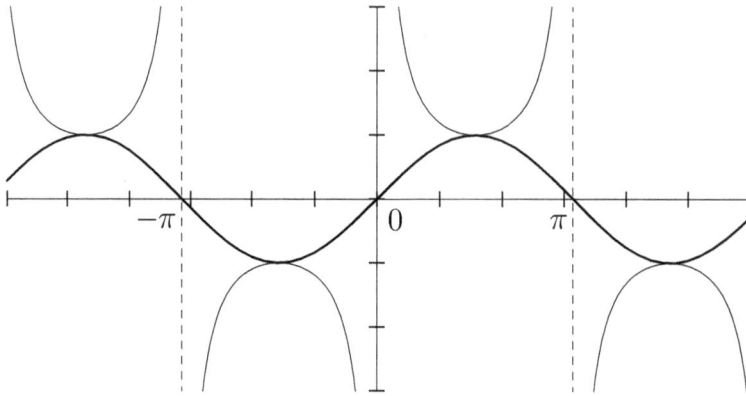

FIGURE 13.5
The sine and cosecant functions.

Remark 13.3.6. The tangent function is odd, as a quotient of an odd function by an even function, and periodic with period π, Figure 13.6:

$$\tan(x + \pi) = \frac{\sin(x + \pi)}{\cos(x + \pi)} = \frac{-\sin x}{-\cos x} = \tan x. \qquad \diamond$$

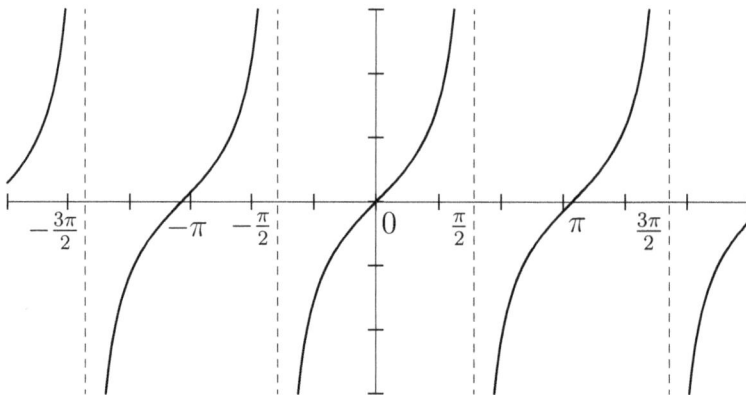

FIGURE 13.6
The tangent function.

Proposition 13.3.7. *The tangent function is differentiable on the "period interval" $(-\pi/2, \pi/2)$, satisfies $\tan' x = \sec^2 x$ for all x in the domain, and maps the period interval bijectively to $\mathbf{R}$.*

For all real x and y,

$$\tan(x + y) = \frac{\tan x + \tan y}{1 - \tan x \tan y}.$$

Proof. The quotient rule gives

$$\tan' = \frac{\cos \sin' - \sin \cos'}{\cos^2} = \frac{\cos^2 + \sin^2}{\cos^2} = \frac{1}{\cos^2} = \sec^2 \geq 1.$$

In particular, tan is strictly increasing (hence injective) on $(-\pi/2, \pi/2)$. Since cos and sin are positive on $(0, \pi/2)$, $\tan x = \sin x / \cos x \to \infty$ as $x \to \pi/2^-$. Since tan is odd, $\tan x \to -\infty$ as $x \to -\pi/2^+$. An intermediate value theorem argument shows tan maps the period interval surjectively to **R**.

For all real x and y,

$$\tan(x + y) = \frac{\sin(x + y)}{\cos(x + y)} = \frac{\sin x \cos y + \cos x \sin y}{\cos x \cos y - \sin x \sin y} = \frac{\tan x + \tan y}{1 - \tan x \tan y};$$

the final equality follows from dividing both the numerator and denominator by $\cos x \cos y$. □

Corollary 13.3.8. *For every integer k, there exists a unique real number x such that $(k - \frac{1}{2})\pi < x < (k + \frac{1}{2})\pi$ and $\tan x = x$.*

Proof. The function $f(x) = \tan x - x$ is differentiable, and has derivative $f'(x) = \sec^2 x - 1 \geq 0$. Since the derivative vanishes at isolated points, namely at $x = k\pi$, f is strictly increasing on $\left((k - \frac{1}{2})\pi, (k + \frac{1}{2})\pi\right)$. On this interval, x is bounded while tan approaches $-\infty$ at the left endpoint and approaches ∞ at the right endpoint. By the intermediate value theorem, f maps $\left((k - \frac{1}{2})\pi, (k + \frac{1}{2})\pi\right)$ bijectively to **R**, and in particular is equal to 0 exactly once in this interval. □

Definition 13.3.9. For x not an integer multiple of π, define the *cotangent* function $\cot x = \cos x / \sin x$.

Remark 13.3.10. The cotangent is periodic with period π. Away from integer multiples of $\pi/2$, $\tan = \sin / \cos$ and $\cot = \cos / \sin$ are reciprocals, and $\cot(x + \pi/2) = -\tan x$. The derivative of cot is $-1/\csc^2$, so cot is decreasing on every interval $\left(k\pi, (k + 1)\pi\right)$ with k an integer. ◇

Example 13.3.11. Power series representations of reciprocals and quotients are not generally easy to calculate. For the circular functions, direct calculation of derivatives quickly becomes onerous. Algebra does allow calculation of any finite germ, however. To illustrate, we'll expand sec and tan as fourth-degree germs at 0. (See also the recursive approach in Exercise 12.3.10.) The starting ingredient is an expansion of the denominator to the desired order, here

$$\cos x = 1 - \tfrac{1}{2}x^2 + \tfrac{1}{24}x^4 + O(x^6) = 1 - e(x), \qquad e(x) = \tfrac{1}{2}x^2 - \tfrac{1}{24}x^4 + O(x^6).$$

Because $e(x) \approx O(x^2)$ is a convergent power series, there is a positive r such that $|e(x)| < 1$ if $|x| < r$. In such an interval, the geometric series formula gives

$$\sec x = \frac{1}{\cos x} = \frac{1}{1 - e(x)} = 1 + e(x) + e(x)^2 + O(x^6).$$

We need not multiply out $e(x)^2$ completely; retaining terms of degree at most 4 suffices:

$$e(x)^2 \approx (\tfrac{1}{2}x^2 - \tfrac{1}{24}x^4)^2 \approx \tfrac{1}{4}x^4 + O(x^6).$$

We conclude that $\sec x = 1 + \tfrac{1}{2}x^2 + \tfrac{5}{24}x^4 + O(x^6)$.

For an expansion of $\tan x$, we multiply the preceding fourth-degree germ by the fourth-degree germ of $\sin x$;

$$\tan x = \frac{\sin x}{\cos x} = [x - \tfrac{1}{6}x^3 + O(x^5)][1 + \tfrac{1}{2}x^2 + \tfrac{5}{24}x^4 + O(x^6)]$$

$$= x - \tfrac{1}{6}x^3 + \tfrac{1}{2}x^3 + O(x^5) = x + \tfrac{1}{3}x^3 + O(x^5).$$

The same ideas handle functions that vanish at the center. For example

$$\cot x = \frac{1}{\tan x} = \frac{1}{x[1 + \tfrac{1}{3}x^2 + O(x^4)]} = \frac{1 - \tfrac{1}{3}x^2 + O(x^4)}{x}. \qquad \diamond$$

Exercises for Section 13.3

Exercise 13.3.1. (★) Prove $\sec' = \sec \cdot \tan$, including that the domains are the same.

Exercise 13.3.2. Prove that $\cot(2x) = \tfrac{1}{2}(\cot x - \tan x)$ if $x \notin (\pi/2)\mathbf{Z}$.

Exercise 13.3.3. (H) Prove that if $|\theta| < \pi$, then

$$\tan\frac{\theta}{2} = \frac{\sin\theta}{1+\cos\theta} = \frac{1-\cos\theta}{\sin\theta}.$$

Exercise 13.3.4. (★) In (a)–(e), SI denotes the *sine integral* defined by

$$s(t) = \frac{\sin t}{t}, \qquad \mathrm{SI}(x) = \int_0^x \frac{\sin t\, dt}{t}.$$

(a) Prove s has a continuous extension to $\mathbf{R}$: We can define $s(0)$ to make s continuous everywhere. Find power series representations for s and SI and show each has infinite radius.

(b) Find the points where $\mathrm{SI}' = 0$, and for each, give the value of SI.

(c) Prove that for each positive integer k, SI has a unique inflection point x_k satisfying $(k-\tfrac{1}{2})\pi < x_k < (k+\tfrac{1}{2})\pi$.

(d) Prove that for every natural number k,

$$\frac{2}{(k+1)\pi} \le \left| \int_{k\pi}^{(k+1)\pi} \frac{\sin t}{t}\, dt \right| = |a_k| \le \frac{2}{k\pi}.$$

Find the absolute maximum of SI as an integral.

(e) Prove

$$\int_0^\infty \frac{\sin t \, dt}{t} = \lim_{x \to \infty} \mathrm{SI}(x)$$

exists, but the integral is not absolutely convergent.

Exercise 13.3.5. Prove that for all positive r,

$$\int_0^\infty \frac{\sin(rt)}{t} \, dt = \int_0^\infty \frac{\sin x}{x} \, dx = \int_0^\infty \frac{\sin^2 x}{x^2} \, dx = \frac{1}{r} \int_0^\infty \frac{\sin^2(rt)}{t^2} \, dt,$$

and the last two integrals converge absolutely.

Exercise 13.3.6. (★) Assume $I \subseteq \mathbf{R}$ and $f : I \to \mathbf{R}$ is a function. The *polar graph* $r = f(\theta)$ is the set of points $(x, y) = \big(f(\theta) \cos \theta, f(\theta) \sin \theta\big)$ for θ in I, namely the set of points $(x, y) = (r \cos \theta, r \sin \theta)$ such that $r = f(\theta)$ and $\theta \in I$.

Describe the hyperbola $x^2 - y^2 = 1$ as a polar graph $r = f(\theta)$ including a domain I for which the hyperbola is traced one time.

Exercise 13.3.7. Describe the parabola $y = x^2$ as a polar graph $r = f(\theta)$ including a domain I for which the parabola is traced one time.

Exercise 13.3.8. (★) Let $X = \mathbf{R}^2 \setminus \{(0,0)\}$. The mapping that sends each point with polar coordinates (r, θ) to $(1/r, \theta)$ is called *inversion in the unit circle*.

Find polar and rectangular equations for the image of the unit hyperbola $x^2 - y^2 = 1$ under inversion in the unit circle.

Exercise 13.3.9. Give rectangular equations for the polar graphs $r = \sec \theta$, $r = 1$, and $r = \cos \theta$, and describe each geometrically. If a is real and positive, describe the image of the line $y = 1/(2a)$ under inversion in the unit circle.

Exercise 13.3.10. Expand $\sec x$ and $\tan x$ as sixth-degree germs at 0.

Exercise 13.3.11. Prove that $\csc x - \dfrac{1}{x^2}$ extends continuously to $(-\pi, \pi)$.

13.4 Inverse Functions

Each circular function is periodic, hence has no "global" inverse. Instead, we restrict each function to an interval on which the function is injective, obtaining a branch of inverse, Figure 13.7.

Remarkably, while the inverse functions are not algebraic, their *derivatives* are algebraic. This is no accident, but a consequence of the differential equations that characterize the elementary trig functions, see also Exercise 11.1.7.

Inverse Sine and Cosine

Definition 13.4.1. The restriction of sin to the interval $[-\pi/2, \pi/2]$ is denoted Sin. The inverse function $\mathrm{Sin}^{-1} : [-1, 1] \to [-\pi/2, \pi/2]$, sometimes denoted arcsin, is called the *principal branch of arcsin*.

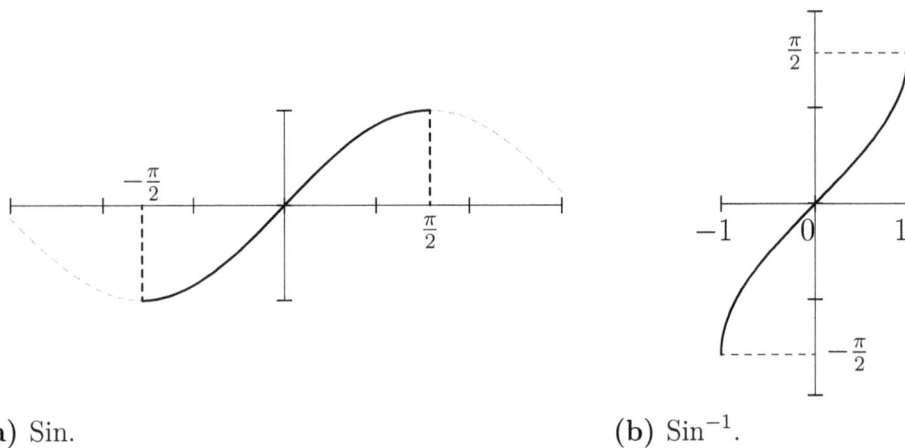

(a) Sin.   **(b)** Sin^{-1}.

FIGURE 13.7
The principal branches of sine and arcsin.

Remark 13.4.2. By definition,

$$\sin(\mathrm{Sin}^{-1} x) = x \quad \text{for all } x \text{ in } [-1, 1],$$
$$\mathrm{Sin}^{-1}(\sin y) = y \quad \text{for all } y \text{ in } [-\pi/2, \pi/2].$$

Sine is *decreasing* on $[\pi/2, 3\pi/2]$ because $\sin(\pi - x) = \sin x$. Periodicity implies sin is monotone on $\left[(k - \frac{1}{2})\pi, (k + \frac{1}{2})\pi\right]$ for every integer k. For each k, there is a corresponding branch of arcsin. On rare occasions when one considers a non-principal branch of arcsin, it is denoted $\sin^{-1}$, and k is supplied by context.   ◇

Proposition 13.4.3. *The function* Sin^{-1} *is continuous on* $[-1, 1]$, *differentiable on* $(-1, 1)$, *and*
$$(\mathrm{Sin}^{-1})'(x) = \frac{1}{\sqrt{1 - x^2}}.$$

Proof. By Theorem 10.2.10, Sin^{-1} is differentiable since $\sin' y = \cos y \neq 0$ if $-\pi/2 < y < \pi/2$, and

$$(\mathrm{Sin}^{-1})'(x) = \frac{1}{\cos(\mathrm{Sin}^{-1} x)} = \frac{1}{\sqrt{1 - \sin^2(\mathrm{Sin}^{-1} x)}} = \frac{1}{\sqrt{1 - x^2}}. \qquad \square$$

Corollary 13.4.4. *If $a > 0$, then $\int_0^a \dfrac{dx}{\sqrt{a^2 - x^2}} = \dfrac{\pi}{2}$.*

Proof. If $0 \le x < a$,

$$\frac{1}{\sqrt{a^2 - x^2}} = \frac{1}{\sqrt{(a + x)(a - x)}} \le \frac{1}{\sqrt{a(a - x)}},$$

and the upper bound is improperly integrable by Proposition 11.3.4. The smaller improper integral therefore converges. Factoring a^2 from the radicand and using the substitution $u = x/a$ gives

$$\int_0^a \frac{dx}{\sqrt{a^2 - x^2}} = \int_0^a \frac{dx/a}{\sqrt{1 - (x/a)^2}} = \int_0^1 \frac{dx}{\sqrt{1 - x^2}}.$$

By Proposition 13.4.3 and Theorem 11.1.2, this is $\mathrm{Sin}^{-1} 1 - \mathrm{Sin}^{-1} 0 = \pi/2$. $\square$

Definition 13.4.5. The restriction of cos to the interval $[0, \pi]$ is denoted Cos. The inverse function $\mathrm{Cos}^{-1} : [-1, 1] \to [0, \pi]$, sometimes denoted arccos, is called the *principal branch of arccos*.

Remark 13.4.6. For each integer k there is a branch of arccos taking values in $[k\pi, (k + 1)\pi]$. The identity $\cos y = x = \sin(\pi/2 - y)$ on $[0, \pi]$ becomes

$$\mathrm{Cos}^{-1} x = y = (\pi/2) - \mathrm{Sin}^{-1} x \quad \text{on } [-1, 1]. \qquad \diamond$$

Inverse Tangent

Definition 13.4.7. The inverse of tan on the interval $(-\pi/2, \pi/2)$ is the *principal branch of arctan*, denoted $\mathrm{Tan}^{-1} : \mathbf{R} \to (-\pi/2, \pi/2)$, Figure 13.8.

Remark 13.4.8. By definition,

$$\tan(\mathrm{Tan}^{-1} x) = x \quad \text{for all real } x,$$
$$\mathrm{Tan}^{-1}(\tan y) = y \quad \text{for all } y \text{ in } (-\pi/2, \pi/2). \qquad \diamond$$

Proposition 13.4.9. *The function Tan^{-1} is differentiable on $\mathbf{R}$, and*

$$(\mathrm{Tan}^{-1})'(x) = \frac{1}{1 + x^2} \quad \text{for all real } x.$$

Proof. Since $\tan' x = \sec^2 x > 0$ for all real x, Theorem 10.2.10 guarantees that Tan^{-1} is differentiable, and

$$(\mathrm{Tan}^{-1})'(x) = \frac{1}{\sec^2(\mathrm{Tan}^{-1} x)} = \frac{1}{1 + \tan^2(\mathrm{Tan}^{-1} x)} = \frac{1}{1 + x^2}. \qquad \square$$

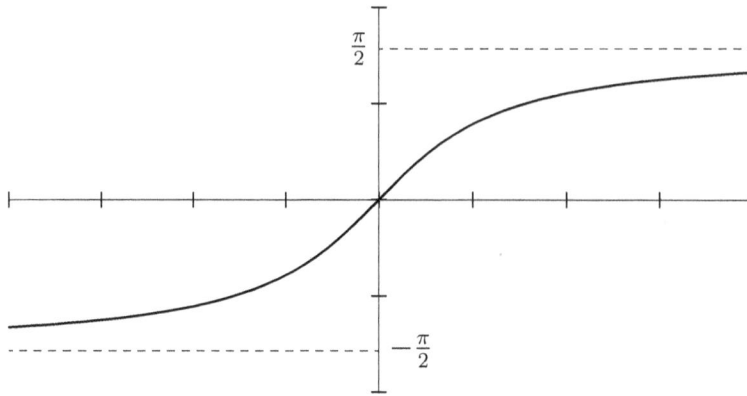

FIGURE 13.8
The principal branch of arctan.

Corollary 13.4.10. *If $a > 0$, then* $\displaystyle\int_0^\infty \frac{dx}{a^2 + x^2} = \frac{\pi}{2a}$.

Proof. The integrand is non-negative and bounded above by $\min(1/a^2, 1/x^2)$, so the improper integral converges by Proposition 11.3.4. If $a = 1$, we have

$$\int_0^\infty \frac{dx}{1 + x^2} = \lim_{b \to \infty} \int_0^b \frac{dx}{1 + x^2} = \lim_{b \to \infty} \mathrm{Tan}^{-1} b - \mathrm{Tan}^{-1} 0 = \frac{\pi}{2}.$$

If $a \neq 1$, use the substitution $u = x/a$:

$$\int_0^\infty \frac{dx}{a^2 + x^2} = \int_0^\infty \frac{a \cdot dx/a}{a^2\left(1 + (x/a)^2\right)} = \frac{1}{a}\int_0^\infty \frac{du}{1 + u^2} = \frac{\pi}{2a}. \qquad \square$$

Corollary 13.4.11. *For x in $(-1, 1)$, $\mathrm{Tan}^{-1} x$ is expanded in a power series*

$$\mathrm{Tan}^{-1} x = \sum_{k=0}^\infty (-1)^k \frac{x^{2k+1}}{2k + 1} = x - \frac{x^3}{3} + \frac{x^5}{5} - \frac{x^7}{7} + \cdots.$$

Proof. The geometric series

$$\frac{1}{1 + t^2} = \sum_{k=0}^\infty (-1)^k t^{2k} = 1 - t^2 + t^4 - t^6 + \cdots$$

converges absolutely for all t in $(-1, 1)$. Integrating term by term,

$$\mathrm{Tan}^{-1} x = \int_0^x \frac{dt}{1 + t^2} = \int_0^x \sum_{k=0}^\infty (-1)^k t^{2k}\, dt$$

$$= \sum_{k=0}^\infty (-1)^k \int_0^x t^{2k}\, dt = \sum_{k=0}^\infty (-1)^k \frac{x^{2k+1}}{2k + 1}. \qquad \square$$

Remark 13.4.12. In fact, at $x = 1$ the series converges to $\mathrm{Tan}^{-1} 1 = \pi/4$, see Exercise 13.4.4. The convergence is too slow to be of practical use, however. ◇

Corollary 13.4.13. $\pi = 2\sqrt{3} \sum_{k=0}^{\infty} (-1)^k \dfrac{1}{3^k(2k+1)}$.

Proof. By Proposition 13.2.6, $\tan(\pi/6) = \sin(\pi/6)/\cos(\pi/6) = 1/\sqrt{3}$, so $\pi = 6\,\mathrm{Tan}^{-1}(1/\sqrt{3})$. Setting $x = 1/\sqrt{3}$ in the power series for Tan^{-1} and noting that $x^{2k+1} = 1/(3^k\sqrt{3})$ and $6/\sqrt{3} = 2\sqrt{3}$,

$$\pi = 6 \sum_{k=0}^{\infty} (-1)^k \frac{1}{3^k\sqrt{3}(2k+1)} = 2\sqrt{3} \sum_{k=0}^{\infty} (-1)^k \frac{1}{3^k(2k+1)}. \qquad \square$$

Remark 13.4.14. This series alternates, so "the tail is bounded by the size of the first omitted term":

$$\left| \frac{\pi}{\sqrt{3}} - 2 \sum_{k=0}^{n-1} (-1)^k \frac{1}{3^k(2k+1)} \right| \leq \frac{2}{3^n(2n+1)}.$$

For example, taking $n = 11$ terms gives a rational estimate of $\pi/\sqrt{3}$ that is accurate to within $\varepsilon = 2/4{,}074{,}382 < 0.5 \times 10^{-6}$. ◇

Exercises for Section 13.4

Exercise 13.4.1. Assume $0 < x < 1$. Draw a right triangle in the first quadrant with angle $\theta = \mathrm{Sin}^{-1} x$ at the origin, use geometry to find the third side, and read off the ratios $\cos(\mathrm{Sin}^{-1} x)$, $\sec(\mathrm{Sin}^{-1} x)$, $\tan(\mathrm{Sin}^{-1} x)$, and $\cot(\mathrm{Sin}^{-1} x)$ as algebraic functions of x.

Similarly, find $\cos(\mathrm{Tan}^{-1} x)$, $\sec(\mathrm{Tan}^{-1} x)$, and $\sin(\mathrm{Tan}^{-1} x)$ for x real.

Exercise 13.4.2. Describe the domain and image of the principal branch of arcsec, formally the inverse of the reciprocal of cosine restricted to $[0, \pi]$.

Exercise 13.4.3. (★) Assume inverse functions are principal branches. Prove that the specification

$$\Theta(x, y) = \begin{cases} \arctan(y/x) & \text{if } x > 0, \\ \mathrm{arccot}(x/y) & \text{if } y > 0, \\ \mathrm{arccot}(x/y) - \pi & \text{if } y < 0, \end{cases}$$

is well-defined, namely, two formulas give the same value if both are defined. Suggestion: Prove that if $-\pi < \theta < \pi$, $r > 0$, and $(x, y) = (r\cos\theta, r\sin\theta)$, then $\Theta(x, y) = \theta$.

Exercise 13.4.4. Prove $\displaystyle\sum_{k=0}^{\infty} \frac{(-1)^k}{(2k+1)} = \lim_{x \to 1^-} \sum_{k=0}^{\infty} \frac{(-1)^k x^{2k+1}}{(2k+1)} = \frac{\pi}{4}.$

Exercise 13.4.5. Use the substitution $u = x - (1/x)$ to evaluate the improper integral

$$\int_0^{\infty} \frac{x^2 + 1}{x^4 + 1}\, dx.$$

14

Complex Numbers

Points in the plane may be viewed as numerical entities in a way that extends the familiar real number line. Exercise 3.2.15 defines complex numbers as ordered pairs of real numbers and establishes the field axioms for the complex numbers. Exercise 4.2.8 introduces the concept of magnitude, which plays a role in analysis analogous to the absolute value function on the real numbers.

Now that we have constructed the exponential and circular functions over the real numbers and established some of their properties, there are substantial (and possibly surprising) benefits to making a close, formal visit to the complex numbers.

14.1   Algebra and Geometry

Definition 14.1.1. A *complex number* is an ordered pair $\alpha = (a, b)$ of real numbers. The set of complex numbers is denoted $\mathbf{C}$.

The real number $\operatorname{Re} \alpha = a$ is called the *real part* of α. We say α is *real* if $b = 0$, and *non-real* if $b \neq 0$. The set of all real complex numbers $(a, 0)$ is the *real axis*.

The real number $\operatorname{Im} \alpha = b$ is called the *imaginary part* of α. We say α is *imaginary* if $a = 0$. The set of imaginary complex numbers $(0, b)$ is the *imaginary axis*.

The complex number $i = (0, 1)$ is the *imaginary unit*.

Remark 14.1.2. A subtle point lurks in this definition: *We have chosen an imaginary unit.* Much of complex analysis rests on a choice of imaginary unit.

In abstract algebra, by contrast, the two imaginary units, which we call $i = (0, 1)$ and $-i = (0, -1)$, are on a more equal footing. If a complex number α can be identified with the real ordered pair (a, b), it can equally well be identified with $(a, -b)$, see also Remark 14.1.9. ◇

Definition 14.1.3. If $\alpha = (a, b)$ and $\alpha' = (a', b')$ are complex numbers, we define their *sum* $\alpha + \alpha'$ and their *product* $\alpha \cdot \alpha'$ to be the complex numbers

$$\alpha + \alpha' = (a + a', b + b'), \qquad \alpha \cdot \alpha' = (aa' - bb', ba' + ab').$$

DOI: 10.1201/9781003601357-14

Remark 14.1.4. To conform with "classical" notation, we write complex numbers not as ordered pairs $(a, b) = a(1, 0) + b(0, 1)$, but as expressions $a + bi$. Viewing the real and imaginary parts of a complex number $\alpha = a + bi$ as rectangular coordinates, we identify $\alpha = a + bi$ with the point (a, b), Figure 14.1.   ◇

Definition 14.1.5. The *conjugate* of $\alpha = a + bi$ is the complex number $\overline{\alpha} = a - bi$, obtained geometrically by reflecting α across the real axis.

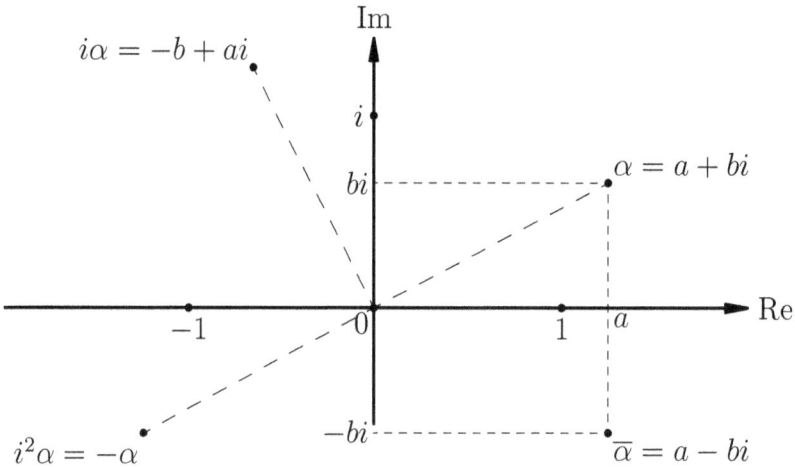

FIGURE 14.1
The complex plane.

Proposition 14.1.6. *The set of complex numbers together with the operations of addition and multiplication is a field.*

Proof. See Exercise 3.2.15, (a)–(c).   □

Remark 14.1.7. The set of *real* complex numbers $(a, 0)$ is a field under *complex* addition and multiplication. In fact, if we define $\phi : \mathbf{R} \to \mathbf{C}$ by $\phi(a) = (a, 0)$, then ϕ is injective, and for all real numbers a and a', we have

$$\phi(a + a') = \phi(a) + \phi(a'), \qquad \phi(a \cdot a') = \phi(a) \cdot \phi(a').$$

In words, complex addition and multiplication reduce to real operations on real complex numbers; we may write a instead of $a + 0i$ without ambiguity.   ◇

Proposition 14.1.8. *For all complex α and α',*

$$\overline{\alpha + \alpha'} = \overline{\alpha} + \overline{\alpha'}, \qquad \overline{\alpha \cdot \alpha'} = \overline{\alpha} \cdot \overline{\alpha'}.$$

Proof. See Exercise 3.2.15 (e).   □

Remark 14.1.9. Conceptually, complex conjugation "respects field operations," or "is a field isomorphism." This is the algebraic sense in which there is no distinguished choice of imaginary unit. ◇

Remark 14.1.10. Imaginary numbers may seem tainted with suspicion, as if they don't really exist but it's mathematically expedient to pretend they do. This sentiment presumably traces back to the Ancient Greeks, who viewed numbers as lengths, or "real numbers." Indeed, no real number has square equal to -1.

The complex number i has a perfectly concrete existence as the point $(0, 1)$ in the plane. By definition of multiplication, $i^2 = (0, 1)^2 = (-1, 0)$. Even the mysterious equation $i^2 = -1$ has a natural geometric interpretation: If $\alpha = (a, b)$, then $i\alpha = (-b, a)$ and $i^2\alpha = i(i\alpha) = (-a, -b)$. Multiplication by i is a counterclockwise quarter-turn of the complex plane about the origin. Multiplying by i twice, a half-turn, multiplies each complex number by -1. ◇

Geometry of Addition and Multiplication

Geometrically, complex addition is the parallelogram law for vector addition in the plane, see Figure 14.2.

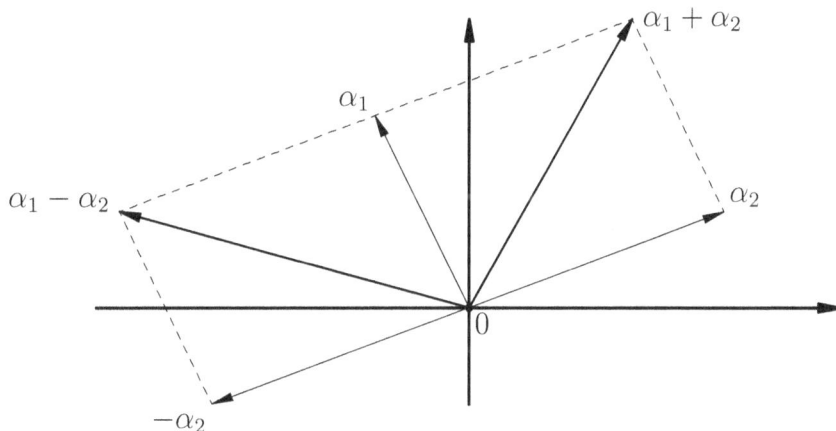

FIGURE 14.2
Adding and subtracting complex numbers.

Definition 14.1.11. Assume r is a non-negative real number, and θ real. The complex number

$$\alpha = r\cos\theta + ir\sin\theta = r(\cos\theta + i\sin\theta)$$

is said to be written in *polar form*.

The radius r is the *magnitude* of α, denoted $|\alpha|$, and θ is the *polar angle* of α. If $-\pi < \theta \leq \pi$, we say θ is the *principal angle* of α.

Remark 14.1.12. The expression $\cos\theta + i\sin\theta$ is convenient to denote $\operatorname{cis}\theta$, short for "cosine plus i sine θ."

The magnitude is given by $|\alpha| = r = \sqrt{a^2 + b^2} = \sqrt{\alpha\overline{\alpha}}$. Combining, the polar form of a complex number is written $\alpha = |\alpha|\operatorname{cis}\theta$. $\diamond$

Proposition 14.1.13. *If $\alpha = |\alpha|\operatorname{cis}\theta$ and $\alpha' = |\alpha'|\operatorname{cis}\theta'$ are in polar form, then $\alpha \cdot \alpha' = |\alpha| \cdot |\alpha'|\operatorname{cis}(\theta + \theta')$.*

Proof. The sum-of-angles formulas for cos and sin, Proposition 13.1.5 (iv), imply

$$\begin{aligned}
\operatorname{cis}\theta \cdot \operatorname{cis}\theta' &= (\cos\theta + i\sin\theta) \cdot (\cos\theta' + i\sin\theta') \\
&= (\cos\theta\cos\theta' - \sin\theta\sin\theta') + i(\cos\theta\sin\theta' + \cos\theta'\sin\theta) \\
&= \cos(\theta + \theta') + i\sin(\theta + \theta') = \operatorname{cis}(\theta + \theta').
\end{aligned}$$

Consequently,

$$\alpha \cdot \alpha' = \left(|\alpha| \cdot \operatorname{cis}\theta\right)\left(|\alpha'| \cdot \operatorname{cis}\theta'\right) = \left(|\alpha| \cdot |\alpha'|\right)\operatorname{cis}(\theta + \theta'). \qquad \square$$

Remark 14.1.14. To multiply two complex numbers geometrically, we *multiply their magnitudes* (compare Exercise 4.2.8) and *add their polar angles*, see Figure 14.3. The action of multiplication by α is the unique scaling-and-rotation of the plane about the origin that carries 1 to α. $\diamond$

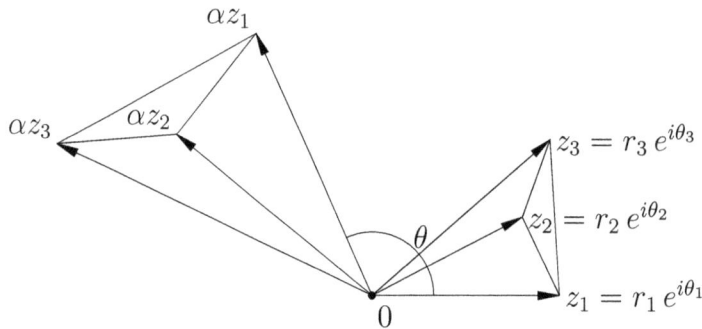

FIGURE 14.3
Complex multiplication by $\alpha = |\alpha|(\cos\theta + i\sin\theta)$.

Remark 14.1.15. Since $i = 0 + 1 \cdot i = \cos(\pi/2) + i\sin(\pi/2)$, the magnitude of i is 1 and the principal angle of i is $\pi/2$. Consequently,

$$i\alpha = i|\alpha|\operatorname{cis}\theta = |\alpha|\operatorname{cis}\left(\theta + (\pi/2)\right);$$

again we see that multiplication by i rotates the plane about the origin by a quarter turn counterclockwise. $\diamond$

Exercises for Section 14.1

Exercise 14.1.1. (★) If $\alpha = a + bi \neq 0$, write $1/\alpha$ in rectangular form.

Exercise 14.1.2. Write $\dfrac{2 - i}{4 + 3i} = a + bi$ with a and b real.

Exercise 14.1.3. Assuming $(\mathbf{C}, +, \cdot)$ is a field, prove the *difference of squares* identity: For all *complex* α and β, $(\alpha + \beta)(\alpha - \beta) = \alpha^2 - \beta^2$.

Exercise 14.1.4. (★) State and prove a binomial theorem for complex numbers.

Exercise 14.1.5. (Geometric sums.) Assume a and r are complex numbers, $r \neq 1$, and n is a natural number. Prove the *geometric sum formula* with first term a, ratio r, and n terms:

$$\sum_{k=0}^{n-1} ar^k = a\frac{r^n - 1}{r - 1}.$$

Exercise 14.1.6. (A) Let i be the complex unit, satisfying $i^2 = -1$. Use the complex binomial theorem, Exercise 14.1.4, to expand the following, and separate the real and imaginary parts.

(a) $(x + iy)^2$.   (b) $(x + iy)^3$.   (c) $(x + iy)^4$.

Exercise 14.1.7. (Square roots.) This exercise establishes that every non-zero complex number w has precisely two complex square roots.

(a) Suppose z_1 and z_2 are complex numbers such that $z_1^2 = w$ and $z_2^2 = w$. Prove that either $z_2 = z_1$ or $z_2 = -z_1$.

(b) Assume u and v are real numbers, not both zero, and $w = u + iv$. If x and y are real, prove that $z = x + iy$ is a square root of w if and only if $u = x^2 - y^2$ and $v = 2xy$.

(c) Solve the equations in (b) for x and y in terms of u and v.

Exercise 14.1.8. (The quadratic formula.) If α, β, and γ are complex and $\alpha \neq 0$, prove that $\alpha z^2 + \beta z + \gamma = 0$ if and only if

$$z = \frac{-\beta \pm \sqrt{\beta^2 - 4\alpha\gamma}}{2\alpha}.$$

Exercise 14.1.9. (★) In each part, define $A = \mathbf{Z} + i\mathbf{Z}$.

(a) Show A is closed under multiplication.

(b) Which elements of A have a reciprocal (multiplicative inverse) in A?

14.2 The Polar Formula

Following mathematical convention, we'll write $z = x + iy$ for a complex variable with real part x and imaginary part y. Assume (α_k) is a complex sequence, z_0 a complex number, and

$$f(z) = \sum_{k=0}^{\infty} \alpha_k(z - z_0)^k = \alpha_0 + \alpha_1(z - z_0) + \alpha_2(z - z_0)^2 + \cdots$$

is the power series in z centered at z_0 with coefficients (α_k). Because the complex numbers form a field and the magnitude satisfies the triangle inequality and reverse triangle inequality (Exercise 4.2.8), convergence properties established earlier for real power series extend without modification. Particularly, if the preceding series converges for some positive real number R, then the series converges absolutely at each z such that $|z - z_0| < R$, and defines a continuous, complex-valued function in the disk $B_R(z_0)$. In many examples, the radius can be computed using the ratio test: If

$$\lim_{k \to \infty} \left| \frac{\alpha_{k+1}(z - z_0)^{k+1}}{\alpha_k(z - z_0)^k} \right| = |z - z_0| \lim_{k \to \infty} \frac{|\alpha_{k+1}|}{|\alpha_k|}$$

exists and is less than 1, then the power series converges absolutely at z. The definition of differentiability makes sense without formal modification:

$$f'(z_0) = \lim_{h \to 0} \frac{f(z_0 + h) - f(z_0)}{h} = \lim_{z \to z_0} \frac{f(z) - f(z_0)}{z - z_0}$$

if the limit exists. The complex chain rule holds for complex-differentiable functions. A convergent complex power series is differentiable on its disk of convergence.

Remark 14.2.1. The integral, by contrast, does not generalize without substantial modification. For real-valued functions we use ordering to split the domain. The "standard" theory of complex integration is not over plane regions, but instead closely resembles line integration in multivariable calculus.   ⋄

Definition 14.2.2. The *complex exponential function* $\exp : \mathbf{C} \to \mathbf{C}$ is defined by

$$\exp(z) = \sum_{k=0}^{\infty} \frac{z^k}{k!}.$$

Definition 14.2.3. The *complex cosine and sine*, denoted $\cos$ and $\sin : \mathbf{C} \to \mathbf{C}$ just as over the real numbers, are defined by

$$\cos z = \sum_{k=0}^{\infty} \frac{(-1)^k z^{2k}}{(2k)!}, \qquad\qquad \sin z = \sum_{k=0}^{\infty} \frac{(-1)^k z^{2k+1}}{(2k+1)!}.$$

Theorem 14.2.4 (The polar formula). *For every complex z,*

$$\exp(iz) = \cos z + i \sin z.$$

Proof. Since $i^2 = -1$, we have $i^{2k} = (-1)^k$ and $i^{2k+1} = i(-1)^k$. Substituting iz into the exponential series and separating even-degree (real) and odd-degree (imaginary) terms, we have

$$\exp(iz) = \sum_{k=0}^{\infty} \frac{(iz)^k}{k!} = \sum_{k=0}^{\infty} \left[\frac{(iz)^{2k}}{(2k)!} + \frac{(iz)^{2k+1}}{(2k+1)!} \right]$$

$$= \sum_{k=0}^{\infty} \left[\frac{(-1)^k z^{2k}}{(2k)!} + \frac{i(-1)^k z^{2k+1}}{(2k+1)!} \right]$$

$$= \sum_{k=0}^{\infty} \frac{(-1)^k z^{2k}}{(2k)!} + i \sum_{k=0}^{\infty} \frac{(-1)^k z^{2k+1}}{(2k+1)!}$$

$$= \cos z + i \sin z. \qquad \square$$

Remark 14.2.5. A formal argument along the lines of characterizing exp and the circular functions using differential equations may be memorable. Although these calculations can be justified, we will not fully do so.

If we define $f(z) = \exp(iz)$ for complex z, then $f'(z) = i\exp(iz) = if(z)$. Differentiating again gives $f''(z) = i^2 f(z) = -f(z)$. This means $f'' + f = 0$, $f(0) = 1$, and $f'(0) = i$. The formal conclusion of Proposition 13.1.4 reads $f(z) = \cos z + i \sin z$ for all complex z. (The conclusion here is "formal" because the proof of Proposition 13.1.4 in Chapter 13 hinged on the implication "$a^2 + b^2 = 0$ implies $a = b = 0$." This implication is true for real numbers but not for complex numbers. Pointedly, $1^2 + i^2 = 0$.) $\diamond$

Remark 14.2.6. Assume θ is a real number. By the polar formula, we have $e^{i\theta} = \operatorname{cis} \theta$. Our earlier calculation using the sum-of-angles formulas implies

$$e^{i\theta_1} \cdot e^{i\theta_2} = \operatorname{cis} \theta_1 \cdot \operatorname{cis} \theta_2 = \operatorname{cis}(\theta_1 + \theta_2) = e^{i(\theta_1 + \theta_2)}.$$

In words, *the law of exponents holds for imaginary exponents.* Because the law of exponents is familiar and simple, many people find it a useful way to remember the sum-of-angle formulas for cosine and sine. $\diamond$

Proposition 14.2.7. *If z and w are complex, then $\exp(z) \cdot \exp(w) = \exp(z+w)$.*

Proof. By the series product formula and binomial theorem,

$$\exp(z) \cdot \exp(w) = \left[\sum_{j=0}^{\infty} \frac{z^j}{j!} \right] \left[\sum_{k=0}^{\infty} \frac{w^k}{k!} \right] = \sum_{n=0}^{\infty} \sum_{j+k=n} \frac{z^j}{j!} \frac{w^k}{k!}$$

$$= \sum_{n=0}^{\infty} \sum_{k=0}^{n} \frac{z^{n-k}}{(n-k)!} \frac{w^k}{k!} = \sum_{n=0}^{\infty} \frac{1}{n!} \sum_{k=0}^{n} \binom{n}{k} z^{n-k} w^k$$

$$= \sum_{n=0}^{\infty} \frac{(z+w)^n}{n!} = \exp(z+w). \qquad \square$$

Corollary 14.2.8. *For every real θ and every integer n, $(e^{i\theta})^n = e^{in\theta}$.*

Proof. See Exercise 14.2.5.   □

Corollary 14.2.9. *If z and z' are complex, then $\exp z = \exp z'$ if and only if there exists an integer k such that $z - z' = 2\pi ki$.*

Proof. If $z = x + iy$ for some real x and y, then by Proposition 14.2.7,

$$\exp z = \exp(x + iy) = \exp(x) \cdot \exp(iy) = e^x(\cos y + i \sin y).$$

This is the polar form of the complex number with magnitude $|\exp(z)| = e^x$ and polar angle y. Similarly, if $z' = x' + iy'$, then $\exp(z') = e^{x'}(\cos y' + i \sin y')$ has magnitude $|\exp(z')| = e^{x'}$ and polar angle y'.

Particularly, if $\exp(z) = \exp(z')$, then $e^x = e^{x'}$, so $x = x'$; and $\operatorname{cis} y = \operatorname{cis} y'$, so $y - y' = 2\pi k$ for some integer k by Corollary 13.2.16.   □

Proposition 14.2.10. $\displaystyle \int_0^\infty \frac{\sin(rt)}{t}\, dt = \frac{\pi}{2}$ *for every positive real r.*

Proof. See Exercise 14.2.13.   □

Roots of Unity

Assume n is a positive integer. In the real numbers, the equation $x^n = 1$ has either one solution (if n is odd) or two (if n is even). The situation over the complex numbers is both algebraically satisfying and geometrically beautiful.

Definition 14.2.11. If n is a positive integer, a complex number ω is an *nth root of unity* if $\omega^n = 1$.

Proposition 14.2.12. *There exist precisely n distinct nth roots of unity:*

$$\omega_n^k = e^{i(2\pi k/n)}, \quad 0 \leq k < n.$$

These are vertices of the regular n-gon inscribed in the unit circle and having 1 as a vertex.

Proof. A complex number $\omega = re^{i\theta}$ is an nth root of unity if and only if

$$e^0 = 1 = \omega^n = (re^{i\theta})^n = r^n e^{in\theta}.$$

(The rightmost equality is Corollary 14.2.8.) Equating magnitudes implies $r = 1$. By Corollary 14.2.9, $e^0 = e^{in\theta}$ if and only if there exists an integer m such that $in\theta = 2\pi mi$, or equivalently, $\theta = 2\pi m/n$. Conversely, for every integer m, the complex number $\omega = e^{i(2\pi m/n)}$ is an nth root of unity by Corollary 14.2.8.

By integer division, for every integer m there exist unique integers d and k such that $m = dn + k$ and $0 \leq k < n$. Since exp is $(2\pi i)$-periodic,

$$e^{i(2\pi m/n)} = e^{i(2\pi(dn+k)/n)} = e^{2\pi di} \cdot e^{i(2\pi k/n)} = e^{i(2\pi k/n)}.$$

That is, every nth root of unity is written uniquely as $\omega = e^{i(2\pi k/n)}$ for some integer k such that $0 \leq k < n$. If $\omega_n = e^{2\pi i/n}$, then $\omega_n^k = e^{i(2\pi k/n)}$. □

Example 14.2.13. For $n = 4$, we have $\omega_4 = e^{i(2\pi/4)} = i$. The fourth roots of unity are $\{\omega_4^0, \omega_4^1, \omega_4^2, \omega_4^3\} = \{1, i, -1, -i\}$, Figure 14.4a. Particularly, in this set we see the non-trivial square root of unity, $\omega_2 = e^{i(2\pi/2)} = -1$.

For $n = 6$, we have $\omega_6 = e^{i(2\pi/6)} = e^{i(\pi/3)} = \frac{1}{2}(1 + i\sqrt{3})$. The sixth roots of unity are

$$\{\omega_6^k\}_{k=0}^5 = \{1, \tfrac{1}{2}(1 + i\sqrt{3}), \tfrac{1}{2}(-1 + i\sqrt{3}), -1, -\tfrac{1}{2}(1 + i\sqrt{3}), \tfrac{1}{2}(1 - i\sqrt{3})\}$$

shown in Figure 14.4b. In this set we see also the non-trivial cube roots of unity, $\omega_3 = e^{i(2\pi/3)} = \frac{1}{2}(-1 + i\sqrt{3})$ and $\omega_3^2 = e^{i(4\pi/3)} = -\frac{1}{2}(1 + i\sqrt{3})$. ◊

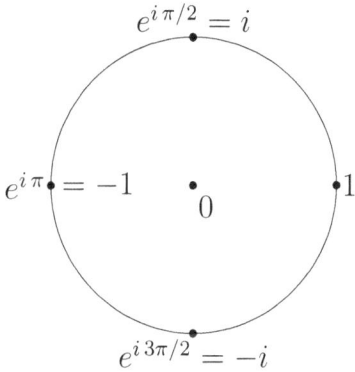

(a) Fourth roots of unity.   (b) Sixth roots of unity.

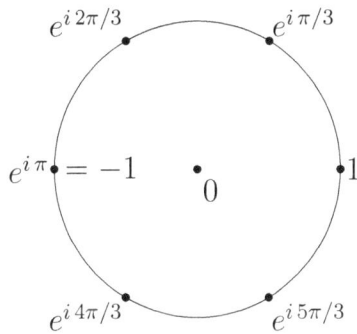

FIGURE 14.4
Complex roots of unity in the unit circle.

Exercises for Section 14.2

Exercise 14.2.1. (★) List the 8th roots of unity in polar and rectangular form.

Exercise 14.2.2. List the 12th roots of unity in polar and rectangular form.

Exercise 14.2.3. Find the two square roots of i in polar and rectangular form.

Exercise 14.2.4. Put $\omega = \omega_3 = \frac{1}{2}(-1 + \sqrt{3})$, and $A = \mathbf{Z} + \omega\mathbf{Z}$. Prove A is closed under addition and under multiplication. Which elements of A have a reciprocal in A?

Exercise 14.2.5. Prove Corollary 14.2.8.

Exercise 14.2.6. Use the polar formula $e^{i\theta} = \cos\theta + i\sin\theta$ to show the angle-sum identities are equivalent to the law of exponents.

Exercise 14.2.7. We define cosh and sinh to be the even and odd parts of exp: For all complex z,

$$\cosh z = \tfrac{1}{2}\Big[\exp(z) + \exp(-z)\Big], \qquad \sinh z = \tfrac{1}{2}\Big[\exp(z) - \exp(-z)\Big].$$

(a) Prove that for all complex z,

$$\cos z = \tfrac{1}{2}\Big[\exp(iz) + \exp(-iz)\Big], \qquad \sin z = \tfrac{1}{2i}\Big[\exp(iz) - \exp(-iz)\Big].$$

(b) Prove $\cos(iz) = \cosh z$ and $\sin(iz) = i\sinh z$ for all complex z.

(c) Prove that up to domain translation and rotation and codomain rotation, the four functions cos, sin, cosh, and sinh are identical.

(d) You may have noticed that various hyperbolic identities look similar to trig identities. Find at least three "analogous pairs" of identities, and use part (b) to explain why the pairs are equivalent.

Exercise 14.2.8.(★) This exercise briefly introduces complex exponentiation, compare Corollary 14.2.8. Take care: The full story is more complex, as it were, than can be detailed in an exercise. Heedless symbolic manipulation guided by "real" expectations leads to devastating errors, such as the four-line "proof"

$$\exp z = e^z = \left(e^{2\pi i}\right)^{z/(2\pi i)} = 1^{z/(2\pi i)} = 1 \quad \text{for all complex } z.$$

Let u and v denote real numbers, and consider the *period strip*, by definition the complex set $H = \{u + iv : -\pi < v \leq \pi\}$. Prove exp maps H bijectively to $\mathbf{C}^\times := \mathbf{C} \setminus \{0\}$, the set of non-zero complex numbers.

By definition, the inverse mapping Log : $\mathbf{C}^\times \to H$ is the *principal logarithm*. If α and β are complex and $\alpha \neq 0$, we define $\alpha^\beta = \exp(\beta \operatorname{Log}\alpha)$.

Prove that if α is real and positive, and β is real, this definition reduces to real exponentiation in Chapter 12.

Calculate $\operatorname{Log} i$, $\operatorname{Log}(-1)$, and $\operatorname{Log}(-i)$. Calculate i^i, $(-1)^i$, and $(-i)^i$. Which laws of exponents hold?

The *principal square root* of a complex number z is $z^{1/2}$. Find $i^{1/2}$, $(-1)^{1/2}$, and $(-i)^{1/2}$.

Exercise 14.2.9. (A) In Exercise 12.3.10 a sequence (b_k) is defined that satisfies

$$t \coth t = \sum_{k=0}^{\infty} b_{2k} \frac{4^k t^{2k}}{(2k)!}.$$

Use (b_k) to express $t \cot t$ as a germ at 0.

Exercise 14.2.10. Prove $\int_0^{\infty} e^{ix^2} \, dx$ converges, but not absolutely. (Do not evaluate.)

Exercise 14.2.11. Under modest technical assumptions, to be studied in Section 17.5, a complex-valued function f on the interval $[-\pi, \pi]$ can be approximated arbitrarily closely by *spectral sums*

$$\frac{\alpha_0}{2} + \sum_{m=1}^{N} \left[\alpha_m \cos(m\phi) + \beta_m \sin(m\phi) \right], \quad \alpha_m, \beta_m \text{ complex scalars.}$$

The identities below are useful in showing that a suitable sequence of spectral sums "converges to f."

Assume ϕ is a real number. Recall that by Exercise 14.2.7,

$$\cos \phi = \frac{e^{i\phi} + e^{-i\phi}}{2}, \qquad \sin \phi = \frac{e^{i\phi} - e^{-i\phi}}{2i}.$$

(a) Prove that if $n \geq 0$, then

$$\sum_{m=-n}^{n} e^{im\theta} = 1 + 2 \sum_{m=1}^{n} \cos(m\theta).$$

(b) Sum the left-hand series in the preceding part, and prove that

$$1 + 2 \sum_{m=1}^{n} \cos(m\theta) = \frac{\sin(n + \frac{1}{2})\theta}{\sin \frac{1}{2}\theta}.$$

Exercise 14.2.12. If n is even, $\int_{-\pi/2}^{\pi/2} e^{inx} \, dx = \pi$ if $n = 0$ and 0 otherwise.

Exercise 14.2.13. This exercise outlines a proof of Proposition 14.2.10, see Chapman, [5]. Throughout, expressions are extended continuously to 0 where necessary. By Exercise 13.3.5 and evenness of the integrand, it suffices to prove

$$\int_{-\infty}^{\infty} \frac{\sin^2 x}{x^2} \, dx = \pi.$$

Establish the following.

(a) For every positive integer n,

$$\sum_{k=0}^{n-1} e^{(2k-n+1)ix} = \frac{\sin(nx)}{\sin x} \quad \text{and} \quad \pi = \frac{1}{n} \int_{-\pi/2}^{\pi/2} \frac{\sin^2(nx)}{\sin^2 x} \, dx.$$

(b) For every positive integer n,

$$\int_{-n\pi/2}^{n\pi/2} \frac{\sin^2 t}{t^2} \, dt = \frac{1}{n} \int_{-\pi/2}^{\pi/2} \frac{\sin^2(nx)}{x^2} \, dx.$$

Subtract from (a) and use Exercise 13.3.11 to finish the proof.

Exercise 14.2.14. Assume $n \geq 2$, and put $\omega = e^{2\pi i/n}$.

(a) Show that the polynomial $z^n - 1$ factors as

$$z^n - 1 = \prod_{k=0}^{n-1} (z - \omega^k) = (z-1)(z-\omega)(z-\omega^2) \cdots (z - \omega^{n-1}).$$

(b) Use part (a) and the geometric sum formula to prove

$$\sum_{j=0}^{n-1} z^j = \prod_{k=1}^{n-1} (z - \omega^k) = 1 + z + z^2 + \cdots + z^{n-1}.$$

(c) By setting $z = 1$ in part (b), prove that

$$n = \prod_{k=1}^{n-1} (1 - \omega^k) = \prod_{k=1}^{n-1} |1 - \omega^k|.$$

This identity has a beautiful geometric interpretation: Inscribe a regular n-gon in the unit circle. Fix a vertex, and consider the $(n-1)$ chords joining that vertex to each of the other vertices. The product of the lengths of these chords is n, the number of sides of the polygon.

15

Linear Spaces

Until now, we have viewed functions as collections of points and used sequences to study the behavior of individual functions. We now take a step of abstraction: A function may be viewed as a *point in a space of functions*. Concepts such as sequences and convergence make sense in this abstraction, and theorems of analysis may be viewed in terms of sequential convergence, such as a sequence of polynomials converging to an arbitrary continuous function.

In this chapter we introduce function spaces and "linear" tools for speaking of distance: inner products generalizing the dot product, and norms generalizing length of vectors.

15.1   Function Spaces

Definition 15.1.1. Assume X is a set. The *vector space of real-valued functions on X* is the set $\mathscr{F}(X) = \mathscr{F}(X, \mathbf{R})$ of functions $f : X \to \mathbf{R}$ equipped with the pointwise operations of addition and scalar multiplication. Specifically, if f and g are functions and c is real, we define $f + g$ and cf by

$$(f + g)(x) = f(x) + g(x), \qquad (cf)(x) = cf(x) \quad \text{for all } x \text{ in } X.$$

The *zero vector* $\mathbf{0} = \mathbf{0}_X$ is the zero function, whose value at each point is 0.

Remark 15.1.2. "The vector space $\mathscr{F}(X)$" implicitly includes pointwise addition and real scalar multiplication. We do not give axioms for a "vector space" here, but merely note that they abstract the algebraic properties of pointwise addition and scalar multiplication. ◇

Remark 15.1.3. There is a "vector space of complex-valued functions on X," comprising the set $\mathscr{F}(X, \mathbf{C})$ of functions $f : X \to \mathbf{C}$ equipped with pointwise addition and pointwise multiplication by *complex* scalars. ◇

Remark 15.1.4. Sometimes in mathematics we also want to consider pointwise multiplication of functions, defined by $(fg)(x) = f(x)g(x)$. In this situation we speak of $\mathscr{F}(X)$ as the *algebra of functions* on X. ◇

DOI: 10.1201/9781003601357-15   309

Example 15.1.5. Assume n is a positive integer, and X is a set with n elements. If we identify X with the initial segment $\underline{\mathbf{n}} = (0, 1, 2, \ldots, n-1)$, each $\mathbf{x} : X \to \mathbf{R}$ becomes the ordered n-tuple $(x_k)_{k=0}^{n-1} = (x_0, x_1, \ldots, x_{n-1})$ whose terms, or *components*, are the values $x_k = \mathbf{x}(k)$. Under this identification, $\mathscr{F}(X) = \mathscr{F}(\underline{\mathbf{n}})$ is the vector space $\mathbf{R}^n$ equipped with the usual operations of vector addition and real scalar multiplication.

Particularly, $\mathscr{F}(\underline{\mathbf{n}})$ comes with a standard basis, the indicators $\mathbf{e}_j = \chi_{\{j\}}$ defined by $\mathbf{e}_j(j) = 1$ and $\mathbf{e}_j(k) = 0$ if $k \in \underline{\mathbf{n}}$ and $k \neq j$. $\quad\diamond$

Remark 15.1.6. By convention, $\mathbf{R}^0 = \mathscr{F}(\varnothing)$ is the real vector space $\{\mathbf{0}\}$ with one element. $\quad\diamond$

Example 15.1.7. A complex-valued function $\mathbf{z} : \underline{\mathbf{n}} \to \mathbf{C}$ may similarly be viewed as the ordered n-tuple $(z_k)_{k=0}^{n-1} = (z_0, z_1, \ldots, z_{n-1})$. The complex vector space of all such functions is the vector space $\mathbf{C}^n$ equipped with the usual operations of vector addition and *complex* scalar multiplication. $\quad\diamond$

Example 15.1.8. Extending Example 15.1.5, $\mathbf{R}^\omega := \mathscr{F}(\mathbf{N})$ denotes the real vector space of real sequences $(x_k)_{k=0}^\infty$, a.k.a., mappings $\mathbf{x} : \mathbf{N} \to \mathbf{R}$. For each n, the space $\mathbf{R}^n$ may be identified with the set of sequences (x_k) satisfying $x_k = 0$ if $k \geq n$. These subspaces are nested outward: $\mathbf{R}^n \subseteq \mathbf{R}^{n+1}$ for each n.

The union $\mathbf{R}^\infty := \bigcup_n \mathbf{R}^n$ is the space of *finite sequences*. While the space of finite sequences is infinite-dimensional, with standard basis $(\mathbf{e}_j)_{j=0}^\infty$, each element of $\mathbf{R}^\infty$ has only finitely many non-zero terms. Particularly, $\mathbf{R}^\infty$ is not all of $\mathbf{R}^\omega$ in the same sense that not all subsets of $\mathbf{N}$ are finite. $\quad\diamond$

Definition 15.1.9. If X is a set of real numbers, the *space of continuous, real-valued functions on* X is the set $\mathscr{C}(X) = \mathscr{C}(X, \mathbf{R})$ of continuous functions $f : X \to \mathbf{R}$ equipped with pointwise operations.

Lemma 15.1.10. *If* $X \subseteq \mathbf{R}$, *then* $\mathscr{C}(X) \subseteq \mathscr{F}(X)$ *is a vector subspace.*

Proof. This is a fancy way of saying "a sum of continuous functions is continuous" and "a scalar multiple of a continuous function is continuous." $\quad\square$

Inner Products

To do analysis in a vector space, we need additional structure to specify "nearness." In this book we consider three forms of structure: inner products, which define length and angle; norms, which define length compatibly with vector space structure; and, in Chapter 16, metrics, which define distance independently of vector space structure.

Definition 15.1.11. Assume $(V, +, \cdot)$ is a real vector space. An *inner product* on V is a function $\langle \, , \, \rangle : V \times V \to \mathbf{R}$ satisfying

(i) Symmetry: For all $\mathbf{u}$ and $\mathbf{v}$ in V, $\langle \mathbf{v}, \mathbf{u} \rangle = \langle \mathbf{u}, \mathbf{v} \rangle$.

(ii) Bilinearity: For all $\mathbf{u}$, $\mathbf{v}$ and $\mathbf{v}'$ in V and all real c,

$$\langle \mathbf{u}, c\mathbf{v} + \mathbf{v}' \rangle = c \langle \mathbf{u}, \mathbf{v} \rangle + \langle \mathbf{u}, \mathbf{v}' \rangle .$$

(iii) Positivity: For all $\mathbf{v}$ in V, $0 \le \langle \mathbf{v}, \mathbf{v} \rangle$, with equality if and only if $\mathbf{v} = \mathbf{0}$.

In an inner product space, the *induced norm* is the function $\|\mathbf{v}\| = \langle \mathbf{v}, \mathbf{v} \rangle^{1/2}$.

Remark 15.1.12. Bilinearity amounts to a formal distributive law:

$$\left\langle \sum_{j=0}^{m-1} a_j \mathbf{u}_j, \sum_{k=0}^{n-1} b_k \mathbf{v}_k \right\rangle = \sum_{j=0}^{m-1} \sum_{k=0}^{n-1} a_j b_k \langle \mathbf{u}_j, \mathbf{v}_k \rangle$$

for all vectors $\mathbf{u}_j$ and $\mathbf{v}_k$, and all real scalars a_j and b_k.

Symmetry amounts, similarly, to a formal commutative law. For example, $\langle \mathbf{u}, \mathbf{v} \rangle + \langle \mathbf{v}, \mathbf{u} \rangle = 2 \langle \mathbf{u}, \mathbf{v} \rangle$ in an inner product space, so

$$\langle \mathbf{u} + \mathbf{v}, \mathbf{u} + \mathbf{v} \rangle = \langle \mathbf{u}, \mathbf{u} \rangle + 2 \langle \mathbf{u}, \mathbf{v} \rangle + \langle \mathbf{v}, \mathbf{v} \rangle ,$$

just as if we were expanding a real binomial. $\diamond$

Definition 15.1.13. For each natural number n, *flat n-space* refers to the vector space $\mathbf{R}^n$ equipped with the *standard inner product*

$$\langle \mathbf{u}, \mathbf{v} \rangle = \sum_{k=0}^{n-1} u_k v_k, \qquad \mathbf{u} = (u_k)_{k=0}^{n-1}, \quad \mathbf{v} = (v_k)_{k=0}^{n-1}.$$

Definition 15.1.14. Assume $a < b$. The *standard inner product* on $\mathscr{C}\big([a,b]\big)$ is

$$\langle f, g \rangle = \frac{1}{b-a} \int_a^b fg = \frac{1}{b-a} \int_a^b f(x)g(x)\, dx.$$

Remark 15.1.15. The scale factor is not essential (or universal), but ensures $\|1\| = 1$ regardless of the interval. In this book, real function spaces have the standard inner product unless explicitly stated otherwise. $\diamond$

Remark 15.1.16. Although the standard inner product is defined for continuous functions, the formula makes sense if f and g are integrable, see Proposition 9.3.6. The issue is positivity: A non-zero function can have integral 0. $\diamond$

Example 15.1.17. Let $V = \mathscr{C}\big([0,1]\big)$. If $p \ge 0$, define f_p by $f_p(x) = x^p$. For all p and q,

$$\langle f_p, f_q \rangle = \int_0^1 x^p x^q \, dx = \left. \frac{x^{p+q+1}}{p+q+1} \right|_{x=0}^{1} = \frac{1}{p+q+1}. \qquad \diamond$$

Theorem 15.1.18 (The cross-term bound). *Assume $(V, +, \cdot)$ is a real vector space and $\langle\ ,\ \rangle$ an inner product. If $\mathbf{u}$ and $\mathbf{v}$ are elements of V, then*

$$|\langle \mathbf{u}, \mathbf{v} \rangle| \leq \|\mathbf{u}\| \, \|\mathbf{v}\|,$$

with equality if and only if one of $\mathbf{u}$ and $\mathbf{v}$ is a multiple of the other.

Proof. If $\mathbf{u} = \mathbf{0}$ or $\mathbf{v} = \mathbf{0}$, the cross-term bound is an equality. Otherwise, for all real s and t, positivity applied to the linear combination $s\mathbf{u} + t\mathbf{v}$ implies

$$0 \leq \langle s\mathbf{u} + t\mathbf{v}, s\mathbf{u} + t\mathbf{v} \rangle = s^2\|\mathbf{u}\|^2 + 2st \langle \mathbf{u}, \mathbf{v} \rangle + t^2\|\mathbf{v}\|^2,$$

with equality if and only if $s\mathbf{u} + t\mathbf{v} = \mathbf{0}$ for some real s and t. Taking $s = \pm\|\mathbf{v}\|$ and $t = \|\mathbf{u}\|$ gives $0 \leq 2\|\mathbf{u}\| \, \|\mathbf{v}\| \big(\|\mathbf{u}\| \, \|\mathbf{v}\| \pm \langle \mathbf{u}, \mathbf{v} \rangle\big)$. Consequently, $0 \leq \|\mathbf{u}\| \, \|\mathbf{v}\| \pm \langle \mathbf{u}, \mathbf{v} \rangle$, or $\mp \langle \mathbf{u}, \mathbf{v} \rangle \leq \|\mathbf{u}\| \, \|\mathbf{v}\|$, or $|\langle \mathbf{u}, \mathbf{v} \rangle| \leq \|\mathbf{u}\| \, \|\mathbf{v}\|$. □

Proposition 15.1.19. *For all $\mathbf{u}$ and $\mathbf{v}$ in an inner product space,*

(i) *(Parallelogram law.)* $2\big(\|\mathbf{u}\|^2 + \|\mathbf{v}\|^2\big) = \|\mathbf{u} + \mathbf{v}\|^2 + \|\mathbf{u} - \mathbf{v}\|^2$.

(ii) *(Polarization identity.)* $\langle \mathbf{u}, \mathbf{v} \rangle = \frac{1}{4}\big(\|\mathbf{u} + \mathbf{v}\|^2 - \|\mathbf{u} - \mathbf{v}\|^2\big)$.

Proof. See Exercise 15.1.4. □

Definition 15.1.20. A real sequence $\mathbf{u} = (u_k)_{k=0}^{\infty}$ is *square-summable* if $\sum_k u_k^2$ converges.

Proposition 15.1.21. *If $\mathbf{u} = (u_k)_{k=0}^{\infty}$ and $\mathbf{v} = (v_k)_{k=0}^{\infty}$ are square-summable, then $(u_k v_k)_{k=0}^{\infty}$ is absolutely summable.*

Proof. If $n \geq 0$, the cross-term bound in flat n-space implies

$$\sum_{k=0}^{n-1} |u_k v_k| \leq \left[\sum_{k=0}^{n-1} u_k^2\right]^{1/2} \cdot \left[\sum_{k=0}^{n-1} v_k^2\right]^{1/2} \leq \left[\sum_{k=0}^{\infty} u_k^2\right]^{1/2} \cdot \left[\sum_{k=0}^{\infty} v_k^2\right]^{1/2} < \infty.$$

Since the partial sums are bounded, $(u_k v_k)_{k=0}^{\infty}$ is absolutely summable. □

Corollary 15.1.22. *The space of square-summable sequences is a real vector space, and*

$$\langle \mathbf{u}, \mathbf{v} \rangle = \sum_{k=0}^{\infty} u_k v_k$$

defines an inner product.

Proof. If $\mathbf{u}$ and $\mathbf{v}$ are square-summable, then

$$\sum_{k=0}^{\infty} (u_k + v_k)^2 = \sum_{k=0}^{\infty} (u_k^2 + 2u_k v_k + v_k^2)$$

is absolutely convergent by Proposition 15.1.21, so $\mathbf{u} + \mathbf{v}$ is square-summable. The conditions for $\langle \mathbf{u}, \mathbf{v} \rangle$ to be an inner product are immediately verified. □

Example 15.1.23. If $s > 1/2$, the real sequence $\mathbf{u} = (u_k)_{k=0}^{\infty}$ with terms $u_k = 1/(1 + k)^s$ is square-summable, and

$$\langle \mathbf{u}, \mathbf{u} \rangle = \sum_{k=0}^{\infty} \frac{1}{(1 + k)^{2s}} = \sum_{k=1}^{\infty} \frac{1}{k^{2s}} = \zeta(2s). \qquad \diamond$$

Complex Inner Products

If $(V, +, \cdot)$ is a *complex* vector space, the definition of an inner product has two small modifications: Symmetry is replaced by *conjugate symmetry*, and bilinearity is replaced by complex linearity in one argument and *conjugate linearity* in the other:

(i)′ For all $\mathbf{u}$ and $\mathbf{v}$ in V, $\langle \mathbf{v}, \mathbf{u} \rangle = \overline{\langle \mathbf{u}, \mathbf{v} \rangle}$.

(ii)′ For all $\mathbf{u}$, $\mathbf{v}$ and $\mathbf{v}'$ in V and all complex c,

$$\langle c\mathbf{v} + \mathbf{v}', \mathbf{u} \rangle = c \langle \mathbf{v}, \mathbf{u} \rangle + \langle \mathbf{v}', \mathbf{u} \rangle,$$
$$\langle \mathbf{u}, c\mathbf{v} + \mathbf{v}' \rangle = \overline{c} \langle \mathbf{u}, \mathbf{v} \rangle + \langle \mathbf{u}, \mathbf{v}' \rangle.$$

Definition 15.1.24. The *standard complex inner product* on $\mathbf{C}^n$ is defined by

$$\langle \mathbf{z}, \mathbf{w} \rangle = \sum_{k=0}^{n-1} z_k \overline{w}_k, \qquad \mathbf{z} = (z_k)_{k=0}^{n-1}, \quad \mathbf{w} = (w_k)_{k=0}^{n-1}.$$

Definition 15.1.25. Assume $a < b$. The *standard complex inner product* on $\mathscr{C}([a, b], \mathbf{C})$ is

$$\langle f, g \rangle = \frac{1}{b - a} \int_a^b f(x) \overline{g(x)} \, dx.$$

Example 15.1.26. Assume $V = \mathscr{C}([0, 2\pi], \mathbf{C})$, and for each integer n, put $e_n(x) = e^{inx}$. Since $\overline{e^{inx}} = e^{-inx}$ for all real x, we have

$$\langle e_n, e_n \rangle = \frac{1}{2\pi} \int_0^{2\pi} e^{inx} e^{-inx} \, dx = \frac{1}{2\pi} \int_0^{2\pi} dx = 1,$$

and, if $m \neq n$,

$$\langle e_m, e_n \rangle = \frac{1}{2\pi} \int_0^{2\pi} e^{i(m-n)x} \, dx = \frac{1}{2\pi i(m - n)} \left(e^{i(m-n)x} \right) \Big|_0^{2\pi} = 1 - 1 = 0. \quad \diamond$$

Proposition 15.1.27. *Assume $(V, +, \cdot)$ is a complex vector space and $\langle \, , \, \rangle$ a complex inner product. If $\mathbf{u}$ and $\mathbf{v}$ are elements of V, then*

$$|\langle \mathbf{u}, \mathbf{v} \rangle| \leq \|\mathbf{u}\| \, \|\mathbf{v}\|,$$

with equality if and only if one of $\mathbf{u}$ and $\mathbf{v}$ is a complex multiple of the other.

Proof. The proof for real inner products requires one modification:

$$\langle \mathbf{u}, \mathbf{v} \rangle + \langle \mathbf{v}, \mathbf{u} \rangle = \langle \mathbf{u}, \mathbf{v} \rangle + \overline{\langle \mathbf{u}, \mathbf{v} \rangle} = 2 \operatorname{Re} \langle \mathbf{u}, \mathbf{v} \rangle \leq 2 |\langle \mathbf{u}, \mathbf{v} \rangle|. \qquad \square$$

Exercises for Section 15.1

Exercise 15.1.1. (★) Assume X and Y are sets. Explain why a mapping $f : X \to Y$ may be viewed as a family of elements of Y indexed by X, and why the notation Y^X is reasonable for the set of all mappings from X to Y, analogous to $\mathbf{R}^n$ for the set of mappings from $\underline{n}$ to $\mathbf{R}$.

Exercise 15.1.2. If $[a, b]$ and $[c, d]$ are real intervals, their ordered product $[a, b] \times [c, d]$ is the rectangle consisting of pairs (x, y) such that $a \leq x \leq b$ and $c \leq y \leq d$. Explain why an element F of $\mathscr{F}([a, b] \times [c, d])$ may be viewed as a mapping $f : [a, b] \to \mathscr{F}([c, d])$, namely, as a path in $\mathscr{F}([c, d])$.

Exercise 15.1.3. (H) Assume $a < b$, and let $R : \mathscr{C}\big([a, b]\big) \to \mathscr{C}\big((a, b)\big)$ denote the restriction map, which sends each function $f : [a, b] \to \mathbf{R}$ to its restriction $Rf = f|_{(a,b)}$. Prove that R is an injective linear map whose image is the subspace of *uniformly* continuous functions.

Exercise 15.1.4. Prove Proposition 15.1.19.

15.2 Geometry of Inner Product Spaces

Throughout, $\big(V, \langle\,,\,\rangle\big)$ denotes a real inner product space. Motivated by the situation in flat n-space, we use the inner product to define length of vectors and angle between non-zero vectors.

Definition 15.2.1. The *length* of a vector $\mathbf{v}$ is its induced norm $\|\mathbf{v}\| = \langle \mathbf{v}, \mathbf{v} \rangle^{1/2}$. A vector $\mathbf{u}$ is *unit* if $\|\mathbf{u}\| = 1$.

Lemma 15.2.2. *If $\mathbf{u}$ and $\mathbf{v}$ are non-zero vectors, there exists a unique θ in $[0, \pi]$ satisfying $\langle \mathbf{u}, \mathbf{v} \rangle = \|\mathbf{u}\| \, \|\mathbf{v}\| \cos\theta$.*

Proof. Since $\|\mathbf{u}\|$ and $\|\mathbf{v}\|$ are positive, $|\langle \mathbf{u}, \mathbf{v} \rangle| \leq \|\mathbf{u}\| \, \|\mathbf{v}\|$ may be written

$$-1 \leq \frac{\langle \mathbf{u}, \mathbf{v} \rangle}{\|\mathbf{u}\| \, \|\mathbf{v}\|} \leq 1.$$

By the proof of Proposition 13.2.13, $\cos : [0, \pi] \to [-1, 1]$ is bijective.   □

Definition 15.2.3. The *angle* between two non-zero vectors $\mathbf{u}$ and $\mathbf{v}$ is the unique θ in $[0, \pi]$ satisfying $\langle \mathbf{u}, \mathbf{v} \rangle = \|\mathbf{u}\| \, \|\mathbf{v}\| \cos\theta$.

Particularly, vectors $\mathbf{u}$ and $\mathbf{v}$ are *orthogonal* if $\langle \mathbf{u}, \mathbf{v} \rangle = 0$. We write $\mathbf{u} \perp \mathbf{v}$ to indicate that $\mathbf{u}$ and $\mathbf{v}$ are orthogonal.

Example 15.2.4. Assume a is a positive real number and $V = \mathscr{C}\big([-a, a]\big)$. Every even function E is orthogonal to every odd function O: The product EO is odd, so Corollary 9.2.9 implies

$$\langle E, O \rangle = \frac{1}{2a} \int_{-a}^{a} E(x)O(x)\, dx = 0. \qquad \Diamond$$

Example 15.2.5. Let $V = \mathscr{C}\big([-1, 1]\big)$. If we define $f_n(x) = x^n$ for non-negative integer n, then

$$\langle f_m, f_n \rangle = \frac{1}{2} \int_{-1}^{1} x^{m+n}\, dx = \begin{cases} \dfrac{1}{m+n+1} & \text{if } m+n \text{ is even,} \\ 0 & \text{if } m+n \text{ is odd.} \end{cases}$$

Thus, $\|f_n\| = 1/\sqrt{2n+1}$ for each n. If $m+n$ is odd, $f_m \perp f_n$. Otherwise, m and n are both even or both odd, and the angle between f_m and f_n is

$$\theta = \arccos \frac{\langle f_m, f_n \rangle}{\|f_m\|\, \|f_n\|} = \arccos \frac{\sqrt{(2m+1)(2n+1)}}{m+n+1}. \qquad \Diamond$$

Theorem 15.2.6 (The hypotenuse theorem). *For all* $\mathbf{u}$ *and* $\mathbf{v}$ *in* V, *we have* $\|\mathbf{u} + \mathbf{v}\|^2 = \|\mathbf{u}\|^2 + \|\mathbf{v}\|^2$ *if and only if* $\mathbf{u} \perp \mathbf{v}$.

Proof. In an inner product space,

$$\|\mathbf{u} + \mathbf{v}\|^2 = \langle \mathbf{u} + \mathbf{v}, \mathbf{u} + \mathbf{v} \rangle = \|\mathbf{u}\|^2 + 2\langle \mathbf{u}, \mathbf{v} \rangle + \|\mathbf{v}\|^2.$$

This is equal to $\|\mathbf{u}\|^2 + \|\mathbf{v}\|^2$ if and only if $\langle \mathbf{u}, \mathbf{v} \rangle = 0$, or $\mathbf{u} \perp \mathbf{v}$. $\qquad \square$

Parallel and Orthogonal Components

Proposition 15.2.7. *Assume* $\mathbf{u}$ *is non-zero. For every* $\mathbf{v}$, *there exist unique vectors* $\mathrm{proj}_{\mathbf{u}}\, \mathbf{v}$ *and* $\mathbf{v}^\perp$ *such that* $\mathrm{proj}_{\mathbf{u}}\, \mathbf{v}$ *is proportional to* $\mathbf{u}$, $\mathbf{v}^\perp$ *is orthogonal to* $\mathbf{u}$, *and* $\mathbf{v} = \mathrm{proj}_{\mathbf{u}}\, \mathbf{v} + \mathbf{v}^\perp$.

Proof. (Uniqueness). Suppose $\mathbf{v} = c\mathbf{u} + \mathbf{v}^\perp$ for some real c and some vector $\mathbf{v}^\perp$ such that $\langle \mathbf{u}, \mathbf{v}^\perp \rangle = 0$. Taking the inner product with $\mathbf{u}$ gives $\langle \mathbf{u}, \mathbf{v} \rangle = c \langle \mathbf{u}, \mathbf{u} \rangle$. Since $\mathbf{u}$ is non-zero, $\langle \mathbf{u}, \mathbf{u} \rangle > 0$ by positivity, so $c = \langle \mathbf{u}, \mathbf{v} \rangle / \langle \mathbf{u}, \mathbf{u} \rangle$. That is, if $\mathrm{proj}_{\mathbf{u}}\, \mathbf{v}$ is proportional to $\mathbf{u}$, $\mathbf{v}^\perp$ is orthogonal to $\mathbf{u}$, and their sum is $\mathbf{v}$, then

$$\mathrm{proj}_{\mathbf{u}}\, \mathbf{v} = \frac{\langle \mathbf{u}, \mathbf{v} \rangle}{\langle \mathbf{u}, \mathbf{u} \rangle}\, \mathbf{u}, \qquad\qquad \mathbf{v}^\perp = \mathbf{v} - \frac{\langle \mathbf{u}, \mathbf{v} \rangle}{\langle \mathbf{u}, \mathbf{u} \rangle}\, \mathbf{u}.$$

(Existence). The preceding vectors satisfy the stated conditions. $\qquad \square$

Definition 15.2.8. Assume $\mathbf{u}$ is non-zero. For every $\mathbf{v}$, we call $\mathrm{proj}_{\mathbf{u}}\, \mathbf{v}$ the *parallel component of* $\mathbf{v}$ *on* $\mathbf{u}$, and we call $\mathbf{v}^\perp = \mathbf{v} - \mathrm{proj}_{\mathbf{u}}\, \mathbf{v}$ the *orthogonal component of* $\mathbf{v}$ *on* $\mathbf{u}$, Figure 15.1.

Remark 15.2.9. If $\mathbf{u} \neq \mathbf{0}$, we have $\mathrm{proj}_{c\mathbf{u}}\, \mathbf{v} = \mathrm{proj}_{\mathbf{u}}\, \mathbf{v}$ for all $\mathbf{v}$ and all non-zero real c. Particularly, if $\mathbf{u}$ is a unit vector, then $\mathrm{proj}_{\mathbf{u}}\, \mathbf{v} = \langle \mathbf{u}, \mathbf{v} \rangle\, \mathbf{u}$. $\qquad \Diamond$

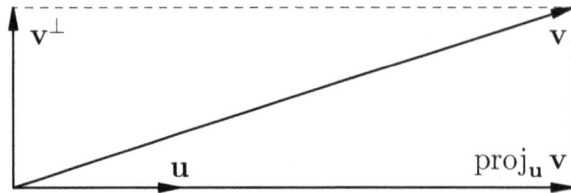

FIGURE 15.1
Parallel and orthogonal components of $\mathbf{v}$ on $\mathbf{u}$.

Orthonormal Sets

Definition 15.2.10. A subset A of V is *orthonormal* if $\|\mathbf{u}\| = 1$ for all $\mathbf{u}$ in A, and $\langle \mathbf{u}, \mathbf{v} \rangle = 0$ for all distinct $\mathbf{u}$ and $\mathbf{v}$ in A.

Example 15.2.11. For each natural number n, the standard basis $(\mathbf{e}_j)_{j=0}^{n-1}$ of $\mathbf{R}^n$ is orthonormal in the standard inner product. Similarly, every finite sequence is square-summable, and the standard basis $(\mathbf{e}_j)_{j=0}^{\infty}$ of $\mathbf{R}^{\infty}$ is orthonormal.

In flat 4-space, the vectors

$$\mathbf{u}_0 = \tfrac{1}{2}(1, \quad 1, \quad 1, \quad 1), \qquad \mathbf{u}_1 = \tfrac{1}{2}(1, \quad 1, -1, -1),$$
$$\mathbf{u}_2 = \tfrac{1}{2}(1, -1, \quad 1, -1), \qquad \mathbf{u}_3 = \tfrac{1}{2}(1, -1, -1, \quad 1),$$

form an orthonormal set: Each has unit norm, and any two agree in two components and agree up to sign in the other two. ◊

Example 15.2.12. In the vector space $\mathscr{C}\big([-\pi, \pi]\big)$ of continuous functions on $[-\pi, \pi]$, the set $(s_n)_{n=1}^{\infty}$ defined by $s_n(x) = \sqrt{2}\sin(nx)$ is orthonormal. Corollary 13.1.6 facilitates the necessary integrals. ◊

Remark 15.2.13. When a vector $\mathbf{v}$ in flat n-space is written $\mathbf{v} = \sum_j v_j \mathbf{e}_j$ in terms of the standard basis, the components of $\mathbf{v}$ are dot products with basis elements, and the squared norm of $\mathbf{v}$ is the sum of the squares of its components: $v_j = \langle \mathbf{v}, \mathbf{e}_j \rangle$ for each j, and $\|\mathbf{v}\|^2 = \sum_j v_j^2$.

Proposition 15.2.14 shows that with a definition of convergence motivated by norms, these properties generalize to countably infinite orthonormal sets. This is the conceptual basis of "spectral decomposition" in Chapter 17. ◊

Proposition 15.2.14. *Assume $(\mathbf{e}_j)_{j=0}^{\infty}$ is an orthonormal set in an inner product space, $(v_j)_{j=0}^{\infty}$ a square-summable real sequence, and define $(\mathbf{v}_n)_{n=0}^{\infty}$ by*

$$\mathbf{v}_n = \sum_{j=0}^{n-1} v_j \mathbf{e}_j.$$

If $\|\mathbf{v} - \mathbf{v}_n\| \to 0$ for some $\mathbf{v}$ in V, then $v_j = \langle \mathbf{v}, \mathbf{e}_j \rangle$ for each j, and $\|\mathbf{v}\|^2 = \sum_j v_j^2$.

Proof. By hypothesis, there exists a $\mathbf{v}$ such that $\|\mathbf{v} - \mathbf{v}_n\| \to 0$. Since $\|\mathbf{e}_j\| = 1$ for all j, bilinearity and the cross-term bound imply

$$|\langle \mathbf{v}, \mathbf{e}_j \rangle - \langle \mathbf{v}_n, \mathbf{e}_j \rangle| = |\langle \mathbf{v} - \mathbf{v}_n, \mathbf{e}_j \rangle| \leq \|\mathbf{v} - \mathbf{v}_n\| \to 0$$

for all j. But if $n > j$, then

$$\langle \mathbf{v}_n, \mathbf{e}_j \rangle = \left\langle \sum_{k=0}^{n-1} v_k \mathbf{e}_k, \mathbf{e}_j \right\rangle = \sum_{k=0}^{n-1} v_k \langle \mathbf{e}_k, \mathbf{e}_j \rangle = v_j.$$

That is, the sequence $|\langle \mathbf{v} - \mathbf{v}_n, \mathbf{e}_j \rangle|$ is eventually constant and converges to both $|\langle \mathbf{v}, \mathbf{e}_j \rangle - v_j|$ and 0. We conclude $v_j = \langle \mathbf{v}, \mathbf{e}_j \rangle$, and also $\mathbf{e}_j \perp (\mathbf{v} - \mathbf{v}_n)$ if $0 \leq j < n$.

Consequently, $\mathbf{v}_n \perp (\mathbf{v} - \mathbf{v}_n)$ for each n. Since $\mathbf{v} = \mathbf{v}_n + (\mathbf{v} - \mathbf{v}_n)$, the hypotenuse theorem implies $\|\mathbf{v}\|^2 = \|\mathbf{v}_n\|^2 + \|\mathbf{v} - \mathbf{v}_n\|^2$ independently of n. Taking the limit as $n \to \infty$,

$$\|\mathbf{v}\|^2 = \lim_{n \to \infty}\left[\sum_{j=0}^{n-1} v_j^2 \right] + \lim_{n \to \infty} \|\mathbf{v} - \mathbf{v}_n\|^2 = \sum_{j=0}^{\infty} v_j^2. \qquad \square$$

In an inner product space, a recursive algorithm converts an ordered basis into an orthonormal set.

Proposition 15.2.15. *If $(\mathbf{v}_k)_{k=0}^{\infty}$ is a linearly independent set in an inner product space, then there exists an orthonormal set $(\mathbf{e}_k)_{k=0}^{\infty}$ such that for each m, the sets $(\mathbf{v}_k)_{k=0}^{m}$ and $(\mathbf{e}_k)_{k=0}^{m}$ have the same span.*

Proof. Define $\mathbf{e}_0 = \mathbf{v}_0/\|\mathbf{v}_0\|$. Recursively, if an orthonormal set $(\mathbf{e}_j)_{j=0}^{m}$ has been constructed with the same linear span as $(\mathbf{v}_j)_{j=0}^{m}$, define

$$\mathbf{u}_{m+1} = \mathbf{v}_{m+1} - \sum_{j=0}^{m} \langle \mathbf{v}_{m+1}, \mathbf{e}_j \rangle \mathbf{e}_j, \qquad \mathbf{e}_{m+1} = \frac{\mathbf{u}_{m+1}}{\|\mathbf{u}_{m+1}\|}.$$

Linear independence of $(\mathbf{v}_k)$ ensures $\mathbf{u}_{m+1} \neq \mathbf{0}$, so this procedure does not involve division by 0. $\qquad \square$

Example 15.2.16. In $\mathscr{C}\big([0,1]\big)$, consider the sequence $(f_k)_{k=0}^{\infty}$ of monomials defined by $f_k(x) = x^k$. Since $f_0(x) = 1$ is a unit vector, we have $e_0(x) = 1$. Proceeding recursively,

$$u_1(x) = x - \langle x, 1 \rangle 1 = x - \int_0^1 x \, dx = x - \tfrac{1}{2}.$$

Since

$$\|u_1\|^2 = \int_0^1 (x - \tfrac{1}{2})^2 \, dx = \int_0^1 (x^2 - x + \tfrac{1}{4}) \, dx = \tfrac{1}{3} - \tfrac{1}{2} + \tfrac{1}{4} = \tfrac{1}{12},$$

we have $1/\|u_1\| = \sqrt{12} = 2\sqrt{3}$, and therefore $e_1(x) = \sqrt{3}(2x - 1)$. Continuing,

$$u_2(x) = x^2 - \langle x^2, 1 \rangle 1 - \langle x^2, e_1 \rangle e_1(x)$$
$$= x^2 - \tfrac{1}{3} - \sqrt{3}\tfrac{1}{6} \cdot \sqrt{3}(2x - 1) = x^2 - x + \tfrac{1}{6}.$$

Calculation gives $1/\|u_2\| = 6\sqrt{5}$, so $e_2(x) = \sqrt{5}(6x^2 - 6x + 1)$, and so forth. $\Diamond$

Exercises for Section 15.2

Exercise 15.2.1. (A) In $\mathscr{C}\big([-1, 1]\big)$ equipped with the L^2 inner product, apply the algorithm of Proposition 15.2.15 to the sequence $(x^k)_{k=0}^4$.

Exercise 15.2.2. ($\bigstar$) Let $\mathbf{1} = (1, 1, \ldots, 1)$ in flat n-space.

(a) Calculate the angle between $\mathbf{1}$ and the standard basis vector $\mathbf{e}_1$. What is this angle when $n = 2$, 3, or 4? What happens to this angle as $n \to \infty$?

(b) Prove that in $\mathbf{R}^{30,000}$, a 1cm cube (30,000-fold product of 1cm intervals) contains a 3-dimensional cube of side length 1m.

Exercise 15.2.3. In flat n-space, put $\mathbf{1} := (1, 1, \ldots, 1)$ and assume $\mathbf{x} = (x_k)_{k=0}^{n-1}$ is arbitrary. We view $\mathbf{x}$ as a data set of n real numbers.

(a) Prove $\mathrm{proj}_1 \mathbf{x} = \overline{x} \mathbf{1}$ is the arithmetic mean of $\mathbf{x}$ multiplied by the vector $\mathbf{1}$.

(b) Prove that $\|\mathbf{x} - \mathrm{proj}_1 \mathbf{x}\|$ is the standard deviation of $\mathbf{x}$.

Exercise 15.2.4. Assume V is the space of polynomials. For f and g in V, define

$$\langle f, g \rangle = \frac{1}{2} \int_{-\infty}^{\infty} f(x)g(x)e^{-|x|} \, dx.$$

(a) Prove that this formula defines an inner product on V, and that the space of even polynomials is orthogonal to the space of odd polynomials.

(b) Calculate the norm of $f_n(x) = x^n$ for each natural number n.

(c) Apply the algorithm of Proposition 15.2.15 to the ordered set $(1, x, x^2, x^3)$ to give an orthonormal basis of the set of cubic polynomials.

Exercise 15.2.5. Assume $\phi : [a, b] \to \mathbf{R}$ is integrable. Find necessary and sufficient conditions on ϕ so that

$$\langle f, g \rangle = \int_a^b f(x)g(x)\phi(x) \, dx$$

defines an inner product on $\mathscr{C}\big([a, b]\big)$.

Exercise 15.2.6. In an inner product space, prove that $\mathbf{u} + \mathbf{v}$ and $\mathbf{u} - \mathbf{v}$ are orthogonal if and only if $\|\mathbf{u}\| = \|\mathbf{v}\|$, and give a geometric interpretation.

Exercise 15.2.7. (★) Fix a positive integer n, and let V be flat n-space. If A is an $n \times n$ real matrix, prove $\langle A\mathbf{u}, A\mathbf{v} \rangle = \langle \mathbf{u}, \mathbf{v} \rangle$ for all $\mathbf{u}$ and $\mathbf{v}$ if and only if $A^t A = I_n$ is the identity matrix, namely, $T_A(\mathbf{v}) = A\mathbf{v}$ is an orthogonal linear transformation.

Exercise 15.2.8. Assume $\mathbf{R}^n$ is equipped with an inner product. Let $(\mathbf{e}_j)_{j=0}^{n-1}$ denote the standard basis, and let B denote the $n \times n$ real matrix with entries $B_{j,k} = \langle \mathbf{e}_j, \mathbf{e}_k \rangle$. Prove B uniquely determines the inner product, and B is symmetric and positive-definite. (Thus, in $\mathbf{R}^n$ an inner product depends on only finitely many real parameters.)

Exercise 15.2.9. (H) A *doubly infinite complex sequence* $\mathbf{z} = (z_k)_{k=-\infty}^{\infty}$ is a mapping $\mathbf{z} : \mathbf{Z} \to \mathbf{C}$. We say $\mathbf{z}$ is *square-summable* if

$$\sum_{k=-\infty}^{\infty} |z_k|^2 := \sum_{k=-\infty}^{-1} |z_k|^2 + \sum_{k=0}^{\infty} |z_k|^2$$

converges. Prove the set of square-summable doubly infinite complex sequences is a complex vector space, and $\langle \mathbf{z}, \mathbf{w} \rangle = \sum_k z_k \overline{w}_k$ is a complex inner product.

Exercise 15.2.10. (H) Assume $(V, \langle \ , \ \rangle)$ is a real inner product space, U a vector subspace, $\mathbf{v}$ an element of V, and $r := \inf_{\mathbf{u} \in U} \|\mathbf{v} - \mathbf{u}\| > 0$. This exercise establishes geometric properties related to minimizing $\|\mathbf{v} - \mathbf{u}\|$ over $\mathbf{u}$ in U.

(a) Assume $\delta > 0$, and $\mathbf{u}_1, \mathbf{u}_2$ are vectors in U such that $\|\mathbf{v} - \mathbf{u}_j\|^2 < r^2 + \delta$ if $j = 1, 2$. Prove $\|\mathbf{u}_2 - \mathbf{u}_1\|^2 < 4\delta$. Conclude, in particular, that there is at most one $\mathbf{u}_0$ in U such that $\|\mathbf{v} - \mathbf{u}_0\| = r$.

(b) Prove that if $\|\mathbf{v} - \mathbf{u}_0\| = r$ for some $\mathbf{u}_0$ in U, then $\langle \mathbf{v} - \mathbf{u}_0, \mathbf{u} \rangle = 0$ for all $\mathbf{u}$ in U.

Exercise 15.2.11. Assume $(V, \langle \ , \ \rangle)$ is an inner product space, and $(\mathbf{e}_k)_{k=0}^{\infty}$ is an orthonormal sequence. Fix a natural number n, and put $V_n = \text{span}(\mathbf{e}_k)_{k=0}^{n-1}$.

(a) For every $\mathbf{v}$ in V, there is a unique $\mathbf{v}_n$ in V_n such that $\mathbf{u} \perp (\mathbf{v} - \mathbf{v}_n)$ for all $\mathbf{u}$ in V_n.

(b) For all $\mathbf{v}'$ in V_n, $\|\mathbf{v} - \mathbf{v}_n\|_2 \leq \|\mathbf{v} - \mathbf{v}'\|_2$, with equality if and only if $\mathbf{v}' = \mathbf{v}_n$.

(c) Assume f is continuous on $[a, b]$. Prove that for each natural number n, there exists a unique polynomial p_n of degree at most n minimizing $\|f - p_n\|_2$.

15.3 Normed Vector Spaces

Inner products allow us to define both length and angle in a vector space. For many applications in analysis, however, no inner product is available. A useful abstraction of length alone is provided by a "norm" in a vector space.

Definition 15.3.1. Assume $(V, +, \cdot)$ is a real vector space. A *norm* on V is a function $\| \, \| : V \to \mathbf{R}$ satisfying

(i) Positivity: For all $\mathbf{v}$ in V, $0 \leq \|\mathbf{v}\|$, with equality if and only if $\mathbf{v} = \mathbf{0}$.

(ii) Homogeneity: For all $\mathbf{v}$ in V and all real c, $\|c\mathbf{v}\| = |c| \, \|\mathbf{v}\|$.

(iii) The triangle inequality: For all $\mathbf{u}$, $\mathbf{v}$ in V, $\|\mathbf{u} + \mathbf{v}\| \leq \|\mathbf{u}\| + \|\mathbf{v}\|$.

If $\| \, \|$ is a norm on V, the pair $(V, \| \, \|)$ is a *normed vector space*.

Proposition 15.3.2. *If* $\left(V, \langle \, , \, \rangle\right)$ *is an inner product space, the induced norm function* $\| \, \| : V \to \mathbf{R}$ *defined by* $\|\mathbf{v}\| = \sqrt{\langle \mathbf{v}, \mathbf{v} \rangle}$ *is a norm.*

Proof. (Positivity). This is condition (iii) in the definition of an inner product.
(Homogeneity). If $\mathbf{v} \in V$ and c is real, then

$$\|c\mathbf{v}\| = \sqrt{\langle c\mathbf{v}, c\mathbf{v} \rangle} = \sqrt{c^2 \langle \mathbf{v}, \mathbf{v} \rangle} = |c| \, \|\mathbf{v}\|.$$

(Triangle inequality). The cross-term bound $\left|\langle \mathbf{u}, \mathbf{v} \rangle\right| \leq \|\mathbf{u}\| \, \|\mathbf{v}\|$ implies

$$\|\mathbf{u} + \mathbf{v}\|^2 = \langle \mathbf{u} + \mathbf{v}, \mathbf{u} + \mathbf{v} \rangle = \|\mathbf{u}\|^2 + 2 \langle \mathbf{u}, \mathbf{v} \rangle + \|\mathbf{v}\|^2$$
$$\leq \|\mathbf{u}\|^2 + 2\|\mathbf{u}\| \, \|\mathbf{v}\| + \|\mathbf{v}\|^2 = \left(\|\mathbf{u}\| + \|\mathbf{v}\|\right)^2.$$

Taking square roots gives $\|\mathbf{u} + \mathbf{v}\| \leq \|\mathbf{u}\| + \|\mathbf{v}\|$. $\qquad\square$

Lemma 15.3.3. *If* $(V, \| \, \|)$ *is a normed vector space, then* $\left|\|\mathbf{u}\| - \|\mathbf{v}\|\right| \leq \|\mathbf{u} + \mathbf{v}\|$ *for all* $\mathbf{u}$ *and* $\mathbf{v}$ *in* V.

Proof. See Exercise 15.3.2. $\qquad\square$

Definition 15.3.4. If $1 \leq p \leq \infty$, the *p-norm* on $\mathbf{R}^\infty$ is defined by

$$\|\mathbf{u}\|_p = \left[\sum_{k=0}^{\infty} |u_k|^p\right]^{1/p}, \quad p < \infty; \qquad\qquad \|\mathbf{u}\|_\infty = \max_{0 \leq k} |u_k|.$$

Remark 15.3.5. This is an idiomatic way of defining the *p*-norm on $\mathbf{R}^n$ for each natural number n. For each $\mathbf{u}$, the sum is finite, and the maximum is taken over a finite set. $\qquad\diamond$

Proposition 15.3.6. *If $1 \le p \le \infty$, the p-norm $\| \; \|_p$ is a norm on $\mathbf{R}^\infty$.*

Proof. It suffices to work in $\mathbf{R}^n$ for some positive integer n. Positivity and homogeneity are immediate. The triangle inequality for the 1-norm follows from the real triangle inequality: If $\mathbf{u} = (u_k)$ and $\mathbf{v} = (v_k)$, then

$$\|\mathbf{u} + \mathbf{v}\|_1 = \sum_{k=0}^{n-1} |u_k + v_k| \le \sum_{k=0}^{n-1} \big(|u_k| + |v_k|\big) = \|\mathbf{u}\|_1 + \|\mathbf{v}\|_1.$$

For the ∞-norm, we have simply

$$\|\mathbf{u} + \mathbf{v}\|_\infty = \max_k |u_k + v_k| \le \max_k |u_k| + \max_j |v_j| = \|\mathbf{u}\|_\infty + \|\mathbf{v}\|_\infty.$$

The triangle inequality if $1 < p < \infty$ is Exercise 15.3.15. $\qquad\square$

Remark 15.3.7. If $0 < p < 1$, the function $\| \; \|_p$ of Definition 15.3.4 is positive and homogeneous, but does not satisfy the triangle inequality. $\qquad\diamond$

Remark 15.3.8. A norm induced by an inner product on $\mathbf{R}^n$ satisfies the parallelogram law of Proposition 15.1.19. If $n > 1$, not every norm does so. For example, the p-norm $\| \; \|_p$ satisfies the parallelogram law only if $p = 2$. $\qquad\diamond$

Proposition 15.3.9. *Let $V = \mathscr{C}\big([a,b]\big)$. If $1 \le p$, the function on V defined by*

$$\|f\|_p = \left[\frac{1}{b-a} \int_a^b |f(x)|^p \, dx\right]^{1/p}$$

is a norm. The function $\|f\|_\infty = \max\limits_{a \le x \le b} |f(x)|$ is a norm.

Proof. See Exercise 15.3.16 $\qquad\square$

Example 15.3.10. If $f(x) = x$ on $[0,1]$, then $\|f\|_p = 1/(p+1)^{1/p}$. Particularly, $\|f\|_1 = 1/2$, $\|f\|_2 = 1/\sqrt{3}$, and $\|f\|_\infty = 1$. $\qquad\diamond$

Definition 15.3.11. If $0 < p$, the set of real sequences $\mathbf{u} = (u_k)_{k=0}^\infty$ with $(|u_k|^p)$ summable is denoted $\ell^p(\mathbf{N})$, and we define

$$\|\mathbf{u}\|_p = \left[\sum_{k=0}^\infty |u_k|^p\right]^{1/p}.$$

On the set $\ell^\infty(\mathbf{N})$ of bounded sequences, we define $\|\mathbf{u}\|_\infty = \sup_k |u_k|$.

Proposition 15.3.12. *If $1 \le p$, then $\ell^p(\mathbf{N})$ is closed under addition, hence is a real vector space, and $\| \; \|_p$ defines a norm.*

Proof. See Exercise 15.3.17. $\qquad\square$

Proposition 15.3.13. *If $1 \le p < q$, then $\ell^p(\mathbf{N}) \subseteq \ell^q(\mathbf{N})$.*

Proof. See Exercise 15.3.18. $\square$

Remark 15.3.14. A sum of norms on a vector space V is a norm on V. A positive real multiple of a norm is a norm. If $(V, \| \ \|)$ and $(V', \| \ \|')$ are normed vector spaces, the following define norms on the ordered product $V \times V'$:

$$\|(\mathbf{v}, \mathbf{v}')\|_1 = \|\mathbf{v}\| + \|\mathbf{v}'\|',$$

$$\|(\mathbf{v}, \mathbf{v}')\|_2 = \left(\|\mathbf{v}\|^2 + \|\mathbf{v}'\|'^2 \right)^{1/2},$$

$$\|(\mathbf{v}, \mathbf{v}')\|_\infty = \max\left(\|\mathbf{v}\|, \|\mathbf{v}'\|' \right).$$ $\diamond$

Definition 15.3.15. Assume V is a real vector space. If $\| \ \|$ and $\| \ \|'$ are norms on V, we say $\| \ \|'$ is *equivalent to* $\| \ \|$ if there exist positive c and C such that $c\|\mathbf{v}\| \le \|\mathbf{v}\|' \le C\|\mathbf{v}\|$ for all $\mathbf{v}$ in V.

Lemma 15.3.16. *On $\mathbf{R}^n$, the 1-, 2-, and ∞-norms are mutually equivalent.*

Proof. See Exercise 15.3.9. $\square$

Exercises for Section 15.3

Exercise 15.3.1. (★) Show that in a normed vector space, every non-zero vector $\mathbf{v}$ is proportional to precisely two vectors of norm 1, and give formulas for these in terms of $\mathbf{v}$ and its norm.

Exercise 15.3.2. (H) Prove Lemma 15.3.3.

Exercise 15.3.3. Assume $a < b$, and $V = \mathscr{C}\left([a, b]\right)$ with the p-norm of Proposition 15.3.9. Define $f_n(x) = x^n$ for each non-negative integer n.

(a) Assuming $0 \le a$, calculate $\|f_n\|_p$.

(b) Not assuming $0 \le a$, calculate $\|f_n\|_1$, $\|f_n\|_2$, and $\|f_n\|_\infty$.

Exercise 15.3.4. (★) Assume V is a vector space with a norm $\| \ \|$. If $r > 0$, the *closed ball of radius r* is the set $\{\mathbf{v} : \|\mathbf{v}\| \le r\}$.

(a) Sketch the closed unit ($r = 1$) ball in the plane with the 1-norm, the 2-norm, and the ∞-norm.

(b) Prove that $\|\mathbf{v}\|_\infty \le \|\mathbf{v}\|_2 \le \|\mathbf{v}\|_1 \le 2\|\mathbf{v}\|_\infty$ for all $\mathbf{v}$, and interpret these inequalities geometrically.

Exercise 15.3.5. Assume $V = \mathbf{R}^2$ with the 2-norm, and $V' = \mathbf{R}$ with the absolute value norm. For each of the three norms in Remark 15.3.14, describe and sketch the closed unit ball in $V \times V'$.

Exercise 15.3.6.(★) Prove that the unit ball in a normed vector space is *convex*: If $\mathbf{u}$ and $\mathbf{v}$ are in the unit ball, then so is $(1-t)\mathbf{u} + t\mathbf{v}$ if $0 \leq t \leq 1$.

Assume $n \geq 2$. Prove the function $\| \ \|_p$ on $\mathbf{R}^n$ is not a norm if $0 < p < 1$.

Exercise 15.3.7. Assume V is the vector space of restrictions of $\mathscr{C}^1$ functions to $[0, 1]$. Does either of the formulas

$$\|f\|_0 = \int_0^1 |f'(x)|\, dx, \qquad \|f\|_1 = \int_0^1 \left(f(x)^2 + f'(x)^2 \right)^{1/2} dx$$

define a norm on V?

Exercise 15.3.8. For each integer k greater than 1, let f_k be the piecewise-affine function whose graph joins $(0,0)$ to $(1/k^2, k)$ to $(2/k^2, 0)$ to $(1, 0)$.

(a) Prove $f_k(x) \to 0$ for every x in $[0, 1]$.

(b) Calculate $\|f_k - \mathbf{0}\|_p$ if $1 \leq p \leq \infty$. For which p does $\|f_k - \mathbf{0}\|_p \to 0$?

Exercise 15.3.9.

(a) Prove that "equivalent to" for norms is an equivalence relation.

(b) Prove Lemma 15.3.16.

Exercise 15.3.10.(★) Assume V is a finite-dimensional real vector space, $\| \ \|$ a norm on V, and $(\mathbf{e}_k)_{k=0}^{n-1}$ an ordered basis. Prove that for each $\mathbf{v}$ in V, the function $N_\mathbf{v} : \mathbf{R}^n \to \mathbf{R}$ defined by

$$N_\mathbf{v}\left((t_k)_{k=0}^{n-1}\right) = \left\| \mathbf{v} + \sum_{k=0}^{n-1} t_k \mathbf{e}_k \right\|$$

has bounded stretch at $\mathbf{0}$ relative to the 2-norm.

Exercise 15.3.11. Assume $0 < a < 1$, and define $\mathbf{a} = (a^k)_{k=0}^{\infty}$. Prove $\mathbf{a} \in \ell^p$ if $1 \leq p \leq \infty$, and calculate its p-norm.

Exercise 15.3.12.(★) Assume $p \geq 1$. Prove that if $\mathbf{v} \in \ell^p$ and $\varepsilon > 0$, there exists a $\mathbf{u}$ in $\mathbf{R}^\infty$ such that $\|\mathbf{v} - \mathbf{u}\|_p < \varepsilon$.

Exercise 15.3.13.(A) Let $V = \mathscr{C}\left([0, 1]\right)$.

(a) If V is equipped with the 1-norm, find two elements of V that do not satisfy the parallelogram law. Conclude that the 1-norm is not induced by an inner product.

(b) If V is equipped with the ∞-norm, find two elements of V that do not satisfy the parallelogram law. Conclude that the ∞-norm is not induced by an inner product.

Exercise 15.3.14. Prove that if $n > 1$, the p-norm on $\mathbf{R}^n$ comes from an inner product only if $p = 2$, see Remark 15.3.8.

Exercise 15.3.15. ($\bigstar$) Suppose $1 < p < \infty$. Prove that if $(a_k)_{k=0}^{n-1}$ and $(b_k)_{k=0}^{n-1}$ are finite real sequences in $\mathbf{R}^n$, then

$$\left[\sum_{k=0}^{n-1} |a_k + b_k|^p\right]^{1/p} \le \left[\sum_{k=0}^{n-1} |a_k|^p\right]^{1/p} + \left[\sum_{k=0}^{n-1} |b_k|^p\right]^{1/p}.$$

Hints: If q is the "dual exponent" defined by $1/p + 1/q = 1$, then

$$|a + b|^p = |a + b|\,|a + b|^{p/q} \le |a|\,|a + b|^{p/q} + |b|\,|a + b|^{p/q}.$$

Sum over k and use Exercise 12.1.15 (d) separately on each piece.

Exercise 15.3.16. (H) Prove Proposition 15.3.9.

Exercise 15.3.17. (H) Prove Proposition 15.3.12.

Exercise 15.3.18. Prove that if $1 \le p < q$, then $\ell^p \subseteq \ell^q$.

16

Metric Spaces

The concepts of open and closed sets, sequences and convergence, and continuity of functions, rely on just a few properties of the real number line and the ordinary distance function. In this chapter we extend concepts of analysis by axiomizing the concept of *distance*. As usual, ε and δ connote arbitrary positive real numbers.

16.1 Metrics and Topology

Definition 16.1.1. A *metric space* is a pair (X, d) comprising a non-empty set X and a function $d : X \times X \to \mathbf{R}$ such that for all x, x', and x'' in X:

(i) Positivity: $0 \leq d(x, x')$, with equality if and only if $x = x'$.

(ii) Symmetry: $d(x', x) = d(x, x')$.

(iii) The triangle inequality: $d(x, x'') \leq d(x, x') + d(x', x'')$.

We call d the *metric*. The value $d(x, x')$ is the *distance* between x and x'.

Remark 16.1.2. If a non-empty set X and prospective metric $d : X \times X \to \mathbf{R}$ are given, the triangle inequality is usually the only non-trivial item to check. ◇

Lemma 16.1.3. *If $\left(V, \| \ \|\right)$ is a normed vector space, then $d(\mathbf{u}, \mathbf{v}) = \|\mathbf{u} - \mathbf{v}\|$ defines a metric on V.*

Proof. Exercise 16.1.1. □

Definition 16.1.4. The *number line* $(\mathbf{R}, d)$ is the set $\mathbf{R}$ equipped with the metric $d(x, x') = |x' - x|$. Generally, if $n \geq 1$, *flat n-space* refers to $\mathbf{R}^n$ with the *flat metric* $d(x, x') = \|x' - x\|_2$, compare Definition 15.1.13.

Lemma 16.1.5. *Assume X is a non-empty set. The function $d(x, x') = 1$ if $x \neq x'$, and $d(x, x) = 0$, is a metric.*

Proof. Positivity and symmetry are immediate. The triangle inequality follows from the observation that if x, x', and x'' are not all equal, then

$$d(x, x'') \leq 1 \leq d(x, x') + d(x', x'').$$ □

DOI: 10.1201/9781003601357-16

Definition 16.1.6. The metric in Lemma 16.1.5 is called the *discrete metric* on X. A set equipped with the discrete metric is a *discrete metric space*.

Example 16.1.7. A two-point discrete space may be viewed as $\{0, 1\}$ in the number line. A three-point discrete space may be viewed as vertices of an equilateral triangle with unit sides. A four-point discrete space may be viewed as vertices of a regular unit tetrahedron. Discrete spaces with more than four points cannot be "accurately represented" in flat 3-space. $\Diamond$

Lemma 16.1.8. *If (X, d) is a metric space, the function $d' = \min(d, 1)$ is a metric.*

Proof. See Exercise 16.1.7. $\square$

Definition 16.1.9. If d is the flat metric in the plane, $d' = \min(d, 1)$ is called the *radar screen metric*.

Lemma 16.1.10 (The reverse triangle inequality)**.** *If (X, d) is a metric space, then for all x, x', and x'' in X,*

$$|d(x, x') - d(x', x'')| \leq d(x, x'').$$

Proof. Assume x, x', and x'' are arbitrary points of X. The triangle inequality $d(x, x') \leq d(x', x'') + d(x, x'')$ implies

$$d(x, x') - d(x', x'') \leq d(x, x'').$$

Similarly, $d(x', x'') \leq d(x, x') + d(x, x'')$ implies

$$d(x', x'') - d(x, x') \leq d(x, x'').$$

But if $a \leq b$ and $-a \leq b$, then $|a| \leq b$. The lemma follows at once. $\square$

In Chapter 4 we often spoke of intervals in terms of center and radius. The same idea makes sense in a metric space, and many concepts and proofs from earlier go through without modification.

Definition 16.1.11. Assume (X, d) is a metric space. For each x_0 in X and each real number $r > 0$, the set

$$B_r(x_0) = \{x \text{ in } X : d(x_0, x) < r\}$$

is called the *open ball of radius r about x_0*. The set

$$B_r^\times(x_0) = \{x \text{ in } X : 0 < d(x_0, x) < r\}$$

is the *punctured open ball of radius r about x_0*. The set

$$\{x \text{ in } X : d(x_0, x) \leq r\}$$

is the *closed ball of radius r about x_0*.

Remark 16.1.12. If multiple metrics are under consideration, we write $B_r^d(x_0)$, and may speak of the *d-open ball*. $\Diamond$

Open and Closed Sets

The next several definitions and statements are, except for replacing the number line with a metric space (X, d), taken verbatim from Chapter 4, with the unit number in parentheses. Statements' proofs are immediate from the corresponding arguments in Chapter 4 and are omitted here, but make good review exercises.

Definition 16.1.13 (cf. 4.4.1). Assume (X, d) is a metric space, $A \subseteq X$, and $A^c = X \setminus A$. For each x_0 in X, precisely one of the following conditions three holds:

(i) There exists an ε such that $B_\varepsilon(x_0) \subseteq A$. In this case we say x_0 is an *interior point* of A.

(ii) There exists an ε such that $B_\varepsilon(x_0) \subseteq A^c$. In this case we say x_0 is an *exterior point* of A.

(iii) For every ε, the sets $B_\varepsilon(x_0) \cap A$ and $B_\varepsilon(x_0) \cap A^c$ are both non-empty. In this case we say x_0 is a *boundary point* of A.

The *interior* of A is the set of interior points of A. The *exterior* and the *boundary ∂A* of A are defined analogously.

Remark 16.1.14 (cf. 4.4.3). The exterior of A is the interior of A^c. The boundary of A is the boundary of A^c. In symbols, $\partial A = \partial(A^c)$. ◇

Remark 16.1.15 (cf. 4.4.4). An interior point of A is an element of A, since $x_0 \in B_\varepsilon(x_0)$ regardless of ε. Similarly, an exterior point of A is not an element of A. However, a boundary point of A may lie in either A or its complement. ◇

Definition 16.1.16 (cf. 4.4.5). Assume (X, d) is a metric space and $A \subseteq X$. If $B_\varepsilon^\times(x_0) \cap A$ is non-empty for every ε, we say x_0 is a *limit point* of A.
The *closure $\overline{A}$* of A is the union of A and its set of limit points.

Remark 16.1.17 (cf. 4.4.6). Contrapositively, x_0 is *not* a limit point of A if and only if there exists an ε such that $B_\varepsilon^\times(x_0) \cap A = \varnothing$. ◇

Definition 16.1.18 (cf. 4.4.7). We say x_0 is an *isolated point of A* if there exists an ε such that $B_\varepsilon(x_0) \cap A = \{x_0\}$, namely, $x_0 \in A$ and x_0 is not a limit point of A.
We say x_0 is a *border point* of A if x_0 is both a limit point of A and boundary point of A.

Proposition 16.1.19 (cf. 4.4.8). *Assume $A \subseteq X$. For every x_0 in X, precisely one of the following conditions holds:*

	Boundary $= F$	*Boundary* $= T$
Limit $= F$	*Exterior*	*Isolated*
Limit $= T$	*Interior*	*Border*

Remark 16.1.20 (cf. 4.4.9). By inspection, the boundary of A is the disjoint union of isolated and border points of A. The limit points of A are the disjoint union of interior and border points of A. The closure of A is the disjoint union of the interior, isolated, and border points of A, namely, the complement of the exterior of A.   ◇

Definition 16.1.21 (cf. 4.4.11). Assume (X, d) is a metric space, and $A \subseteq X$. We say A is an *open set* if every element of A is an interior point of A, that is, if A contains *none* of its boundary points.

   We say A is a *closed set* if A contains all of its limit points, that is, $\overline{A} = A$, or A contains *all* of its boundary points.

Proposition 16.1.22 (cf. 4.4.13). *Assume (X, d) is a metric space. For every subset A of X, the following are equivalent:*

   (i) *A is closed.*   (ii) $\overline{A} = A$.   (iii) $\partial A \subseteq A$.   (iv) A^c *is open.*

Proposition 16.1.23 (cf. 4.4.14). *In a metric space (X, d), an arbitrary open ball $B_r(x_0)$ is an open set. An arbitrary closed ball is a closed set.*

Proposition 16.1.24 (cf. 4.4.15). *Assume (X, d) is a metric space. A union of open sets is open, and a finite intersection of open sets is open. Precisely:*

   (i) *If $\{O_i\}_{i \in \mathscr{I}}$ is a collection of open subsets of (X, d), then $\bigcup_i O_i$ is open.*

   (ii) *If $\{O_i\}_{i=0}^{n}$ is a finite collection of open subsets of (X, d), then $\bigcap_i O_i$ is open.*

Remark 16.1.25 (cf. 4.4.16). By Proposition 16.1.22 and the complement laws, an intersection of closed sets is closed, and a finite union of closed sets is closed.   ◇

   This concludes our parallel development from Chapter 4.

Definition 16.1.26. Assume X is a set. A collection $\mathscr{T}$ (script T) of subsets of X is a *topology* on X if $\varnothing$ and X are in $\mathscr{T}$, and if $\mathscr{T}$ is closed under arbitrary unions and finite intersections.

   A property of metric spaces is *topological* if it depends only on open sets.

Remark 16.1.27. Just as a metric furnishes an axiomization of "distance," a topology furnishes an axiomization of "openness" or "nearness."

   Metric topologies have a "countable flavor" due to finitude and reciprocal finitude. General non-metric topologies can be distinctly different, ranging from $\mathscr{T} = \{\varnothing, X\}$ to topologies with unavoidably uncountable local and/or global structure, and are beyond the scope of this book. Munkres, [21] is an excellent introduction.   ◇

Definition 16.1.28. If X is a non-empty set and $Y \subseteq X$, a collection $\{X_i\}_{i \in \mathscr{I}}$ (script I) of subsets of X *covers* Y if $Y \subseteq \bigcup_i X_i$.

Relative Openness

If (X, d) is a metric space and $Y \subseteq X$, the restriction of d to Y is a metric. If $A \subseteq Y$, we may ask about "openness" or "closedness" of A in two senses: Viewing A as a subset of (Y, d), or as a subset of (X, d).

Definition 16.1.29. If (X, d) is a metric space and $Y \subseteq X$, a subset $A \subseteq Y$ is *relatively open in Y* if there exists an open set $G \subseteq X$ such that $A = G \cap Y$.

Similarly, $A \subseteq Y$ is *relatively closed in Y* if there exists a closed set $F \subseteq X$ such that $A = F \cap Y$.

Example 16.1.30. Assume (X, d) is the number line, and $Y = [-1, 1]$. The half-open interval $(0, 1] = (0, 2) \cap Y$ is relatively open in Y, though not open in X.

If instead $Y = (0, \infty)$, the half-open interval $(0, 1] = [0, 1] \cap Y$ is relatively closed in Y, though not closed in X. $\diamond$

Example 16.1.31. In the number line, assume $Y = \mathbf{Q}$ is the set of rational numbers. The set $A = \{x \text{ in } \mathbf{Q} : x^2 < 2\} = \{x \text{ in } \mathbf{Q} : x^2 \leq 2\}$ is both relatively open in Y (because $A = (-\sqrt{2}, \sqrt{2}) \cap \mathbf{Q}$) and relatively closed in Y (because $A = [-\sqrt{2}, \sqrt{2}] \cap \mathbf{Q}$). $\diamond$

Example 16.1.32. Let $(X, d) = (\mathbf{R}^2, d)$ be the flat plane, and Y the horizontal axis, identified with the real number line $\mathbf{R}$. An open interval (a, b) is relatively open in Y, though no subset of Y is open in X. $\diamond$

Lemma 16.1.33. *If (X, d) is a metric space and $Y \subseteq X$, then $A \subseteq Y$ is relatively open in Y if and only if A is open in the metric space (Y, d).*

Consequently, $A \subseteq Y$ is relatively closed in Y if and only if A is closed in the metric space (Y, d).

Proof. For every point a in Y and every positive r, the intersection $B_r(a) \cap Y$ is, by definition, the open ball of radius r about a in (Y, d).

Assume A is relatively open in Y, so that $A = G \cap Y$ for some open set G. If $a \in A$, there exists a positive real number r such that $B_r(a) \subseteq G$. Consequently, $B_r(a) \cap Y \subseteq G \cap Y = A$, so A is open in Y.

Conversely, assume A is open in Y. For each a in A, there exists an r such that $B_r(a) \cap Y \subseteq A$. The union of this collection of balls over all points of a is an open subset $G \subseteq X$, and $A = G \cap Y$, so A is relatively open in Y. $\square$

Sequences in Metric Spaces

In an arbitrary metric space (X, d), for every point x_0, there exists a "countable base" $\{B_{1/n}(x_0)\}_{n=1}^{\infty}$ of neighborhoods such that *every* neighborhood of x_0 contains some set of the base. Sequences consequently play an important role in the study of metric spaces. Convergence-related properties of sequences generalize immediately to metric spaces.

Definition 16.1.34. A sequence (x_k) in X is *condensing* if for every ε, there exists an index N such that $n, m \geq N$ implies $d(x_n, x_m) < \varepsilon$.

A sequence $(x_k)_{k=k_0}^{\infty}$ in X *converges* to a point x_∞ in X if for every ε, there exists an index N such that $k \geq N$ implies $d(x_k, x_\infty) < \varepsilon$.

If $\mathbf{x}$ is a sequence in X and $\nu : \mathbf{N} \to \mathbf{N}$ is strictly increasing, the mapping $\overline{\mathbf{x}} = \mathbf{x} \circ \nu$, namely, $\overline{x}_k = x_{\nu(k)}$, is a *subsequence* of $\mathbf{x}$.

Definition 16.1.35. Assume (X, d) is a metric space. We say (X, d) is *complete* if every condensing sequence in X converges to a point of X.

Continuity of Mappings

Like sequential convergence, continuity of mappings generalizes immediately from the real definition.

Definition 16.1.36 (cf. 8.1.4). Assume (X, d) and (Y, e) are metric spaces, $f : X \to Y$ is a mapping, and x is a point of X. We say f is *continuous* at x (with respect to the given metrics) if the following condition holds:

$$\big(f(x_k)\big) \to f(x) \text{ in } (Y, e) \text{ for every sequence } (x_k) \to x \text{ in } (X, d).$$

If f is continuous at x for every x in X, we say f is *continuous on* X.

The ε-δ criterion for continuity holds without modification:

Proposition 16.1.37 (cf. 8.2.7). *Assume (X, d) and (Y, e) are metric spaces. A mapping $f : X \to Y$ is continuous at x_0 if and only if:*

For every ε, there is a δ such that $d(x_0, x) < \delta$ implies $e\big(f(x_0), f(x)\big) < \varepsilon$.

As for one-variable functions, continuity has a topological characterization:

Proposition 16.1.38 (cf. Exercise 8.2.11). *Assume (X, d) and (Y, e) are metric spaces. A mapping $f : X \to Y$ is continuous on X if and only if for every open subset $V \subseteq Y$, the preimage $f^*(V)$ is open in X.*

Proof. Suppose f is continuous on X, and that $V \subseteq Y$ is open. If $f^*(V) = \varnothing$ there is nothing to prove. Otherwise, assume x_0 is an arbitrary element of $f^*(V)$, and write $y_0 = f(x_0)$. Fix ε so that $B_\varepsilon(y_0) \subseteq V$. Since f is continuous at x_0, there exists a δ such that if $d(x_0, x) < \delta$, then $e\big(y_0, f(x)\big) < \varepsilon$, or $f(x) \in B_\varepsilon(y_0) \subseteq V$. That is, $B_\delta(x_0) \subseteq f^*(V)$, so x_0 is an interior point. Since x_0 was an arbitrary point of $f^*(V)$, $f^*(V)$ is open.

Conversely, fix x_0 in X arbitrarily, put $y_0 = f(x_0)$, and fix ε arbitrarily. The ball $V = B_\varepsilon(y_0)$ is open, so by hypothesis $f^*(V)$ is open. Since $x_0 \in f^*(V)$, there exists a δ such that $B_\delta(x_0) \subseteq f^*(V)$. In other words, if $d(x_0, x) < \delta$, then $f(x) \in V = B_\varepsilon(y_0)$, or $e\big(y_0, f(x)\big) < \varepsilon$. $\square$

Equivalent Metrics

Some properties of metric spaces depend not on the metric, but only on the induced topology of the metric. For topological purposes including convergence of sequences and continuity of mappings, we may replace a metric by an "equivalent" metric if doing so is convenient.

Definition 16.1.39. Assume X is a non-empty set. Two metrics d and d' on X are *equivalent* if for every subset $O \subseteq X$, O is d-open if and only if O is d'-open.

Lemma 16.1.40. *Assume X is a non-empty set, and d, d' are metrics on X.*

(i) *The metrics d and d' are equivalent if and only if every d-open ball is d'-open and every d'-open ball is d-open.*

(ii) *If d and d' are equivalent, then a sequence (x_k) in X converges to x_∞ with respect to d if and only if it converges to x_∞ with respect to d'.*

Proof. Exercise 16.1.8. □

Lemma 16.1.41. *If V is a real vector space with equivalent norms $\| \ \|$ and $\| \ \|'$, the induced metrics are equivalent.*

Proof. Exercise 16.1.9. □

Example 16.1.42. If (X, d) is a metric space, the metric $d' = \min(d, 1)$ is equivalent to d because every d-open set may be written as a union of open balls of radius less than 1, and *vice versa*. ◇

Example 16.1.43. On $\mathbf{R}$, the flat metric d and discrete metric d' are not equivalent: Every singleton is d'-open, since $\{x_0\} = B_1(x_0)$, so *every subset of $\mathbf{R}$ is d'-open*, while "most" subsets of $\mathbf{R}$ are not d-open. ◇

Products of Metric Spaces

Definition 16.1.44. If (X, d) and (X', d') are metric spaces, their *product* is the ordered product $X \times X'$ equipped with the metric

$$D\big((x, x'), (y, y')\big) = \big(d(x, y)^2 + d'(x', y')^2\big)^{1/2}.$$

Remark 16.1.45. The product of the number line with itself is the flat plane. Generally, our definition ensures the product of flat spaces is a flat space. This choice is not universal; take care when consulting other sources. ◇

Lemma 16.1.46. *Assume (X, d) and (X', d') are metric spaces. If $r > 0$, x_0 in X, and x_0' in X', then*

$$B_{r/2}^d(x_0) \times B_{r/2}^{d'}(x_0') \subseteq B_r^{d \times d'}(x_0, x_0') \subseteq B_r^d(x_0) \times B_r^{d'}(x_0').$$

Proof. If u and v are real and $\max(|u|, |v|) < r/2$, then $|u|^2 + |v|^2 < r^2/2 < r^2$. Further, if $|u|^2 + |v|^2 < r^2$, then $\max(|u|, |v|) < r$. The lemma follows at once if we let x be an arbitrary point of X, x' be an arbitrary point of X', and we put $u = d(x, x_0)$ and $v = d'(x', x_0')$. $\qquad\qquad\square$

Exercises for Section 16.1

Exercise 16.1.1. Prove Lemma 16.1.3.

Exercise 16.1.2. (★) Suppose A is an open set. For each x in A, there exists a positive r_x such that $B_{r_x}(x) \subseteq A$. Prove that for every such family of choices, A is the union $\bigcup_x B_{r_x}(x) \subseteq A$.

Exercise 16.1.3. Sketch the following sets, and determine whether or not each is open in $\mathbf{R}^2$ with the flat metric:

(a) $H^+ = \{(x, y) \text{ in } \mathbf{R}^2 : x > 0\}$.

(b) $\overline{H^-} = \{(x, y) \text{ in } \mathbf{R}^2 : x \leq 0\}$.

(c) $A = \{(x, y) \text{ in } \mathbf{R}^2 : x \neq 0\}$.

Exercise 16.1.4. (★) Assume (X, d) is a metric space. Prove the *reverse quadrangle inequality*: If x, x', y, and y' are elements of X, then

$$|d(x, y) - d(x', y')| \leq d(x, x') + d(y, y').$$

In words, the difference between two sides of a quadrangle is no larger than the sum of the other two sides.

Exercise 16.1.5. Referring to the sets in Exercise 16.1.3:

(a) Is $\overline{H^-} \cap B_1(0, 0)$ relatively open in $\overline{H^-}$?

(b) Is $\overline{H^-} \cap B_1(0, 0)$ open in $\mathbf{R}^2$?

(c) Can A be partitioned into two relatively open subsets?

Exercise 16.1.6. (★) Assume (X, d) is a metric space and $A \subseteq X$. Prove that $\overline{\overline{A}} = \overline{A}$: The closure of the closure of A is the closure of A.

Exercise 16.1.7. Assume (X, d) is a metric space. Prove that the function $\bar{d}(x, x') = \min\big(d(x, x'), 1\big)$ is a metric on X.

Exercise 16.1.8.(★) Prove Lemma 16.1.40.

Exercise 16.1.9. Prove Lemma 16.1.41.

Exercise 16.1.10.(H) Assume $f : [0, \infty) \to [0, \infty)$ is a concave, non-decreasing function such that $f(0) = 0$ and $f(x) > 0$ if $x > 0$.

(a) (H) Prove that if d is a metric on X, then $d_f = f \circ d$ is a metric on X.

(b) Let d be the absolute value metric on $\mathbf{R}$. Find a concave, non-decreasing function f such that d_f is the radar screen metric.

(c) Prove that if d is a metric, then $d/(1 + d)$ is a metric.

Exercise 16.1.11. Assume $a < b$, and $V = \mathscr{C}([a, b])$ is the vector space of continuous functions on $[a, b]$.

(a) Prove that $\|f\|_1 \leq \|f\|_\infty$ for all f in V.

(b) Prove the 1-norm and ∞-norm are not equivalent on V.

Exercise 16.1.12.(★) Assume n and m are positive integers, and equip $\mathbf{R}^n$ and $\mathbf{R}^m$ with the flat metrics. Assume $U \subseteq \mathbf{R}^n$ is open and $f = (f_k)_{k=0}^{m-1} : U \to \mathbf{R}^m$ a mapping. Prove f is continuous if and only if each component f_k is continuous.

Exercise 16.1.13. To state this question, we'll introduce *multi-index notation* in $\mathbf{R}^n$. A *multi-index* is an ordered n-tuple of natural numbers, $I = (i_k)_{k=0}^{n-1}$. If $x = (x_k)_{k=0}^{n-1}$ is a point of $\mathbf{R}^n$, we define

$$x^I = \prod_{k=0}^{n-1} x_k^{i_k} = x_0^{i_0} \cdot x_1^{i_1} \cdots x_{n-1}^{i_{n-1}}.$$

A *polynomial function in n variables* is a function p defined by a formula $p(x) = \sum_I a_I x^I$ with only finitely many non-zero coefficients a_I. A *rational function in n variables* is a function defined by a quotient of polynomials functions having no common factor, defined on the complement of the zero set of the denominator.

Prove that every rational function in n variables is continuous. Suggestion: Use induction on the number of variables to prove polynomials are continuous.

Exercise 16.1.14. (★) Define *inverse stereographic projection* $f : \mathbf{R}^2 \to \mathbf{R}^3$ by

$$f(u, v) = \left(\frac{2u}{u^2 + v^2 + 1}, \frac{2v}{u^2 + v^2 + 1}, \frac{u^2 + v^2 - 1}{u^2 + v^2 + 1} \right).$$

Prove f is a continuous injection whose image is the punctured unit sphere, $\{(x, y, z) \text{ in } \mathbf{R}^3 : x^2 + y^2 + z^2 = 1\}$ with the point $(0, 0, 1)$ removed.

Suggestion: Write $(x, y, z) = f(u, v)$, solve for (u, v), and prove the resulting mapping inverts f on a suitable domain.

16.2   Separability and Total Boundedness

The intermediate and extreme value theorems in one variable are set on closed, bounded intervals of real numbers. We naturally seek to generalize these theorems to arbitrary metric spaces. As we have just seen, continuity of mappings generalizes to metric spaces as easily as the definitions and properties of open sets. Generalizing "closed and bounded" from the real numbers to arbitrary metric spaces is another matter, and takes three entire sections.

"Closed and bounded" makes sense in an arbitrary metric space, but generally imposes no conditions on continuous mappings: An arbitrary set with the discrete metric is closed and bounded. Because every point in a discrete metric space is isolated, every mapping on a discrete metric space is continuous, compare Proposition 8.2.6 (i).

This section introduces properties that either name familiar conditions in the number line, or else suitably generalize number line concepts to arbitrary metric spaces. Illustrating definitions and proofs with sketches as you read may provide helpful assistance in assimilating these concepts.

Distance Between Subsets

Definition 16.2.1. If (X, d) is a metric space and if A, B are non-empty subsets, the *distance* between A and B is

$$d(A, B) = \inf\{d(a, b) : a \in A,\ b \in B\}.$$

If $x \in X$, then $d(x, A) = \inf\{d(x, a) : a \in A\}$.

Example 16.2.2. If A and B are not disjoint, then $d(A, B) = 0$. In particular, despite the notation d is not a metric on the power set $\mathscr{P}(X)$.

In $(\mathbf{R}, |\ |)$, $A = (-1, 0)$ and $B = (0, 1)$ are disjoint but $d(A, B) = 0$. $\Diamond$

Example 16.2.3. In the number line, $d(x, \mathbf{Z}) \le 1/2$ for every real x, since for every real x, $\lfloor x \rfloor \in \mathbf{Z}$ and Corollary 4.3.2 implies $\lfloor x \rfloor \le x \le 1 + \lfloor x \rfloor$. $\Diamond$

Denseness and Separability

For the remainder of the book, metric spaces (X, d) are assumed to be *infinite*, namely to contain infinitely many points, unless explicitly stated otherwise.

Definition 16.2.4. Assume (X, d) is a metric space. A subset $A \subseteq X$ is *dense* if the closure of A in X is X, namely, if the exterior is empty.

We say (X, d) is *separable* if X contains a countable dense subset.

Example 16.2.5. The set $\mathbf{Q}$ of rational numbers is dense in the number line. Since $\mathbf{Q}$ is countable, the number line is separable. $\Diamond$

Example 16.2.6. No proper subset of a discrete metric space is dense, because every singleton is open. A discrete metric space is separable if and only if it is countable. ◇

Lemma 16.2.7. *Assume (X, d) is a metric space. A subset $A \subseteq X$ is dense if and only if $d(x, A) = 0$ for every x in X.*

Proof. See Exercise 16.2.1. □

When (X, d) contains a countable dense subset A, the collection of open balls centered at a point of A and with rational radius is itself countable. For later use, we establish an "encapsulation" property of balls with center in a dense set and having rational radius.

Lemma 16.2.8. *Assume (X, d) is a metric space and $A \subseteq X$ a dense subset. If $U \subseteq X$ is open and x is an arbitrary point of U, there exists a point x_0 of A and a positive rational r such that $x \in B_r(x_0) \subseteq U$.*

Proof. Since U is open, there is an ε such that $B_{3\varepsilon}(x) \subseteq U$. Since A is dense, there is a point x_0 of A in $B_\varepsilon(x)$. Assume r is a rational number in $(\varepsilon, 2\varepsilon)$. Since $d(x_0, x) < \varepsilon < r$, we have $x \in B_r(x_0)$. Further, if $x' \in B_r(x_0)$, then $d(x, x') \leq d(x_0, x) + d(x_0, x') < \varepsilon + r < 3\varepsilon$, so $x' \in B_{3\varepsilon}(x) \subseteq U$. □

Boundedness

Definition 16.2.9. Assume (X, d) is a metric space and A a non-empty subset of X. The *diameter* of A is the extended real number

$$\operatorname{diam} A = \sup\{d(x, x') : x, \ x' \in A\} \in [0, \infty].$$

We say A is *bounded* if $\operatorname{diam} A < \infty$.

Example 16.2.10. In the number line, an interval with endpoints a and b has diameter $|b - a|$. The set of natural numbers has diameter ∞. ◇

Example 16.2.11. The flat plane is unbounded. Generally, a non-trivial vector space V with metric coming from a norm is unbounded.

The plane with the radar screen metric has diameter 1. ◇

Example 16.2.12. A discrete metric space is bounded. Every singleton set has diameter 0, and every set containing at least two points has diameter 1. ◇

Example 16.2.13. In a metric space, every ball $B_r(x_0)$ has diameter at most $2r$, Exercise 16.2.2 (d).

In a discrete metric space, open unit balls have diameter 0. ◇

Lemma 16.2.14. *Assume (X, d) is a metric space. A subset $A \subseteq X$ is bounded if and only if for every x_0 in X there exists a positive r such that $A \subseteq B_r(x_0)$.*

Proof. Exercise 16.2.6. □

Total Boundedness

Discrete spaces and the radar screen metric show that boundedness in an arbitrary metric space is weaker than we might like. A stronger condition, total boundedness, better captures number line intuition.

Definition 16.2.15. Assume (X, d) is a metric space and $A \subseteq X$. We say A is *totally bounded* if for every positive r, there exists a finite set $C = \{c_j\}_{j=0}^{N}$ in X (C for "centers") such that $A \subseteq \bigcup_{j=0}^{N} B_r(c_j)$.

Example 16.2.16. An infinite discrete space is bounded (every open ball of radius $1 + \varepsilon$ is the entire space) but not totally bounded (the space is not covered by finitely many open balls of radius 1). $\diamond$

Example 16.2.17. The vector space $\mathbf{R}^n$ equipped with the radar-screen metric d' is bounded but not totally bounded: The diameter is 1, but finitely many d'-balls of radius $1/2$, which are balls of radius $1/2$ in the flat metric, do not cover. $\diamond$

Proposition 16.2.18. *A totally bounded metric space (X, d) is separable.*

Proof. By hypothesis, for each positive integer n there exists a finite set C_n such that the balls of radius $1/n$ centered at points of C_n cover X, which implies $d(x, C_n) < 1/n$ for every x in X.

The set $A = \bigcup_n C_n$ is a countable union of finite sets, hence countable, and $d(x, A) < 1/n$ for every positive integer n, so $d(x, A) = 0$, for all x in X. By Lemma 16.2.7, A is dense in X. $\square$

Proposition 16.2.19. *Assume (X, d) is a metric space. A totally bounded subset A is bounded.*

Proof. If A is totally bounded, there exists a finite set $C = \{c_j\}_{j=0}^{N}$ in X such that $A \subseteq \bigcup_{j=0}^{N} B_1(c_j)$. Put $r = 1 + \operatorname{diam} C$. By the triangle inequality, $B_1(c_j) \subseteq B_r(c_0)$ for each j, so $A \subseteq \bigcup_{j=0}^{N} B_1(c_j) \subseteq B_r(c_0)$. $\square$

Proposition 16.2.20. *In $(\mathbf{R}, | \ |)$, a bounded set is totally bounded.*

Proof. Assume $A \subseteq \mathbf{R}$ is a bounded set, namely, there exists a real M such that $A \subseteq [-M, M]$, and that $r > 0$. By the accretion principle, Corollary 4.3.5, there exists a positive integer n such that $M < nr$. Consequently, the set $r\mathbf{Z} \cap [-M, M] \subseteq \{jr : |j| \leq n\}$ is finite, and $A \subseteq [-M, M] \subseteq \bigcup_{j=-n}^{n} B_r(jr)$. $\square$

Remark 16.2.21. If you are in the habit of emphasizing your utterances with *totally*, it is safe to do so when describing bounded sets in the number line. $\diamond$

Proposition 16.2.22. *Assume (X, d) and (X', d') are metric spaces. If $A \subseteq X$, $A' \subseteq X'$ are totally bounded, then $A \times A'$ is totally bounded in $(X \times X', d \times d')$.*

Proof. Assume $r > 0$. Since A is totally bounded, there exists a finite set $C = \{c_i\}_{i=0}^{N}$ in X such that $A \subseteq \bigcup_{i=0}^{N} B_{r/2}^{d}(c_i)$. (Note the usage of balls of radius $r/2$.) Similarly there is a finite set $C' = \{c'_j\}_{j=0}^{N'}$ in X' such that $A' \subseteq \bigcup_{j=0}^{N'} B_{r/2}^{d'}(c'_j)$. By Lemma 16.1.46, $B_{r/2}^{d}(c_i) \times B_{r/2}^{d'}(c'_j) \subseteq B_{r}^{d \times d'}(c_i, c'_j)$ for all (i, j). Consequently,

$$
A \times A' \subseteq \left[\bigcup_{i=0}^{N} B_{r/2}^{d}(c_i) \right] \times \left[\bigcup_{j=0}^{N'} B_{r/2}^{d'}(c'_j) \right]
$$

$$
= \bigcup_{i=0}^{N} \bigcup_{j=0}^{N'} B_{r/2}^{d}(c_i) \times B_{r/2}^{d'}(c'_j) \subseteq \bigcup_{i=0}^{N} \bigcup_{j=0}^{N'} B_{r}^{d \times d'}(c_i, c'_j). \qquad \square
$$

Corollary 16.2.23. *A bounded set in flat n-space is totally bounded.*

Proof. Immediate from Propositions 16.2.20 and 16.2.22 by induction on n. $\square$

Exercises for Section 16.2

Exercise 16.2.1.(★) Prove Lemma 16.2.7.

Exercise 16.2.2. Assume (X, d) is a metric space and $A \subseteq X$.

(a) Prove $\operatorname{diam} A = 0$ if and only if A is a singleton.

(b) In the number line, prove $\operatorname{diam} B_r(x_0) = 2r$.

(c) In a discrete metric space, what is $\operatorname{diam} B_r(x_0)$?

(d) In an arbitrary metric space, prove that $\operatorname{diam} B_r(x_0) \leq 2r$.

Exercise 16.2.3. Assume (X, d) is a metric space, $x_0 \in X$, and $r > 0$. Is the closure of $B_r(x_0)$ equal to the closed ball of radius r about x_0? If so, give a proof. If not, does either inclusion hold in general?

Exercise 16.2.4. Assume (X, d) is a metric space, $A \subseteq X$ a dense subset, and $B \subseteq A$ dense in (A, d). Prove B is dense in (X, d).

Exercise 16.2.5. Assume $f : (X, d) \to (Y, e)$ is continuous and A is dense in (X, d). Prove the image $f(A)$ is dense in the image $f(X)$.

Exercise 16.2.6.(★) Prove Lemma 16.2.14: A subset A of a metric space is bounded if and only if A is contained in some open ball.

Exercise 16.2.7. Prove the closure of a bounded set is bounded.

Exercise 16.2.8. An infinite-dimensional inner product space $\left(V, \langle \ , \ \rangle\right)$ is *separable* if there exists a countable, dense subset of V in the 2-norm metric. An orthonormal set $B \subseteq V$ is an L^2-*basis* if the algebraic span of B is dense.
    Prove V is separable if and only if V has a countable L^2-basis.

Exercise 16.2.9. Assume (Y, e) is a metric space and $f : X \to Y$ is a bijection. Prove that the function $d(x, x') = e\big(f(x), f(x')\big)$ defines a metric on X. (We call $d = f^*e$ the *pullback* of e by f. Conceptually, (X, d) "is" (Y, e), except the underlying sets are different.)

Exercise 16.2.10. Let $X = \mathbf{R} \setminus \{0\}$ be the set of non-zero real numbers. Prove that the function $d(x, x') = |x' - x|/|xx'|$ defines a metric on X.

Exercise 16.2.11. Let n denote a positive integer. In this exercise we'll use geometry and calculus to find the volume of a ball in flat n-space without systematically developing a theory of volume. The closed ball of radius r in $\mathbf{R}^n$ is the set of $(x_j)_{j=0}^{n-1}$ such that $\sum_j x_j^2 \leq r^2$. The key fact we assume is, scaling a ball in flat n-space by a factor r scales the volume by r^n: If $V^n(r)$ denotes the volume of the closed ball of radius r in $\mathbf{R}^n$, then $V^n(r) = r^n V^n(1)$.

(a) Show that if $-1 \leq t \leq 1$, the intersection of the unit ball in $\mathbf{R}^{n+1}$ with $\mathbf{R}^n \times \{t\}$ is (congruent to) a ball of radius $(1 - t^2)^{1/2}$ in $\mathbf{R}^n$. Use this to express $V^{n+1}(1)$ as an integral of n-dimensional volumes in terms of $V^n(1)$.

(b) Use Proposition 13.2.10 to express the recursion from (a) in terms of the Γ function, and prove that $V^n(1) = \pi^{n/2}/\Gamma(\frac{n}{2} + 1)$. Check that this formula agrees with your geometric knowledge if $n = 2$ and $n = 3$.

(c) In flat $\mathbf{R}^n$, the complement of the closed ball of radius $r - dr$ (about $\mathbf{0}$) in the closed ball of radius r is a thin shell whose volume is approximately dr times the $(n-1)$-dimensional volume of the sphere of radius r. Use this to find the $(n-1)$-dimensional volume of the sphere in $\mathbf{R}^n$.

(d) Prove that the volume of the unit ball converges to 0 as $n \to \infty$. Prove that the volume concentrates at the boundary.

16.3 Connectedness and Compactness

This section introduces *intrinsic* (namely, non-relative), *topological* conditions in an arbitrary metric space: connectedness for generalizing "intervals," and compactness for generalizing "closed and bounded." Section 16.4 establishes that subsets of flat n-space are compact if and only if they are closed and bounded, and characterizes compact subsets of arbitrary metric spaces as being complete and totally bounded. Section 16.5 contains sweeping yet simple generalizations of the extreme value theorem and intermediate value theorem.

Connectedness

Definition 16.3.1. Assume (X, d) is a metric space, $A \subseteq X$. A *disconnection* of A is a partition of A into two relatively open subsets, namely, a pair of disjoint open sets U and V such that $A \subseteq U \cup V$ and the intersections $A \cap U$ and $A \cap V$ are non-empty.

If A admits a disconnection, A is *disconnected*. Otherwise A is *connected*.

Example 16.3.2. Every singleton is trivially connected. Every set with more than one element in a discrete space is disconnected: every partition into two subsets is a disconnection.

In the number line, the set $A = [-1, 0) \cup (0, 1]$ is disconnected, as is $\mathbf{Q}$, the set of rational numbers, and K, the ternary set. ◇

Remark 16.3.3. The standard idiom for proving A is connected is to assume U and V are disjoint, non-empty open sets whose union contains A, to assume $A \cap U$ is non-empty, and to prove $A \subseteq U$, or equivalently, $A \cap V = \varnothing$. ◇

Remark 16.3.4. Connectedness is topological by definition. Further, as noted above, connectedness is intrinsic: If $A \subseteq X \subseteq (X', d)$, then A is connected as a subspace of (X, d) if and only if A is connected as a subspace of (X', d). ◇

Proposition 16.3.5. *If J is a subset of the number line, then J is connected if and only if J is an interval.*

Proof. If J is not an interval, there exist real numbers a, b, and x such that $a < x < b$, $a \in J$ and $b \in J$, but $x \notin J$. The open sets $U = (-\infty, x)$ and $V = (x, \infty)$ are a disconnection of J.

The converse is proven in Exercises 16.3.3 and 16.3.4. □

Compactness

Definition 16.3.6. Assume (X, d) is a metric space and $A \subseteq X$. An *open-cover* of A indexed by a set $\mathscr{I}$ is a cover $\{O_i\}_{i \in \mathscr{I}}$ of A by open subsets of (X, d).

A *subcover* of A from $\{O_i\}_{i \in \mathscr{I}}$ is a subcollection $\{O_i\}_{i \in \mathscr{F}}$ for some $\mathscr{F} \subseteq \mathscr{I}$ whose union contains A. If the set $\mathscr{F}$ is finite, we speak of a *finite subcover*.

Remark 16.3.7. In mathematics, the *red herring principle* states, "A red herring is in general neither 'red' nor a 'herring.'" The hyphenated noun *open-cover* reminds us that an "open cover" is not an "open" "cover," but a cover consisting of open sets. ◇

Example 16.3.8. In the number line, the sets $O_x = B_1(x) = (x - 1, x + 1)$ with radius 1 and arbitrary center comprise an open-cover $\{O_x\}_{x \in \mathbf{R}}$ indexed by real numbers. The subcollection $\{O_x\}_{x \in \mathbf{Z}}$ centered at integers is a subcover.

Generally, if $r > 0$, the collection $\{B_r(x)\}_{x \in A}$ of balls of radius r centered at points of A is an open-cover of A.   ◇

Example 16.3.9. In the number line, the sets $O_r = B_r(0) = (-r, r)$ of fixed center and arbitrary radius comprise an open-cover $\{O_r\}_{r>0}$ of $\mathbf{R}$. The subcollection $\{O_r\}_{r \in \mathbf{Z}^+}$ is a subcover.

Generally, if $x_0 \in X$, the collection $\{B_r(x_0)\}_{r>0}$ of open balls of radius r centered at x_0 is an open-cover of X: An arbitrary point x is contained in the ball of radius $1 + d(x_0, x)$.   ◇

Definition 16.3.10. Assume (X, d) is a metric space. A set $K \subseteq X$ is *compact* if every open-cover of K has a finite subcover.

Remark 16.3.11. The definition of compactness may be viewed as an adversarial game. The referee specifies a metric space (X, d) and the subset K. Player O picks an open-cover of K, a collection $\{O_i\}$ of open sets whose union contains K. Player S tries to find a finite subcover, finitely many sets *among the* $\{O_i\}$ whose union contains K. The set K is compact if and only if Player S has a winning strategy against a perfect opponent.   ◇

Remark 16.3.12. Compactness is topological by definition. Further, compactness is intrinsic, not relative: If $K \subseteq X \subseteq (X', d)$, then K is compact as a subspace of (X, d) if and only if K is compact as a subspace of (X', d).   ◇

Example 16.3.13. If (X, d) is an arbitrary metric space, then every finite subset of X is compact.

In a discrete space, inversely, every infinite set is non-compact: The singleton subsets constitute an open-cover with no finite subcover.   ◇

Example 16.3.14. In the number line, the open interval $(0, 1)$ is non-compact: The intervals $O_n = (2^{-(n+1)}, 1)$ form an open-cover $\{O_n\}_{n \in \mathbf{N}}$ having no finite subcover.

The number line is non-compact. For instance, if $O_n = (-n, n)$ for each positive integer n, the collection $\{O_n\}$ is an open-cover by finitude, but there is no finite subcover.   ◇

Proposition 16.3.15. *Assume (X, d) is a metric space, $K \subseteq X$ a compact set.*

(i) *The set K is totally bounded, hence bounded.*

(ii) *The set K is closed in X.*

(iii) *If $F \subseteq X$ closed, the intersection $K \cap F$ is compact.*

Proof. (i). Assume $r > 0$. The open-cover $\{B_r(x)\}_{x \in K}$ has a finite subcover by compactness of K. Since K is covered by finitely many r-balls for every r, K is totally bounded by definition.

(ii). Assume x_0 is an arbitrary point of $X \setminus K$. It suffices to show x_0 is interior to $X \setminus K$, or that $X \setminus K$ is open.

If $x \in K$, put $r_x = \frac{1}{2}d(x_0, x)$. The open balls $B_{r_x}(x_0)$ and $B_{r_x}(x)$ are disjoint, and the collection $\{B_{r_x}(x)\}_{x \in K}$ is an open-cover of K. Since K is compact, this covering has a finite subcover, say $\{B_{r_j}(x_j)\}_{j=0}^N$. The finite intersection $\bigcap_j B_{r_j}(x_0)$ is open, contains x_0, and is disjoint from the union $\bigcup_j B_{r_j}(x_j)$, hence disjoint from K.

(iii). Assume $\{O_i\}$ is an arbitrary open-cover of $K \cap F$. Since F is closed, the complement $X \setminus F$ is open. The collection $\{O_i\} \cup \{X \setminus F\}$ is an open-cover of K. By compactness of K, this new open-cover has a finite subcover. Deleting $X \setminus F$ if necessary, we have extracted a finite subcover of $K \cap F$ from $\{O_i\}$. $\quad\square$

Lemma 16.3.16 (Uniform radius). *Assume (X, d) is a compact metric space and $\{O_i\}_{i \in \mathscr{I}}$ an open-cover of X. There exists a positive real number r, independent of x, such that for every x in X, the ball $B_r(x)$ is contained in some open set O_i of the covering.*

Proof. For each x in X, there is a positive r_x such that $B_{2r_x}(x)$ is contained in some set of the covering. By compactness, there exist finitely many points $\{x_j\}_{j=0}^N$ with corresponding radii r_j such that the balls $\{B_{r_j}(x_j)\}_{j=0}^N$ cover X. (Note carefully: We choose r_x so that the balls of radius r_x cover X but each ball of radius $2r_x$ is contained in some covering set.) Let $r = \min\{r_j\}_{j=0}^N$.

If $x \in X$, then $x \in B_{r_j}(x_j)$ for some j. But by the triangle inequality, if $d(x', x) < r$, then $d(x', x_j) \leq d(x', x) + d(x, x_j) < r + r_j \leq 2r_j$. That is, $B_r(x) \subseteq B_{2r_j}(x_j)$, which is contained in some covering set. $\quad\square$

Theorem 16.3.17 (Finite intersection property). *Assume (X, d) is a metric space and $(K_n)_{n=0}^\infty$ a collection of non-empty compact sets such that $K_n \supseteq K_{n+1}$ for each n. The intersection $\bigcap_n K_n$ is empty if and only if $K_N = \varnothing$ for some N.*

Proof. See Exercise 16.3.6. $\quad\square$

Products of Compact Sets

Lemma 16.3.18. *Assume (X, d) and (X', d') are metric spaces and x' an arbitrary point of X'. If $K \subseteq X$ is compact, and if $\{O_i\}_{i \in \mathscr{I}}$ is an arbitrary open-cover of $K \times \{x'\}$ in the product metric, then there exists a positive r' and a finite subcover $\{O_j\}_{j=0}^N$ of $K \times B_{r'}(x')$.*

Proof. For each x in K, the point (x, x') is contained in O_i for some i in $\mathscr{I}$, so there exists a positive r_x such that $B_{r_x}^d(x) \times B_{r_x}^{d'}(x') \subseteq O_i$. The collection $\{B_{r_x}^d(x)\}_{x \in K}$ is an open-cover of K, so there exists a finite subcover, say $\{B_{r_j}^d(x_j)\}_{j=0}^N$. Put $r' = \min\{r_j : 0 \le j \le N\}$. Since $0 < r' \le r_j$ for each j, we have $B_{r'}^{d'}(x') \subseteq B_{r_j}^{d'}(x')$ for each j.

By construction, each of the finitely many products $B_{r_j}^d(x_j) \times B_{r'}^{d'}(x')$ is contained in some open set O_j from the original covering. The collection $\{O_j\}_{j=0}^N$ is a finite subcover of $K \times B_{r'}(x')$. $\qquad\square$

Proposition 16.3.19. *Assume (X, d) and (X', d') are metric spaces. If $K \subseteq X$ and $K' \subseteq X'$ are compact, then $K \times K'$ is compact in the product metric space.*

Proof. Assume $\{O_i\}_{i \in \mathscr{I}}$ is an arbitrary open-cover of $K \times K'$. By Lemma 16.3.18, for each x' in K', there exists a positive $r'(x')$ and a finite collection $\{O_j\}_{j=0}^N$ whose union contains $K \times \{B_{r'(x')}^{d'}(x')\}$.

The collection $\{B_{r'(x')}^{d'}(x')\}_{x' \in K'}$ is an open-cover of K'. Since K' is compact, there exists a finite subcover, say $\{B_{r_k'}^{d'}(x_k')\}_{k=0}^{N'}$. To each set in this finite subcover is associated a finite covering of $K \times B_{r_k'}^{d'}(x_k')$ from the collection $\{O_i\}_{i \in \mathscr{I}}$. This finite union of finite collections is a finite subcover of $K \times K'$. $\qquad\square$

Exercises for Section 16.3

Exercise 16.3.1. (★) True or false: An intersection of two connected sets is connected.

Exercise 16.3.2. (The hub lemma.) Let (X, d) be a metric space and x_0 a point of X. Assume $\{C_i\}_{i \in \mathscr{I}}$ is a collection of connected subsets of X, and that $x_0 \in C_i$ for all i in $\mathscr{I}$. Prove $\bigcup_{i \in \mathscr{I}} C_i \subseteq X$ is connected.

Exercise 16.3.3. If $a < b$, prove that the interval $[a, b]$ is connected in the number line. Suggestion: Use interval induction.

Exercise 16.3.4. (H) Prove that every interval (closed, open, or half-open; bounded or unbounded) is connected in the number line.

Exercise 16.3.5. Assume (X, d) is a metric space, and define a binary relation $\sim$ on X by $x \sim x'$ if and only if there exists a connected subset $A \subseteq X$ such that $\{x, x'\} \subseteq A$. Prove that $\sim$ is an equivalence relation. (The equivalence classes are called the (*connected*) *components* of (X, d).)

Exercise 16.3.6. (★) Prove the finite intersection property, Theorem 16.3.17.

Exercise 16.3.7. Assume (X, d) is a metric space and $(a_k)_{k \in \mathbf{N}}$ a convergent sequence in (X, d). Prove the set of terms $A = \{a_k\}_{k \in \mathbf{N}}$ has compact closure.

16.4 Characterizations of Compactness

Proving compactness from the definition involves handling arbitrary open-covers. In this section we prove a subset of flat n-space is compact if and only if it is closed and bounded. We also prove a metric space is compact if and only if every sequence has a convergent subsequence, if and only if the space is complete and totally bounded.

Theorem 16.4.1 (Closed intervals are compact). *Let $\left(\mathbf{R}, |\ |\right)$ be the number line. If $a < b$, the closed interval $[a, b]$ is compact.*

Proof. Fix an open-cover $\{O_i\}_{i \in \mathscr{I}}$ of $[a, b]$. We'll say a subset A of $[a, b]$ is *finitely covered* if A is contained in the union of finitely many of the O_i. We will proceed by interval induction, Theorem 4.2.14, using the set

$$J = \{t \text{ in } [a, b] : [a, t] \text{ is finitely covered}\} \subseteq [a, b].$$

(Priming). Since $a \in O_i$ for some i, $a \in J$.

(Climbing). If $t \in J$, then there exist finitely many sets $\{O_j\}_{j=0}^{n-1}$ such that $[a, t] \subseteq O_n' := \bigcup_{j=0}^{n-1} O_j$. Since O_n' is open, $B_r(t) \subseteq O_n' \subseteq J$ for some r.

(Capping). Since $\sup J \in [a, b]$, we have $\sup J \in O_k$ for some index k. Since O_k is open, there is an ε such that $B_\varepsilon(\sup J) \subseteq O_k$. By definition of a supremum, there is a t in J such that $(\sup J) - \varepsilon < t$. Since $[a, t]$ is finitely covered, appending O_k to a finite covering collection shows $[a, t']$ is finitely covered for all t' in $B_\varepsilon(\sup J) \cap [a, b]$. Particularly, $\sup J \in J$. $\qquad\square$

Remark 16.4.2. To be sure you understand, determine where the argument fails for the open interval (a, b) and for the half-open interval $[a, b)$, neither of which is compact. $\qquad\diamond$

Example 16.4.3. The ternary set, Example 4.1.19, is compact as a closed subset of the compact interval $[0, 1]$. $\qquad\diamond$

Theorem 16.4.4 (Closed, bounded theorem). *Let $(\mathbf{R}^n, d)$ be flat n-space. If $K \subseteq \mathbf{R}^n$, then the following are equivalent:*

(i) *K is closed and bounded.*

(ii) *K is compact.*

Proof. ((i) implies (ii)). If K is bounded, there exist a point x_0 of $\mathbf{R}^n$ and an $R > 0$ such that $K \subseteq B_R(x_0) \subseteq x_0 + [-R, R]^n$. But $x_0 + [-R, R]^n$ is a product of compact intervals, hence compact by Proposition 16.3.19 and induction on the number of factors. Proposition 16.3.15 implies K itself is compact as a closed subset of a compact set.

((ii) implies (i)). Immediate from Proposition 16.3.15. $\qquad\square$

Example 16.4.5. If x_0 is an arbitrary point of flat n-space, and if $r > 0$, the closed ball $\overline{B_r(x_0)}$ is closed and bounded, hence compact. Its boundary, the *sphere* $S_r^{n-1}(x_0) = \{x \text{ in } \mathbf{R}^n : d(x_0, x) = r\}$, is also closed and bounded, hence compact.   ◇

Example 16.4.6. Assume $n \geq 1$, and $(\mathbf{R}^n, d)$ is the n-dimensional vector space with the discrete metric. Every infinite set is closed and bounded, but not compact in this metric space.   ◇

Example 16.4.7. Assume $n \geq 1$, and $\mathbf{R}^n \setminus \{\mathbf{0}\}$ is flat n-space with the origin removed. The set $\{x : \|x\| \leq 1\}$ is closed and bounded, but not compact. In Theorem 16.4.4, "closed" means *closed in flat n-space*.   ◇

Sequential Compactness

In the number line, sequences detect compactness via existence of convergent subsequences. In this section we generalize to arbitrary metric spaces.

Definition 16.4.8. A metric space (X, d) is *sequentially compact* if every sequence in K has a subsequence convergent to a point of K.

Proposition 16.4.9. *If (K, d) is a sequentially compact metric space, then (K, d) is totally bounded.*

Proof. We'll argue contrapositively. If (K, d) is not totally bounded, there exists a positive r such that no finite collection of r-balls covers K. Pick x_0 in K arbitrarily. Inductively, use the fact that no finite collection of r-balls covers K to pick x_{k+1} in $K \setminus \bigcup_{j=0}^{k} B_r(x_j)$. By construction, $d(x_m, x_n) \geq r$ for all m and n, so the sequence $(x_n)_{n=0}^{\infty}$ has no convergent subsequence.   □

Corollary 16.4.10. *If (K, d) is sequentially compact, then (K, d) is separable.*

Proof. A totally bounded metric space is separable by Proposition 16.2.18.   □

Proposition 16.4.11. *If (X, d) is separable, then every open-cover $\{O_i\}_{i \in \mathscr{I}}$ of X has a countable subcover.*

Proof. Assume $A \subseteq X$ is a countable, dense set and consider the countable collection $\mathscr{N} := \{B_r(x_0)\}_{x_0 \in A, \ r \in \mathbf{Q}^+}$ of open balls centered at a point of A and having rational radius. Suppose $\{O_i\}_{i \in \mathscr{I}}$ is an arbitrary open-cover of X. For each x in X, there exists an index i in $\mathscr{I}$ such that $x \in O_i$. By Lemma 16.2.8, there exists a ball $B_r(x_0)$ in $\mathscr{N}$ such that $x \in B_r(x_0) \subseteq O_i$. The set of balls arising this way covers X, and is countable as a subset of $\mathscr{N}$, so can be written $\{B_j\}_{j=0}^{\infty}$. For each j, pick an index $i(j)$ in $\mathscr{I}$ such that $B_j \subseteq O_{i(j)}$. The collection $\{O_{i(j)}\}_{j=0}^{\infty}$ is a countable subcover.   □

Theorem 16.4.12 (Convergent subsequence theorem). *A metric space (K, d) is compact if and only if it is sequentially compact.*

Proof. (Compactness implies sequential compactness). Assume K is compact and $\mathbf{x} = (x_k)$ is a sequence in K. Since K is totally bounded, finitely many closed $1/2$-balls (diameter at most 1) cover K. At least one of these balls, $K_0 \subseteq K$ say, contains infinitely many terms of $\mathbf{x}$. Let x_{k_0} be a term of $\mathbf{x}$ in K_0.

Inductively, suppose K_m is a compact set of diameter at most 2^{-m} containing infinitely many terms of $\mathbf{x}$ and we have picked a term x_{k_m} in K_m. Use total boundedness to cover K_m with finitely many closed balls of radius $2^{-(m+2)}$. At least one of these balls, B_{m+1} say, contains infinitely many terms of $\mathbf{x}$. Pick an index k_{m+1} greater than k_m such that $x_{k_{m+1}} \in K_{m+1} := B_{m+1} \cap K_m$.

We have constructed a sequence of compact sets, $K_0 \supseteq K_1 \supseteq \ldots$ with the property that $\operatorname{diam}(K_m) \leq 2^{-m}$ for each m. The finite intersection property, Theorem 16.3.17, implies $K_\infty := \bigcap_m K_m$ is non-empty. For every natural number m, $\operatorname{diam} K_\infty \leq \operatorname{diam} K_m \leq 2^{-m}$. Thus $\operatorname{diam} K_\infty = 0$, so $K_\infty = \{x_\infty\}$ for some x_∞ in K. But $d(x_{n_m}, x_\infty) \leq 2^{-m}$ for each m, so $(x_{n_m}) \to x_\infty$.

(Sequential compactness implies compactness). Assume $\{O_i\}_{i \in \mathscr{I}}$ is an open-cover of K. By Proposition 16.4.9, K is separable. By Proposition 16.4.11, there exists a countable subcover $\{O_{i(j)}\}_{j=0}^{\infty}$ of $\{O_i\}_{i \in \mathscr{I}}$, and therefore a countable *nested* cover $O_n' := \bigcup_{j=0}^{n} O_{i(j)}$. It suffices to prove there is a natural number N such that $K \subseteq O_N'$, namely, the sets $\{O_{i(j)}\}_{j=0}^{N}$ cover K. But the closed sets $K_n = K \setminus O_n' \subseteq K$ are compact and nested inward, and their intersection is empty. By the finite intersection property, $K_N = \varnothing$ for some N, or $K \subseteq O_N'$. $\square$

Corollary 16.4.13. *A metric space (K, d) is compact if and only if it is complete and totally bounded.*

Proof. (Complete and totally bounded implies compact). Suppose (K, d) is complete and totally bounded, and (x_k) is an arbitrary sequence in K. The argument in the forward direction of the convergent subsequence theorem shows (x_k) has a *condensing* subsequence. Since (K, d) is complete, the subsequence converges. Thus every sequence in K has a convergent subsequence; that is, (K, d) is sequentially compact, hence compact by Theorem 16.4.12.

(Compact implies complete and totally bounded). By Proposition 16.3.15, a compact metric space is totally bounded. If (x_k) is a condensing sequence in K, Theorem 16.4.12 implies (x_k) has a convergent subsequence, which implies the sequence (x_k) itself converges. That is, (K, d) is complete. $\square$

Remark 16.4.14. In a topological space, compactness is usually defined by every open-cover having a finite subcover. Sequences, namely *countable* ordered lists, do not characterize compactness in this more general setting. $\diamond$

Exercises for Section 16.4

Exercise 16.4.1. (★) Assume $f : \mathbf{R}^n \to \mathbf{R}^m$ is continuous relative to the flat metrics. For each c in $\mathbf{R}^m$, the *level of f at c* is the preimage $f^*(\{c\})$. Prove that if the level at c is bounded, it is compact.

Exercise 16.4.2. (H) Assume (X, d) is a metric space and $C \subseteq X$ a closed subset. Prove there exists a continuous function $f : X \to \mathbf{R}$ whose zero set is precisely C.

Exercise 16.4.3. Which of the following subsets of the flat plane are compact?

(a) The unit circle $x^2 + y^2 = 1$.   (d) A rectangle $(a, b) \times (c, d)$.

(b) The unit hyperbola $x^2 - y^2 = 1$.   (e) The disk $\{(x, y) : x^2 + y^2 \le 1\}$.

(c) The parabola $y = x^2$.   (f) The polar graph $r = e^{-\theta^2}$.

Exercise 16.4.4. (★) Let n be a positive integer, $\mathbf{R}^{n \times n}$ the set of real $n \times n$ matrices, A^T the transpose of a matrix, and $\operatorname{tr} A$ the trace of a square matrix, the sum of the diagonal entries. Prove that the formula $\langle A, B \rangle = \operatorname{tr}(A^\mathsf{T} B)$ defines an inner product on $\mathbf{R}^{n \times n}$, which amounts to the standard inner product on $\mathbf{R}^{n^2}$.

Exercise 16.4.5. (★) Let n be a positive integer. Is the set $SL(n, \mathbf{R}) \subseteq \mathbf{R}^{n \times n}$ of $n \times n$ real matrices of determinant 1 compact? Closed? Use the metric defined by the inner product of Exercise 16.4.4.

Exercise 16.4.6. Let n be a positive integer. Recall that an $n \times n$ real matrix A is *orthogonal* if $A^\mathsf{T} A = I_n$, the identity matrix. Is the set $O(n) \subseteq \mathbf{R}^{n \times n}$ of $n \times n$ real orthogonal matrices compact? Use the metric defined by the inner product of Exercise 16.4.4.

Exercise 16.4.7. Equip $\mathbf{R}^\infty$ with the metric induced by the 2-norm. Is the closed unit ball compact?

Exercise 16.4.8. A set X of real numbers has *measure zero* if for every ε, there exist countably many intervals (a_j, b_j) covering X whose total length $\sum_j (b_j - a_j)$ is at most ε. Prove that:

(a) A countable set has measure zero. The ternary set has measure zero.

(b) A countable union of sets of measure zero has measure zero.

(c) A compact set of measure zero can be covered by *finitely many* intervals of arbitrarily small total length.

(d) If $a < b$ and (a_j, b_j) is a collection of open intervals whose union contains $[a, b]$, then finitely many of these intervals cover $[a, b]$. Conclude that $[a, b]$ does not have measure zero.

Exercise 16.4.9. (H) Assume $X \subseteq \mathbf{R}$ is a set of real numbers and f is a real-valued function on X. For each c in X, define functions on the set of positive reals by

$$U_c f(\delta) = \sup\{f(x) : x \in X, |x - c| < \delta\},$$
$$L_c f(\delta) = \inf\{f(x) : x \in X, |x - c| < \delta\}.$$

(a) Prove $U_c f$ is non-decreasing and bounded above, and $L_c f$ is non-increasing and bounded below.

(b) The quantity

$$\operatorname{osc}_c f := \lim_{\delta \to 0^+} \left(U_c f(\delta) - L_c f(\delta) \right)$$

is the *oscillation* of f at c. Prove that $\operatorname{osc}_c f \geq 0$, with equality if and only if f is continuous at c.

(c) (H) If $r > 0$, prove the set $D_r := \{c \text{ in } I : \operatorname{osc}_c f \geq r\}$ is closed in I.

(d) Calculate the oscillation at each point for $\chi_{\mathbf{Q}}$ and for the denominator function of Exercise 8.1.5.

Exercise 16.4.10. (H) Assume $a < b$. Prove a function f on $[a, b]$ is integrable if and only if f is bounded and the set D of discontinuities of f has measure zero.

16.5 Properties of Continuous Mappings

Now that we have found a suitable abstract characterization of closed, bounded sets, the extreme value theorem generalizes to arbitrary metric spaces with gratifying simplicity.

Continuous Images of Compact Spaces

Theorem 16.5.1 (The compact image theorem). *Assume (X, d) and (Y, e) are metric spaces, and $f : X \to Y$ is continuous. If $K \subseteq X$ is compact, the image $f(K)$ is compact.*

Proof. Assume $\{O_i\}_{i \in \mathscr{I}}$ is an arbitrary open-cover of $f(K)$. Since f is continuous, each preimage $f^*(O_i)$ is open in X by Proposition 16.1.38. The collection

$\{f^*(O_i)\}_{i \in \mathscr{I}}$ is an open-cover of K, so by compactness there exists a finite subcover $\{f^*(O_{i(j)})\}_{j=0}^N$. The images $\{O_{i(j)}\}_{j=0}^N$ are a finite subcover of the image $f(K)$. $\qquad\square$

Corollary 16.5.2 (The extreme value theorem). *Suppose (X, d) is a metric space and $f : X \to \mathbf{R}$ is continuous with respect to the flat metric on $\mathbf{R}$. If $K \subseteq X$ is compact, then f achieves minimum and maximum values on K. That is, there exist elements $x_{\min}$ and $x_{\max}$ in K such that $f(x_{\min}) \le f(x) \le f(x_{\max})$ for all x in K.*

Proof. By Theorem 16.5.1, the image $f(K) \subseteq \mathbf{R}$ is compact, hence closed and bounded in the number line. But a closed, bounded set of real numbers contains its supremum $y_{\max}$ and infimum $y_{\min}$. Since these are in the image of f, there exist elements $x_{\min}$ and $x_{\max}$ in K such that $y_{\min} = f(x_{\min})$ and $y_{\max} = f(x_{\max})$. $\qquad\square$

Proposition 16.5.3. *If V is a finite-dimensional real vector space, then any two norms on V are equivalent.*

Proof. Exercise 16.5.8. $\qquad\square$

Definition 16.5.4. Assume (X, d) and (Y, e) are metric spaces. A bijective mapping $f : X \to Y$ is a *homeomorphism* if f is continuous and the inverse mapping $f^{-1} : Y \to X$ is continuous.

Proposition 16.5.5. *Assume (X, d) and (Y, e) are metric spaces, and that $f : X \to Y$ is continuous. If $K \subseteq X$ is compact and f is injective on K, then f is a homeomorphism from (K, d) to its image $\left(f(K), e\right)$.*

Proof. It suffices to prove that the preimage of an arbitrary closed set in (X, d) under f^{-1} is closed in (Y, e), namely, that f maps closed subsets of K to closed subsets of $f(K)$. But a closed subset of a compact set is compact, so its image is compact, hence closed. $\qquad\square$

Example 16.5.6. The formula $f(t) = (\cos t, \sin t)$ defines a continuous bijection from the (non-compact) half-open interval $[0, 2\pi)$ with the flat metric to the (compact) unit circle in the flat plane. The inverse mapping is discontinuous at $(1, 0)$, so f is not a homeomorphism.

This does not contradict Proposition 16.5.5 because the half-open interval is not compact. $\qquad\diamond$

Continuous Images of Connected Spaces

Theorem 16.5.7 (The connected image theorem). *Assume (X, d) and (Y, e) are metric spaces, $f : X \to Y$ continuous. If $A \subseteq X$ is connected, the image $f(A)$ is connected.*

Proof. We prove the contrapositive. If $\{U, V\}$ is a disconnection of $f(A)$, the preimages $\{f^*(U), f^*(V)\}$ are disjoint, have non-empty intersections with A, and their union is A; that is, they constitute a disconnection of A. $\quad\square$

Corollary 16.5.8 (The intermediate value theorem). *Suppose (X, d) is a metric space and $f : X \to \mathbf{R}$ is continuous with respect to the flat metric on $\mathbf{R}$. If $A \subseteq X$ is connected, then $f(A)$ is an interval.*

Particularly, if a and b are elements of A, and if y is a real number satisfying $f(a) < y < f(b)$, there exists an x in A such that $y = f(x)$.

Proof. By Proposition 16.3.5, connected subsets of the number line are precisely intervals. $\quad\square$

Uniform Continuity

Uniform continuity generalizes immediately to metric spaces.

Definition 16.5.9. Assume (X, d) and (Y, e) are metric spaces, $f : X \to Y$ a mapping. We say f is *uniformly continuous* on X if the following condition holds:

For every ε, there exists a δ such that for all x and x' in X, $d(x, x') < \delta$ implies $e\big(f(x), f(x')\big) < \varepsilon$.

Proposition 16.5.10. *Assume (X, d) is a compact metric space, (Y, e) a metric space. A continuous mapping $f : X \to Y$ is uniformly continuous.*

Proof. A contrapositive proof can be given along the lines in Chapter 8. Here we give a direct proof using the uniform radius lemma, Lemma 16.3.16.

Fix ε arbitrarily. For each x in X, use continuity to pick δ_x such that if $d(x, x') < \delta_x$, then $e\big(f(x), f(x')\big) < \varepsilon/2$. The collection $\mathcal{N} = \{B_{\delta_x}(x)\}_{x \in X}$ is an open-cover of X. By the uniform radius lemma, there exists a δ such that for every x in X, the ball $B_\delta(x)$ is contained in some element of $\mathcal{N}$.

If x and x' are arbitrary elements of X such that $d(x, x') < \delta$, then $x' \in B_\delta(x)$, which is contained in some some element $B_{\delta_j}(x_j)$ of $\mathcal{N}$. That is, $d(x, x_j) < \delta_j$ and $d(x_j, x') < \delta_j$. If we write $y = f(x)$, $y' = f(x')$, and $y_j = f(x_j)$ for simplicity, the triangle inequality gives

$$e(y, y') \leq e(y, y_j) + e(y_j, y') < \varepsilon/2 + \varepsilon/2 = \varepsilon. \qquad \square$$

Bounded stretch for real-valued functions generalizes immediately to metric spaces, and implies uniform continuity.

Definition 16.5.11. Assume (X, d) and (Y, e) are metric spaces, and that $f : X \to Y$ is a mapping. We say f has *bounded stretch* if there exists a real M

such that $e\big(f(x), f(x')\big) \le M d(x, x')$ for all x and x' in X. The infimum of all such M is called the *stretch* of f.

We say f has *locally bounded stretch* if for every x_0 in X, there is an open ball $B_r(x_0)$ such that f has bounded stretch in $B_r(x_0)$.

Isometry

Definition 16.5.12. Assume (X, d) and (Y, e) are metric spaces. We say $f : X \to Y$ is *distance-preserving* if $e\big(f(x), f(x')\big) = d(x, x')$ for all x, x' in X.

A distance-preserving *surjection* $i : (X, d) \to (Y, e)$ is an *isometry*. If there exists an isometry $i : (X, d) \to (Y, e)$, we say (X, d) and (Y, e) are *isometric*.

Remark 16.5.13. The noun *isometry* is usually accented on the second syllable. The adjective *isometric* is usually accented on the third.   ◇

Remark 16.5.14. An isometry and its inverse both have stretch 1. Isometric spaces are "the same" in regard to properties of metric spaces.   ◇

Proposition 16.5.15. *Assume (X, d) is a metric space. The set of isometries of (X, d) is a group under composition, namely: A composition of isometries is an isometry, every isometry is invertible as a mapping, and the inverse mapping is an isometry.*

Proof. See Exercise 16.5.11.   □

Remark 16.5.16. An isometry is uniformly continuous, hence a homeomorphism by Proposition 16.5.15.

The set of isometries of (X, d), viewed as a group under mapping composition, called the *isometry group* of (X, d).   ◇

Example 16.5.17. If $b > 0$, the mapping $f : (0, \infty) \to (0, \infty)$ defined by $f(x) = b + x$ is distance-preserving and an isometry onto its image, but not surjective, so not an isometry of $(0, \infty)$.   ◇

Theorem 16.5.18 (The rigid motion theorem). *A mapping $f : \mathbf{R}^n \to \mathbf{R}^n$ is an isometry of flat n-space if and only if there exist a real orthogonal $n \times n$ matrix A and a $\mathbf{b}$ in $\mathbf{R}^n$ such that $f(\mathbf{x}) = A\mathbf{x} + \mathbf{b}$ for all x in $\mathbf{R}^n$.*

Proof. See Exercise 16.5.12.   □

Corollary 16.5.19. *Let $O(n)$ denote the set of $n \times n$ real orthogonal matrices. The isometry group of flat n-space is the set $O(n) \times \mathbf{R}^n$, equipped with the binary operation $(A', \mathbf{b}') \circ (A, \mathbf{b}) = (A'A, A'\mathbf{b} + \mathbf{b}')$.*

Proof. For all $\mathbf{x}$ in $\mathbf{R}^n$, $A'(A\mathbf{x} + \mathbf{b}) + \mathbf{b}' = A'A\mathbf{x} + (A'\mathbf{b} + \mathbf{b}')$.   □

Contraction

Definition 16.5.20. Assume (X, d) is a metric space. A mapping $T : X \to X$ is a *contraction* if T has stretch λ for some $\lambda < 1$.

Remark 16.5.21. A contraction moves points closer by fixed ratio $\lambda < 1$. It is *not* enough that $d\big(T(x), T(x')\big) < d(x, x')$ for all x, x' in X. $\diamond$

Definition 16.5.22. If X is a set and $T : X \to X$ a mapping, a *fixed point* of T is an element x of X such that $T(x) = x$.

Theorem 16.5.23 (The contraction theorem, cf. Exercise 8.6.1). *Assume (X, d) is a complete metric space. If $T : X \to X$ is a contraction, then T has a unique fixed point x_∞ in X.*

Proof. Pick x_0 in X arbitrarily and recursively define a sequence $(x_k)_{k=0}^\infty$ by $x_{k+1} = T(x_k)$. Set $d_0 = d(x_0, x_1)$. Since T is a contraction,

$$d(x_k, x_{k+1}) = d\big(T(x_{k-1}), T(x_k)\big) \leq \lambda\, d(x_{k-1}, x_k)$$

if $k \geq 1$. Induction on k shows $d(x_k, x_{k+1}) \leq \lambda^k d_0$.

Fix ε arbitrarily. Since $0 \leq \lambda < 1$, there exists a positive integer N such that $\lambda^N d_0 / (1 - \lambda) < \varepsilon$. If m and n are integers such that $N \leq m < n$, the triangle inequality gives

$$d(x_m, x_n) \leq \sum_{k=m}^{n-1} d(x_k, x_{k+1}) \leq \sum_{k=m}^{n-1} \lambda^k d_0 \leq \sum_{k=m}^{\infty} \lambda^k d_0 = \frac{\lambda^m d_0}{1 - \lambda} < \varepsilon.$$

Thus (x_k) is a condensing sequence. Since (X, d), is complete, (x_k) converges to some point x_∞ of X. Since T is continuous and $x_{k+1} = T(x_k)$ for each k,

$$T(x_\infty) = \lim_{k \to \infty} T(x_k) = \lim_{k \to \infty} x_{k+1} = x_\infty.$$

Finally, a contraction has at most one fixed point: If $T(x') = x'$, then

$$d(x, x') = d\big(T(x), T(x')\big) \leq \lambda d(x, x').$$

Since $\lambda < 1$, we have $d(x, x') = 0$, or $x = x'$. $\square$

Exercises for Section 16.5

Exercise 16.5.1. ($\bigstar$) Assume n is a positive integer, $\ell > 0$, and $f : \mathbf{R} \to \mathbf{R}^n$ an ℓ-periodic continuous mapping. Prove the image $f(\mathbf{R}) \subseteq \mathbf{R}^n$ is compact.

Exercise 16.5.2. Let $A \subseteq \mathbf{R}^2$ be image of the mapping $f : (-\pi, \pi) \to \mathbf{R}^2$ defined by $f(t) = (\sin t, \sin 2t)$. Determine whether A is compact.

Exercise 16.5.3. Assume (X, d) is a metric space, $A \subseteq X$ non-empty. Define $f : X \to \mathbf{R}$ by $f(x) = d(x, A)$. Prove f is uniformly continuous on X, and $f(x) = 0$ if and only if $x \in \overline{A}$.

Exercise 16.5.4. Assume (X, d) is a metric space, and A, B are disjoint, closed, non-empty subsets of X.

(a) If A is compact, prove $d(A, B) > 0$.

(b) Assume $f : A \cup B \to \mathbf{R}$ is a function whose restrictions $f|_A$ and $f|_B$ are uniformly continuous. Must f be uniformly continuous? Does it matter whether or not A is compact?

Exercise 16.5.5. Assume (X, d) and (X, d') are metric spaces with the same underlying set.

(a) Prove that d and d' are equivalent if and only if the identity mapping $i : (X, d) \to (X, d')$ is a homeomorphism (i.e., continuous in both directions).

(b) Assume d and d' are equivalent and (X, d) is compact. If $f : (X, d) \to (Y, e)$ is uniformly continuous, prove f is uniformly continuous when viewed as a mapping from (X, d') to (Y, e).

(c) In part (b), can the compactness hypothesis be omitted?

Exercise 16.5.6. Assume (X, d), (Y, e) are metric spaces, and $f : X \to Y$ a mapping.

(a) Prove that if f is uniformly continuous and (x_k) is a condensing sequence in (X, d), then the image sequence $\big(f(x_k)\big)$ in (Y, e) is condensing.

(b) Weaken the hypothesis in (a) to continuity of f, or find a continuous mapping that maps some condensing sequence to a non-condensing sequence.

(c) Suppose conversely that f maps condensing sequences to condensing sequences. Prove f is continuous. Caution: In general, a condensing sequence in Y need not converge.

(d) Strengthen the conclusion in (c) to uniform continuity, or find a continuous mapping that is not uniformly continuous, yet maps condensing sequences to condensing sequences.

Exercise 16.5.7. ($\bigstar$) Assume $I = [a, b]$ is an interval of real numbers and $\mathscr{C}^\infty(I)$ is equipped with the metric $d(f, g) = \|f - g\|_\infty$.

(a) Prove the definite integration operator $I : V \to V$, $I(f)(x) = \int_a^x f(t)\,dt$, has stretch $(b - a)$.

(b) Prove that the derivative operator $D : V \to V$, $Df = f'$, is discontinuous. Hint: A function of small absolute value can have arbitrarily large slopes.

Exercise 16.5.8. (H) Prove Proposition 16.5.3.

Exercise 16.5.9. (H) Assume $(V, \langle\,,\,\rangle)$ is a real inner product space, viewed as a metric space with the 2-norm. Prove:

(a) For each $\mathbf{v}$ in V, the linear function $\lambda_{\mathbf{v}} : V \to \mathbf{R}$ defined by $\lambda_{\mathbf{v}}(\mathbf{u}) = \langle \mathbf{v}, \mathbf{u} \rangle$ is continuous.

(b) If $U \subseteq V$ is a vector subspace, its *orthogonal complement*

$$U^{\perp} = \{\mathbf{v} \text{ in } V : \langle \mathbf{u}, \mathbf{v} \rangle = 0 \text{ for all } \mathbf{u} \text{ in } U\}$$

is a closed subspace.

(c) (H) If U is a closed subspace and V is complete with respect to the 2-norm, then $(U^{\perp})^{\perp} = U$.

Exercise 16.5.10. (H) Assume g is continuous on $[-1, 1]$, and that for all $\mathscr{C}^1$ functions ϕ satisfying $\phi(-1) = \phi(1) = 0$, we have $\int_{-1}^{1} g\phi' = 0$. Prove g is constant.

Exercise 16.5.11. (★) Prove Proposition 16.5.15.

Exercise 16.5.12. (H) Prove Theorem 16.5.18.

Exercise 16.5.13. Assume (X, d) is a metric space. We say (X, d) is *path-connected* if for all x, x' in X, there exists a continuous mapping $c : [0, 1] \to X$ such that $c(0) = x$ and $c(1) = x'$. (That is, x and x' can be *joined by a path in X*.)

(a) Prove that a path-connected space is connected.

(b) Define a binary relation on X by $x \sim x'$ if and only if x and x' can be joined by a path in X. Prove $\sim$ is an equivalence relation. (The equivalence classes are the *path components* of (X, d).)

(c) Prove that an *open* connected subset X of flat n-space $(\mathbf{R}^n, d)$ is path-connected. Hint: If x can be joined to x' by a path in X, there exists an open ball about x' whose points can be joined to x by a path in X.

(d) Prove that the set $X = \big(\{0\} \times [-1, 1]\big) \cup \{(x, \sin(1/x)) : 0 < x\}$ is connected but not path-connected in the flat plane.

Exercise 16.5.14. (H) In this exercise, we construct a metric on a particular quotient of the number line, $\mathbf{R}/(2\pi\mathbf{Z})$, obtaining the simplest non-trivial "compact manifold" and a motivating example of a "covering space."

Let $S^1 \subseteq \mathbf{R}^2$ denote the unit circle, viewed as a metric space (S^1, d) in the flat plane. Define a relation $\sim$ on the number line by $t \sim t'$ if and only if $t - t' \in 2\pi\mathbf{Z}$, and define a mapping $f : \mathbf{R} \to S^1$ by $f(t) = (\cos t, \sin t)$.

(a) Prove $\sim$ is an equivalence relation, and f factors through the quotient.

(b) (H) If d_1 denotes the number line distance, prove that

$$\bar{d}([t], [t']) = d_1(t + 2\pi\mathbf{Z}, t' + 2\pi\mathbf{Z})) \quad \text{for all } [t] \text{ and } [t']$$

defines a metric on the quotient $\mathbf{R}/\sim$, the *angular separation* metric.

(c) Is the induced mapping $\bar{f} : (\mathbf{R}/\sim, \bar{d}) \to (S^1, d)$ a homeomorphism? An isometry?

Exercise 16.5.15. (H) Let Λ (Lambda, for *lattice*) denote the set $2\pi(\mathbf{Z} \times \mathbf{Z})$ in the flat plane. Define an equivalence relation on $\mathbf{R}^2$ by $(s, t) \sim (s', t')$ if and only if $(s' - s, t' - t) \in \Lambda$. (Check this *is* an equivalence relation if the claim is not apparent, paying attention to the properties of Λ that correspond to reflexivity, symmetry, and transitivity.)

Let $S^1 \subseteq \mathbf{R}^2$ denote the unit circle equipped with the angular separation metric of Exercise 16.5.14 (b).

(a) Define $f : \mathbf{R}^2 \to S^1 \times S^1$ by $f(s, t) = (\cos s, \sin s, \cos t, \sin t)$. Prove there is an induced mapping $\bar{f} : (\mathbf{R}^2/\sim) \to S^1 \times S^1$, namely, that f factors through the quotient.

Prove that

$$\bar{d}\big([(s, t)], [(s', t')]\big) = d\big((s, t) + \Lambda, (s', t') + \Lambda\big)$$

defines a metric on the quotient $\mathbf{R}^2/\sim$, and the induced mapping is an isometry. (Thus, our use of $\bar{d}$ to denote both is practically unambiguous.) The metric space $(S^1 \times S^1, \bar{d})$ is called the *flat Λ-torus*, or (here) simply "the torus."

(b) For each real α, let $\ell_\alpha = \{(s, t) \text{ in } \mathbf{R}^2 : t = \alpha s\}$ denote the line of slope α through the origin. Prove $f^*\big(f(\ell_\alpha)\big) = \ell_\alpha + \Lambda$, the set of translates of ℓ_α by elements of Λ.

(c) Prove $f(\ell_\alpha)$ is compact if and only if α is rational, and is dense in the torus if and only if α is irrational. (If α is irrational, the image $f(\ell_\alpha)$ is called an *irrational winding* on the torus.)

(d) Prove that the complement of an irrational winding is connected but has *uncountably many* path components, see Exercise 16.5.13.

17

Approximation Theorems

Metric spaces provide a conceptual framework for studying "limits of well-understood objects." In this chapter, the language and machinery of sequences is applied toward approximation problems: embedding an arbitrary metric space as a dense subset of a complete metric space, approximating continuous functions using polynomials and periodic functions with trigonometric polynomials, solving first-order ordinary differential equations. This small selection only hints at the variety of applications illuminated by metric spaces.

17.1 Completion of a Metric Space

Rational numbers, ratios of integers, are as simple and concrete as the ordered field axioms. Unfortunately, the rational number system is not complete: a condensing sequence of rationals need not converge in the rationals. In Chapter 3 we started from axioms for the real number system that included completeness. (In the axioms, completeness referred to existence of suprema, which is equivalent to convergence of arbitrary condensing sequences.) Now, in the final chapter, we'll use condensing sequences to prove an arbitrary metric space has a completion, unique up to isometry. As a special case, we construct the real number system as a completion of the rationals.

Definition 17.1.1. Assume (X, d) is a metric space. A *completion* of (X, d) is a complete metric space $(\overline{X}, \overline{d})$ together with a distance-preserving mapping $i : X \to \overline{X}$ whose image is dense.

Theorem 17.1.2 (The completion theorem). *Assume (X, d) is a metric space. There exists a completion $(\overline{X}, \overline{d})$. Moreover, any two completions of (X, d) are isometric.*

Remark 17.1.3. Despite the sweeping generality, the conceptual strategy is straightforward. Loosely, we'll represent "limits from X" as condensing sequences (x_k) from X. Precisely, we'll view two condensing sequences (x_k) and (x'_k) as equivalent if $d(x_k, x'_k) \to 0$, and we'll define points of $\overline{X}$ to be equivalence classes of condensing sequences. The key point is, "a condensing

DOI: 10.1201/9781003601357-17

sequence of condensing sequences is represented by a condensing sequence," or a condensing sequence in $\overline{X}$ converges to a point of $\overline{X}$. The approximation idiom is a "diagonal sequence."

A sequence in X is a mapping $x : \mathbf{N} \to X$. A sequence *of sequences in* X is, in effect, a mapping $\mathbf{x} : \mathbf{N} \times \mathbf{N} \to X$. For each n, we get a sequence $\mathbf{x}_n = \mathbf{x}(n, \cdot)$ by letting the second variable run from 0 to ∞. The *diagonal sequence* has kth term $\overline{x}_k = \mathbf{x}(k, k)$, the kth term of the kth sequence.   ◇

Proposition 17.1.4 (Existence of a completion). *Let CX denote the set of all condensing sequences in (X, d). Define a relation $\equiv$ on CX by $(x_k) \equiv (x'_k)$ if and only if $\lim d(x_k, x'_k) = 0$.*

(i) *If (x_k) and (y_k) are in CX, then $d\big((x_k), (y_k)\big) = \lim d(x_k, y_k)$ exists as a real number. The relation $\equiv$ is an equivalence in CX and the distance function d on CX is well-defined mod $\equiv$, namely, if $(x_k) \equiv (x'_k)$ and $(y_k) \equiv (y'_k)$, then $\lim d(x_k, y_k) = \lim d(x'_k, y'_k)$.*

(ii) *If $\overline{X} = CX/\equiv$ is the set of equivalence classes, and if $[x_k]$ denotes the equivalence class of a condensing sequence (x_k), then the function $\overline{d}\big([x_k], [y_k]\big) = \lim d(x_k, y_k)$ is a complete metric on $\overline{X}$.*

(iii) *If $x \in X$ and (x) denotes the constant sequence, the mapping $i : X \to \overline{X}$ defined by $i(x) = (x)$ is an isometry with dense image.*

Proof. Exercise 17.1.3.   □

Proposition 17.1.5 (Uniqueness of a completion). *Assume (X, d) is a metric space and $(\overline{X}, \overline{d})$ is a complete metric space such that $X \subseteq \overline{X}$ is dense.*

(i) *If (Y, e) is a complete metric space and $f : (X, d) \to (Y, e)$ is uniformly continuous, there exists a unique continuous extension $\overline{f} : \overline{X} \to Y$ of f.*

(ii) *If $(\overline{X}', \overline{d}')$ is an arbitrary completion of (X, d), then there exists a unique isometry $i : (\overline{X}, \overline{d}) \to (\overline{X}', \overline{d}')$ acting as the identity mapping on X.*

Proof. Exercise 17.1.4.   □

Example 17.1.6. If $X = \mathscr{C}\big([a, b]\big)$ equipped with the (incomplete) metric induced by the 1-norm,

$$d(f, g) = \int_a^b |f(x) - g(x)| \, dx,$$

the completion is denoted $L^1\big([a, b]\big)$. Elements of this space are equivalence classes of limits of integrable functions. The integral, viewed as a real-valued function on X, extends to an integral built on "measurability," a generalized theory of length for subsets of the number line. Instead of using indicators of intervals, the extended integral uses indicators of measurable sets.   ◇

To construct the real number system, namely, the ordered field structure on the completion of $(\mathbf{Q}, |\ |)$, we must further show that sums, products, and ordering pass from the rational numbers to equivalence classes of condensing sequences. This amounts to material from Chapter 6.

Corollary 17.1.7 (Construction of the real number system). *Let $(\mathbf{Q}, d)$ be the ordered rational field equipped with the number line metric, and let $(\mathbf{R}, |\ |)$ denote the completion of Theorem 17.1.2.*

(i) *Addition and multiplication are well-defined mod $\equiv$, and induce field operations on $\mathbf{R}$.*

(ii) *Ordering is well-defined mod $\equiv$, and defines a subset $P \subseteq \mathbf{R}$ satisfying the order axioms.*

(iii) *If $A \subseteq \mathbf{R}$ is non-empty and bounded above in $\mathbf{R}$, then A has a supremum in $\mathbf{R}$.*

Proof. See Exercise 17.1.5. □

Continuous Extension

Corollary 17.1.8. *Assume (X, d) is a metric space, $A \subseteq X$ a set with compact closure, and (Y, e) complete. A mapping $f : (A, d) \rightarrow (Y, e)$ is uniformly continuous if and only if there exists a continuous extension $\overline{f} : \overline{A} \rightarrow Y$.*

Proof. If $f : (A, d) \rightarrow (Y, e)$ is uniformly continuous, part (i) of Proposition 17.1.5 guarantees there exists a continuous extension $\overline{f} : \overline{A} \rightarrow Y$.

Conversely, if there exists a continuous extension $\overline{f} : \overline{A} \rightarrow Y$ of f, the extension is uniformly continuous by Proposition 16.5.10 because $\overline{A}$ is compact, so the restriction f is *a fortiori* uniformly continuous. □

Example 17.1.9. The signum function $\operatorname{sgn}(x) = x/|x|$ on $B_1^\times(0)$ does not extend continuously to $B_1(0)$, and therefore is not uniformly continuous on the punctured unit ball. The same is true of $f(x) = \sin(1/x)$. ◊

Example 17.1.10. If $0 < \delta < 1$, the signum function on $[-1, 1] \setminus [-\delta, \delta]$ does extend continuously to the closure $[-1, -\delta] \cup [\delta, 1]$, and therefore is uniformly continuous on the complement of $B_\delta(0)$ in the unit ball. The same is true of $f(x) = \sin(1/x)$. ◊

Example 17.1.11. If $0 < r < 1$, the power function $f(x) = x^r$ is uniformly continuous on $(0, 1)$. Note that f' is unbounded on this interval. ◊

Example 17.1.12. The polar angle function θ defined on the slit plane $\mathbf{R}^2 \setminus (-\infty, 0]$ is not uniformly continuous: There is a jump discontinuity across the negative x-axis, hence no continuous extension to the negative axis. ◊

Exercises for Section 17.1

Exercise 17.1.1. Assume I is a bounded real interval and f a bounded monotone function on I. Prove f is uniformly continuous. If f is strictly monotone, can we drop the hypothesis I is an interval?

Exercise 17.1.2. Define $f : (0,1)^2 \to \mathbf{R}$ by $f(x,y) = x/(x+y)$. Prove f is bounded, and is uniformly continuous in each variable (give a careful definition for a function defined on an open rectangle), but not uniformly continuous.

Exercise 17.1.3. (H) Prove Proposition 17.1.4.

Exercise 17.1.4. (H) Prove Proposition 17.1.5.

Exercise 17.1.5. (H) Prove Corollary 17.1.7.

Exercise 17.1.6. Assume (X,d) is a metric space, and fix a point x_0 in X arbitrarily. For each x' in X, define a function $f_{x'} : X \to \mathbf{R}$ by

$$f_{x'}(x) = d(x,x') - d(x,x_0) \quad \text{for all } x \text{ in } X.$$

Prove $f_{x'}$ is uniformly continuous for each x', that $|f_{x'}(x)| \le d(x',x_0)$ for all x in X, and that $\|f_{x''} - f_{x'}\|_\infty = d(x',x'')$ for all x' and x'' in X.

(If we equip $\mathscr{C}(X,\mathbf{R})$ with the metric $d(f,g) = \sup|f(x) - g(x)|$, then the mapping $\Phi : X \to \mathscr{C}(X,\mathbf{R})$ defined by $\Phi(x') = f_{x'}$ is an isometry onto its image. The closure of the image is therefore a completion of (X,d).)

17.2 The Uniform Metric

Metrics on function spaces are nearly as varied as approximation problems themselves. In this section we'll study convergence in the "uniform metric" generalizing the ∞-norm. This metric avoids technical complications that occur with most metrics on function spaces. Particularly, convergent sequences in the uniform metric converge pointwise, so limits are mappings rather than equivalence classes of mappings. Separately, a uniform limit of continuous mappings is continuous.

Definition 17.2.1. Assume (X,d) and (Y,e) are metric spaces, and $\mathscr{B}(X,Y)$ is the set of bounded mappings $f : X \to Y$, namely, mappings whose image is bounded. The *uniform metric* is defined, for f and g in $\mathscr{B}(X,Y)$, by

$$d_\infty(f,g) = \sup_{x \in X} e\big(f(x), g(x)\big).$$

If (X, d) is compact, the uniform metric defines a metric on $\mathscr{C}(X, Y)$, the set of continuous mappings $f : X \to Y$.

Remark 17.2.2. The elements (or "points") of $\mathscr{B}(X, Y)$ are *mappings*. In the uniform metric, the distance between two mappings f and g is the supremum of the distances between the elements $f(x)$ and $g(x)$ in Y. To say "$d_\infty(f, g) \le \varepsilon$" is to say $e\big(f(x), g(x)\big) \le \varepsilon$ for all x in X. ◇

Lemma 17.2.3 (Fundamental idiom of uniform approximation). *If f and g are bounded mappings from (X, d) to (Y, e), then*

$$e\big(f(x), f(x')\big) \le 2 d_\infty(f, g) + e\big((g(x), g(x')\big) \quad \text{for all } x, \, x' \text{ in } X.$$

Proof. Consider the chain of points $f(x)$, $g(x)$, $g(x')$, $f(x')$. Apply the triangle inequality to the ends, noting that $e\big(f(x), g(x)\big)\big) \le d_\infty(f, g)$ for all x. □

Remark 17.2.4. Assume (X, d) and (Y, e) are metric spaces, $\mathscr{X} = \mathscr{B}(X, Y)$ equipped with the uniform metric d_∞. Classically, a condensing sequence (f_k) in $(\mathscr{X}, d_\infty)$ is said to be *uniformly condensing* on X. Similarly, a convergent sequence in $(\mathscr{X}, d_\infty)$ is *uniformly convergent* on X.

If (f_k) is condensing in $(\mathscr{X}, d_\infty)$, then for each x in X, $\big(f_k(x)\big)$ is a condensing sequence in (Y, e). "Uniform" refers to "at the same rate for all x": For every ε, there exists an N *independent of x* such that if $N \le k < n$, then $e\big(f_k(x), f_n(x)\big) < \varepsilon$ for all x in X. ◇

Proposition 17.2.5. *Suppose (X, d) is a metric space and (Y, e) is complete.*

(i) *The metric space $\big(\mathscr{B}(X, Y), d_\infty\big)$ is complete.*

(ii) *The set $\mathscr{C}(X, Y) \cap \mathscr{B}(X, Y)$ of bounded continuous mappings is a closed subset. Particularly, if (X, d) is compact, then $\mathscr{C}(X, Y)$ is closed.*

Proof. Assume (f_k) is a condensing sequence in $\big(\mathscr{B}(X, Y), d_\infty\big)$. Claim (i) says there exists a bounded mapping f in $\mathscr{B}(X, Y)$ such that $(f_k) \to f$. Claim (ii) says if each f_k is continuous and bounded, then f is continuous and bounded. That is, $\mathscr{C}(X, Y) \cap \mathscr{B}(X, Y)$ contains all its limit points.

(Existence of the pointwise limit). For each x in X, the sequence $\big(f_k(x)\big)$ in Y is condensing. Since (Y, e) is complete, $\big(f_k(x)\big) \to y$ for some y in Y. Let $f : X \to Y$ be the mapping defined by $y = f(x)$ for each x in X.

(Convergence). Although $(f_k) \to f$ pointwise, we must prove convergence relative to the uniform metric, a stronger condition. Fix ε arbitrarily. Since (f_k) is condensing, there exists an N such that if $N \le k < n$, then

$$d_\infty(f_k, f_n) = \sup_{x \in X} e\big(f_k(x), f_n(x)\big) < \varepsilon/2.$$

Fixing k and letting $n \to \infty$, we have

$$d_\infty(f_k, f) = \sup_{x \in X} e\big(f_k(x), f(x)\big) \leq \varepsilon/2 < \varepsilon$$

if $k \geq N$, so (f_k) converges to f in the uniform metric. It remains to prove the mapping $f : (X, d) \to (Y, e)$ is bounded. But in the preceding notation, if x and x' are arbitrary points of X, the fundamental idiom (Lemma 17.2.3) with $g = f_N$ implies

$$e\big(f(x), f(x')\big) \leq 2(\varepsilon/2) + e\big(f_N(x), f_N(x')\big).$$

Taking the supremum over x and x', $\operatorname{diam}\big(f(X)\big) \leq \varepsilon + \operatorname{diam}\big(f_N(X)\big)$. Since f_N is bounded, f is bounded.

(Continuity of the limit). Now assume each f_k is bounded and continuous. Fix x in X and ε arbitrarily. Since $(f_k) \to f$, there exists a positive integer N such that $d_\infty(f_N, f) < \varepsilon/3$.

Since f_N is continuous at x, there exists a positive δ such that if $x' \in X$ and $d(x, x') < \delta$, then $e\big(f_N(x), f_N(x')\big) < \varepsilon/3$. The fundamental idiom with $g = f_N$ implies

$$e\big(f(x), f(x')\big) \leq 2(\varepsilon/3) + e\big(f_N(x), f_N(x')\big) < \varepsilon.$$

Since ε was arbitrary, f is continuous at x. Since x was an arbitrary element of X, f is continuous on X.   □

A striking application of these ideas is existence of *space-filling curves*:

Proposition 17.2.6. *For each positive integer n, there exist a continuous surjection $c : [0, 1] \to [0, 1]^n$, a uniformly continuous surjection $c : \mathbf{R} \to \mathbf{R}^n$, and a continuous surjection $c : (0, 1) \to \mathbf{R}^n$. If $n \geq 2$, no such mapping is injective.*

Proof. See Exercises 17.2.2, 17.2.4, and 17.2.5.   □

Remark 17.2.7. Visually, if $n = 2$, it may be tempting to imagine scribbling with a pen to fill in a region on a piece of paper. If so, that picture is technically inadequate. A real pen draws sets of positive width, while an ideal path has width 0. Though we will not prove this, it turns out the image of a *differentiable* path has measure zero, so does not contain any non-empty open set.   ◇

Equicontinuity

In the same way the closed, bounded theorem gives an easily verified criterion for compactness in flat n-space, Theorem 17.2.11 gives a sufficient condition for compactness in an infinite-dimensional function space.

Definition 17.2.8. Assume (X, d) and (Y, e) are metric spaces. A family $\mathscr{F}$ of mappings $f : X \to Y$ is *equicontinuous* if for every ε and x in X, there exists a $\delta(x)$ such for all x' in $B_{\delta(x)}(x)$, $e\big(f(x), f(x')\big) < \varepsilon$ for all f in $\mathscr{F}$.

Remark 17.2.9. Loosely, not only is each f in $\mathscr{F}$ continuous on X, but at each x, for a given challenge ε, we can pick the same response δ for all f in $\mathscr{F}$. ◇

Lemma 17.2.10. *If $\mathscr{F}$ is an equicontinuous family of mappings from a compact metric space (X, d) to a metric space (Y, e), then for every ε, there exists a δ such that for all x, x' in X, and for all f in $\mathscr{F}$, $d(x, x') < \delta$ implies $e\big(f(x), f(x')\big) < \varepsilon$.*

Proof. Fix ε arbitrarily. By equicontinuity of $\mathscr{F}$, for each x_0 in X, there exists a $\delta(x_0) > 0$ such that if $x \in B_{\delta(x_0)}(x_0)$, then $e\big(f(x_0), f(x)\big) < \varepsilon/2$ for all f in $\mathscr{F}$. By the uniform radius lemma (Lemma 16.3.16), there exists a δ such that for all x and x' in X, $d(x, x') < \delta$ implies x and x' are in some ball $B_{\delta(x_0)}(x_0)$. For all f in $\mathscr{F}$, the triangle inequality implies,

$$e\big(f(x), f(x')\big) \leq e\big(f(x_0), f(x)\big) + e\big(f(x_0), f(x')\big) < \varepsilon. \qquad \square$$

Theorem 17.2.11 (The equicontinuity theorem). *Assume (X, d) is compact and (Y, e) is flat n-space. If $\mathscr{F}$ is a bounded, equicontinuous family of mappings from X to Y, then $\mathscr{F}$ has compact closure in the uniform metric.*

Proof. It suffices to prove every sequence $(f_m)_{m=0}^{\infty}$ in $\mathscr{F}$ has a condensing subsequence in the uniform metric: Since (Y, e) is complete, Proposition 17.2.5 implies a condensing sequence has a bounded, continuous limit.

The compact space (X, d) is separable. Fix a countable, dense subset $A = \{a_j\}_{j=0}^{\infty}$. We first construct a sequence of subsequences of (f_m) that converge at points of A. For clarity, write $\nu(k, m)$ to denote the index of the mth term of the kth subsequence, so that the kth subsequence is $(f_{\nu(k,m)})_{m=0}^{\infty}$.

Since $\big(f_m(a_0)\big)$ is a bounded sequence in flat n-space, there is a subsequence $(f_{\nu(0,m)})$ such that $\big(f_{\nu(0,m)}(a_0)\big)$ converges. Inductively, if $(f_{\nu(k,m)})$ is a sequence such that $\big(f_{\nu(k,m)}(a_j)\big)$ converges for all j such that $0 \leq j \leq k$, pick a subsequence $(f_{\nu(k+1,m)})$ of $(f_{\nu(k,m)})$ such that $\big(f_{\nu(k+1,m)}(a_{k+1})\big)$ converges. The diagonal sequence $\phi_m := f_{\nu(m,m)}$ converges at each point of A. It suffices to prove (ϕ_m) condenses in the uniform metric.

Fix ε arbitrarily. By Lemma 17.2.10, there exists a δ such that for all x and x' in X, and for all f in $\mathscr{F}$, $d(x, x') < \delta$ implies $e\big(f(x), f(x')\big) < \varepsilon/3$. Because (X, d) is totally bounded, there exist finitely many $\delta/2$-balls, say $\{O_j\}_{j=0}^{J}$, that cover X. For each j, pick a point $a_{k(j)}$ in O_j and an index N_j such that if $N_j \leq m < n$, then $e\big(\phi_m(a_{k(j)}), \phi_n(a_{k(j)})\big) < \varepsilon/3$. Let $N = \max\{N_j\}_{j=0}^{J}$. If

$N \le m < n$, then for all x in X, we have $d(x, a_{k(j)}) < \delta$ for some j. The triangle inequality applied to the chain $\phi_m(x)$, $\phi_m(a_{k(j)})$, $\phi_n(a_{k(j)})$, $\phi_n(x)$ as in the proof of Lemma 17.2.3 implies $e\big(\phi_m(x), \phi_n(x)\big) < \varepsilon$ independently of x. $\quad\square$

Exercises for Section 17.2

Exercise 17.2.1. Assume (a_k) is an arbitrary real sequence. Find a sequence (f_k) of continuous functions on $[0, 1]$ that converges to 0 uniformly on every compact subset of $(0, 1)$, but for which $\int_0^1 f_k = a_k$ for all k.

Qualitatively, uniform convergence on arbitrary compact subsets of the interior gives no control over integrals. What, if anything, can we guarantee if the sequence (f_k) is bounded?

Exercise 17.2.2. Put $X = [0, 1]$ and $Y = [0, 1]^2$, each with the usual metric.

Use the sketch provided to prove there exists a continuous surjection $c : X \to Y$. We'll construct c recursively. Let c_0 be the constant-speed piecewise-affine path from $(0, 0)$ to $(1, 0)$ on the left in Figure 17.1, and let c_1 be the piecewise-affine path with 8 segments in the middle diagram. If we bisect each edge of the square, c_1 comprises four pieces that up to reflection and translation are c_0 scaled by a factor $1/2$.

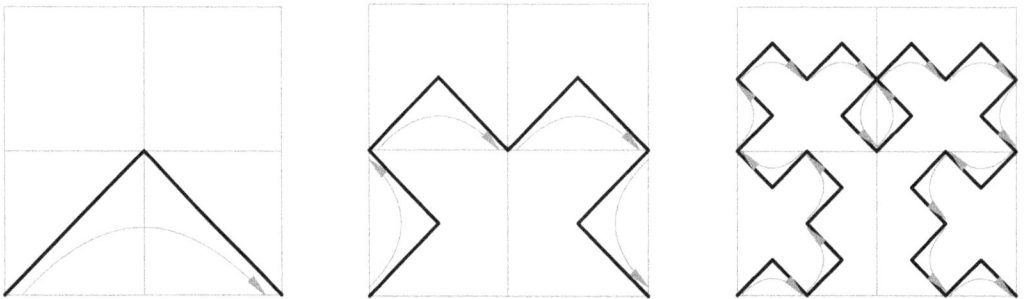

FIGURE 17.1
Recursively constructing a space-filling path.

Now suppose inductively that c_m has been constructed for some m, and that c_m consists of finitely many $\wedge$-shaped pieces parameterized at constant speed over intervals $I_i = (i/4^m, (i + 1)/4^m)$, $0 \le i < 4^m$. Let c_{m+1} be the piecewise-affine path that results from replacing $c_m|_{I_i}$ with four $\wedge$s, each half the size and four times as fast, but collectively defined over the same interval. The diagram on the right shows the image of c_2.

Prove that $c_{m+1}(i/4^m) = c_m(i/4^m)$ if $0 \le i \le 4^m$, and

$$\|c_{m+2} - c_{m+1}\|_\infty \le \tfrac{1}{2}\|c_{m+1} - c_m\|_\infty \quad \text{for all } m.$$

Conclude that $(c_n)_{n=0}^{\infty}$ is condensing in $\mathscr{C}(X, Y)$ with the uniform metric, so the pointwise limit c exists and is continuous. Finally, prove the image of c is dense in Y, and conclude the image is all of Y.

Exercise 17.2.3. (H) Prove that no continuous surjection $c : [0,1] \to [0,1]^2$ is injective.

Exercise 17.2.4. (H) Use Exercise 17.2.2 to prove that if $N \geq 2$, there exists a continuous surjection $c_N : [0,1] \to [0,1]^N$.

Exercise 17.2.5. ($\bigstar$) Use Exercise 17.2.4 to prove there exists a uniformly continuous surjection $c : \mathbf{R} \to \mathbf{R}^n$, and consequently a continuous surjection $c : (0,1) \to \mathbf{R}^n$.

Prove there is no continuous surjection $c : [0,1] \to \mathbf{R}^n$. Prove there is no uniformly continuous surjection $c : (0,1) \to \mathbf{R}^n$. See how many different proofs you can give.

17.3   Uniform Approximation by Polynomials

Assume $a < b$, and the vector space $V = \mathscr{C}\big([a,b]\big)$ is equipped with the uniform metric, $d(f,g) = \|f - g\|_{\infty}$. In Chapter 8 we saw that a typical element of V is "infinitely jagged." Polynomials, by contrast, are as smooth as functions come. In this section we prove that polynomials are dense in $(V, \| \ \|_{\infty})$. Our approach is, loosely, to "blur" an arbitrary continuous function f suitably to obtain polynomials, but so that as we sharpen our view, f emerges.

Convolution

In finite-dimensional linear algebra, a vector in $\mathbf{R}^n$ is an ordered n-tuple $\mathbf{v} = (v_k)$ of real numbers. A matrix is an $n \times n$ array $A = [A_j^i]$ with row index i and column index j. Matrix multiplication defines a linear operator $T_A : \mathbf{R}^n \to \mathbf{R}^n$ given by $T_A(\mathbf{v}) = A\mathbf{v} = \sum_i A_j^i v_i$.

In Chapter 15, we viewed $\mathbf{R}^n$ as the *space of real-valued functions* on a set $\underline{\mathbf{n}}$ of n elements. In the same spirit, a matrix is a "function of two variables," with domain $\underline{\mathbf{n}} \times \underline{\mathbf{n}}$.

This perspective generalizes to infinite-dimensional linear algebra. For simplicity we will work in a space of functions unconnected to any particular real interval.

Definition 17.3.1. A function $f : \mathbf{R} \to \mathbf{R}$ is *rapidly decreasing* if for every positive integer n, $|x|^n f(x) \to 0$ as $|x| \to \infty$.

Example 17.3.2. If $f = 0$ outside some bounded set, then f is rapidly decreasing. If f is continuous on some interval $[a, b]$ and $f(a) = f(b) = 0$, we may extend f by 0, obtaining a continuous, rapidly decreasing function. $\Diamond$

Example 17.3.3. The functions $f(x) = e^{-x^2}$ and $g(x) = e^{-|x|}$ are rapidly decreasing. The product of a rapidly decreasing function and an arbitrary polynomial is rapidly decreasing. A sum or product of rapidly decreasing functions is rapidly decreasing. $\Diamond$

Returning to our matrix multiplication analogy, we may view a continuous, rapidly decreasing function f as a vector. Our analog of a matrix is a continuous function of two variables $K : \mathbf{R}^2 \to \mathbf{R}$ that is rapidly decreasing in each variable separately. The linear operator motivated by matrix multiplication is

$$T_K(f)(x) = \int_{-\infty}^{\infty} K(x, y) f(y) \, dy.$$

(The integral converges absolutely for each x; why?) That is, $T_K(f)$ is a function, whose value at x is given by this formula. For our needs we can specialize to $K(x, y) = g(x - y)$ for some continuous, rapidly decreasing function g.

Definition 17.3.4. If f and g are continuous and rapidly decreasing, their *convolution* is the function

$$(f * g)(x) = \int_{-\infty}^{\infty} g(x - t) f(t) \, dt.$$

Example 17.3.5. A discontinuous example conveys the intuition of convolution. For each positive integer n, let $\chi_{[-1/n, 1/n]}$ be the indicator of $[-1/n, 1/n]$, and define $g_n = (n/2)\chi_{[-1/n, 1/n]}$. If f is integrable, then

$$(f * g_n)(x) = \frac{n}{2} \int_{-\infty}^{\infty} f(t) \chi_{[x-(1/n), x+(1/n)]} \, dt = \frac{1}{2/n} \int_{x-(1/n)}^{x+(1/n)} f(t) \, dt$$

is the average of f over the interval $[x - (1/n), x + (1/n)]$. Particularly, if f is continuous at x, then $(f * g_n)(x) \to f(x)$ as $n \to \infty$. If f is rapidly decreasing and uniformly continuous on $\mathbf{R}$, then $(f * g_n) \to f$ in the uniform metric, see Proposition 17.3.11.

Generally, we think of g as a "filter" and the convolution $f * g$ as a weighted average of f as we translate g. $\Diamond$

Lemma 17.3.6. *The convolution product is commutative on the space of rapidly decreasing functions.*

Proof. The substitution $u = x - t$, or $t = x - u$, gives

$$(f * g)(x) = \int_{-\infty}^{\infty} g(x - t) f(t) \, dt = \int_{-\infty}^{\infty} g(u) f(x - u) \, du = (g * f)(x). \quad \square$$

Unit Spikes

Can we represent the identity operator $I(f) = f$ as convolution, say $\mathsf{I} * f = f$? If so, then

$$f(x) = \int_{-\infty}^{\infty} \mathsf{I}(x - y)f(y)\,dy$$

for all x. Taking $f = 1$, we have $\int \mathsf{I} = 1$. Taking f positive in some short interval $(-\delta, \delta)$ and identically 0 outside, we find that $\mathsf{I}(y) = 0$ if $|y| > 2\delta$. Since δ is arbitrary, $\mathsf{I}(y) = 0$ except at $y = 0$, and integrates to 1 over the real line. Formally, the graph of I is a spike of area 1 and width 0; or I is the "derivative of the unit step function." Mathematically no such function exists. Using sequences of functions, however, we can get as close to this ideal behavior as we like.

Definition 17.3.7. A *unit spike* is a sequence (I_n) of non-negative, continuous, rapidly decreasing functions satisfying the following conditions:

(i) For each n, $\displaystyle\int_{-\infty}^{\infty} \mathsf{I}_n = 1$.

(ii) For every δ, $\displaystyle\int_{|x| \geq \delta} |\mathsf{I}_n| \to 0$ as $n \to \infty$.

Remark 17.3.8. Intuitively, condition (ii) guarantees that the integral of I_n concentrates in an arbitrarily small interval about 0. ◇

Example 17.3.9. If ϕ is non-negative, continuous, rapidly decreasing, and has integral 1, then the functions $\mathsf{I}_n(x) = n\phi(nx)$ are a unit spike. Particularly, the functions $\mathsf{I}_n(x) = (n/\sqrt{2\pi})e^{-(nx)^2/2}$ are a unit spike. ◇

Example 17.3.10. If n is a positive integer, define

$$\frac{1}{c_n} = \int_{-1}^{1} (1 - x^2)^n\,dx = 2\int_{0}^{1} (1 - x^2)^n\,dx.$$

Note that

$$\frac{1}{c_n} > 2\int_{0}^{1/\sqrt{n}} (1 - x^2)^n\,dx > 2\int_{0}^{1/\sqrt{n}} (1 - nx^2)\,dx = \frac{4}{3\sqrt{n}}.$$

The piecewise-polynomials

$$\mathsf{I}_n(x) = \begin{cases} c_n(1 - x^2)^n, & |x| \leq 1, \\ 0 & 1 < |x|, \end{cases}$$

constitute a unit spike. Condition (i) is immediate. For (ii), the preceding estimate gives $0 \leq \mathsf{I}_n(x) \leq (3\sqrt{n}/4)(1 - x^2)^n$. Thus, if $0 < \delta \leq 1$ and $|x| \geq \delta$, we have $0 \leq \mathsf{I}_n(x) \leq (3\sqrt{n}/4)(1 - \delta^2)^n \to 0$. ◇

Proposition 17.3.11. *If* $f : \mathbf{R} \to \mathbf{R}$ *is bounded and uniformly continuous, and* (I_n) *is a unit spike, then* $(f * \mathsf{I}_n) \to f$ *in the uniform metric on* $\mathbf{R}$.

Proof. For all n, we have

$$f_n(x) = \int_{-\infty}^{\infty} f(x-t)\mathsf{I}_n(t)\,dt, \qquad f(x) = \int_{-\infty}^{\infty} f(x)\mathsf{I}_n(t)\,dt.$$

If $\delta > 0$ and x is real, we have

$$\left| f(x) - f_n(x) \right| = \left| \int_{-\infty}^{\infty} \big(f(x) - f(x-t) \big)\, \mathsf{I}_n(t)\,dt \right|$$

$$\leq \left| \int_{-\delta}^{\delta} \big(f(x) - f(x-t) \big)\, \mathsf{I}_n(t)\,dt \right| + \left| \int_{|t| \geq \delta|} \big(f(x) - f(x-t) \big)\, \mathsf{I}_n(t)\,dt \right|$$

$$\leq \int_{-\delta}^{\delta} \left| f(x) - f(x-t) \right| \mathsf{I}_n(t)\,dt + \int_{|t| \geq \delta|} \left| f(x) - f(x-t) \right| \mathsf{I}_n(t)\,dt.$$

Conceptually, for small t, the increment of f is small (so the first term is small) because f is uniformly continuous on $\mathbf{R}$, while for large t the second term is small because f is bounded and the I_n concentrate at 0.

Fix ε arbitrarily. Because f is uniformly continuous on $\mathbf{R}$, there exists a positive δ such that $|t| \leq \delta$ implies $|f(x) - f(x-t)| < \varepsilon/2$. Because f is bounded, there exists an M such that $|f(x) - f(x-t)| \leq M$ for all x and t in $\mathbf{R}$. By property (ii) of a unit spike, there is an N such that

$$\int_{|t| \geq \delta} \mathsf{I}_n(t)\,dt < \frac{\varepsilon}{2M+1} \quad \text{if } n \geq N.$$

If $n \geq N$, then we have, independently of x,

$$\underbrace{\int_{-\delta}^{\delta} \left| f(x) - f(x-t) \right|}_{<\varepsilon/2} \mathsf{I}_n(t)\,dt + \int_{|t| \geq \delta} \underbrace{\left| f(x) - f(x-t) \right|}_{\leq M} \mathsf{I}_n(t)\,dt$$

$$< \frac{\varepsilon}{2} \int_{-\delta}^{\delta} \mathsf{I}_n(t)\,dt + \int_{|t| \geq \delta} M\mathsf{I}_n(t)\,dt < \frac{\varepsilon}{2} + M \cdot \frac{\varepsilon}{2M+1} < \varepsilon. \quad \square$$

Theorem 17.3.12 (Uniform polynomial approximation). *If* $f : [a,b] \to \mathbf{R}$ *is continuous, there exists a sequence* (p_n) *of polynomials such that* $(p_n) \to f$ *in the uniform metric on* $[a,b]$.

Proof. We can approximate f by polynomials if and only if we can approximate

$$g(x) = f(x) - f_{a,b}(x) = f(x) - f(a) - \frac{f(b) - f(a)}{b-a}(x-a)$$

by polynomials. Further, we may substitute $x = a + (b-a)u$, reducing to the case $[a,b] = [0,1]$. In more detail, if we write $\tilde{\phi}(u) = \phi(x)$, then $(p_n) \to f$

on $[a, b]$ if and only if $(\widetilde{p_n}) \to \tilde{f}$ on $[0, 1]$. Thus it suffices to prove the theorem for a continuous function $f : [0, 1] \to \mathbf{R}$ satisfying $f(0) = f(1) = 0$.

Let (I_n) be the unit spike of Example 17.3.10, and set $p_n = f * I_n$. Because $f = 0$ outside $[0, 1]$,

$$p_n(x) = \int_{-\infty}^{\infty} f(t) I_n(x - t)\, dt = \int_0^1 f(t) I_n(x - t)\, dt.$$

If $x \in [0, 1]$, then $x - t \in [-1, 1]$ for all t in $[0, 1]$. Consequently, the integrand $f(t) I_n(x - t) = c_n \big(1 - (x - t)^2\big)^n f(t)$ is a polynomial in x whose coefficients are continuous functions of t. Integrating from $t = 0$ to $t = 1$ shows p_n is a polynomial in x. By Proposition 17.3.11, $(p_n) \to f$ in the uniform metric. $\square$

Exercises for Section 17.3

Exercise 17.3.1. Assume $\gamma : [0, 1] \to \mathbf{R}^n$ is a continuous path in the flat metric. Prove there exists a sequence (p_k) of polynomial mappings that uniformly approximates γ on $[0, 1]$.

Exercise 17.3.2. In a well-defined sense, "most" continuous functions do not have a power series expansion. Why does this not contradict Theorem 17.3.12?

Exercise 17.3.3. Prove $\mathscr{C}([a, b])$ is separable with respect to the uniform metric.

Exercise 17.3.4. (H) Assume I is an open real interval, and $f : I \to \mathbf{R}$ is a continuous function. Can we extend Theorem 17.3.12, namely, can f be uniformly approximated by polynomials on I? What if the interval is bounded, and/or f is assumed to be uniformly continuous?

17.4 Differential Equations

Historically, the study of differential equations is motivated by classical physics, where the acceleration of a point mass is proportional to the net forces acting on the mass, and the forces typically depend on position and velocity of the mass. Throughout this section, $\mathbf{R}^n$ denotes flat n-space.

Example 17.4.1. A point mass m attached to a thin, rigid, massless pendulum of length ℓ in a constant, vertical gravitational field with acceleration g is free to swing. According to Newton's second law of motion, the angle θ from vertical obeys the equation $\theta'' = (g/\ell) \sin \theta$.

To convert this "second-order" equation into two "coupled first-order" equations, physicists introduce "phase space" variables $x_0 = \theta$ and $x_1 = \theta'$, so $x_0' = x_1$ and $x_1' = \theta'' = (g/\ell)\sin x_0$. ◇

Remark 17.4.2. In a physical system with N interacting particles in ordinary flat 3-space, such as planets, the positions and velocities of the particles may be described by a system of $6N$ equations in $6N$ unknowns: Three spatial coordinates and three velocity components for each particle.

Classically, a *first-order ordinary differential equation* is a set of n conditions on a set of n functions of one variable:

$$x_k' = f_k(x_0, x_1, \ldots, x_{n-1}), \quad 0 \le k < n.$$

The modern definition packages these unknown functions into a single vector-valued function $\mathbf{x} = (x_k)_{k=0}^{n-1}$ of one variable, and views $\mathbf{f} = (f_k)_{k=0}^{n-1}$ as a *vector field*, which assigns to each location a vector in $\mathbf{R}^n$. The preceding system of conditions becomes $\mathbf{x}' = \mathbf{f} \circ \mathbf{x}$ as vector-valued functions on some real interval. Naturally, to prove theorems we must specify properties of $\mathbf{f}$. ◇

Definition 17.4.3. Assume $U \subseteq \mathbf{R}^n$ is a non-empty open set. A *vector field* on U is a continuous mapping $\mathbf{f} : U \to \mathbf{R}^n$. If I is a non-empty open interval, a *flow line* of $\mathbf{f}$ on I is a $\mathscr{C}^1$ mapping $\mathbf{x} : I \to U$ satisfying the *first-order differential equation* $\mathbf{x}' = \mathbf{f} \circ \mathbf{x}$.

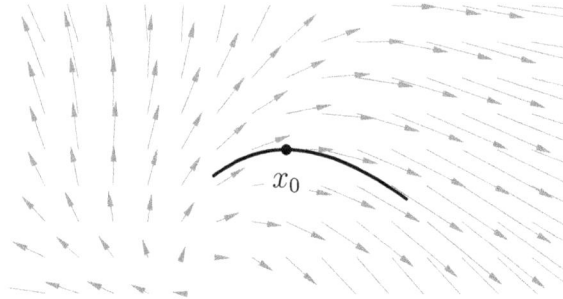

FIGURE 17.2
A vector field (gray) and a flow line.

Definition 17.4.4. Assume $U \subseteq \mathbf{R}^n$ is a non-empty open set, $\mathbf{f} : U \to \mathbf{R}^n$ a continuous mapping. For each $\mathbf{x}_0$ in U, the conditions $\mathbf{x}' = \mathbf{f} \circ \mathbf{x}$ and $\mathbf{x}(0) = \mathbf{x}_0$ constitute an *initial-value problem*. A *solution* is a $\mathscr{C}^1$ mapping $\mathbf{x} : I \to U$ on some open interval containing 0 and satisfying $\mathbf{x}' = \mathbf{f} \circ \mathbf{x}$ and $\mathbf{x}(0) = \mathbf{x}_0$.

In this section we give sufficient conditions for a first-order system to have, for each initial condition, a unique solution.

First-Order Differential Equations

Theorem 17.4.5 (The existence-uniqueness theorem). *Assume $U \subseteq \mathbf{R}^n$ is a non-empty open set, $\mathbf{f} : U \to \mathbf{R}^n$ a mapping with locally bounded stretch, and $\mathbf{x}_0$ an arbitrary point of U. There exist a positive real number a and a unique mapping $\mathbf{x} : (-a, a) \to U$ satisfying the initial-value problem*

$$\mathbf{x}' = \mathbf{f} \circ \mathbf{x}, \qquad \mathbf{x}(0) = \mathbf{x}_0.$$

Proof. Integrating the differential equation from 0 to t gives

$$\mathbf{x}(t) - \mathbf{x}(0) = \int_0^t \mathbf{x}'(s)\, ds = \int_0^t (\mathbf{f} \circ \mathbf{x})(s)\, ds,$$

or after substituting the initial condition and rearranging,

$$\mathbf{x}(t) = (T\mathbf{x})(t) := \mathbf{x}_0 + \int_0^t (\mathbf{f} \circ \mathbf{x})(s)\, ds.$$

Conversely, any function $\mathbf{x}$ satisfying $\mathbf{x} = T\mathbf{x}$ is a solution of the initial-value problem by Theorem 11.1.1. Technically, solving this "integral equation" is preferable: Differentiation is a discontinuous operator by Exercise 16.5.7, while under the hypotheses of the theorem, integration on a sufficiently small interval is a contraction. If we construct a complete function space on which the operator T is a contraction, Theorem 16.5.23 guarantees the existence of a unique fixed point, namely, a solution of our initial-value problem.

Since $\mathbf{f}$ has locally bounded stretch in U, there exists a positive r and a non-negative real M such that $\|\mathbf{f}(\mathbf{v}_2) - \mathbf{f}(\mathbf{v}_1)\| \leq M\|\mathbf{v}_2 - \mathbf{v}_1\|$ for all $\mathbf{v}_1$ and $\mathbf{v}_2$ in $B_{2r}(\mathbf{x}_0)$. Let $Y = \overline{B_r(\mathbf{x}_0)} \subseteq \mathbf{R}^n$ be the closed ball of radius r about $\mathbf{x}_0$. Since Y is compact and $\|\mathbf{f}\|$ is continuous, $L := \max\{\|\mathbf{f}(\mathbf{v})\| : \mathbf{v} \in Y\}$ exists. Set $a = \min\big(r/(L+1), 1/(M+1)\big)$, and put $I = [-a, a]$.

Let $\mathscr{C}(I, Y)$ be the space of continuous mappings equipped with the uniform metric d_∞. By Proposition 17.2.5, $\mathscr{C}(I, Y)$ is d_∞-complete. The subset X of mappings satisfying $\mathbf{x}(0) = \mathbf{x}_0$ is a closed subset (why?), hence also complete. It suffices to prove $T : X \to X$ is a contraction.

Since $\mathbf{x}$ is continuous and $\mathbf{f}$ has locally bounded stretch, $\mathbf{f} \circ \mathbf{x}$ is continuous. By Proposition 9.4.4, $T\mathbf{x}$ is continuous in I. The triangle inequality gives

$$\|T\mathbf{x}(t) - \mathbf{x}_0\| = \left\| \int_0^t (\mathbf{f} \circ \mathbf{x})(s)\, ds \right\| \leq \left| \int_0^t \|(\mathbf{f} \circ \mathbf{x})(s)\|\, ds \right| \leq aL < r$$

for all t in I. This means $T\mathbf{x} : I \to Y$, or $T\mathbf{x} \in X$, so $T : X \to X$.

To complete the argument, it suffices to prove T is a contraction. By the choice of a, $\lambda := Ma < 1$. If $\mathbf{x}_1$ and $\mathbf{x}_2$ are elements of X and $|t| \leq a$, then

$$\|T\mathbf{x}_2(t) - T\mathbf{x}_1(t)\| = \left\| \int_0^t (\mathbf{f} \circ \mathbf{x}_2)(s) - (\mathbf{f} \circ \mathbf{x}_1)(s)\, ds \right\|$$

$$\leq \left| \int_0^t \|(\mathbf{f} \circ \mathbf{x}_2)(s) - (\mathbf{f} \circ \mathbf{x}_1)(s)\| \, ds \right|$$

$$\leq M \left| \int_0^t \|\mathbf{x}_2(s) - \mathbf{x}_1(s)\| \, ds \right|$$

$$\leq M \left| \int_0^t d_\infty(\mathbf{x}_1, \mathbf{x}_2) \, ds \right|$$

$$\leq (Ma) \, d_\infty(\mathbf{x}_1, \mathbf{x}_2) = \lambda d_\infty(\mathbf{x}_1, \mathbf{x}_2).$$

Taking the supremum over t, we have $d_\infty(T\mathbf{x}_1, T\mathbf{x}_2) \leq \lambda d_\infty(\mathbf{x}_1, \mathbf{x}_2)$ that is, T is a contraction. $\qquad\square$

Corollary 17.4.6. *If $U \subseteq \mathbf{R}^n$ is non-empty and open, $\mathbf{f} : U \to \mathbf{R}^n$ has locally bounded stretch, and $\mathbf{x}_0$ is an arbitrary point of U, viewed as a constant mapping, then there exists an a such that the sequence $(\mathbf{x}_k)_{k=0}^\infty$ defined by*

$$\mathbf{x}_{k+1}(t) = \mathbf{x}_0 + \int_0^t (\mathbf{f} \circ \mathbf{x}_k)(s) \, ds$$

converges in the uniform metric on $[-a, a]$ to the solution of the initial-value problem $\mathbf{x}' = \mathbf{f} \circ \mathbf{x}$, $\mathbf{x}(0) = \mathbf{x}_0$.

Example 17.4.7. Consider the differential equation $x' = x^2$, which we might view as the equation of motion of a point particle whose velocity is x^2 when the particle is at x. Since $f(x) = x^2$ has locally bounded stretch, Theorem 17.4.5 guarantees that for each real x_0, there is a unique solution of the corresponding initial-value problem on some interval. In this case we can find the solutions explicitly. If $x(0) = 0$, then $x(t) = 0$ for all t. Otherwise, $x_0 := x(0) \neq 0$, and

$$\left[-\frac{1}{x} \right]'(t) = \frac{x'(t)}{x(t)^2} = 1.$$

Integrating from 0 to t and solving for x gives

$$-\frac{1}{x(t)} + \frac{1}{x_0} = t \quad \text{or} \quad x(t) = \frac{x_0}{1 - x_0 t}.$$

Although the vector field f is smooth on $\mathbf{R}$, the solutions are not defined for all t. Particularly, if $x_0 > 0$, then the solution *blows up in finite time*. Specifically, $x \to \infty$ as $t \to 1/x_0$. The maximal interval containing 0 on which the solution exists is $(-\infty, 1/x_0)$. For large x_0, the interval of positive time for which the solution exists is small. $\qquad\diamond$

Example 17.4.8. Consider $x' = x^{1/3}$. Separating variables as above gives $x(t) = (\frac{2}{3}t + x_0^{2/3})^{3/2}$. If $x_0 = 0$ we have $x(t) = (\frac{2}{3}t)^{3/2}$.

Note that $x(t) \equiv 0$ is also a solution with initial data $x_0 = 0$. That is, this initial-value problem has non-unique solutions. There is no contradiction: The function $f(x) = x^{1/3}$ does not have locally bounded stretch at 0. $\qquad\diamond$

Global Solutions

Definition 17.4.9. A *continuous dynamical system* is a metric space (X, d) together with a family of homeomorphisms $\{f_t\}_{t \in \mathbf{R}}$ such that $f_{t'} \circ f_t = f_{t+t'}$ for all real t and t'.

Remark 17.4.10. The prototypical example is a non-empty open set in flat n-space with a vector field whose flow lines are defined for all time. ◇

Theorem 17.4.11 (Global existence-uniqueness). *Assume* $\mathbf{f} : \mathbf{R}^n \to \mathbf{R}^n$ *has finite stretch. For every* $\mathbf{x}_0$ *in* $\mathbf{R}^n$, *there is a unique* $\mathbf{x} : \mathbf{R} \to \mathbf{R}^n$ *satisfying* $\mathbf{x}' = \mathbf{f} \circ \mathbf{x}$ *and* $\mathbf{x}(0) = \mathbf{x}_0$.

We separate the proof into several steps.

Lemma 17.4.12. *Assume* $t_0 > 0$. *If* u *is continuous and non-negative on* $[0, t_0]$, *and there are non-negative numbers* U_0 *and* M *such that*

$$u(t) \leq U(t) := U_0 + \int_0^t M u(s) \, ds, \quad \text{on } [0, t_0],$$

then $u(t) \leq U_0 e^{Mt}$ *on* $[0, t_0]$.

Proof. We have $U'(t) = M u(t) \leq M U(t)$, so

$$\frac{d}{dt}\left(e^{-Mt} U(t)\right) = e^{-Mt}\left(U'(t) - M U(t)\right) \leq 0.$$

Integrating from 0 to t gives $e^{-Mt} U(t) - U(0) \leq 0$, and therefore

$$u(t) \leq U(t) \leq U(0) e^{Mt} = U_0 e^{Mt}. \qquad \square$$

Lemma 17.4.13. *Assume* $\mathbf{f}$ *has stretch* M. *If* $\mathbf{x}$ *and* $\mathbf{y}$ *are solutions of* $\mathbf{x}' = \mathbf{f} \circ \mathbf{x}$ *on some interval* $[a, b]$ *containing* 0, *with respective initial data* $\mathbf{x}_0$ *and* $\mathbf{y}_0$, *then*

$$\|\mathbf{x}(t) - \mathbf{y}(t)\| \leq \|\mathbf{x}_0 - \mathbf{y}_0\| e^{M|t|} \quad \text{on } [a, b].$$

If $\mathbf{x}(t_0) = \mathbf{y}(t_0)$ *for some* t_0 *in* $[a, b]$, *then* $\mathbf{x}(t) = \mathbf{y}(t)$ *for all* t *in* $[a, b]$.

Remark 17.4.14. Everyday experience suggests predictions become less accurate the farther ahead they go. In a discrete setting, such as the space of binary sequences, exponential growth has a natural interpretation: At "each step" there is some fixed set of possibilities, so each successive step of prediction expands the space of possibility by some factor. Lemma 17.4.13 may be interpreted as a continuous version: Solutions of a differential equation move apart exponentially rapidly over time intervals of fixed length, with the exponential coefficient equal to the stretch of the vector field f. ◇

Proof. Define $u(t) = \|\mathbf{x}(t) - \mathbf{y}(t)\|$. Using the integral equations that $\mathbf{x}$ and $\mathbf{y}$ satisfy,

$$
\begin{aligned}
u(t) &= \left\| \mathbf{x}_0 - \mathbf{y}_0 + \int_0^t \big((\mathbf{f} \circ \mathbf{x})(s) - (\mathbf{f} \circ \mathbf{y})(s) \big)\, ds \right\| \\
&\le \|\mathbf{x}_0 - \mathbf{y}_0\| + \left| \int_0^t \|(\mathbf{f} \circ \mathbf{x})(s) - (\mathbf{f} \circ \mathbf{y})(s)\|\, ds \right| \\
&\le \|\mathbf{x}_0 - \mathbf{y}_0\| + \left| \int_0^t M \|\mathbf{x}(s) - \mathbf{y}(s)\|\, ds \right| \\
&= \|\mathbf{x}_0 - \mathbf{y}_0\| + \left| \int_0^t M u(s)\, ds \right|.
\end{aligned}
$$

Hence u satisfies the hypotheses of Lemma 17.4.12, and therefore

$$
u(t) \le \|\mathbf{x}_0 - \mathbf{y}_0\| e^{Mt}. \qquad \square
$$

Lemma 17.4.15. *Assume* $\mathbf{f} : \mathbf{R}^n \to \mathbf{R}^n$ *has stretch* M *and put* $a = 1/(M+1)$. *For all* $\mathbf{x}_0$ *in* $\mathbf{R}^n$, *the initial-value problem* $\mathbf{x}' = \mathbf{f} \circ \mathbf{x}$, $\mathbf{x}(0) = \mathbf{x}_0$, *has a unique solution* $\mathbf{x} : [-a, a] \to \mathbf{R}^n$.

Proof. Using notation in the proof of Theorem 17.4.5, if $a \le 1/(M+1)$, then for all $\mathbf{x}_0$ in $\mathbf{R}^n$, the mapping T is a contraction on $\mathscr{C}\big([-a, a], \mathbf{R}^n\big)$. $\qquad \square$

Proof of Theorem 17.4.11. Assume $\mathbf{f}$ has stretch M, and put $a = 1/(M+1)$. For each $\mathbf{x}_0$ in $\mathbf{R}^n$, Lemma 17.4.15 guarantees there is some $\mathbf{x}_1$ defined on $[-a, a]$ such that $\mathbf{x}_1' = \mathbf{f} \circ \mathbf{x}_1$ and $\mathbf{x}_1(0) = \mathbf{x}_0$.

Now put $\mathbf{y}_0 = \mathbf{x}(a)$. Lemma 17.4.15 implies there is a unique $\mathbf{y}$ so that $\mathbf{y}' = \mathbf{f} \circ \mathbf{y}$ if $-a \le t \le a$ and $\mathbf{y}(0) = \mathbf{y}_0$. Define $\mathbf{x}_2(t) = \mathbf{y}(t - a)$. We have $\mathbf{x}_2' = \mathbf{f} \circ \mathbf{x}_2$ if $0 \le t \le 2a$, and $\mathbf{x}_2(a) = \mathbf{y}(0) = \mathbf{y}_0 = \mathbf{x}_1(a)$. Hence $\mathbf{x}_1$ and $\mathbf{x}_2$ are solutions on $[0, a]$ that agree at $t = a$. By the last part of Lemma 17.4.13, $\mathbf{x}_1(t) = \mathbf{x}_2(t)$ if $0 \le t \le a$.

The solution $\mathbf{x}_1$, defined on $[-a, a]$, may be extended to a solution on $[-a, 2a]$ by defining $\mathbf{x}_1 = \mathbf{x}_2$ on $[a, 2a]$. Inductively, we may extend to a unique solution on arbitrarily large intervals, hence on $(-\infty, \infty)$. $\qquad \square$

Matrix Exponentiation

In this section we study systems of first-order differential equations with constant coefficients, namely having the form $\mathbf{x}' = A\mathbf{x}$ for some real $n \times n$ matrix $A = [A_j^i]$.

Definition 17.4.16. If A is an $n \times n$ real matrix, its *exponential series* is

$$
\exp(tA) = \sum_{k=0}^{\infty} \frac{(tA)^k}{k!} = I + tA + t^2 \frac{A^2}{2!} + t^3 \frac{A^3}{3!} + \cdots.
$$

Lemma 17.4.17. *For every $n \times n$ real matrix $A = [A^i_j]$, its exponential series converges absolutely in each entry.*

Proof. Since there are only finitely many entries of A, there is a real number M such that $|A^i_j| \leq M$ for all i and j. It suffices to prove that for every k, $|(A^k)^i_j| \leq (Mn)^k$ for all i and j. The base case $(k = 0)$ is true since $A^0 = I_n$ has all entries 0 or 1. Assume inductively that the stated inequality holds for some m. By definition of matrix multiplication,

$$|(A^{m+1})^i_j| = \left| \sum_{\ell=0}^{n-1} (A^m)^i_\ell A^\ell_j \right| \leq \sum_{\ell=0}^{n-1} |(A^m)^i_\ell A^\ell_j| \leq (Mn)^m \sum_{\ell=0}^{n-1} |A^\ell_j| \leq (Mn)^{m+1}.$$

By induction, the stated inequality holds for all k. Consequently, each entry in the exponential series is bounded in absolute value by the convergent series

$$\sum_{k=0}^{\infty} \frac{(tMn)^k}{k!} = e^{tMn}. \qquad \square$$

Example 17.4.18. If $A = \operatorname{diag}[d^1, d^2, \ldots, d^n]$, then A^m is the diagonal matrix whose entries are the mth powers of the entries of A, so

$$\exp(tA) = \operatorname{diag}[e^{td^1}, e^{td^2}, \ldots, e^{td^n}]. \qquad \Diamond$$

Example 17.4.19. If N is the 4×4 nilpotent block, its fourth power is $\mathbf{0}^{4\times 4}$, so N, N^2, N^3, and $\exp(tN)$ are

$$\begin{bmatrix} 0 & 1 & 0 & 0 \\ 0 & 0 & 1 & 0 \\ 0 & 0 & 0 & 1 \\ 0 & 0 & 0 & 0 \end{bmatrix}, \begin{bmatrix} 0 & 0 & 1 & 0 \\ 0 & 0 & 0 & 1 \\ 0 & 0 & 0 & 0 \\ 0 & 0 & 0 & 0 \end{bmatrix}, \begin{bmatrix} 0 & 0 & 0 & 1 \\ 0 & 0 & 0 & 0 \\ 0 & 0 & 0 & 0 \\ 0 & 0 & 0 & 0 \end{bmatrix}, \begin{bmatrix} 1 & t & \frac{t^2}{2!} & \frac{t^3}{3!} \\ 0 & 1 & t & \frac{t^2}{2!} \\ 0 & 0 & 1 & t \\ 0 & 0 & 0 & 1 \end{bmatrix}. \qquad \Diamond$$

Example 17.4.20. If $N = B_n(0)$ is the $n \times n$ nilpotent block, then N^m has a diagonal of 1s lying m rows above the main diagonal, and $N^n = \mathbf{0}^{n\times n}$. The exponential series is consequently a polynomial, and

$$\exp(tN) = \begin{bmatrix} 1 & t & \frac{t^2}{2!} & \cdots & \frac{t^{n-2}}{(n-2)!} & \frac{t^{n-1}}{(n-1)!} \\ 0 & 1 & t & \cdots & \frac{t^{n-3}}{(n-3)!} & \frac{t^{n-2}}{(n-2)!} \\ \vdots & \vdots & \vdots & \ddots & \vdots & \vdots \\ 0 & 0 & 0 & \cdots & t & \frac{t^2}{2!} \\ 0 & 0 & 0 & \cdots & 1 & t \\ 0 & 0 & 0 & \cdots & 0 & 1 \end{bmatrix}. \qquad \Diamond$$

Proposition 17.4.21. *Assume A and B are $n \times n$ matrices.*

(i) *If P is an invertible $n \times n$ matrix, then $\exp(tP^{-1}AP) = P^{-1}\exp(tA)P$.*

(ii) *If $BA = AB$, then $\exp\big(t(A + B)\big) = \exp(tA)\exp(tB)$.*

(iii) *If A is diagonalizable, then $\det\exp(tA) = e^{t\,\mathrm{tr}(A)}$.*

(iv) $\exp(tA^{\mathsf{T}}) = \big(\exp(tA)\big)^{\mathsf{T}}$.

(v) *If $A^{\mathsf{T}} = -A$, then $\exp(tA)$ is orthogonal and has determinant 1.*

Proof. (i). This is a consequence of the identity $(P^{-1}AP)^k = P^{-1}A^k P$:

$$\exp(tP^{-1}AP) = \sum_{k=0}^{\infty} \frac{(tP^{-1}AP)^k}{k!} = \sum_{k=0}^{\infty} P^{-1}\frac{(tA)^k}{k!}P = P^{-1}\exp(tA)P.$$

(ii). This is formally identical to the proof for complex numbers. Since A and B commute, there is a binomial theorem

$$\frac{(A + B)^k}{k!} = \sum_{i=0}^{k} \frac{A^i}{i!}\frac{B^{k-i}}{(k-i)!} = \sum_{i+j=k} \frac{A^i}{i!}\frac{B^j}{j!}.$$

The product formula for absolutely convergent double series gives

$$\exp\big(t(A + B)\big) = \sum_{k=0}^{\infty} \frac{\big(t(A + B)\big)^k}{k!} = \sum_{k=0}^{\infty}\sum_{i+j=k} \frac{(tA)^i}{i!}\frac{(tB)^j}{j!}$$

$$= \Big[\sum_{i=0}^{\infty} \frac{(tA)^i}{i!}\Big]\Big[\sum_{j=0}^{\infty} \frac{(tB)^j}{j!}\Big] = \exp(tA)\exp(tB).$$

(iii). Suppose $P^{-1}AP = A' = \mathrm{diag}[d^1, \ldots, d^n]$ for some invertible matrix P. By (i) and Example 17.4.18,

$$P^{-1}\exp(tA)P = \exp(tA') = \mathrm{diag}[e^{td^1}, e^{td^2}, \ldots, e^{td^n}].$$

Taking determinants,

$$\det\exp(tA) = \det\exp(tA') = \prod_{k=1}^{n} e^{td^k} = e^{t\sum_k d^k} = e^{t\,\mathrm{tr}(A')} = e^{t\,\mathrm{tr}(A)}.$$

(iv). This is an immediate consequence of $(A^k)^{\mathsf{T}} = (A^{\mathsf{T}})^k$.

(v). We have $\exp(\mathbf{0}^{n\times n}) = I_n$. By (ii), $\exp(-tA) = \exp(tA)^{-1}$. If $A^{\mathsf{T}} = -A$. then

$$\exp(tA)^{-1} = \exp(-tA) = \exp(tA^{\mathsf{T}}) = \exp(tA)^{\mathsf{T}},$$

which proves $\exp(tA)$ is orthogonal.

Since the diagonal entries of a skew-symmetric matrix are all 0, $\mathrm{tr}(A) = 0$. By (iii), $\det\exp(tA) = e^{t\,\mathrm{tr}(A)} = e^0 = 1$. $\qquad\square$

Remark 17.4.22. An orthogonal $n \times n$ matrix of determinant 1 is, by definition, a *rotation*. Part (v) of the theorem says that a skew-symmetric matrix is an "infinitesimal" rotation. ◇

Example 17.4.23. A matrix version of the polar formula holds: If

$$J = \begin{bmatrix} 0 & -1 \\ 1 & 0 \end{bmatrix}, \quad \text{then } \exp(tJ) = \begin{bmatrix} \cos t & -\sin t \\ \sin t & \cos t \end{bmatrix}.$$

To prove this, note that $J^2 = -I_2$, from which it follows immediately that $J^{2k} = (-1)^k I_2$ and $J^{2k+1} = (-1)^k J$. Splitting the exponential series into terms of even and odd degree and recalling the power series for the circular functions,

$$\cos t = \sum_{k=0}^{\infty} (-1)^k \frac{t^{2k}}{(2k)!}, \qquad \sin t = \sum_{k=0}^{\infty} (-1)^k \frac{t^{2k+1}}{(2k+1)!},$$

we have

$$\begin{aligned} \exp(tJ) &= \sum_{k=0}^{\infty} \frac{(tJ)^{2k}}{(2k)!} + \sum_{k=0}^{\infty} \frac{(tJ)^{2k+1}}{(2k+1)!} \\ &= \Big[\sum_{k=0}^{\infty} (-1)^k \frac{t^{2k}}{(2k)!}\Big] I_2 + \Big[\sum_{k=0}^{\infty} (-1)^k \frac{t^{2k+1}}{(2k+1)!}\Big] J \\ &= (\cos t) I_2 + (\sin t) J. \end{aligned}$$ ◇

Example 17.4.24. Consider the linear system $\mathbf{x}' = A\mathbf{x}$, $\mathbf{x}(0) = \mathbf{x}_0$. Iteration as in Corollary 17.4.6, with initial function $\mathbf{x}_0(t) \equiv \mathbf{x}_0$ gives

$$\mathbf{x}_1(t) = \mathbf{x}_0 + \int_0^t A\mathbf{x}_0 \, ds = (I + tA)\mathbf{x}_0,$$

$$\mathbf{x}_2(t) = \mathbf{x}_0 + \int_0^t (I + tA)\mathbf{x}_0 \, ds = \Big(I + tA + \tfrac{1}{2}t^2 A^2\Big)\mathbf{x}_0,$$

and generally, by induction on n,

$$\mathbf{x}_{n-1}(t) = \Big[\sum_{k=0}^{n-1} \frac{t^k A^k}{k!}\Big] \mathbf{x}_0,$$

the partial sums of the series for the unique solution, $\mathbf{x}(t) = e^{tA}\mathbf{x}_0$. ◇

Exercises for Section 17.4

Exercise 17.4.1. (H) Assume $\phi : (-1, 1) \to \mathbf{R}$ is continuous and positive.

(a) Prove that

$$x = \int_0^{u(x)} \frac{dt}{\phi(t)}$$

defines an increasing, $\mathscr{C}^1$ function u satisfying $u' = \phi \circ u$.

(b) Prove that

$$f(x) = \int_0^{u(x)} \frac{t \, dt}{\phi(t)}$$

defines a $\mathscr{C}^2$ function f satisfying $f' = u$.

(c) If $\phi(t) = 1 - t^2$, find u and f as explicit functions of x.

(d) If $\phi(t) = 1 + t^2$, find u and f as explicit functions of x.

Exercise 17.4.2. Throughout, a, b, and c denote real numbers, and put

$$A = \begin{bmatrix} a & 0 \\ 0 & -a \end{bmatrix} \qquad B = \begin{bmatrix} 0 & b \\ b & 0 \end{bmatrix}, \qquad C = \begin{bmatrix} 0 & -c \\ c & 0 \end{bmatrix} \qquad X = \begin{bmatrix} a & c \\ b & -a \end{bmatrix}.$$

Calculate $\exp(tA)$, $\exp(tB)$, $\exp(tC)$, and $\exp(tX)$. Hint: Start by calculating powers of each matrix.

Exercise 17.4.3. Let $x(t)$ represent the position of a point particle at time t. A simple model of one-dimensional oscillation has a point mass m attached to a spring with spring constant k, subject to friction proportional to velocity. If no outside forces act, the law of motion $F = ma$ reads $mx'' = -bx' - kx$, or $mx'' + bx' + kx = 0$.

Find all *complex* r such that $x(t) = e^{rt}$ is a solution. Under what conditions on m, b, and k are the values of r real?

If r is not real, prove that the real and imaginary parts of $x(t) = e^{rt}$ solve the same differential equation.

If there exists a unique (necessarily real) r such that $x(t) = e^{rt}$ solves the differential equation, prove that te^{rt} is also a solution.

17.5   Spectral Series

For every positive integer n, the circular functions $\cos(nx)$ and $\sin(nx)$ are 2π-periodic, as is the constant function $1 = \cos(0)$, and every convergent series in these functions. Conversely, if $f : \mathbf{R} \to \mathbf{R}$ is 2π-periodic, can f be written as a trigonometric series? In this section we introduce "piecewise $\mathscr{C}^1$" functions, and prove the answer for these is "yes (except possibly at jump discontinuities)."

Definition 17.5.1. Assume $a < b$. If I is an interval (open, closed, or half-open) with endpoints a and b, a function $f : I \to \mathbf{R}$ has $\mathscr{C}^1$ *extension on I* if there exists an open interval $J \supseteq [a, b]$ and a $\mathscr{C}^1$ extension of f to J.

Lemma 17.5.2. *Assume $a < b$. If I is an interval with endpoints a and b, then $f : I \to \mathbf{R}$ has $\mathscr{C}^1$ extension on I if and only if f is of class $\mathscr{C}^1$ on (a, b), extends continuously to $[a, b]$, and the one-sided limits $f'(a^+)$ and $f'(b^-)$ exist.*

Proof. If f has a $\mathscr{C}^1$ extension to an open interval J containing $[a, b]$, then the restriction to $[a, b]$ of the extension is of class $\mathscr{C}^1$ on (a, b), continuous on $[a, b]$, and the one-sided limits of f' exist at the endpoints.

Conversely, if f is of class $\mathscr{C}^1$ on (a, b), extends continuously to $[a, b]$, and the one-sided limits of f' exist at the endpoints, we can extend f to a larger open interval using line segments of matching height and slope at each endpoint.

In more detail, define $f(x) = f(a) + f'(a^+)(x - a)$ if $x < a$, and define $f(x) = f(b) + f'(b^-)(x - b)$ if $b < x$. The extended function is differentiable on $\mathbf{R}$ by smooth patching, and the derivative is continuous. $\quad\square$

Lemma 17.5.3. *Assume I is a bounded real interval. If f and $g : I \to \mathbf{R}$ have $\mathscr{C}^1$ extension on I, then $f + g$ and fg have $\mathscr{C}^1$ extension on I. If in addition g is bounded away from 0, then f/g has $\mathscr{C}^1$ extension on I.*

Definition 17.5.4. If $a < b$, a function $f : [a, b] \to \mathbf{R}$ is *piecewise $\mathscr{C}^1$* if there exists a splitting $\Pi = \{t_i\}_{i=0}^n$ of $[a, b]$ such that for each i, the restriction of f to $I_i = (t_i, t_{i+1})$ has $\mathscr{C}^1$ extension on I_i.

Remark 17.5.5. If f is piecewise $\mathscr{C}^1$, then at all but finitely many splitting points we have $f(x) = \frac{1}{2}\big(f(x^-) + f(x^+)\big)$. $\quad\diamond$

Lemma 17.5.6. *Assume I is a bounded real interval. If f and g are piecewise $\mathscr{C}^1$ on I, then $f + g$ and fg are piecewise $\mathscr{C}^1$. If in addition g is bounded away from 0, then f/g is piecewise $\mathscr{C}^1$.*

Proof. If Π_f is a splitting for f in the sense of Definition 17.5.4 and similarly Π_g is a splitting for g, then $\Pi = \Pi_f \cup \Pi_g$ is a splitting for both. The claims follow immediately from Lemma 17.5.3. $\quad\square$

Corollary 17.5.7. *If $a < b$, the set of piecewise $\mathscr{C}^1$ functions on $[a, b]$ is a vector subspace of $\mathscr{F}\big([a, b]\big)$, and a subalgebra if we consider pointwise multiplication.*

Without further mention, we work in $\mathscr{C}\big([-\pi, \pi]\big)$ with the standard inner product

$$\langle f, g \rangle = \frac{1}{2\pi} \int_{-\pi}^{\pi} f(x)g(x)\, dx.$$

Lemma 17.5.8. *The functions*

$$c_0 = 1, \qquad c_m(x) = \sqrt{2}\cos(mx), \quad s_m(x) = \sqrt{2}\sin(mx) \quad (m \geq 1)$$

form an orthonormal set.

Proof. See Exercise 17.5.1.   □

Definition 17.5.9. Assume f is integrable and 2π-periodic. The *spectral amplitudes* of f are the inner products $\langle f, c_m \rangle$ and $\langle f, s_m \rangle$. The *circular amplitudes* of f are the more conveniently scaled numbers

$$a_m = \frac{1}{\pi} \int_{-\pi}^{\pi} f(t) \cos mt \, dt, \qquad\qquad b_m = \frac{1}{\pi} \int_{-\pi}^{\pi} f(t) \sin mt \, dt.$$

If n is a natural number, the *nth spectral sum* of f is

$$f_n(x) = \frac{a_0}{2} + \sum_{m=1}^{n} \Big[a_m \cos(mx) + b_m \sin(mx) \Big].$$

The *spectral series* of f is

$$f_\infty(x) = \frac{a_0}{2} + \sum_{m=1}^{\infty} \Big[a_m \cos(mx) + b_m \sin(mx) \Big].$$

Remark 17.5.10. The summand $\frac{1}{2}a_0 = \langle f, c_0 \rangle c_0$ is the component of f along c_0, while

$$\left.\begin{array}{l} a_m \cos(mx) = \langle f, c_m \rangle c_m(x), \\ b_m \sin(mx) = \langle f, s_m \rangle s_m(x), \end{array}\right\} \quad (m \geq 1)$$

are components of f along c_m and s_m. The spectral series is the reconstruction of the "waveform" f from "pure harmonics" $\cos(mx)$ and $\sin(mx)$ of respective amplitudes a_m and b_m.   ◇

Theorem 17.5.11 (Piecewise-$\mathscr{C}^1$ spectral series). *If f is 2π-periodic and piecewise $\mathscr{C}^1$, the spectral series of f converges to $\frac{1}{2}\big(f(x^-) + f(x^+)\big)$ pointwise, and converges to f in the 2-norm:* $\|f_n - f\|_2 \to 0$.

Remark 17.5.12. In words, if f is piecewise $\mathscr{C}^1$, the spectral series at x converges to $f(x)$ if f is continuous at x, and to the average of the one-sided limits at each jump discontinuity of f.

For functions that are not piecewise-$\mathscr{C}^1$, the spectral series does not generally converge pointwise at all, much less to f. Nonetheless, the 2-norm $\|f - f_n\|_2$ converges to 0 in great generality: If f is measurable and square-integrable, then $f_\infty = f$ in the 2-norm completion of the space of integrable functions, see for example Dym and McKean, [6].   ◇

Remark 17.5.13. The proof of Theorem 17.5.11 is essentially a (lengthy) calculation. We separate off two steps: Lemma 17.5.15 shows the tails of the spectral series converge to 0, while Lemma 17.5.16 calculates the spectral sums as the convolution of f with a formal (not everywhere-positive) unit spike in a form to which Lemma 17.5.15 can be applied. ◇

Example 17.5.14. Let f be the 2π-periodic extension of $f(x) = x$ on $(-\pi, \pi)$, defined to be 0 at each integer multiple of π. The cosine amplitudes vanish because f is odd. The sine amplitudes are given by

$$b_m = \frac{1}{\pi} \int_{-\pi}^{\pi} x \sin(mx)\, dx = \frac{2}{\pi} \int_0^{\pi} x \sin(mx)\, dx = (-1)^{m-1} \frac{2}{m}.$$

Figure 17.3 shows the spectral sums with 4, 8, 24, and 48 terms. ◇

Lemma 17.5.15 (Rapid oscillation lemma). *Assume f is an integrable function on $[a, b]$. If β is real, then*

$$\lim_{\alpha \to \infty} \int_a^b f(t) \sin(\alpha t + \beta)\, dt = 0.$$

Proof. If χ is the indicator of $[c, d] \subseteq [a, b]$, then

$$\left| \int_a^b \chi \sin(\alpha t + \beta)\, dt \right| = \left| \int_c^d \sin(\alpha t + \beta)\, dt \right| = \left| \frac{1}{\alpha} \cos(\alpha t + \beta) \right|_c^d \right| \leq \frac{2}{\alpha},$$

which goes to 0 as $\alpha \to \infty$. Fix ε arbitrarily. Since f is integrable, there exists a step function g such that

$$\int_a^b \left| f(t) - g(t) \right| dt < \frac{\varepsilon}{2},$$

see Remark 9.1.11. Since g is a finite linear combination of indicators, there exists an R such that if $\alpha > R$, then

$$\int_a^b \left| g(t) \sin(\alpha t + \beta) \right| dt < \frac{\varepsilon}{2}.$$

The ordinary triangle inequality gives

$$\left| f(t) \sin(\alpha t + \beta) \right| \leq \left| \left(f(t) - g(t) \right) \sin(\alpha t + \beta) \right| + \left| g(t) \sin(\alpha t + \beta) \right|$$
$$\leq \left| f(t) - g(t) \right| + \left| g(t) \sin(\alpha t + \beta) \right|.$$

If $\alpha > R$, the triangle inequality for integrals guarantees

$$\left| \int_a^b f(t) \sin(\alpha t + \beta)\, dt \right| \leq \int_a^b \left| f(t) \sin(\alpha t + \beta) \right| dt$$
$$\leq \int_a^b \left| f(t) - g(t) \right| dt + \int_a^b \left| g(t) \sin(\alpha t + \beta) \right| dt < \varepsilon. \quad \square$$

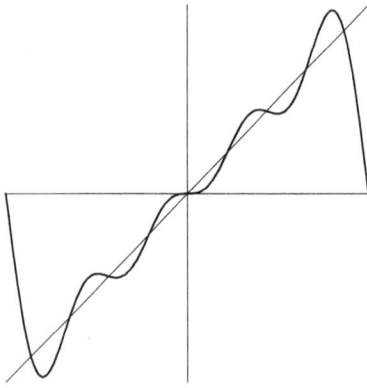

$$2\sum_{k=1}^{4}(-1)^n\frac{\sin nx}{n}$$

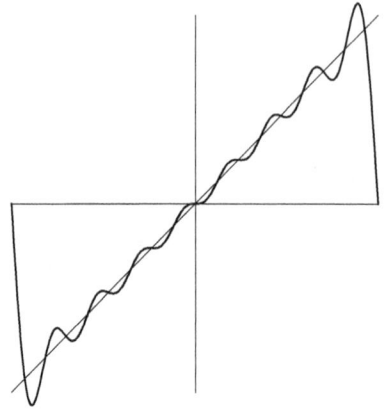

$$2\sum_{k=1}^{8}(-1)^n\frac{\sin nx}{n}$$

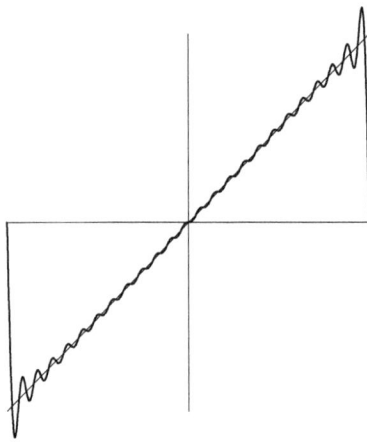

$$2\sum_{k=1}^{24}(-1)^n\frac{\sin nx}{n}$$

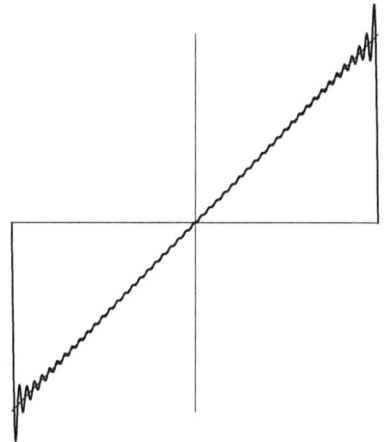

$$2\sum_{k=1}^{48}(-1)^n\frac{\sin nx}{n}$$

FIGURE 17.3

Spectral sums of $f(x) = x$ on $(-\pi, \pi)$.

Lemma 17.5.16. *If f is 2π-periodic and piecewise $\mathscr{C}^1$, and $n \geq 1$, then*

$$f_n(x) = \frac{2}{\pi}\int_0^{\pi}\frac{f(x+t)+f(x-t)}{2}\frac{\sin(n+\frac{1}{2})t}{2\sin\frac{t}{2}}\,dt.$$

Proof. For each positive integer m, we have

$$a_m\cos(mx) = \left[\frac{1}{\pi}\int_{-\pi}^{\pi}f(t)\cos(mt)\,dt\right]\cos(mx),$$

$$b_m\sin(mx) = \left[\frac{1}{\pi}\int_{-\pi}^{\pi}f(t)\sin(mt)\,dt\right]\sin(mx),$$

and therefore

$$a_m \cos(mx) + b_m \sin(mx)$$
$$= \frac{1}{\pi} \int_{-\pi}^{\pi} f(t) \Big[\cos(mt) \cos(mx) + \sin(mt) \sin(mx) \Big] dt$$
$$= \frac{1}{\pi} \int_{-\pi}^{\pi} f(t) \cos\big(m(x - t)\big) dt \quad \text{(Proposition 13.1.5 (iv))}$$
$$= \frac{1}{\pi} \int_{x-\pi}^{x+\pi} f(x - t) \cos(mt) dt$$
$$= \frac{1}{\pi} \int_{-\pi}^{\pi} f(x - t) \cos(mt) dt \quad \text{(the integrand is } 2\pi\text{-periodic)}$$
$$= \frac{1}{\pi} \left[\int_{0}^{\pi} f(x + t) \cos(mt) dt + \int_{0}^{\pi} f(x - t) \cos(mt) dt \right]$$
$$= \frac{2}{\pi} \int_{0}^{\pi} \frac{f(x + t) + f(x - t)}{2} \cos(mt) dt.$$

Substituting into the spectral sum,

$$f_n(x) = \frac{2}{\pi} \int_{0}^{\pi} \frac{f(x + t) + f(x - t)}{2} \left[\frac{1}{2} + \sum_{m=1}^{n} \cos(mt) \right] dt.$$

To complete the proof, it suffices to show

$$\frac{1}{2} + \sum_{m=1}^{n} \cos(mt) = \frac{\sin(n + \frac{1}{2})t}{2 \sin \frac{t}{2}}.$$

This formula is immediate if $n = 0$. The inductive step follows from the identity $2 \sin \frac{t}{2} \cos(n + 1)t = \sin(n + \frac{3}{2})t - \sin(n + \frac{1}{2})t$, see Corollary 13.1.6. (Exercise 14.2.11 gives a "complex" proof using the polar formula.) $\square$

Proof of Theorem 17.5.11. If $0 < \delta < \pi$, then

$$f_n(x) = \frac{2}{\pi} \int_{0}^{\delta} \frac{f(x + t) + f(x - t)}{2} \frac{\sin(n + \frac{1}{2})t}{2 \sin \frac{t}{2}} dt$$
$$+ \frac{2}{\pi} \int_{\delta}^{\pi} \frac{f(x + t) + f(x - t)}{2} \frac{\sin(n + \frac{1}{2})t}{2 \sin \frac{t}{2}} dt.$$

By rapid oscillation (Lemma 17.5.15), the second integral vanishes in the limit as $n \to \infty$ since $\sin \frac{t}{2}$ is bounded away from 0. Thus $0 < \delta < \pi$ implies

$$f_\infty(x) = \lim_{n \to \infty} \frac{2}{\pi} \int_{0}^{\delta} \frac{f(x + t) + f(x - t)}{2} \frac{\sin(n + \frac{1}{2})t}{2 \sin \frac{t}{2}} dt$$

in the sense that the spectral series converges at x to the limit on the right if and only if this limit exists. Introduce

$$g_x(t) := \frac{f(x + t) + f(x - t)}{2} \frac{t}{2 \sin \frac{t}{2}},$$

which is piecewise $\mathscr{C}^1$ by Lemma 17.5.6 and satisfies $g_x(0^+) = \frac{1}{2}\big(f(x^+) + f(x^-)\big)$. Substituting $g_x(t) = \big(g_x(t) - g_x(0^+)\big) + g_x(0^+)$, we have

$$
\begin{aligned}
f_\infty(x) &= \lim_{n\to\infty} \frac{2}{\pi} \int_0^\delta g_x(t) \frac{\sin(n+\frac{1}{2})t}{t}\, dt \\
&= \lim_{n\to\infty} \frac{2}{\pi} \int_0^\delta \left[\frac{g_x(t) - g_x(0^+)}{t} \sin(n+\tfrac{1}{2})t + g_x(0^+) \cdot \frac{\sin(n+\frac{1}{2})t}{t} \right] dt.
\end{aligned}
$$

The one-sided derivative

$$
g_x'(0^+) = \lim_{t\to 0^+} \frac{g_x(t) - g_x(0^+)}{t}
$$

exists, so the first integral vanishes in the limit by rapid oscillation. Consequently, the spectral series converges to

$$
g_x(0^+) \cdot \lim_{\alpha\to\infty} \frac{2}{\pi} \int_0^\delta \frac{\sin(\alpha t)}{t}\, dt = g_x(0^+) \cdot \frac{2}{\pi} \int_0^\infty \frac{\sin t}{t}\, dt = g_x(0^+)
$$

by Proposition 14.2.10.

The claim $\|f - f_n\|_2 \to 0$ is Exercise 17.5.3. $\square$

Spectral Isometry

Lemma 17.5.17. *If*

$$
f_n(x) = \frac{a_0}{2} + \sum_{m=1}^n \Big[a_m \cos(mx) + b_m \sin(mx) \Big]
$$

is a spectral sum, then

$$
\|f_n\|_2^2 = \frac{1}{2\pi} \int_{-\pi}^\pi |f_n(x)|^2\, dx = \frac{1}{2} \left[\frac{a_0^2}{2} + \sum_{m=1}^n (a_m^2 + b_m^2) \right].
$$

Proof. The functions $\{1, \cos(mx), \sin(mx)\}_{m=1}^n$ are mutually orthogonal on $[-\pi, \pi]$ with respect to the standard inner product, and

$$
\frac{1}{2\pi} \int_{-\pi}^\pi dx = 1, \qquad \frac{1}{2\pi} \int_{-\pi}^\pi \cos^2(mx)\, dx = \frac{1}{2\pi} \int_{-\pi}^\pi \sin^2(mx)\, dx = \frac{1}{2}
$$

if $m \geq 1$. The lemma follows at once from the hypotenuse theorem. $\square$

Theorem 17.5.18 (Spectral isometry). *If f is a 2π-periodic and piecewise $\mathscr{C}^1$ function, then*

$$
\|f\|_2^2 = \frac{1}{2\pi} \int_{-\pi}^\pi |f(x)|^2\, dx = \frac{1}{2} \left[\frac{a_0^2}{2} + \sum_{m=1}^\infty (a_m^2 + b_m^2) \right].
$$

Proof. Theorem 17.5.11 guarantees $\|f - f_n\|_2 \to 0$. By Proposition 15.2.14,

$$\|f\|_2^2 = \|f_n\|_2^2 + \|f - f_n\|_2^2 \to \|f_\infty\|_2^2. \qquad \square$$

Example 17.5.19. For the function $f(x) = x$, Example 17.5.14 gives $a_m = 0$ for all non-negative m and $b_m = (-1)^{m-1} 2/m$ for all positive m. By Theorem 17.5.18,

$$\frac{\pi^2}{3} = \frac{1}{2\pi} \int_{-\pi}^{\pi} x^2 \, dx = \frac{1}{2} \sum_{m=1}^{\infty} \frac{4}{m^2}, \quad \text{or} \quad \zeta(2) = \sum_{m=1}^{\infty} \frac{1}{m^2} = \frac{\pi^2}{6}. \qquad \diamondsuit$$

Exercises for Section 17.5

Exercise 17.5.1. ($\bigstar$) Prove Lemma 17.5.8.

Exercise 17.5.2. (A) As in Example 17.5.14, find the spectral decomposition of the indicated functions, and determine the conclusion of Theorem 17.5.18.

(a) $f(x) = \operatorname{sgn} x.$ \qquad (b) $f(x) = |x|,$ \qquad (c) $f(x) = x^2.$

In each part, the formula holds on $(-\pi, \pi)$ and the function is extended by periodicity. If you have a plotting program, plot each function and several of its spectral sums.

Exercise 17.5.3. ($\bigstar$) Prove that if f is integrable on $[-\pi, \pi]$ and f_n is the nth spectral sum, then $\|f - f_n\|_2 \to 0$. Suggestion: Approximate f by a step function and use Lemma 17.5.15.

Exercise 17.5.4. Assume f is of class $\mathscr{C}^1$ on $\mathbf{R}$, 2π-periodic, and has circular amplitudes a_m and b_m. Find the circular amplitudes a'_m and b'_m of f'.

Exercise 17.5.5. Each part (a)–(c) refers to the complex inner product space V of functions $f = u + iv$, u and v piecewise $\mathscr{C}^1$, with standard complex inner product

$$\langle f, g \rangle = \frac{1}{2\pi} \int_0^{2\pi} f(x)\overline{g(x)} \, dx.$$

(a) Prove that the collection $e_n(x) = e^{inx}$ is orthonormal.

(b) If $f \in V$ and n is an arbitrary integer, define the *complex spectral amplitude* by $c_n = \langle f, e_n \rangle$, and the *complex spectral series* of f by

$$s(x) = \lim_{N \to \infty} s_N(x) = \lim_{N \to \infty} \sum_{n=-N}^{N} c_n e^{inx}.$$

Prove that $s(x) = \frac{1}{2}\big(f(x^-) + f(x^+)\big)$. Suggestions: See Exercise 14.2.11, and apply Theorem 17.5.11 to the real and imaginary parts of f.

(c) Prove that $\|f\|_2^2 = \sum\limits_{n=-\infty}^{\infty} |c_n|^2$.

Exercise 17.5.6. In the spirit of a "capstone" to your travels through this book, this exercise draws together three threads: The rational sequence (b_k) introduced recursively in Exercise 12.3.10 by

$$b_0 = 0, \qquad b_k = -\frac{1}{k+1}\sum_{j=0}^{k-1}\binom{k+1}{j}b_j,$$

which satisfies

$$t\coth t = \sum_{m=0}^{\infty}\frac{b_{2m}2^{2m}t^{2m}}{(2m)!};$$

the ζ function of Exercise 12.4.10; and complex spectral series of Exercise 17.5.5. The end result is the evaluation, for each positive integer m,

$$\zeta(2m) = \sum_{n=1}^{\infty}\frac{1}{n^{2m}} = \frac{(-1)^{m-1}b_{2m}(2\pi)^{2m}}{2\cdot(2m)!}.$$

The first two of these, $\zeta(2) = \pi^2/6$ and $\zeta(4) = \pi^4/90$, are calculated somewhat laboriously in Exercise 17.5.2. Generally, $\zeta(2m)$ is a rational multiple of π^{2m}.

Let t denote a real parameter, and define $f(x) = e^{tx}$ on $[0, 2\pi]$. Calculate the complex spectral amplitudes c_n of f, and apply Exercise 17.5.5 (c). You will need to change the order of a double series, and should look to apply the identity theorem for power series in t.

A

Solutions to Selected Exercises

Solutions: Logic and Sets

Exercises for Section 1.1

Ex. 1.1.1. We have

P	Q	$\neg P$	$\neg P$ or Q	P implies Q
T	T	F	T	T
T	F	F	F	F
F	T	T	T	T
F	F	T	T	T

Ex. 1.1.3. One powerful intuitive cognitive mode humans have is the ability to detect "cheating." Here is the relevant insight about the cards:

(a) The legal drinking age in a certain state is 21. Your job at a gathering is to ensure that no one under 21 years of age is drinking alcohol, and to report those that are. A group of four people consists of a 20 year old who is drinking, a 46 year old who is drinking, a 16 year old who is not drinking, and a 25 year old who is not drinking. Which of these people is/are violating the law?

(b) After reporting this incident, you find four people at the bar: An 18 year old and a 35 year old with their backs to you, and two people of unknown age, one of whom is drinking. From which people do you need further information to see whether or not they are violating the law?

One major goal of this book is to help you team up your powerful social intuition with your analytical ability to think about mathematics. Each mode has complementary strengths. Together they are a superpower duo.

Ex. 1.1.4. To be deeply philosophical for a moment, *finding the answer to a question* and *verifying a proposed solution* are different. The first is looking for a needle in a haystack. The second is verifying you are holding a needle.

"Solving" is a relatively loose process where we search for prospective needles. The "absolute, ironclad" part is the proof that the prospective needle really is a needle, and perhaps that there are no other needles to be found.

This is not a casual analogy, but a structural feature of mathematics. Discovering and writing proofs are nearly opposite activities, logically speaking. To discover a proof (and before that, to guess what is true, or *formulate a conjecture*), we must permit ourselves to make partial, open-ended, possibly unjustified guesses and see where they lead. Ultimately, most of the work we do in discovering mathematics does not need to be written up; it's the process we used to sift the hay, and pieces of hay we examined that were not needles.

(a) The requested hypothesis is P: "The real number x is equal to [such and such]." The requested conclusion is Q "$0 = x^3 - 4x^2 - 4x + 16$." At the start we do not know which numbers are [such and such]. Our first task is to make an educated guess by "solving the equation."

The calculation proceeds by assuming Q and deducing P: If $0 = x^3 - 4x^2 - 4x + 16$, *then* $0 = (x - 4)(x^2 - 4) = (x - 4)(x - 2)(x + 2)$, so $x = 4$ or $x = 2$ or $x = -2$. This is the converse of what was asked, but adds crucial information: *We now know exactly which numbers are such and such.*

A literal answer to the original question should show that (i) each of $x = 4$, 2, and -2 satisfies the polynomial equation, and (ii) no other numbers do. It remains to see why in practice we do not do this.

In algebra, teachers sometimes speak of "allowable" operations, ways we can transform an equation into a new equation *without losing or gaining solutions*. Logically, these are if and only if deductions. Factoring and expanding are allowable in this sense. Adding expressions to both sides and multiplying by non-zero numbers are similarly allowable. If a calculation consists entirely of logically reversible steps, the solution reached by calculating is logically a solution. The calculation is a "trail of bread crumbs" from P back to Q. Since verification that we have a solution is often tedious, it tends to get omitted.

(b) Each step in the argument is a true deduction, so we have a valid proof of "If $-1 = 1$ (Q), then $1 = 1$ (P)." Treating this argument the way we treat a calculation with logically reversible steps, however, is an error: The statement $1 = 1$ is tautological, but $-1 = 1$ is false. Abstractly, "Q implies P" is (vacuously) true, but "P implies Q" is false.

What went wrong? Here, the first step, squaring, is not logically reversible: Two different real numbers may have the same square. Specifically, -1 and 1 *do* have the same square. When we write proofs, it's best to avoid this two-column style. For one thing, it can introduce errors. As a result, it looks unprofessional even if correct.

The preferred style for reporting calculations is chains of equalities, or of inequalities from smaller to larger. This book makes a special effort to instill good habits.

(c) If you are trying to "debug" a calculation, to find the numerical and/or logical errors, it may help to consider specific cases. The whole reason algebra works, after all, is that it allows us to calculate with unknown quantities. Setting variables to specific numbers in a correct calculation must give a correct chain of deductions. If it does not, that is an error of the original.

So, let's try our favorite real numbers, $a = b = 0$. Every step until the last reads $0 = 0$; the last reads $2 = 1$. The last step, canceling a, is not logically correct if $a = 0$. What if we try our second-favorite real numbers, $a = b = 1$? Now the chain of statements starts $1 = 1$, becomes $0 = 0$ for two steps, then becomes $2 = 1$. The canceled common factor after the third step is $b - a = 0$, leading to the false implication "If $0 = 0$, then $2 = 1$." Arguably this is the heart of the problem: By hypothesis, $a = b$, so no matter what numerical value we pick, this step is division by 0.

The remaining steps, incidentally, are readily checked to be logically reversible, so we have found the two questionable steps, one wrong if $a \neq 0$ and one wrong if $a = 0$. Also, your first three favorite real numbers, in order, are 0, 1, and -1. After that you can pick whatever idiosyncratic choice you like.

Ex. 1.1.6. The maximum, $10,000 plus the larger of $500 and the value of the Hotel Resort Platinum Getaway, is unknown. The minimum is the value of the two scratch tickets, effectively $0.

Exercises for Section 1.2

Ex. 1.2.1. Each quantifier ("for every" and "there exists") specifies conditions on a variable. If this variable does not appear in the predicate (as in (a)), then the quantifier is misused. Separately, quantifiers come before predicates. Any condition referring to variables before they have been quantified (as in (b) and (c)) is potentially anomalous. Finally, placing quantifiers inside an implication conveys that the quantification is part of the hypothesis. That may be the intent, but can likely be reworded to improve clarity.

(a) The predicate "$2 + 2 = 4$" does not depend on x, so the entire statement is insensitive to quantifying x. The statement is true, but merely means "$2 + 2 = 4$."

(b) Rephrasing to place the quantifier first gives the implication, "If, for every (real) x such that $1 < x$, (we have) $0 < x$, then $0 < 1 < x^2$." This is true but a logical disconnect: The hypothesis $0 < x$ is guaranteed by the universal quantification, as is the conclusion $0 < 1 < x^2$, but $0 < x$ does not imply $0 < 1 < x^2$.

The presumed meaning is "For all (real) x, if $1 < x$ then $1 < x^2$." This is both true, and the conclusion really does "follow" from the hypothesis and properties of inequalities.

(c) The sentence "If $y = x^2$ for every $x > 0$, then $y > 0$," refers to x and y before x is quantified, y is not explicitly quantified, and colloquial meaning suggests, in particular, that $y = 1^2$ and $y = 2^2 = 4$, making "$y = x^2$ for every $x > 0$" a false hypothesis, so the implication is vacuously true. All these features are anomalous.

Rephrasing to put the quantifier first gives the presumed meaning, "For every (positive real) x (and every real y), if $y = x^2$ then $y > 0$." This is true, but again for different reasons than the original wording is true.

Ex. 1.2.3. For every x in $[0, 1]$, there exists a y in $[0, 1]$ such that $x < y$. This is false (if $x = 1$, no y exists), so the first payer has a winning strategy.

Exercises for Section 1.3

Ex. 1.3.1. For each x in X let $P(x)$ be the statement $x \in X$ and let $Q(x)$ be $x \in Y$. We have

$$X \subseteq Y \text{ if and only if for every } x, P(x) \text{ implies } Q(x)$$
$$\text{if and only if for every } x, \neg Q(x) \text{ implies } \neg P(x)$$
$$\text{if and only if for every } x, Y^c \subseteq X^c.$$

The middle step replaces an implication by its contrapositive.

Ex. 1.3.3. In each part we tabulate truth values using all combinations of truth values for P and Q. Equivalence of the indicated statement in encoded by equality of the fourth and seventh columns.

(a) We have

P	Q	P or Q	$\neg(P$ or $Q)$	$\neg P$	$\neg Q$	$\neg P$ and $\neg Q$
T	T	T	F	F	F	F
T	F	T	F	F	T	F
F	T	T	F	T	F	F
F	F	F	T	T	T	T

(b) Similarly,

P	Q	P and Q	$\neg(P$ and $Q)$	$\neg P$	$\neg Q$	$\neg P$ or $\neg Q$
T	T	T	F	F	F	F
T	F	F	T	F	T	T
F	T	F	T	T	F	T
F	F	F	T	T	T	T

(c) For each x in the universe, let $P(x)$ be the statement $x \in X$ and $Q(x)$ the statement $x \in Y$. Since complements correspond to negation, unions to "or," and intersections to "and," the logical equivalences in (a) and (b) translate directly to the set identities here.

Ex. 1.3.4. For all x in $\mathcal{U}$, if $x \in (X \cup Y) \cap Z$, then $x \in (X \cup Y)$ and $x \in Z$. Consequently, $x \in X$ or $x \in Y$, and $x \in Z$.

If $x \in X$, then $x \in (X \cap Z) \subseteq (X \cap Z) \cup (Y \cap Z)$. Similarly, if $x \in Y$, then $x \in (Y \cap Z) \subseteq (X \cap Z) \cup (Y \cap Z)$. Since x was arbitrary, we have shown $(X \cup Y) \cap Z \subseteq (X \cap Z) \cup (Y \cap Z)$.

Conversely, if $x \in (X \cap Z) \cup (Y \cap Z)$, then $x \in X \cap Z \subseteq (X \cup Y) \cap Z$, or $x \in Y \cap Z \subseteq (X \cup Y) \cap Z$. Since x was arbitrary, $(X \cup Y) \cap Z \supseteq (X \cap Z) \cup (Y \cap Z)$.

The proof of the other identity is entirely similar.

Ex. 1.3.6. We can argue "inductively" using Example 1.3.26: Every subset of X either does not contain 2 or does. This gives, respectively,

$$A' : \quad \varnothing \quad \{0\} \quad \{1\} \quad \{0,1\}$$
$$A : \quad \{2\} \quad \{0,2\} \quad \{1,2\} \quad \{0,1,2\}.$$

In words, there are four subsets not containing 2, namely elements of the power set of $X' := X \setminus \{2\}$. Further, for each such set A', there is a subset $A := A' \cup \{2\}$. Finally, if A' and B' are subsets of X', then $A' = B'$ if and only if $A = B$. Consequently, there are four subsets of X that contain 2, making eight subsets of X, or eight elements of $\mathscr{P}(X)$.

Exercises for Section 1.4

Ex. 1.4.1. In each part, we may visualize $X \times X$ as a 3×3 array of os. We are counting how many ways we can fill in three os, one in each column.

(a) There are three choices for $f(0)$, three for $f(1)$, and three for $f(2)$. These choices are independent, so the total number of mappings is their product, $3 \cdot 3 \cdot 3 = 27$.

(b) There are three choices for $f(0)$, but only two for $f(1)$ since $f(1) \neq f(0)$, and only one choice for $f(2)$. The total number of mappings is their product, $3 \cdot 2 \cdot 1 = 6$. Listing them as suggested gives $(0,1,2)$, $(0,2,1)$, $(1,0,2)$, $(1,2,0)$, $(2,0,1)$, and $(2,1,0)$.

Ex. 1.4.2. Without loss of generality, assume $X = \{0,1\}$.

(a) There are five choices for $f(0)$ and five for $f(1)$. These choices are independent, so the total number of mappings is their product, $5 \cdot 5 = 25$.

If instead the mapping takes distinct values at the two inputs, there are only four choices for $f(1)$, so the total number of mappings is $5 \cdot 4 = 20$.

(b) Every partition of Y into two subsets comprises either a singleton and a set of four elements, or a set of two elements and a set of three. There are five singleton subsets of Y, one for each element.

To count subsets of two elements, we can put the elements of Y in order, say $Y = \{0,1,2,3,4\}$, to avoid double-counting. There are four subsets containing 0, three subsets of smallest element 1, two subsets of smallest element 2, and one subset of smallest element 3, ten in all. Consequently, there are 15 partitions of Y into two subsets.

Ex. 1.4.5. Assume $\{\{x\}, \{x,y\}\} = \{\{x'\}, \{x',y'\}\}$. If each is a singleton, then in particular, $\{x\} = \{x'\}$, so $x = x'$, and $\{x,y\} = \{x',y'\} = \{x,y'\}$, so $y = y'$.

Otherwise $\{x\} \neq \{x,y\}$, so $y \neq x$. By definition of set equality, the singleton $\{x'\}$, which is an element of $\{\{x\}, \{x,y\}\}$, is equal to $\{x\}$ or to $\{x,y\}$. The second is false since $\{x,y\}$ is not a singleton. We deduce $\{x'\} = \{x\}$, so $x = x'$, and $\{x,y\} = \{x',y'\} = \{x,y'\}$, so $y = y'$.

Ex. 1.4.6. Hint: This amounts to the observation that if $A \subseteq X$, there is a unique truth function f_A that asks each element x of X, "Are you an element of A?"

Ex. 1.4.7. Answers:

	$m \neq n$	$m < n$	$m \leq n$	$0 < mn$	$0 \leq mn$
Reflexive	N	N	Y	N	Y
Symmetric	Y	N	N	Y	Y
Transitive	N	Y	Y	Y	N

Ex. 1.4.9. A deck of playing cards has four suits, $S = \{\clubsuit, \diamondsuit, \heartsuit, \spadesuit\}$.

(a) $[\clubsuit] = [\spadesuit] = \{\clubsuit, \spadesuit\}$ and $[\diamondsuit] = [\heartsuit] = \{\diamondsuit, \heartsuit\}$.

(b) No: A card's color does not determine its suit.

Ex. 1.4.12. First check whether R is reflexive, symmetric, and/or transitive.

(a) R is an equivalence relation. For transitivity, if $x - x'$ and $x' - x''$ are even, so is their sum $(x - x') + (x' - x'') = x - x''$. The equivalence classes are the set $[0] = 2\mathbf{Z}$ of even integers and the set $[1] = 2\mathbf{Z} + 1$ of odd integers.

(b) This relation is symmetric but neither reflexive nor transitive.

Solutions: Induction

Exercises for Section 2.1

Ex. 2.1.4. For each natural number n, let $P(n)$ be the statement $2^n < n!$.

If $2^k < k!$ for some positive integer k, then by commutativity of multiplication and the definitions of exponentiation and factorials,

$$2^{k+1} = 2^k \cdot 2 < k! \cdot 2 = 2 \cdot k! \leq (k+1) \cdot k! = (k+1)!.$$

That is, $P(k)$ implies $P(k+1)$ for all k.

The statement $P(4)$ reads $2^4 < 4!$, or $16 < 24$, which is true. By induction and Remark 2.1.3, $P(n)$ is true if $n \geq 4$.

Exercises for Section 2.2

Ex. 2.2.1. The number of ways is

$$\binom{52}{5} = \frac{52!}{5!\,47!} = \frac{52 \cdot 51 \cdot 50 \cdot 49 \cdot 48}{5 \cdot 4 \cdot 3 \cdot 2}.$$

Canceling $5 \cdot 2 = 10$ from 50 and $4 \cdot 3 = 12$ from 48 gives

$$52 \cdot 51 \cdot 5 \cdot 49 \cdot 4 = (20 \cdot 52) \cdot (51 \cdot 49) = 1040 \cdot (50^2 - 1) = 1040 \cdot 2499.$$

Since $40 \cdot 2499 = 100,000 - 40 = 99960$ and $1000 \cdot 2499 = 2499000$, we have $\binom{52}{5} = 2,598,960$.

Ex. 2.2.3. To a laughable extent, *no*. The number of orderings of a deck of 52 playing cards is $52!$. Without a calculator, this is greater than 10^{43} since 43 of the factors are 10 or larger. (With a calculator, the actual value is a bit less than 8.1×10^{67}.) No macroscopic event on earth has occurred 10^{43} times. For example, if 100 raindrops fell every second on every square foot of the earth non-stop for 4.6 billion years, the total number of drops would be on the order of 10^{35}, one hundred-millionth of the target. The total number of earthly card shuffles surely does not exceed 10^{21} (a trillion dealers each shuffling one billion times), so in every practical sense *zero percent* of possible orderings have been seen.

Ex. 2.2.4. Answers: We have $9!! = 945$, $10!! = 3840$ (if these are correct, your earlier values probably are as well), and $(2n)!! = 2^n n!$, so $(2n+1)!! = (2n+1)!/(2n)!! = (2n+1)!/(2^n n!)$.

Ex. 2.2.5. Geometrically, B^n may be viewed as the set of vertices of a cube with unit sides in n-space. Figure A.1 uses light gray lines to indicate the edges.

(a) For each natural number n, let $P(n)$ be the statement, "If b and b' are arbitrary elements of B^n, then $|b' - b|$ is the number of components in which b and b' differ." Since B^0 has only one element and the value of an empty sum is 0, the statement $P(0)$ is true: (0) differs from (0) in 0 components. The "real" base case is $P(1)$: The distance between two elements of $B^1 = B$ is 1 if the elements differ and 0 if the elements are the same; that is, $|b' - b|$ is the number of components in which b and b' differ.

Assume inductively that $P(m)$ is true for some positive integer m. If $b = (b_k)_{k=0}^m$ and $b' = (b'_k)_{k=0}^m$ are arbitrary elements of B^{m+1}, then by definition of distance and the recursive definition of summation,

$$|b' - b| = \sum_{k=0}^m |b'_k - b_k| = \left[\sum_{k=0}^{m-1} |b'_k - b_k|\right] + |b'_m - b_m|.$$

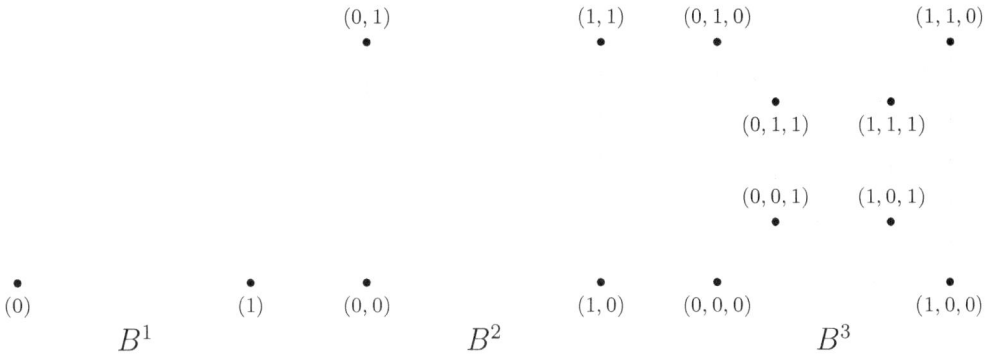

$(0,1)$ $(1,1)$ $(0,1,0)$ $(1,1,0)$

$(0,1,1)$ $(1,1,1)$

$(0,0,1)$ $(1,0,1)$

(0) (1) $(0,0)$ $(1,0)$ $(0,0,0)$ $(1,0,0)$

B^1 B^2 B^3

FIGURE A.1
One way to depict the sets B^1, B^2, and B^3.

The sum in brackets is the distance between $(b_k)_{k=0}^{m-1}$ and $(b_k')_{k=0}^{m-1}$. By the inductive hypothesis, this is the number of components in which these elements differ. The final summand "counts" whether the final components differ. Consequently, $|b' - b|$ is the number of components in which b and b' differ. Since $P(0)$ and $P(1)$ are true and $P(m)$ implies $P(m+1)$ for every positive integer m, induction guarantees $P(n)$ is true for every natural number n.

Consequently, $|b' - b| = 0$ if and only if b and b' agree in all components, if and only if $b = b'$ (positivity); the distance $|b - b'|$ is the number of components in which b' and b differ, a concept independent of which tuple is which, so $|b - b'| = |b' - b|$ (symmetry); and $|b'' - b|$, the number of components that must be "flipped" to change b into b'', does not exceed $|b'' - b'| + |b' - b|$, the number of components that must be flipped to change b into b' and then to change b' into b'' (the triangle inequality).

TABLE A.1
Points at specified distance from $\mathbf{0}$ in an n-cube.

	$m = 1$		$m = 2$		$m = 3$	
B^1	$\binom{1}{1} = 1$	(1)				
B^2	$\binom{2}{1} = 2$	$(1,0)$ $(0,1)$	$\binom{2}{2} = 1$	$(1,1)$		
B^3	$\binom{3}{1} = 3$	$(1,0,0)$ $(0,1,0)$ $(0,0,1)$	$\binom{3}{2} = 3$	$(0,1,1)$ $(1,0,1)$ $(1,1,0)$	$\binom{3}{3} = 1$	$(1,1,1)$

(b) If $b \in B^n$, then for all b' in B^n and all m such that $0 \le m \le n$, $|b' - b| = m$ if and only if b and b' differ in precisely m components of n. There are $\binom{n}{m}$ ways to pick those m components, so there are precisely $\binom{n}{m}$ elements of B^n at distance m from b. These sets of points partition B^n. Since B^n has 2^n elements, compare Exercise 2.1.5, we have

$$2^n = \text{number of elements of } B^n = \sum_{m=0}^{n} \binom{n}{m}.$$

In Figure A.1 we may take $b = \mathbf{0}$ to be the n-tuple whose components are all 0. Table A.1 shows the numbers of points at distance m from b.

The $m = 0$ column is omitted; $\binom{n}{0} = 1$ if $n \ge 0$, and the unique point at distance 0 is $\mathbf{0}$ itself.

Exercises for Section 2.3

Ex. 2.3.1. Hint: Think of $X \times X$ as an $n \times n$ array, with a diagonal comprising pairs (j, j). For each question, count how many choices there are.

Ex. 2.3.2. Assume a, b, and c denote arbitrary integers.

(a) By definition of $*$,

$$(a * b) * c = (a + b - 1) * c = (a + b - 1) + c - 1$$
$$= a + (b + c - 1) - 1 = a + (b * c) - 1 = a * (b * c).$$

Commutativity is immediate from $b + a - 1 = a + b - 1$.

(b) It suffices to check $a * 1 = a$ and $1 * a = a$ for all a.

To locate the "candidate" identity element $e = 1$, consider the condition $a * e = a$ for all a, namely, $a + e - 1 = a$ for all a, and if possible solve for e.

For inverses, it suffices to prove that for every a, the integer $b = 2 - a$ satisfies $a * b = 1$ and $b * a = 1$. As above, solving $a * b = e = 1$ for b gives the "candidate" inverse.

Ex. 2.3.4. Assume A, B, and C are arbitrary subsets of X.

(a) For all x in X, $x \in A \cup B$ if and only if $x \in A$ or $x \in B$, if and only if $x \in B$ or $x \in A$, if and only if $x \in B \cup A$. That is, $B \cup A = A \cup B$.

Similarly for associativity, $x \in A \cup (B \cup C)$ if and only if $x \in A$ or $x \in B \cup C$, if and only if $x \in A$ or $x \in B$ or $x \in C$, if and only if $x \in A \cup B$ or $x \in C$, if and only if $x \in (A \cup B) \cup C$.

(b) If E is an identity element for $\cup$, then $E \cup A = A$ for all A. Particularly, $E \subseteq A$ for all A, so the only candidate is $\varnothing$. Since $\varnothing \cup A = A$ for all A, $\varnothing$ is the identity element for $\cup$. A set A is invertible if and only if there exists a set B such that $A \cup B = \varnothing$, if and only if $A = \varnothing$.

Ex. 2.3.6. Suggestion: Fix a "standard grouping" inductively, such as $p_2 = (x_0 * x_1) * x_2$ and $p_n = p_{n-1} * x_n$, and show inductively that an arbitrary grouping is equal to the standard grouping.

Exercises for Section 2.4

Ex. 2.4.2. By hypothesis, there exists a natural number $k \neq 0$ such that $m + k = n$. By associativity of addition, $(\ell + m) + k = \ell + (m + k) = \ell + n$, so $\ell + m < \ell + n$ by definition.

Since ℓk is a natural number, distributivity implies $(\ell m) + (\ell k) = \ell(m+k) = \ell n$, so $\ell m \leq \ell n$. If $\ell = 0$, both sides are zero and equality holds. If $\ell \neq 0$, then $\ell k \neq 0$, and strict inequality holds.

Ex. 2.4.3. Hint for transitivity: First prove if ℓ and m are natural numbers, and $\ell + n = m + n$ for some n, then $\ell = m$.

Ex. 2.4.4. Hint: It suffices to prove that if $a_1 = [(m_1, n_1)]$, $a_1' = [(m_1', n_1')]$, $a_2 = [(m_2, n_2)]$, $a_2' = [(m_2', n_2')]$ are integers such that $a_1 = a_1'$ and $a_2 = a_2'$, then $a_1 + a_2 = a_1' + a_2'$.

Ex. 2.4.7. Let ℓ, m, and n denote arbitrary natural numbers.

(a) As avatars, $(m, 0) + (n, 0) = (m + n, 0)$ and $(m, 0)(n, 0) = (mn, 0)$.

(b) Assume (ℓ, m) is an arbitrary integer avatar. Precisely one of the following is true: $\ell = m$, or $m < \ell$, or $\ell < m$.

If $\ell = m$, then $(\ell, m) = (\ell, \ell) \equiv (0, 0)$.

If $m < \ell$, then $\ell = m+k$ for some non-zero k, so $(\ell, m) = (m+k, m) \equiv (k, 0)$.

If $\ell < m$, then $m = \ell+k$ for some non-zero k, so $(\ell, m) = (\ell, \ell+k) \equiv (0, k)$.

Ex. 2.4.9. If $r_j = (p_j, q_j)$ and $r_j' = (p_j', q_j')$, $j = 0, 1$, are equivalent rational avatars, then

$$
\begin{aligned}
(p_0, q_0)(p_1, q_1) &= (p_0 p_1, q_0 q_1) && \text{definition of multiplication} \\
&\equiv (p_0 p_1 q_0' q_1', q_0 q_1 q_0' q_1') && \text{equivalent representation} \\
&= (p_0' p_1' q_0 q_1, q_0 q_1 q_0' q_1') && p_0 q_0' = p_0' q_0, \quad p_1 q_1' = p_1' q_1 \\
&\equiv (p_0' p_1', q_0' q_1') && \text{equivalent representation} \\
&= (p_0', q_0')(p_1', q_1'),
\end{aligned}
$$

so multiplication is well-defined. Commutativity is immediate from the formula:

$$(p_0, q_0)(p_1, q_1) = (p_0 p_1, q_0 q_1) = (p_1 p_0, q_1 q_0) = (p_1, q_1)(p_0, q_0).$$

Solutions: Real Numbers

Exercises for Section 3.1

Ex. 3.1.1. We have

$$
\begin{aligned}
(x + y)^2 &= (x + y)(x + y) && \text{def. of squaring} \\
&= (x + y)x + (x + y)y && \text{distributive law} \\
&= (x \cdot x + yx) + (xy + y \cdot y) && \text{distributive law} \\
&= (x^2 + yx) + (xy + y^2) && \text{def. of squaring} \\
&= (x^2 + xy) + (xy + y^2) && \text{commutativity of } \cdot \\
&= \left((x^2 + xy) + xy\right) + y^2 && \text{associativity of } + \\
&= \left(x^2 + (xy + xy)\right) + y^2 && \text{associativity of } + \\
&= \left(x^2 + \left((1 + 1)xy\right)\right) + y^2 && \text{distributive law} \\
&= \left(x^2 + ((2)xy)\right) + y^2 && \text{distributive law} \\
&= x^2 + 2xy + y^2.
\end{aligned}
$$

Ex. 3.1.2. Hint: The three parts are set up so (a) can be used to establish (b), which can be used for (c).

Ex. 3.1.3. If $x = u + v$ and $y = u - v$, then

$$
\begin{aligned}
x + y &= (u + v) + (u - v) = 2u, \\
x - y &= (u + v) - (u - v) = 2v.
\end{aligned}
$$

Conversely, if $2u = x + y$ and $2v - x - y$, then

$$
\begin{aligned}
u + v &= \tfrac{1}{2}[2u + 2v] = \tfrac{1}{2}[(x + y) + (x - y)] = x, \\
u - v &= \tfrac{1}{2}[2u - 2v] = \tfrac{1}{2}[(x + y) - (x - y)] = y.
\end{aligned}
$$

These identities play a wide variety of roles throughout the book.

Ex. 3.1.6. For each natural number n, let $P(n)$ be the statement to be shown, $p_n/q_n = [a_0, a_1, a_2, \ldots, a_n]$. Since $p_0 = a_0$ and $q_0 = 1$, the base case $P(0)$ reads $p_0/q_0 = a_0$, which is true.

Assume inductively that $P(m)$ is true for some natural number m:

$$\frac{p_m}{q_m} = [a_0, a_1, \ldots, a_m].$$

The recursion relations for (p_n) and (q_n) imply

$$\frac{a_m p_{m-1} + p_{m-2}}{a_m q_{m-1} + q_{m-2}} = [a_0, a_1, \ldots, a_m].$$

This formula is true for *all* positive, real a_m since the finite sequences $(p_n)_{n=0}^{m-1}$ and $(q_n)_{n=0}^{m-1}$ do not depend on a_m. Particularly,

$$\begin{aligned}
\frac{a_{m+1} p_m + p_{m-1}}{a_{m+1} q_m + q_{m-1}} &= \frac{a_{m+1}(a_m p_{m-1} + p_{m-2}) + p_{m-1}}{a_{m+1}(a_m q_{m-1} + q_{m-2}) + q_{m-1}} \\
&= \frac{[a_m + (1/a_{m+1})]p_{m-1} + p_{m-2}}{[a_m + (1/a_{m+1})]q_{m-1} + q_{m-2}} \\
&= [a_0, a_1, \ldots, a_{m-1}, a_m + (1/a_{m+1})] \\
&= [a_0, a_1, \ldots, a_m, a_{m+1}].
\end{aligned}$$

(The algebra may read more naturally from bottom to top.) Since $P(0)$ is true and $P(m)$ implies $P(m+1)$ for all m, $P(n)$ is true for all n by induction.

Exercises for Section 3.2

Ex. 3.2.3. Hint: In (c), $x = \dfrac{-b \pm \sqrt{b^2 - 4ac}}{2a}$, the famous *quadratic formula*.

Ex. 3.2.4. Suggestion for part (i): If x and y are real, then $-|x| \leq x \leq |x|$ and $-|y| \leq y \leq |y|$. Add these inequalities and apply Proposition 3.2.10.

Suggestion for part (ii): Apply (i) to the identities $y = x + (y - x)$ and $x = y + (x - y)$.

Ex. 3.2.5. (Conceptual sketch) For all real y, we have $|-y| = |y|$. Corollary 3.2.12 asserts four inequalities. Two are Proposition 3.2.11 verbatim, the other two follow by replacing y with $-y$ in Proposition 3.2.11.

Ex. 3.2.10. Suggestion: Mimic the proof of Proposition 3.2.25, replacing "largest" with "smallest" and "max" with "min" throughout. The point is to coax you into reading the proof carefully.

Ex. 3.2.11. Hint: Add and subtract xy_0, factor the four terms in pairs, and apply the triangle inequality.

Ex. 3.2.12. Hint: To get an upper bound on a positive reciprocal, establish a lower bound on the denominator. The reverse triangle inequality is set up for this.

Ex. 3.2.13. Fix n and argue inductively on m. When $m = 0$ or 1, the inequality is an equality. Inductively, if $(n+1)^k n! \leq (n+k)!$ for some positive integer k, then

$$\begin{aligned}
(n+1)^{k+1} n! &= (n+1) \cdot (n+1)^k n! \\
&\leq (n+1) \cdot (n+k)! \\
&< (n+k+1) \cdot (n+k)! \\
&= (n+k+1)!
\end{aligned}$$

so strict inequality holds if $m = k + 1$.

Exercises for Section 3.3

Ex. 3.3.1. Hint: To handle induction with two variables, assume m and n are non-negative integers, and let $P(n)$ be the universally quantified statement

$$x^{m+n} = x^m \cdot x^n \quad \text{for all } m \text{ in } \mathbf{N}.$$

Ex. 3.3.4. Hint: Introduce

$$s_n = a + ar + ar^2 + \cdots + ar^{n-1} = \sum_{k=0}^{n-1} ar^k.$$

Handle the cases $r = 1$ and $r \neq 1$ separately.

Ex. 3.3.5. The expression is the geometric sum with first term $a = x^{n-1}$, ratio $r = y/x$, and n terms. (If this is unclear, factor out x^{n-1}.) By Proposition 3.3.8, if $x = y$ the sum is nx. Otherwise,

$$\sum_{k=0}^{n-1} x^{n-k-1} y^k = x^{n-1} \frac{1 - (y/x)^n}{1 - y/x} = x^n \frac{1 - (y/x)^n}{x - y} = \frac{x^n - y^n}{x - y}.$$

Ex. 3.3.6. Since x^n is the $k = 0$ term in the binomial theorem,

$$\frac{1}{h}\left((x+h)^n - x^n\right) = \frac{1}{h} \sum_{k=1}^{n} \binom{n}{k} x^{n-k} h^k = \sum_{k=1}^{n} \binom{n}{k} x^{n-k} h^{k-1}.$$

Setting $h = 0$ leaves only the $k = 1$ term, nx^{n-1}.

Ex. 3.3.7. Conceptually, $(1 - x)^n$ has "the same terms as $(1 + x)^n$ except the signs alternate." Adding doubles the even-index terms and cancels the odd-index terms. In symbols, $\frac{1}{2}(1 + (-1)^k) = 1$ if k is even and 0 if k is odd.

By contrast, subtracting cancels the even-index terms and doubles the odd-index terms. In symbols, $\frac{1}{2}(1 - (-1)^k) = 0$ if k is even and 1 if k is odd.

Setting $n/2 = m$ if $n = 2m$ is even or if $n = 2m + 1$ is odd, we have

$$\frac{1}{2}\left((1 + x)^n + (1 - x)^n\right) = \frac{1}{2}\sum_{k=0}^{n}(1 + (-1)^k)\binom{n}{k}x^k = \sum_{k=0}^{n/2}\binom{n}{2k}x^{2k},$$

$$\frac{1}{2}\left((1 + x)^n - (1 - x)^n\right) = \frac{1}{2}\sum_{k=0}^{n}(1 - (-1)^k)\binom{n}{k}x^k = \sum_{k=0}^{n/2}\binom{n}{2k+1}x^{2k+1}.$$

For $n = 2$ we have $(1 + x)^2 = 1 + 2x + x^2$ and $(1 - x)^2 = 1 - 2x + x^2$, so

$$\frac{1}{2}\left((1 + x)^2 + (1 - x)^2\right) = 1 + x^2,$$

$$\frac{1}{2}\left((1 + x)^2 - (1 - x)^2\right) = 2x.$$

For $n = 3$, $(1 + x)^3 = 1 + 3x + 3x^2 + x^3$ and $(1 - x)^3 = 1 - 3x + 3x^2 - x^3$, so

$$\frac{1}{2}\left((1 + x)^3 + (1 - x)^3\right) = 1 + 3x^2,$$

$$\frac{1}{2}\left((1 + x)^3 - (1 - x)^3\right) = 3x + x^3.$$

For $n = 4$, $(1 \pm x)^4 = 1 \pm 4x + 6x^2 \pm 4x^3 + x^4$, so

$$\frac{1}{2}\left((1 + x)^4 + (1 - x)^4\right) = 1 + 6x^2 + x^4,$$

$$\frac{1}{2}\left((1 + x)^4 - (1 - x)^4\right) = 4x + 4x^3.$$

Ex. 3.3.10.

(a) Assume inductively that $3^k < k!$ for some integer k such that $k \geq 2$. Then

$$3^{k+1} = 3 \cdot 3^k < 3 \cdot k! \leq (k + 1)k! = (k + 1)!.$$

That is, $P(k)$ implies $P(k + 1)$ if $k \geq 2$.

(b) It suffices to exhibit an n_0 such that $3^{n_0} < n_0!$.

Since $3^6 = 9^3 = 729 < 6! = 720$ is false but $3^7 = 2187 < 5040 = 7!$ is true, we may take $n_0 = 7$ or any larger integer

Ex. 3.3.13. Hint: To establish the inductive step, apply the distributive law to $(x + y)^{m+1} = (x + y)(x + y)^m$, and shift the indices in one sum so terms can be combined.

Solutions: Sets of Real Numbers

Exercises for Section 4.1

Ex. 4.1.1. For every real x, $x \in B_r^\times(x_0)$ if and only if $0 < |x - x_0| < r$, if and only if $x - x_0 \in B_r^\times(0)$, if and only if $x \in x_0 + B_r^\times(0)$.

Ex. 4.1.5. Assume contrapositively that $J \cap J' = \varnothing$, and without loss of generality assume $a < a'$. Because the intersection is empty, we have $b < a'$. Assume $x = \frac{1}{2}(b + a')$ is the midpoint. Since $b < x < a'$, $x \notin J \cup J'$. But since b and a' are in $J \cup J'$, the union is not an interval.

Ex. 4.1.6. Suggestion: Prove the intersection is non-empty using the idea of Exercise 4.1.5. Then use the fact J and J' are intervals to prove the intersection is an interval. Finally, use Exercise 4.1.4 to prove the intersection is an open interval.

Exercises for Section 4.2

Ex. 4.2.1. If $x \in (a, b)$, then $a < x < b$, so again b is an upper bound. Instead of showing the second condition directly, it's easier to work with the contrapositive. Assume $\varepsilon > 0$, and put $x = b - \frac{1}{2}\varepsilon$. Certainly we have $b - \varepsilon < x < b$, so if $x \in (a, b)$, namely, if $a < x$, then we have shown $b - \varepsilon$ is not an upper bound of (a, b).

While our choice of x "works" for small ε, it doesn't work if $\varepsilon > 2(b - a)$. The standard workaround is to make a two-case definition. The conventional choice is this: If $\varepsilon > 0$, define $x = \max\left(b - \frac{1}{2}\varepsilon, \frac{1}{2}(a + b)\right)$. For this choice, we have $b - \varepsilon < x$ and $a < \frac{1}{2}(a + b) \leq x < b$. Since $x \in (a, b)$, we have shown $b - \varepsilon$ is not an upper bound of (a, b). Since ε was arbitrary, $b = \sup(a, b)$.

Ex. 4.2.6. Suggestion: Consider the sets $A = \{a_n\}_{n=0}^\infty$ and $B = \{b_n\}_{n=0}^\infty$. Prove that $a \leq b$ for all a in A and all b in B, and conclude that $\sup A \leq \inf B$ and $[\sup A, \inf B]$ is contained in $\bigcap_{n=0}^\infty I_n$.

Ex. 4.2.7. Suggestion: Use interval induction together with Exercises 4.1.4 and 4.1.7.

Exercises for Section 4.3

Ex. 4.3.1. Qualitatively, no real number is more than 1 unit away from some integer. By scaling, if q is a positive integer then no real number is further than $1/q$ from the set $(1/q)\mathbf{Z}$. Since $(1/N)\mathbf{Z} \subseteq \mathbf{Q}_N$, the "maximum distance" from $\mathbf{Q}_N$ is at most $1/N$.

Using Corollary 4.3.2, we can both make this argument rigorous and improve the bound. If x is an arbitrary real number, then $n := \lfloor Nx \rfloor \leq Nx < n+1$. Consequently, $n/N \leq x < (n+1)/N$, so the distance from x to some point of $(1/N)\mathbf{Z}$, and therefore to some point of $\mathbf{Q}_N$, is no larger than $1/(2N)$. (Why?)

There is no smaller bound: $(0, 1/N)$ is disjoint from $\mathbf{Q}_N$, and has length $1/N$.

Ex. 4.3.5. Suggestion: The outline below establishes existence of $\lfloor x \rfloor$ if $x > 0$, and then for general x, and then establishes uniqueness.

(a) Assume $x > 0$. Use finitude, well-ordering of $\mathbf{N}$, and a well-chosen set of natural numbers depending on x to prove there exists a natural number n and real number x' such that $x = n + x'$ and $0 \leq x' < 1$.

(b) Assume x is real. If $x \leq 0$, prove there exists a natural number N such that $0 < N + x$, then apply part (a).

(c) Prove that the representation in (b) is unique: If $x = n_1 + x_1' = n_2 + x_2'$ for integers n_1 and n_2, and real numbers x_1' and x_2' such that $0 \leq x_k' < 1$, then $n_1 = n_2$ and $x_1' = x_2'$.

Ex. 4.3.6. Hints: In (a), argue contrapositively. For part (b), if $x < y$, then $x/\sqrt{2} < y/\sqrt{2}$.

Ex. 4.3.8. By Proposition 3.3.8, there is a closed formula

$$s_n = \frac{1 - (1/2)^n}{1 - (1/2)} = 2 - (1/2^{n-1}).$$

Since $2 - (1/n) < 2 - (1/2^{n-1}) = s_n < 2$ for all n, we have $\sup s = 2$.

Ex. 4.3.10. For each m,

$$H_{m+1} - H_m = \sum_{k=2^m+1}^{2^{m+1}} \frac{1}{k}$$

is a sum of 2^m terms, each no smaller than $1/2^{m+1}$, so the sum is at least $1/2$. Consequently,

$$H_{2^{m+1}} = 1 + \sum_{j=0}^{m} \sum_{k=2^j+1}^{2^{j+1}} \frac{1}{k} \geq 1 + \sum_{j=0}^{m} \frac{1}{2} \geq 1 + \frac{m+1}{2}.$$

Since m is arbitrary, $\{H_m\}$ is not bounded above.

Exercises for Section 4.4

Ex. 4.4.1. No: The intersection is $\{0\}$, which is not open. To prove the claim about the intersection, note that $0 \in O_n$ for each n. Inversely, if $x \neq 0$, then $\varepsilon = |x| > 0$. By reciprocal finitude (Corollary 4.3.7), there is a positive integer n such that $1/n < |x|$. Consequently, $x \notin O_n$, so x is not in the intersection.

Ex. 4.4.3. Throughout, $(A_k)_{k \in \mathbf{N}}$ is a sequence of sets, each having no limit points.

(a) Assume x_0 is a real number. Since x_0 is not a limit point of A_0, there exists an ε_0 such that $B_{\varepsilon_0}^\times(x_0) \cap A_0 = \varnothing$. Since x_0 is not a limit point of A_1, there exists an ε_1 such that $B_{\varepsilon_1}^\times(x_0) \cap A_1 = \varnothing$. If $\varepsilon = \min(\varepsilon_0, \varepsilon_1)$, then $\varepsilon > 0$ and $B_\varepsilon^\times(x_0) \cap (A_0 \cup A_1) = \varnothing$, so x_0 is not a limit point of the union.

(b) Immediate from (a) by induction on n.

(c) For example, if $A_k = \{k\}$, the union is the set of natural numbers, which has no limit points. At the other extreme, $A_k = 2^{-k}\mathbf{Z}$ has no limit points, but the union $\bigcup_k A_k$ is the set $\mathbf{Z}[\frac{1}{2}]$ of dyadic rationals, which is dense by Exercise 4.3.7.

Ex. 4.4.4. Hint: A direct approach is not difficult, but the perspective of Exercise 4.4.3 may also help.

Ex. 4.4.7. Hint: Put $J = \{x \text{ in } [a, b] : [a, x] \text{ is finitely covered from } \{O_i\}_{i \in \mathscr{I}}\}$ and use interval induction.

Ex. 4.4.9. If A is the set from Exercise 4.4.4, then $\overline{A} = A \cup \{0\}$ is the closure, and O is the complement of $\overline{A}$, hence is open as the complement of an open set. The open intervals $O_{-1} := (-\infty, 0)$, $O_0 := (1, \infty)$, and $O_n := (1/(n+1), 1/n)$ if $n \geq 1$ are non-empty, mutually disjoint, and their union is the complement of $\overline{A}$ (compare Corollary 4.3.2), so they partition O and are consequently the components by Exercise 4.4.8.

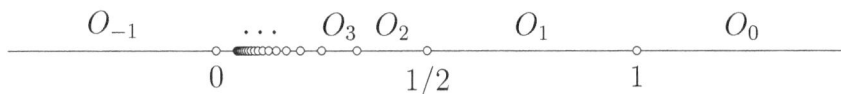

FIGURE A.2

The components of O in Exercise 4.4.9.

Solutions: Functions

Exercises for Section 5.1

Ex. 5.1.1. Assume $p(x) = \sum_{k=0}^{n} a_k x^k$ and $q(x) = \sum_{k=0}^{n} b_k x^k$ are polynomials of degree at most n.

(a) The sum $(p+q)(x) = \sum_{k=0}^{n} (a_k + b_k)x^k$ is a polynomial of degree at most n. The degree is n if and only if $a_n + b_n \neq 0$.

(b) If c is real, then $cp(x) = \sum_{k=0}^{n} (ca_k)x^k$, is a polynomial of degree at most n. If $\deg p = n$, namely, $a_n \neq 0$, the product is either of degree n if $c \neq 0$, or identically 0 if $c = 0$.

(c) By Corollary 5.1.19, $p(x_j) = 0$ if and only if $(x - x_j)$ divides p. If $p(x_j) = 0$ for all j, then $p(x) = \prod_{j=0}^{n}(x - x_j)q(x)$ for some polynomial q. By hypothesis, p has degree at most n, while if q is not identically 0, the product on the right has degree at least $(n+1)$. We conclude q, and hence p, is identically 0.

Ex. 5.1.2. Since $0 \le (x-1)^2$ for all real x,

$$f^+(x) = \begin{cases} 0 & x < 0, \\ x(x-1)^2 & 0 \le x; \end{cases} \qquad f^-(x) = \begin{cases} x(x-1)^2 & x < 0, \\ 0 & 0 \le x. \end{cases}$$

Ex. 5.1.4. Hint: For existence, use induction on "If p is a polynomial of degree at most n and if q is a polynomial, there exist polynomials d and r such that $p(x) = d(x)q(x) + r(x)$ and $\deg r < \deg q$."

Ex. 5.1.5. We have

$$\frac{p(x)}{q(x)} = \frac{x^4 - x^3 + 3x^2 - 6x + 1}{x^3 - 2x^2 + x} = \frac{x^4 - x^3 + 3x^2 - 6x + 1}{x(x-1)^2}.$$

Polynomial division gives $d(x) = x + 1$ and $r(x) = 4x^2 - 7x + 1$, so it suffices to decompose

$$\frac{r(x)}{q(x)} = \frac{4x^2 - 7x + 1}{x(x-1)^2} = \frac{c_{0,1}}{x} + \left[\frac{c_{1,2}}{(x-1)^2} + \frac{c_{1,1}}{x-1}\right].$$

Theorem 5.1.27 guarantees the form of the right-hand side: For each root x_j of the denominator of order m_j, there is a summand $c_{j,k}/(x - x_j)^k$ for each k such that $1 \leq k \leq m_j$. The double indices signify the root and the "degree of singularity." Our task is to evaluate the $c_{j,k}$.

There are two roots, 0 of order 1 (the first summand), and 1 of order 2 (the remaining summands). Pick a root x_j arbitrarily, and multiply the entire equation by the highest power $(x - x_j)^{m_j}$ that divides q. Starting with the root 1, we multiply by $(x-1)^2$ and evaluate at 1:

$$\frac{4x^2 - 7x + 1}{x} = \frac{c_{0,1}}{x}(x-1)^2 + c_{1,2} + c_{1,1}(x-1), \quad \text{or } c_{1,2} = 4 - 7 + 1 = -2.$$

Substituting and rearranging gives

$$\frac{4x^2 - 7x + 1}{x(x-1)^2} + \frac{2}{(x-1)^2} = \frac{c_{0,1}}{x} + \frac{c_{1,1}}{x-1}.$$

Combining terms on the right, factoring, and canceling gives

$$\frac{4x^2 - 5x + 1}{x(x-1)^2} = \frac{(4x-1)(x-1)}{x(x-1)^2} = \frac{4x-1}{x(x-1)} = \frac{c_{0,1}}{x} + \frac{c_{1,1}}{x-1}.$$

Now repeat the process: Multiply by $(x - 1)$ and evaluate at 1 to obtain $c_{1,1}$; multiply by x and evaluate at 0 to obtain $c_{0,1}$. The end result is $c_{1,1} = 3$ and $c_{0,1} = 1$. Substituting all these,

$$\frac{x^4 - x^3 + 3x^2 - 6x + 1}{x^3 - 2x^2 + x} = (x + 1) + \frac{1}{x} - \frac{2}{(x-1)^2} + \frac{3}{x-1}, \quad x \notin \{0, 1\}.$$

Ex. 5.1.6. For each natural number N, let $P(N)$ be the statement, "For every real polynomial p, and for every real polynomial q of degree N that factors completely, $p(x)/q(x)$ has a partial fractions decomposition as in Theorem 5.1.27."

The base case $P(0)$ is true: If $\deg q = 0$, then q is a constant, so $d = p/q$ is a polynomial. Incidentally, we may as well assume q is monic: If not, divide both p and q by the leading coefficient of q.

Assume inductively that $P(N)$ is true for some N, and let q be a completely factored polynomial of degree $(N + 1)$, say

$$q(x) = \prod_{j=0}^{n} (x - x_j)^{m_j} = (x - x_0)^{m_0} \cdot \prod_{j=1}^{n} (x - x_j)^{m_j}.$$

Calling the product on the right $q_0(x)$, we have $q_0(x_0) \neq 0$. Since p and q have no common root, $p(x_0) \neq 0$ as well. Subtracting the "highest-order pure singularity" gives the rational function

$$\frac{p(x)}{q(x)} - \frac{p(x_0)/q_0(x_0)}{(x-x_0)^{m_0}} = \frac{1}{(x-x_0)^{m_0}}\left[\frac{p(x)}{q_0(x)} - \frac{p(x_0)}{q_0(x_0)}\right]$$
$$= \frac{p(x)q_0(x_0) - p(x_0)q_0(x)}{q(x)q_0(x_0)}.$$

The numerator vanishes at x_0, so by Corollary 5.1.19, $(x-x_0)$ is a factor of the numerator. Writing $p(x)q_0(x_0) - p(x_0)q_0(x) = (x-x_0)p_0(x)$, we have

$$\frac{p(x)}{q(x)} - \frac{p(x_0)/q_0(x_0)}{(x-x_0)^{m_0}} = \frac{p_0(x)}{(x-x_0)^{m_0-1}q_0(x)}.$$

The denominator of the right-hand side has degree at most N, so the inductive hypothesis guarantees it decomposes as in Theorem 5.1.27. This establishes the inductive step, and completes the proof of the theorem.

Ex. 5.1.7. Answers (may be verified by algebra):

(a) $\dfrac{1}{a-b}\left[\dfrac{1}{x-a} - \dfrac{1}{x-b}\right].$

(b) $\dfrac{1/a^2}{x^2} - \dfrac{1/(2a)}{x-a} + \dfrac{1/(2a)}{x+a}.$

(c) $\dfrac{1/(ab)}{x^2} + \dfrac{(a+b)/(ab)^2}{x} + \dfrac{1/[a^2(a-b)]}{x-a} + \dfrac{1/[b^2(b-a)]}{x-b}.$

Ex. 5.1.8. Conceptually, each f_j vanishes at "the other points," and e_j is normalized to have value 1 at x_j.

(a) By direct calculation, $f_0(x_1) = f_0(x_2) = 0$, so $e_0(x_1) = e_1(e_2) = 0$, while $f(x_0) = (x_0 - x_1)(x_0 - x_2) \neq 0$, so $e_0(x_0) = f(x_0)/f(x_0) = 1$. Analogous calculations show $e_i(x_j) = 1$ if $i = j$ and $e_i(x_j) = 0$ if $i \neq j$.

(b) The polynomial $p(x) = y_0e_0(x) + y_1e_1(x) + y_2e_2(x)$ is a linear combination (sum of constant multiples) of polynomials of degree at most 2, so has degree at most 2. By (a),

$$p(x_0) = y_0e_0(x_0) + y_1e_1(x_0) + y_2e_2(x_0) = y_0 \cdot 1 + y_1 \cdot 0 + y_2 \cdot 0 = y_0,$$

and similarly $p(x_j) = y_j$ for $j = 1, 2$. That is, the points (x_0, y_0), (x_1, y_1), and (x_2, y_2) satisfy $y = p(x)$.

(c) Setting $x_0 = -1$, $x_1 = 0$, and $x_2 = 1$,

$$f_0(x) = x(x-1), \qquad\qquad e_0(x) = \frac{x(x-1)}{2},$$

$$f_1(x) = (x+1)(x-1), \qquad e_1(x) = \frac{(x+1)(x-1)}{-1},$$

$$f_2(x) = (x+1)x, \qquad\qquad e_2(x) = \frac{(x+1)x}{2}.$$

Consequently,

$$p(x) = y_0 \cdot \tfrac{1}{2}(x^2 - x) + y_1 \cdot (1 - x^2) + y_2 \cdot \tfrac{1}{2}(x^2 + x)$$
$$= \tfrac{1}{2}(y_0 - 2y_1 + y_2)x^2 + \tfrac{1}{2}(y_2 - y_0)x + y_1.$$

The degree is less than 2 if and only if $y_0 - 2y_1 + y_2 = 0$, if and only if the three points are collinear.

Ex. 5.1.9. Hint for (c): Apply Exercise 5.1.1 (c) to the polynomial $p(x) - q(x)$.

Ex. 5.1.11. Hint: Fix x_0 in I arbitrarily. For each x in I, subdivide the interval with endpoints x_0 and x.

Exercises for Section 5.2

Ex. 5.2.1. The domain of f is $\mathbf{R} \setminus \{1\}$. Formal substitution and canceling if $x \neq 1$ gives

$$(f \circ f)(x) = \frac{1 + f(x)}{1 - f(x)} = \left[1 + \frac{1+x}{1-x}\right] \Big/ \left[1 - \frac{1+x}{1-x}\right] = \frac{(1-x) + (1+x)}{(1-x) - (1+x)} = -\frac{1}{x}.$$

This formula is true if $x \notin \{0, 1\}$. Formal substitution in $f \circ f \circ f = (f \circ f) \circ f$ gives

$$(f \circ f \circ f)(x) = -\frac{1}{f(x)} = -\frac{1-x}{1+x}.$$

This formula is true if $x \notin \{-1, 0, 1\}$. Finally, formal substitution in the fourth iterate $f \circ f \circ f \circ f = (f \circ f) \circ (f \circ f)$ gives

$$(f \circ f \circ f \circ f)(x) = -\frac{1}{(f \circ f)(x)} = \frac{-1}{-1/x} = x.$$

This is not the identity function on $\mathbf{R}$, but only on $\mathbf{R} \setminus \{-1, 0, 1\}$.

Ex. 5.2.2. Assume $g : Y' \to Z$ is composable with $f : X \to Y$ and $h : Z' \to W$ is composable with g. For all x in X,

$$\left[(h \circ g) \circ f\right](x) = (h \circ g)\left(f(x)\right) = h\left[g\left(f(x)\right)\right]$$
$$= h \circ \left[(g \circ f)(x)\right] = \left[h \circ (g \circ f)\right](x).$$

Since $(h \circ g) \circ f$ and $h \circ (g \circ f)$ map X to W and take the same values at each x, they are the same mapping.

Ex. 5.2.3. Multiplying out, $p(x) = x^3 - 2x^2 + x$. On substituting $-x$ for x, the odd-degree terms change sign and the even-degree terms are preserved: $p(-x) = -x^3 - 2x^2 - x$. The even part, obtained by adding these and dividing by 2, is the sum of the even-degree terms. The odd part, obtained by subtracting the second from the first and dividing by 2, is the sum of the odd-degree terms: $f_{\text{even}}(x) = -2x^2$, $f_{\text{odd}}(x) = x^3 + x$.

Ex. 5.2.11. For each n in $\mathbf{N}$, let $P(n)$ be the statement "$f(x + n\ell) = f(x)$ for all real x." The base case $P(0)$ is a tautology. Assume inductively that $P(k)$ is true for some k. For all real x, we have

$$f\left(x + (k+1)\ell\right) = f(x + k\ell) \qquad \ell\text{-periodicity}$$
$$= f(x) \qquad \text{inductive hypothesis.}$$

By induction, for every n we have $f(x + n\ell) = f(x)$ for all real x. Now let $Q(n)$ be the statement "$f(x - n\ell) = f(x)$ for all real x." An entirely similar induction shows $Q(n)$ is true for all n. Together, $P(n)$ and $Q(n)$ imply that for every integer n, $f(x + n\ell) = f(x)$ for all real x.

Exercises for Section 5.3

Ex. 5.3.1. Throughout we refer to Proposition 3.2.1 freely.

(a) For all real x, we have $0 < x < b$ if and only if $0 < 1/b < 1/x$. Consequently, $0 < a < x < b$ if and only if $0 < 1/b < 1/x < 1/a$, and we may interpret this as encompassing the first if we interpret $1/0 = \infty$. With this understanding, the reciprocal image of (a, b) is $(1/b, 1/a) \subseteq (0, \infty)$.

(b) For all real x, we have $a < x < 0$ if and only if $1/x < 1/a < 0$. Consequently, $a < x < b < 0$ if and only if $1/b < 1/x < 1/a < 0$, and we may interpret this as encompassing the first if we interpret $1/0 = -\infty$. With this understanding, the reciprocal image of (a, b) is $(1/b, 1/a) \subseteq (-\infty, 0)$.

(c) Since $X = (a, 0) \cup (0, b)$, Proposition 5.3.12 (i) together with the results of (a) and (b) implies the reciprocal image of X is $(-\infty, 1/a) \cup (1/b, \infty)$.

Ex. 5.3.5.

(a) (Injectivity of $g \circ f$). If $x_1 \neq x_2$ are distinct points of X, then $y_1 = f(x_1)$ and $y_2 = f(x_2)$ are distinct points of Y because f is injective. Consequently, $z_1 = g(y_1) = (g \circ f)(x_1)$ and $z_2 = g(y_2) = (g \circ f)(x_2)$ are distinct points of Z since g is injective. Since x_1 and x_2 were arbitrary, $g \circ f$ is injective.

(Surjectivity of $g \circ f$). Assume z is an arbitrary point of Z. Since g is surjective, there exists a y in Y such that $z = g(y)$. Since f is surjective, there exists an x in X such that $y = f(x)$. By definition, $z = (g \circ f)(x)$. Since z was arbitrary, $g \circ f$ is surjective.

(Inverse of $g \circ f$). We have $z = (g \circ f)(x)$ if and only if $g^{-1}(z) = f(x)$, if and only if $(f^{-1} \circ g^{-1})(z) = x$. That is, $(g \circ f)^{-1} = f^{-1} \circ g^{-1}$.

(b) If the composition $g \circ f : X \to Z$ is a bijection, f must be injective and g surjective. Contrapositively, if f is not injective, then f identifies some pair $x_1 \neq x_2$. Since g is a mapping, $(g \circ f)$ identifies the same pair, so $g \circ f$ is not injective. And, if g is not surjective, then $(g \circ f)$ is not surjective, since $(g \circ f)(X) \subseteq g(Y)$.

We cannot generally conclude f is surjective or g is injective. The simplest example is $X = Z = \{0\}$ and $Y = \{0, 1\}$, with $f(0) = 0$ and $g(y) = 0$ for each y. The most general statement is, $g \circ f$ is bijective if and only if f is injective and the restriction of g to the image of f is bijective.

Ex. 5.3.7. Answers:

(a) $\Pi(t) = (x, y) = \dfrac{(2t, t^2 - 1)}{t^2 + 1}$.   (b) $\Pi^{-1}(x, y) = t = \dfrac{x}{1 - y}$ inverts Π.

(c) Each of Π and Π^{-1} is a quotient of polynomials with rational coefficients, so each maps rational numbers to rational numbers.

(d) and (e) are now just computations, but may require care.

Exercises for Section 5.4

Ex. 5.4.3. To start, let's tabulate values of f as shown in Figure A.3.

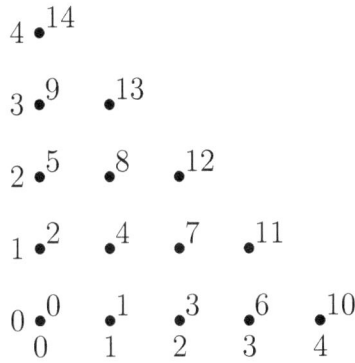

$$4 \bullet^{14}$$

$$3 \bullet^{9} \quad \bullet^{13}$$

$$2 \bullet^{5} \quad \bullet^{8} \quad \bullet^{12}$$

$$1 \bullet^{2} \quad \bullet^{4} \quad \bullet^{7} \quad \bullet^{11}$$

$$0 \bullet^{0} \quad \bullet^{1} \quad \bullet^{3} \quad \bullet^{6} \quad \bullet^{10}$$

$$\quad\quad 0 \quad 1 \quad 2 \quad 3 \quad 4$$

FIGURE A.3
Counting points of a countable union of countable sets.

To prove f is a bijection, the evidence in Figure A.3 leads us to "walk northwest along successive diagonals," namely, to introduce the natural number $N = m + n$, which is constant along these diagonals, and to "parameterize" each diagonal by letting n (the vertical coordinate) run from 0 to N. Here is one way to make this precise.

For each natural number N, consider the sets

$$A_N = \{(N - n, n) : 0 \leq n \leq N\} \subseteq \mathbf{N} \times \mathbf{N},$$
$$B_N = \{k : f(N, 0) \leq k < f(N + 1, 0)\} \subseteq \mathbf{N}.$$

The collection $\{A_N\}_{N=0}^{\infty}$ partitions $\mathbf{N} \times \mathbf{N}$: A pair (m, n) is in A_N if and only if $m + n = N$, so every pair is in some set, and distinct sets are disjoint.

Further, the collection $\{B_N\}_{N=0}^{\infty}$ partitions $\mathbf{N}$. To see this, recall that $f(N, 0) = \frac{1}{2}(N^2 + N)$ is the sum of the natural numbers from 0 to N. Particularly, $f(N, 0) + N + 1 = f(N + 1, 0)$, or $f(N, 0) + N = f(N + 1, 0) - 1$. Informally, the pattern of the first few sets,

$$\mathbf{N} = B_0 \cup B_1 \cup B_2 \cup B_3 \cup \cdots = \{0\} \cup \{1, 2\} \cup \{3, 4, 5\} \cup \{6, 7, 8, 9\} \cup \cdots,$$

continues forever. Specifically, B_N contains precisely $N + 1$ elements, and successive Bs abut but do not overlap.

Finally, the restriction $f : A_N \to B_N$ is a bijection for each N: As n runs from 0 to N, the values

$$f(N - n, n) = \frac{1}{2}(N^2 + N + 2n) = \frac{1}{2}(N^2 + N) + n$$

step through all of B_N. We have partitioned the domain $\mathbf{N} \times \mathbf{N}$ and the codomain $\mathbf{N}$, and shown that f maps each set A_N in the partition of $\mathbf{N} \times \mathbf{N}$ bijectively to the corresponding set B_N in our partition of $\mathbf{N}$. It follows that f is a bijection.

Ex. 5.4.4. Pick a bijection $f_k : \mathbf{N} \to X_k$ for each natural number k. The mapping $f : \mathbf{N} \times \mathbf{N} \to \bigcup_k X_k$ defined by $f(k, \ell) = f_k(\ell)$ is surjective. But $\mathbf{N} \times \mathbf{N}$ is countable by Proposition 5.4.9, so the union is countable by Corollary 5.4.8.

Ex. 5.4.8. Hint: What data define a mapping $f : \mathbf{N} \to \{0, 1\}$?

Solutions: Sequences

Exercises for Section 6.1

Ex. 6.1.1. We first do the scratch work of "solving" for N in terms of ε:

$$\left| \frac{4k - 5}{3k + 2} - \frac{4}{3} \right| = \left| \frac{-23}{3(3k + 2)} \right| \le \frac{23/9}{k} < \varepsilon$$

if $k > 23/(9\varepsilon)$. Now we're ready to write the proof:

Assume $\varepsilon > 0$. By finitude, there exists an integer $N > 23/(9\varepsilon)$. If $k \ge N$, then

$$|a_k - a_\infty| = \left| \frac{4k - 5}{3k + 2} - \frac{4}{3} \right| = \left| \frac{-23}{3(3k + 2)} \right| \le \frac{23/9}{k} \le \frac{23/9}{N} < \varepsilon.$$

Since ε was arbitrary, $(a_k) \to a_\infty$.

Ex. 6.1.3. There exists an index N and a real number c such that if $k \ge N$, then $a_k = c$. (Other wordings are possible.)

Ex. 6.1.4. Hint: Think in terms of adversarial games.

Exercises for Section 6.2

Ex. 6.2.1. In each part, factor out the highest power of k from the numerator and from the denominator.

(a) If $k > 0$,

$$\frac{4k^2 - 5k + 7}{5k^2 + 1} = \frac{k^2(4 - 5/k + 7/k^2)}{k^2(5 + 1/k^2)} = \frac{4 - 5/k + 7/k^2}{5 + 1/k^2}.$$

By Example 6.1.13 and Proposition 6.2.1, this approaches $4/5$ as $k \to \infty$.

(b) If $k > 0$,

$$\frac{4k^5 - 5k + 7}{5k^4 + 1} = \frac{k^5(4 - 5/k^4 + 7/k^5)}{k^4(5 + 1/k^4)} = k \cdot \frac{(4 - 5/k^4 + 7/k^5)}{5 + 1/k^4}.$$

The fraction approaches $4/5$ as $k \to \infty$, so this sequence has unbounded terms, hence is divergent.

Ex. 6.2.2. Hint: The claim is immediate if p is constant or $p(x) = x$. Use induction on the degree and Proposition 6.2.1 to establish the general case.

Ex. 6.2.3. Largely, the point is the introduce the idea of a "sequence of functions."

(a) Proposition 6.1.16 says, more specifically, that $(f_k(x)) \to 0$ if $0 \le x < 1$ and $(f_k(1)) \to 1$.

(b) It suffices to prove

$$\sup\{|f_k(x) - f(x)| : 0 \le x \le 1\} = \sup\{x^k : 0 \le x < 1\} = 1 \quad \text{for each } k.$$

Since $x^k < 1$ if $0 \le x < 1$, the supremum is no larger than 1. On the other hand, the sequence with terms $x_m = 1 - 2^{-m} < 1$ strictly increases with limit 1, and $f_k(x_m) = (1 - 2^{-m})^k \to 1$, so $1 \le \sup\{x^k : 0 \le x < 1\}$.

We have shown $\sup\{|f_k(x) - f(x)| : 0 \le x \le 1\} = 1$ for every k, which implies the sequence of suprema does not converge to 0.

Ex. 6.2.5. Hint: Each general term is a product of k factors. As in Example 6.2.4, taking limits "factor by factor" does not strictly apply. Nonetheless, examining factors can help if there are "useful trends."

Exercises for Section 6.3

Ex. 6.3.1. In words, if each term of a sequence is no smaller than the preceding term, then all subsequent terms are no smaller than a given term (and conversely).

If (ii) holds, then the particular case $m = 1$ says $a_k \le a_{k+1}$ for all k, so (i) holds.

Conversely, assume (i) holds, and if $m \ge 0$ let $P(m)$ be the statement "$a_k \le a_{k+n}$ for all k." The statement $P(0)$, $a_k \le a_k$ for all k, is immediate, and $P(1)$ is our hypothesis (i).

Assume inductively that $P(m)$ is true for some m, namely, that $a_k \leq a_{k+m}$ if $k \geq 0$. By (i), we have $a_k \leq a_{k+m} \leq a_{(k+m)+1} = a_{k+(m+1)}$ for all k. That is, $P(m)$ implies $P(m+1)$. Since the base cases $P(0)$ and $P(1)$ are true, and since $P(m)$ implies $P(m+1)$ if $m \geq 1$, induction guarantees $P(n)$ is true for all n; that is, (ii) is true.

Ex. 6.3.2.

(a) If $2k - 13 > 0$, namely, if $k \geq 7$, then increasing k makes the denominator larger and positive, so a_k decreases.

More formally, if $k \geq N = 7$ and $m > 0$, then

$$a_k - a_{k+m} = \frac{1}{2k-13} - \frac{1}{2k+2m-13}$$
$$= \frac{(2k+2m-13) - (2k-13)}{2k-13} = \frac{2m}{2k+2m-13} > 0,$$

since the numerator and denominator are both positive. This means (a_k) is eventually decreasing.

(b) Since $8k - 3 = 4(2k - 13) + 49$, we have

$$b_k = \frac{8k-3}{2k-13} = 4 + \frac{49}{2k-13}.$$

By (a), this is eventually decreasing, specifically, provided $k \geq 7$.

Ex. 6.3.4. It suffices to "shuffle" two sequences approaching 0 at different rates. Since $2 + (-1)^k$ is alternately 3 and 1, for example, $a_k = (2 + (-1)^k)/2^k$ is not eventually monotone. To prove this we can note that the ratio of consecutive terms a_{k+1}/a_k is $3/2 > 1$ if k is odd and $1/6 < 1$ if k is even. (If the sequence were eventually monotone, these ratios would be eventually no larger than 1.)

Ex. 6.3.6. Assume $\varepsilon > 0$ arbitrarily. Because $(a_k) \to L$, there is an N_1 such that if $k \geq N_1$, then $|a_k - L| < \varepsilon$, or equivalently, $-\varepsilon < a_k - L < \varepsilon$, which implies $L - \varepsilon < a_k$. Similarly, there is an N_2 such that if $k \geq N_2$, then $|b_k - L| < \varepsilon$, or $L - \varepsilon < b_k < L + \varepsilon$, which implies $b_k < L + \varepsilon$.

Put $N = \max(N_1, N_2)$. If $k \geq N$, then $k \geq N_1$ and $k \geq N_2$, so both inequalities above hold: We have $L - \varepsilon < a_k \leq c_k \leq b_k < L + \varepsilon$. This implies $|c_k - L| < \varepsilon$. Since ε was arbitrary, $(c_k) \to L$.

Ex. 6.3.9. Let M be an arbitrary real number. Since $(a_k) \to 0$, there exists an index N such that if $k \geq N$, then $0 < a_k < 1/(1 + |M|)$, and therefore $b_k = 1/a_k > 1 + |M| > M$. Since M was arbitrary, $(b_k) \to \infty$.

Ex. 6.3.11. Hint: One challenge is working with the (scant) tools developed so far. Properties of the function $f(x) = \sqrt{bx}$, particularly its monotonicity, whether or not it is larger or smaller than x, and how $|b - f(x)|$ compares to $|b - x|$, will help.

Exercises for Section 6.4

Ex. 6.4.1. The subsequence of even-index terms is constant (equal to 1) and the subsequence of odd-index terms is constant (equal to -1). Since $-1 \neq 1$, this sequence diverges.

Ex. 6.4.5. We can accomplish (b), and implicitly therefore (a), by considering the reciprocal sequence $b_k = 1/a_k$ and applying Exercise 6.4.4. Alternatively, we can proceed in a self-contained way:

(a) For each m, we can use the positive minimum $\varepsilon = \min\{a_k : k \leq m\}$ as challenge in the limit game. Since $(a_k) \to 0$, there exists an n such that $k \geq n$ implies $a_k = |a_k - 0| < \varepsilon$. In particular, $a_n < \varepsilon = \min\{a_k : k \leq m\}$, so $n > m$ and $a_n < a_m$.

(b) Proceed inductively using (a): Let $\nu(0) = 0$. If we have chose $\nu(k)$, use (a) to pick $\nu(k+1) > \nu(k)$ such that $a_{\nu(k+1)} < a_{\nu(k)}$. The resulting subsequence $(a_{\nu(k)})_{k=0}^{\infty}$ is strictly decreasing by construction.

Ex. 6.4.6. Hint: Proceed along the same lines as Exercise 6.4.5.

Ex. 6.4.7. In light of density of $\mathbf{Q}$ in $\mathbf{R}$, a standard idiom is to look at intervals centered at α with shrinking radii, or with right endpoint α.

(a) For each natural number k, density of $\mathbf{Q}$ in $\mathbf{R}$ guarantees there exists a rational number a_k in $B_{2^{-k}}(\alpha)$. Since $|a_k - \alpha| < 2^{-k}$ for each k, $(a_k) \to \alpha$. (We can, with minor modifications left to you, arrange that $a_k \neq \alpha$ for all k, and/or that $|a_{k+1} - \alpha| < |a_k - \alpha|$ for all k.)

(b) Let $A = \{r \text{ in } \mathbf{Q} : r < \alpha\}$. Since A is non-empty and bounded above by α, it has a real supremum $\sup A \leq \alpha$. It suffices to show that for every ε, $\alpha - \varepsilon$ is not an upper bound of A. But because $\mathbf{Q}$ is dense in $\mathbf{R}$, there is a rational number r in $(\alpha - \varepsilon, \alpha)$. Particularly, $r < \alpha$, so $r \in A$.

Ex. 6.4.12. Hints: For (a) part of the question is copying the definitions for real sequences, and part is recognizing that we need additional structure to measure distance between points in the plane. For definiteness, use the function d_2 of Exercise 6.4.11.

For (c) you'll need the triangle inequality. By (b), it suffices to work with d_∞ instead of d_2, for which the triangle inequality is easily verified.

Solutions: Infinite Series

Exercises for Section 7.1

Ex. 7.1.1.

(a) This is a geometric series with first term $a = 4$ and ratio $r = -3/5$. Since $|r| < 1$, the series converges, with sum $a/(1 - r) = 5/2$.

(b) This is a geometric series with first term $a = 0.01$ and ratio $r = 7/5$. Since $|r| > 1$, the series diverges

(c) This is a geometric series with first term $a = 0$, so the geometric series formula as stated does not apply. Although the ratio $r = 1000$ is "in the divergence range," all the terms are 0, so the series converges and the sum is 0.

Ex. 7.1.2. Following the suggestion, we have

$$1000x = 123.45345\overline{345},$$
$$x = \quad 0.12345\overline{345}, \qquad x = \frac{12333}{999 \cdot 100} = \frac{12333}{99900} = \frac{4111}{33300}.$$
$$999x = 123.33;$$

Ex. 7.1.4. Hint: If you are stuck, see Exercise 2.1.3.

Ex. 7.1.6. Hint: First decompose the summands in partial fractions.

Ex. 7.1.8. Each series is geometric, so converges if and only if the absolute ratio is smaller than 1. The respective sums are:

(a) $\dfrac{1}{1 - x^2}$ if $|x| < 1$; (b) $\dfrac{1}{(1/2) - x^2}$ if $|x| < \sqrt{1/2}$; (c) $\dfrac{1}{x^2}$ if $0 < |x| < 1$.

Ex. 7.1.13. Hint for (c): Assume (a_k) and (b_k) are distinct digit sequences defining the same real number, and assume (i) $a_k = b_k$ if $1 \le k < n$; (ii) $a_n < b_n$. How much smaller than b_n can a_n be? What can you say about all subsequent digits of each sequence?

Ex. 7.1.14. Answer: The image of f is the closed unit interval, $[0, 1]$. The function f is not injective. If $f(A) = f(A')$ for distinct sets A and A', then one set is non-empty and finite, say A, and the other results from removing the largest element and appending *all* subsequent integers. In symbols, if $N = \max A$, then $A' = (A \setminus \{N\}) \cup \{j\}_{j=N+1}^{\infty}$.

Exercises for Section 7.2

Ex. 7.2.2. The series converges if $p > 0$ by the alternating series test, and converges absolutely if and only if $p > 1$ by the p-series test.

Ex. 7.2.6. Answer: $c_k = \sum_{j=0}^{k} \binom{k}{j} a_j b_{k-j}$.

Exercises for Section 7.3

Ex. 7.3.1. If $a_k = 1/k^2$, the alternating series bounds give

$$\frac{1}{n^2} - \frac{1}{(n+1)^2} = \frac{2n+1}{n^2(n+1)^2} \leq \left| \sum_{k=n}^{\infty} \frac{(-1)^k}{k^2} \right| \leq \frac{1}{n^2}.$$

Ex. 7.3.3. In the notation of the proof, and under the assumption $a_k \geq 0$ for all k, we have

$$s_{2m} \leq s_{2m+2} \leq \ell \leq s_{2m+3} \leq s_{2m+1}$$

for all m. Consequently, if $n = 2m$ is even, then

$$a_n - a_{n+1} = s_{2m+2} - s_{2m} \leq \ell - s_n = \sum_{k=n}^{\infty} (-1)^k a_k \leq s_{2m+1} - s_{2m} = a_n.$$

If instead $n = 2m + 1$ is odd, then

$$a_n - a_{n+1} = s_{2m+1} - s_{2m+3} \leq s_{2m+1} - \ell = -\sum_{k=n}^{\infty} (-1)^k a_k \leq s_{2m+1} - s_{2m+2} = a_n.$$

Ex. 7.3.4.

(a) Since $(2k+2)! = (2k+2)(2k+1)(2k)!$, the ratio test gives

$$\lim_{k \to \infty} \left| \frac{x^{2k+2}}{(2k+2)!} \cdot \frac{(2k)!}{x^{2k}} \right| = \lim_{k \to \infty} \left| \frac{x^2}{(2k+2)(2k+1)} \right| = 0 \quad \text{for all real } x.$$

Since this is less than 1, the series converges absolutely for all real x.

(b) If $|x| \leq 2$, the terms are decreasing in absolute value (with k) if $k \geq 1$. The error estimate for an alternating series gives

$$|C(x) - C_n(x)| = \left| \sum_{k=n}^{\infty} \frac{(-1)^k x^{2k}}{(2k)!} \right| = \frac{|x|^{2n}}{(2n)!} \leq \frac{4^n}{(2n)!}$$

This bound is smaller than 0.5×10^{-6} if and only if $(2n)!/4^n > 2,000,000$. For $n = 5$ we have $10!/4^5 = 3,628,800/1024 > 3000$. Multiplying by successive ratios,

$$\underbrace{\frac{10!}{4^5}}_{>3000} \cdot \underbrace{\frac{12 \cdot 11}{4}}_{=33} \cdot \underbrace{\frac{14 \cdot 13}{4}}_{>45},$$

which suffices. That is, $n = 6$ is not enough, but $n = 7$ suffices.

Exercises for Section 7.4

Ex. 7.4.5. If $\phi = \frac{1}{2}(1 + \sqrt{5})$, then

$$\phi^2 = \left(\tfrac{1}{2}(1 + \sqrt{5})\right)^2 = \tfrac{1}{4}(1 + 2\sqrt{5} + 5) = \tfrac{1}{2}(3 + \sqrt{5}) = 1 + \phi.$$

(a) Let $P(n)$ be the *pair* of inequalities $F_n \le \phi^n$ and $F_{n+1} \le \phi^{n+1}$. The base case $P(0)$ is true: $F_0 = 1 \le \phi^0$ and $F_1 = 1 \le \phi$. Assume inductively that $P(k)$ is true for some k. Since $F_{k+1} \le \phi^{k+1}$ by hypothesis, it suffices to prove $F_{k+2} \le \phi^{k+2}$. But since $1 + \phi = \phi^2$,

$$F_{k+2} = F_k + F_{k+1} \le \phi^k + \phi^{k+1} = \phi^k(1 + \phi) = \phi^{k+2},$$

so $P(k+1)$ is true. Since $P(0)$ is true and $P(k)$ implies $P(k+1)$ for all k, $P(n)$ is true for all n by mathematical induction. Particularly, $F_n \le \phi^n$ for all n.

(b) The sequence starts 1, 1, 2, 3, 5, 8, 13, so

$$f(x) = 1 + x + 2x^2 + 3x^3 + 5x^4 + 8x^5 + 13x^6 + \cdots.$$

Since

$$|f(x)| = \left| \sum_{k=0}^{\infty} F_k x^k \right| \le \sum_{k=0}^{\infty} F_k |x|^k \le \sum_{k=0}^{\infty} \phi^k |x|^k$$

and the rightmost series is geometric with radius $1/\phi$, the series for $f(x)$ converges absolutely on the interval $(-\phi, \phi)$.

(c) By splitting off terms of degree less than 2 and shifting indices,

$$(1 - x - x^2)f(x) = (1 - x - x^2) \sum_{k=0}^{\infty} F_k x^k$$

$$= \sum_{k=0}^{\infty} F_k x^k - \sum_{k=0}^{\infty} F_k x^{k+1} - \sum_{k=0}^{\infty} F_k x^{k+2}$$

$$= \left[(1+x) + \sum_{k=2}^{\infty} F_k x^k\right] - \left[x + \sum_{k=1}^{\infty} F_k x^{k+1}\right] - \sum_{k=0}^{\infty} F_k x^{k+2}$$

$$= 1 + \sum_{k=0}^{\infty} F_{k+2} x^{k+2} - \sum_{k=0}^{\infty} F_{k+1} x^{k+2} - \sum_{k=0}^{\infty} F_k x^{k+2}$$

$$= 1 + \sum_{k=0}^{\infty} (F_{k+2} - F_{k+1} - F_k) x^{k+2} = 1,$$

the last because every coefficient in parentheses is 0 by the recursion for F_k. This calculation is meaningful on every interval where the series $f(x)$ converges. Particularly, $f(x) = 1/(1 - x - x^2)$ if $|x| < \phi$.

Ex. 7.4.7. Hint for (c): Suppose $E(1) = p/q$ with p and q integers in lowest terms. Show that $E(1)$ is not an integer, namely, that $q > 1$. Now use (a) and (b) to show there exists an integer between 0 and 1.

Solutions: Continuous Functions

Exercises for Section 8.1

Ex. 8.1.1. The graph is shown with compressed vertical scale in Figure A.4.

(a) Since $x_k < 0$ for all k we have $f(x_k) = x_k = -1/k$; the "image sequence" converges to $f(0) = 0$.

(b) Here, by contrast, we have $x_k > 0$, so $f(x_k) = 1/x_k = k$ diverges to ∞.

(c) Because there exists a sequence $(x_k) \to 0$ whose image under f does not approach $f(0)$, f is discontinuous at 0.

Ex. 8.1.2. The graph is shown in Figure A.5.

(a) Suppose $\ell \neq 0$, and put $\varepsilon = |\ell| > 0$. The open ball $B_\varepsilon(\ell)$ does not contain 0 (by the reverse triangle inequality, for example), so f is constant in this ball, either 1 (if $\ell > 0$) or -1 (if $\ell < 0$).

Suppose $\ell > 0$. If $(x_k) \to \ell$, there exists an N such that if $k \geq N$, then $|x_k - \ell| < \varepsilon$, which implies $f(x_k) = 1$. That is, the image sequence is eventually constant, hence converges to $1 = f(\ell)$. Since (x_k) was arbitrary, f is continuous at ℓ. An entirely similar argument handles the case $\ell < 0$.

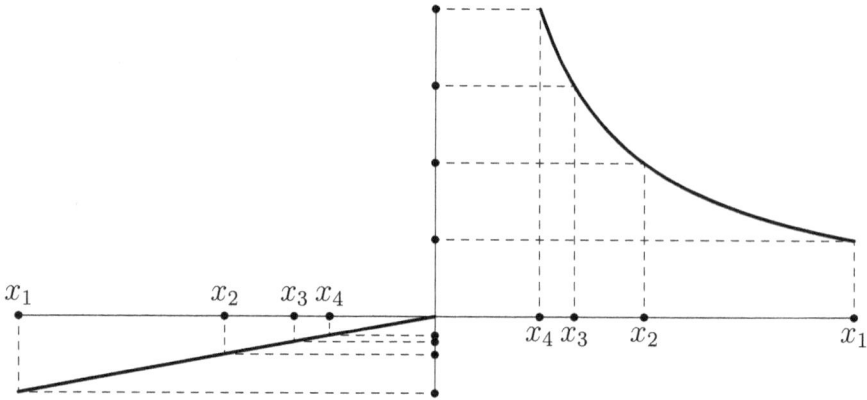

FIGURE A.4
The function of Exercise 8.1.1.

(b) There is no way to define $f(0)$ to make f continuous at 0. That is, f *has no continuous extension to* 0. If there were, then for every sequence (x_k) converging to 0, the image sequence $\big(f(x_k)\big)$ would converge to $f(0)$. But the sequence $x_k = (-1)^{k-1}/k$ converges to 0, while $f(x_k) = (-1)^{k-1}$, which by Example 6.1.15 does not converge.

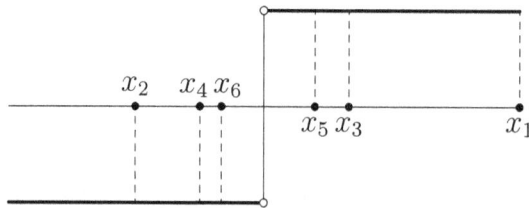

FIGURE A.5
The signum function of Exercise 8.1.2.

Ex. 8.1.4. Figure 5.6 shows the graph of $\chi_{\mathbf{Q}}$.

(a) The set of rationals is dense in $\mathbf{R}$, Theorem 4.3.12, and by Exercise 4.3.6 the set of irrationals is dense. If ℓ is an arbitrary real number, then for each positive integer k there exist a rational number x_k such that $|x_k - \ell| < 1/k$ and an irrational number y_k such that $|y_k - \ell| < 1/k$.

By construction, $(x_k) \to \ell$ and $(y_k) \to \ell$. Also by construction, $f(x_k) = 1$ for all k, and $f(y_k) = 0$ for all k. It follows that f is discontinuous at ℓ.

(b) Write $g = f \cdot \chi_{\mathbf{Q}}$. For all real x, we have $|\chi_{\mathbf{Q}}(x)| \leq |1|$, and consequently $|g(x)| = |f(x)\chi_{\mathbf{Q}}(x)| \leq |f(x)|$.

Assume $f(x_\infty) = 0$ for some x_∞ and (x_k) is an arbitrary sequence converging to x_∞. Fix ε arbitrarily and pick an index N such that if $k \geq N$ then

$|f(x_k)| = |f(x_k) - f(x_\infty)| < \varepsilon$, and consequently $|g(x_k)| \leq |f(x_k)| < \varepsilon$. Since ε was arbitrary, we have shown $\big(g(x_k)\big) \to g(x_\infty)$.

An alternative argument can be given using the squeeze theorem (Exercise 6.3.6) by noting $0 \leq g(x) \leq |f(x)|$ for all real x.

It remains to show g is discontinuous at every real ℓ such that $f(\ell) \neq 0$. In the notation of part (a), pick a rational sequence (x_k) converging to ℓ, and an irrational sequence (y_k) converging to ℓ. Since $g(x_k) = 0$ for all k while $g(y_k) = f(y_k) \to f(\ell) \neq 0$, g is discontinuous at ℓ.

Ex. 8.1.5. Hint: If n is a positive integer, then $f(x) \geq 1/n$ if and only if $x \in \mathbf{Q}_n$. By Exercise 4.4.3 (b), the set $\mathbf{Q}_n$ has no limit points.

Exercises for Section 8.2

Ex. 8.2.1. We say $\lim(f, c) = L$ if: For every ε, there exists a δ such that if $0 < |x - c| < \delta$, then $|f(x) - L| < \varepsilon$. But since $f(x) = g(x)$ if $x \neq c$, and since $0 < |x - c|$ implies $x \neq c$, the preceding condition implies if $0 < |x - c| < \delta$, then $|g(x) - L| < \varepsilon$. Thus $\lim(g, c) = L$, in the sense that the limit exists and is equal to L.

Ex. 8.2.2. Assume $\varepsilon > 0$, and put $\delta = \varepsilon/3$. If $|x - c| < \delta$, then

$$|f(x) - f(c)| = |(3x + 5) - (3c + 5)| = |3(x - c)| = 3|x - c| < 3\delta = \varepsilon.$$

Ex. 8.2.3. Hint: As needed, re-read the proof of Proposition 6.2.1 (ii).

Ex. 8.2.4. Hint: As needed, re-read the proof of Proposition 6.2.1 (iii).

Ex. 8.2.14. (Sketch) Suppose $a < b$, and put $\ell = b - a$. The mean value property determines $f(b+\ell)$ (in terms of $f(a)$ and $f(b)$) since b is the midpoint of $[a, b + \ell]$. By induction, $f(b + n\ell)$ is determined if $n \geq 0$, and similarly $f(a - n\ell)$ is determined if $n \geq 0$. In terms of sets, the values $f(a)$ and $f(b)$ uniquely determine $f(x)$ for all x in $a + \ell\mathbf{Z}$.

Further, f is determined at every midpoint $\frac{1}{2}(a+b)$, hence is determined on $a + (\ell/2)\mathbf{Z}$. Another induction shows f is determined on $a + 2^{-n}\ell\mathbf{Z}$ for every n in $\mathbf{N}$, and therefore on the union. The resulting countable set of real numbers is "closed under taking midpoints," so in general the mean value property gives no additional information.

(A *dyadic rational* is a real number of the form $m2^{-k}$ for some integer m and non-negative integer k, see Exercise 4.3.7. The set of dyadic rationals is denoted $\mathbf{Z}[\frac{1}{2}]$. The preceding remarks amount to the claim that if $f(a)$ and $f(b)$ are specified, and if $\ell = b - a$, then f is determined on $a + \ell\mathbf{Z}[\frac{1}{2}]$.)

In general (assuming the axiom of choice), we may freely specify $f(0)$, and $f(x_i)$ for some (uncountable) set $\{x_i\}_{i \in \mathscr{I}}$ of real numbers such that (i) every real number is a dyadic multiple of some x_i and (ii) no two points x_i and $x_{i'}$ have dyadic ratio.

Because the dyadic rationals are dense in $\mathbf{R}$, however, if f is continuous, then f is affine, given by $f(x) = f(0) + x(f(1) - f(0))$.

Exercises for Section 8.3

Ex. 8.3.1. By the binomial theorem or multiplying out,

$$
\begin{aligned}
(x_0 + h)^2 &= x_0^2 + 2x_0 h + h^2 \\
&= x_0^2 + h(2x_0 + h) \quad \left[\approx x_0^2 + O(h)\right] \\
&= x_0^2 + 2x_0 h + h^2 \quad \left[\approx x_0^2 + 2x_0 h + O(h^2)\right].
\end{aligned}
$$

Exercises for Section 8.4

Ex. 8.4.1. Hint: By definition, $x^{m/n}$ is an *integer* power of $x^{1/n}$.

Ex. 8.4.3. Since $\rho \geq 1$, we have $1 \leq \rho^{1/n}$ for all positive n, or $\rho^{1/n} = 1 + u_n$ for some non-negative u_n. By Proposition 3.3.5, $1 + nu_n \leq (1 + u_n)^n = \rho$ for all positive n, from which we deduce $0 \leq u_n < \rho/n$. By the squeeze theorem or reciprocal finitude, $(u_n) \to 0$ and therefore $(\rho^{1/n}) \to 1$.

If instead $0 < \rho < 1$, then $1 < 1/\rho$. The argument of the preceding paragraph shows $1/(\rho^{1/n}) = (1/\rho)^{1/n} \to 1$, so $\rho^{1/n} \to 1$ as well.

Ex. 8.4.6. Each sign change of $f(x)$ guarantees a solution of $f(x) = 0$ by the intermediate value theorem. There is a sign change on $[-1, 0]$, a sign change on $[0, 1]$, and $f(2) = 0$, so there are at least three zeros of f.

To analyze -6, look for values that "bracket" -6, namely, for sign changes of $f(x) - (-6) = f(x) + 6$. Sign changes occur in $[-2, -1]$ and $[0, 1]$, so there are at least two solutions of $f(x) = -6$.

Ex. 8.4.7. (Injectivity). Since odd integer powers of x with positive coefficient are increasing, and a sum of increasing functions in increasing, f is increasing, hence injective.

(Surjectivity). Let y be an arbitrary real number. It suffices to find inputs $-M$ and M whose values bracket y, and in fact to ensure $-3M+1 < y < 3M+1$. Let $M = |y| + 1$. We have

$$f(-M) = -M^5 - 3M + 1 < -3M + 1 = -3|y| - 2$$
$$< -|y| \le y \le |y| \le 3|y| + 4 = 3M + 1 < f(M).$$

Since $f(-M) < y < f(M)$, the intermediate value theorem guarantees there is an x in $(-M, M)$ such that $y = f(x)$. Since y was arbitrary, f is surjective.

Ex. 8.4.10.

(a) Assume y_0 is real. If $x^n = y_0$, then $|x|^n = |x^n| = |y_0|$, so $|x| = \sqrt[n]{|y_0|}$. There are at most two real numbers with any given absolute value, so at most two x satisfying $x^n = y_0$. (By contrast, as we will see in Chapter 14, every non-zero complex number has precisely n distinct complex nth roots.)

(b) If $n = 2m + 1$ is odd, then $x < 0$ if and only if $x^n < 0$. (Technically, we need an induction on m together with $x^2 > 0$ for all real x.)

If $y_0 > 0$, there exists a unique positive real x such that $x^n = y_0$, and there are no other real nth roots since $0^n = 0$ and $x^n < 0$ if $x < 0$.

If $y_0 < 0$, there is a unique real nth root x of $-y_0 > 0$ by the preceding argument, and $(-x)^n = y_0$ is therefore the unique real nth root of y_0. Finally, 0 has a unique real nth root, namely 0.

Since every real y_0 has a unique real nth root, the nth power mapping is a bijection.

Ex. 8.4.11. Hint: Apply the intermediate value theorem to $g(x) = f(x) - x$.

Exercises for Section 8.5

Ex. 8.5.1. Since $p(5)$ is positive, $\varepsilon = f(5)$ is positive. Because the denominator is positive and has degree greater than the degree of p, $|f(x)| \to 0$ as $|x| \to \infty$ by Proposition 8.5.8 (vi) and Example 8.5.9.

Consequently, there is an R such that $|x| > R$ implies $|f(x)| < \varepsilon/2 < f(5)$. Without loss of generality we may assume $R > 5$. Apply the extreme value theorem to f on $[-R, R]$: There exists a point $x_{\max}$ such that $f(x) \le f(x_{\max})$ for all x in $[-R, R]$. But if $|x| > R$, then

$$f(x) < \varepsilon/2 < f(5) \le f(x_0);$$

that is, $f(x_0)$ is an absolute maximum for f on $\mathbf{R}$.

Ex. 8.5.7. No to both. A counterexample to (b), and therefore to (a), is to let ψ be a continuous, periodic function that achieves both positive and negative values, and to put $f(x) = (1/x)\psi(1/x)$ on $(0, 1]$. To prove such a ψ exists, it suffices to define $\psi(x) = -1 + 2|x|$ on $[-1, 1]$, and to invoke Proposition 5.2.17.

Exercises for Section 8.6

Ex. 8.6.2. Assume $f(x) = x + (1/x)$.

(a) If x, $x' \in (1, \infty)$, then

$$|f(x') - f(x)| = \left| x - x' + \frac{1}{x} - \frac{1}{x'} \right| = \left| x - x' + \frac{x' - x}{xx'} \right| = |x - x'| \cdot \left| 1 - \frac{1}{xx'} \right|,$$

which is smaller than $|x - x'|$. Despite this, f is not a contraction on $(1, \infty)$: The factor in the absolute value is not bounded above by any real number less than 1.

(b) Contrapositively, if the recursive sequence with seed $x_0 = 2$ is condensing, it has a real limit x_∞ by Proposition 6.4.18. Exercise 8.6.1 implies $f(x_\infty) = x_\infty$, or $1/x_\infty = 0$. No such real number exists, so (x_k) is not condensing.

Solutions: Integration

Exercises for Section 9.1

Ex. 9.1.1. Using the "one moving target" principle, put $h = g - f$, so h is integrable and non-negative. By linearity of the integral, it suffices to prove the integral of h is non-negative.

For every splitting Π of $[a, b]$, and every piece I_i associated to Π, we have $m_i = \inf\{h(t) : t \in I_i\} \geq 0$ since $h(t) \geq 0$ for all t. Consequently, $L(h, \Pi) = \sum_i m_i \Delta t_i \geq 0$, so the integral, the supremum of the lower sums, is non-negative.

Ex. 9.1.3. If $0 \leq i \leq n$, the ith piece $I_i = [t_i, t_{i+1}] = [ib/n, (i+1)b/n]$ has length $\Delta t = b/n$. Since f is increasing, the infimum is the left-hand endpoint value and the supremum is the right-hand value: $m_i = f(t_i) = ib/n$ and $M_i = f(t_{i+1}) = (i+1)b/n$. The lower and upper sums are

$$L(f, \Pi) = \sum_{i=0}^{n-1} m_i \, \Delta t = \sum_{i=0}^{n-1} \frac{ib}{n} \cdot \frac{b}{n} = \frac{b^2}{n^2} \sum_{i=0}^{n-1} i = \frac{b^2}{n^2} \frac{n(n-1)}{2} = \frac{b^2}{2} \frac{n-1}{n},$$

$$U(f, \Pi) = \sum_{i=0}^{n-1} M_i \, \Delta t = \sum_{i=0}^{n-1} \frac{(i+1)b}{n} \cdot \frac{b}{n} = \frac{b^2}{n^2} \sum_{i=0}^{n-1} (i+1) = L(f, \Pi) + \frac{b^2}{n}.$$

The supremum over *all* lower sums is no smaller than the supremum over *these* lower sums, which is $b^2/2$. Similarly, the infimum over *all* upper sums is no larger than the infimum over *these* upper sums, which is $b^2/2$. Consequently, the integral is equal to $b^2/2$.

Ex. 9.1.5. Fix ε arbitrarily. The lower sums are all 0, so it suffices to prove there exists a splitting Π for which $U(f, \Pi) < \varepsilon$.

If $\varepsilon \geq 1$, take $\Pi = \{0, 1\}$. Otherwise, we'll use half our "error budget" to cover all but finitely many $1/k$ near 0 and the other half to cover any remaining $1/k$.

Consider the closed intervals $I_0 = [0, \varepsilon/2]$, and for each positive i the intersection of $[0, 1]$ with the closed interval I_i of length $\varepsilon 2^{-(i+1)}$ centered at $1/i$. The total length of these interval is at most $3\varepsilon/4$, because we have clipped off the right half of I_1. Moreover, all but finitely many of the I_i with i greater than 1 are contained in I_0 since $(1/i) + 2^{-(i+2)} \to 0$. The union $\bigcup_{i=0}^{\infty} I_i$ is therefore a finite union of disjoint closed intervals, say $\{J_j\}_{j=0}^{n-1}$. Let Π be the splitting consisting of the endpoints of these intervals. Since $f \leq 1$ on $[0, 1]$, $U(f, \Pi)$ is no larger than the sum of the lengths of the J_j, which is less than ε.

We have shown that for every ε, there exists a splitting for which $L(f, \Pi) = 0$ and $U(f, \Pi) < \varepsilon$. By Proposition 9.1.9, f is integrable, and the integral is $\sup L(f, \Pi) = 0$.

Exercises for Section 9.2

Ex. 9.2.1. Throughout, assume $b > 0$ and write $f(t) = t^k$.

(f is integrable on $[0, b]$). Assume n is a positive integer and $\Pi = \{t_i\}_{i=0}^{n}$ is the equal-length splitting of $[0, b]$ with n pieces. Since f is strictly increasing on $[0, b]$, we have $m_i = f(t_i)$ and $M_i = f(t_{i+1})$ for each i. Consequently,

$$U(f, \Pi) - L(f, \Pi) = \sum_{i=0}^{n-1} (M_i - m_i) \, \Delta t$$

$$= \sum_{i=0}^{n-1} \Big(f(t_{i+1}) - f(t_i) \Big) \Delta t = \Big(f(b) - f(0) \Big) \frac{b}{n} = \frac{b^{k+1}}{n}.$$

If ε is arbitrary, reciprocal finitude guarantees that there exists an n such that $U(f, \Pi) - L(f, \Pi) = b^{k+1}/n < \varepsilon$. By Proposition 9.1.9, f is integrable on $[0, b]$.

(f is integrable on $[-b, 0]$). Immediate from the preceding argument and Proposition 9.2.8 (ii) with $\mu = -1$.

Now assume a and b are arbitrary and $a < b$.

(f is integrable on $[a, b]$). The cocycle property gives

$$\int_a^b f = \int_a^0 f + \int_0^b f,$$

including existence of the integral on the left given existence of the integrals on the right.

It remains to evaluate the integral. We first evaluate the integral from 0 to b assuming $b > 0$. Since f is non-negative, we have, for every δ in $(0, b)$,

$$\frac{b^{k+1} - \delta^{k+1}}{k+1} = \int_\delta^b t^k \, dt \le \int_0^b t^k \, dt \le \delta^{k+1} + \frac{b^{k+1} - \delta^{k+1}}{k+1}.$$

Taking limits as $\delta \to 0$ gives $\int_0^b t^k \, dt = b^{k+1}/(k+1)$. The cocycle condition implies

$$\int_a^b t^k \, dt = \int_a^0 t^k \, dt + \int_0^b t^k \, dt = -\int_0^a t^k \, dt + \int_0^b t^k \, dt = \frac{b^{k+1} - a^{k+1}}{k+1}.$$

Ex. 9.2.4.

(a) Since the set of rational numbers is countable, there exists an injective sequence $(a_j)_{j=0}^\infty$ whose set of terms is $\mathbf{Q}$. For each k, define $A_k = \{j\}_{j=0}^{k-1}$ and $f_k = \chi_{A_k}$. Since each f_k is non-zero only at finitely many points, each is integrable, with integral equal to 0 over an arbitrary interval $[a, b]$.

(b) If $I_k = [a_k - \varepsilon 2^{-(k+2)}, a_k + \varepsilon 2^{-(k+2)}]$ is the "closed ball" of length $\varepsilon 2^{-(k+1)}$ about a_k, then $a_k \in I_k$ for each k, and the total length of the I_k is ε by the geometric series formula.

(c) The sets in (b) tell us that if the integral were defined in a way that permitted "countable splittings," then the lower sums of f would all be 0 and the upper sums could be made arbitrarily small. It would then be reasonable to define the integral of f over an arbitrary interval to be 0.

Ex. 9.2.5. Hints: The hypothesis might remind you of uniform convergence. Earlier proofs about continuous functions converging to a continuous limit may provide structural guidance.

Exercises for Section 9.3

Ex. 9.3.1. For definiteness, extend f to $[a, b]$ by 0; that is, continue to denote the extended function by f, and put $f(a) = f(b) = 0$. Assume $|f| \leq M$ on (a, b), and $\varepsilon > 0$ arbitrarily. Pick points $a' < b'$ in (a, b) so that $a' - a < \varepsilon/(8M)$ and $b - b' < \varepsilon/(8M)$.

Since f is continuous on $[a', b']$, there exists a splitting Π' whose upper sum minus lower sum is less than $\varepsilon/2$. Let $\Pi = \Pi' \cup \{a, b\}$. On the first interval $[a, a']$, $\sup f - \inf \leq 2M$. Similarly on the last interval $[b', b]$, $\sup f - \inf \leq 2M$. Consequently,

$$U(f, \Pi) - L(f, \Pi) \leq \frac{2M\varepsilon}{8M} + U(f, \Pi') - L(f, \Pi') + \frac{2M\varepsilon}{8M} < \frac{\varepsilon}{4} + \frac{\varepsilon}{2} + \frac{\varepsilon}{4} = \varepsilon.$$

Since ε was arbitrary, f is integrable. By Corollary 9.2.2, the endpoint values of f do not affect the value of the integral.

Ex. 9.3.2. By hypothesis, there exists a t_0 in (a, b) such that $f(t_0) > 0$. (If f is identically zero in the open interval, continuity implies $f = 0$ at the endpoints.) Corollary 8.2.8 (ii) implies f is locally bounded away from zero: there exists a δ such that if $|t - t_0| < \delta$, then $f(t) > f(t_0)/2$. Shrinking δ if necessary, we may assume $[t_0 - \delta, t_0 + \delta] \subseteq [a, b]$. Since f is integrable, monotonicity implies

$$0 < (2\delta)\big(f(t_0)/2\big) = \int_{t_0-\delta}^{t_0+\delta} f(t_0)/2\, dt \leq \int_{t_0-\delta}^{t_0+\delta} f(t)\, dt \leq \int_{a}^{b} f(t)\, dt.$$

Ex. 9.3.3. Hint: Use Proposition 9.2.8 (ii) and Exercise 9.3.2.

Ex. 9.3.6. We have $\Delta t = b - a$, $f(t_0) = a^2$, $f(t_1) = b^2$, and $f(\bar{t}_0) = \frac{1}{4}(a+b)^2$. Consequently, $\text{LEFT}(f, \Pi) = a^2(b - a)$, $\text{RIGHT}(f, \Pi) = b^2(b - a)$,

$$\text{TRAP}(f, \Pi) = \frac{(a^2 + b^2)(b - a)}{2}, \qquad \text{MID}(f, \Pi) = \frac{(a+b)^2(b-a)}{4}.$$

The parabolic sum is

$$\frac{1}{3}\Big(\text{TRAP}(f, \Pi) + 2\,\text{MID}(f, \Pi)\Big) = \frac{1}{3}\left[\frac{(a^2 + b^2)}{2} + \frac{(a+b)^2}{2}\right](b-a)$$
$$= \frac{(a^2 + ab + b^2)(b - a)}{3}$$
$$= \frac{b^3 - a^3}{3} = \int_{a}^{b} t^2\, dt.$$

Ex. 9.3.8. Answer for part (c): $\overline{t^k} = \dfrac{1}{b-a}\dfrac{b^{k+1} - a^{k+1}}{k+1} = \dfrac{1}{k+1}\displaystyle\sum_{j=0}^{k} a^{k-j}b^{j}.$

Ex. 9.3.15. Hint: Use Exercise 9.3.14.

Exercises for Section 9.4

Ex. 9.4.1. The signum function is non-decreasing on $\mathbf{R}$, hence integrable on every closed, bounded interval $[a, b]$. (Other proofs are possible.)

If $0 \leq x$, then on $[0, x]$, $\operatorname{sgn}(t) = 1$ except at one point. Corollary 9.2.2 guarantees

$$\int_0^x \operatorname{sgn}(t)\, dt = \int_0^x 1\, dt = x - 0 = x = |x|.$$

If $x < 0$, then on $[x, 0]$, $\operatorname{sgn}(t) = -1$ except at one point. Corollary 9.2.2 guarantees

$$\int_0^x \operatorname{sgn}(t)\, dt = \int_0^x (-1)\, dt = -\int_0^x dt = \int_x^0 dt = 0 - x = -x = |x|.$$

Ex. 9.4.6. Hints: For (a), use induction to handle natural numbers p, then show $\log(x^{-p}) = -p \log x$ by algebra. For (b), write $b = x^{p/q}$, so that $b^q = x^p$, and use (a).

Ex. 9.4.13. Hint: First expand $1/t$ in a geometric series about x_0.

Solutions: Differentiation

Exercises for Section 10.1

Ex. 10.1.1. Suggestion: Write $f(x) = ax^2 + bx + c$, calculate the derivative in terms of a, b, and c, and only at the end substitute the given numerical values.

Ex. 10.1.2. Since $f_{1/3}(0) = 0$, the difference quotient is

$$\Delta f_{1/3}(0, h) = \frac{f_{1/3}(h) - 0}{h} = \frac{h^{1/3}}{h} = h^{-2/3} \to \infty.$$

Ex. 10.1.6. Periodicity of f implies that for all x,

$$f'(x + \ell) = \lim_{h \to 0} \frac{f(x + \ell + h) - f(x + \ell)}{h} = \lim_{h \to 0} \frac{f(x + h) - f(x)}{h} = f'(x).$$

The periodic function 1 does not have a periodic primitive.

Ex. 10.1.8. In a word, *no*. Perhaps the simplest counterexample is $f(x) = x$ and $g(x) = 0$. The equation $f(x) = g(x)$ has solution $x = 0$, but $f'(x) = g'(x)$ reads $1 = 0$. A general explanation is given in [9].

Ex. 10.1.9. Taking $x = 0$ gives $f(0) \leq 0$, which implies $f(0) = 0$.

(a) The absolute value of the difference quotient of f at 0 is

$$\left| \lim_{h \to 0} \frac{f(h) - f(0)}{h} \right| = \left| \lim_{h \to 0} \frac{f(h)}{h} \right| \leq \left| \lim_{h \to 0} \frac{h^2}{h} \right| = 0.$$

This simultaneously proves $f'(0)$ exists and is equal to 0.

(b) The function $f(x) = x^2 \cdot \chi_{\mathbf{Q}}(x)$ satisfies $|f(x)| \leq x^2$ for all real x, hence is differentiable at 0 and $f'(0) = 0$ by (a), but is discontinuous at x for every non-zero x by Exercise 8.1.4 (b).

Ex. 10.1.11. Let $(u, v) = (x - x_0, y - y_0)$ denote displacement from (x_0, y_0) as seen in the viewing window. Zooming in with factor c maps (u, v) to $(u/c, v/c)$, transforming the graph from

$$y_0 + v = y = f(x) = f(x_0 + u),$$

or $v = f(x_0 + u) - f(x_0)$, to $v/c = f(x_0 + u/c) - f(x_0)$, or

$$v = \frac{f(x_0 + u/c) - f(x_0)}{1/c} = \frac{f(x_0 + u/c) - f(x_0)}{u/c} \cdot u.$$

This is a line in the limit if and only if

$$\lim_{c \to \infty} \frac{f(x_0 + u/c) - f(x_0)}{u/c} = \lim_{h \to 0} \frac{f(x_0 + h) - f(x_0)}{h} = f'(x_0)$$

exists, if and only if f is differentiable at x_0.

Exercises for Section 10.2

Ex. 10.2.1. In each part, write $f^{(k)}$ to denote the kth derivative of a function f.

(a) $s'(x) = 1 - (x^2/2!)$, $s''(x) -- -x$, $s^{(3)}(x) = -1$, $s^{(k)}(x) = 0$ if $k \geq 4$.

(b) $c'(x) = -s(x)$, so $c^{(k+1)}(x) = -s^{(k)}(x)$.

(c) The derivative of each term is the preceding term, and the derivative of the first term is 0, so $e'(x) = 1 + x + (x^2/2!)$, $e''(x) = 1 + x$, $e^{(3)}(x) = 1$, and $e^{(k)}(x) = 0$ if $k \geq 4$.

Ex. 10.2.2. Differentiating successively, $f'(x) = nx^{n-1}$, $f''(x) = n(n-1)x^{n-2}$, $f^{(3)}(x) = n(n-1)(n-2)x^{n-3}$, and generally

$$f^{(k)}(x) = n(n-1)(n-2)\cdots(n-k+1)x^{n-k} = \frac{n!}{(n-k)!}x^{n-k}, \quad \text{if } 0 \le k \le n.$$

Particularly, $f^{(n)}(x) = n!$ is constant, so $f^{(k)}(x) = 0$ if $k > n$. These formulas may be established using induction on k.

Ex. 10.2.5. Write $r = p/q$ with q positive. The monomial function $g(y) = y^q$ is increasing (hence invertible) and differentiable on $(0, \infty)$, with derivative $g'(y) = qy^{q-1}$. The inverse function is the qth root function, $g^{-1}(x) = x^{1/q}$. By Theorem 10.2.10, g^{-1} is differentiable at $x = y^q$, and

$$(g^{-1})'(x) = \frac{1}{g'(y)} = \frac{1}{qy^{q-1}} = \frac{1}{q}x^{(1/q)-1}.$$

Since $f(x) = x^{p/q} = \left(g^{-1}(x)\right)^p$ for all positive x, the preceding calculation and the chain rule imply $f'(x) = p(x^{1/q})^{p-1} \cdot (1/q)x^{(1/q)-1} = p/qx^{(p/q)-1} = rx^{r-1}$.

Ex. 10.2.8. Throughout, intuition from calculus guides us.

(a) Taking $u = 1+t$ gives $F(t) = (1+t)^{n+1}/(n+1)$ as a primitive of f. Since the derivative of $F(t^2)$ is $2tF'(t^2) = 2t(1+t^2)^n = 2g(t)$, $G(t) = (1/2)F(t^2)$ is a primitive of g. For h there is no such recourse. Instead, the binomial theorem gives

$$h(t) = \sum_{k=0}^{n} \binom{n}{k} t^{2k}, \qquad H(t) = \sum_{k=0}^{n} \binom{n}{k} \frac{t^{2k+1}}{2k+1}.$$

(b) Similarly, $F(t) = 2(1+t)^{(n+2)/2}/(n+2)$ is a primitive of f, $G(t) = F(t^2)$ is a primitive of g, and $H(t) = (2+t^2)^{(2-n)/2}/(2-n)$ is a primitive of h if $n \ne 2$.

Ex. 10.2.9. Respectively, these are linearity of the derivative, the product rule, the chain rule, and the derivative of an inverse function.

Exercises for Section 10.3

Ex. 10.3.1. The value of c in $(0, b)$ given by the mean value theorem satisfies

$$nc^{n-1} = f'(c) = \frac{f(b) - f(0)}{b-0} = \frac{b^n}{b} = b^{n-1},$$

or $c = b/(n^{1/n})$. (Note that "existence and uniqueness" comes from existence and uniqueness of root of positive real numbers, not from anything specific to the mean value theorem, and is demonstrated by the fact of a formula.)

Ex. 10.3.2. Answer: When calculating derivatives, expanded form is usually easiest to differentiate, but factored form is easiest for determining the sign. There are four maximal branches of inverse, $x = \pm\sqrt{1 \pm \sqrt{y}}$ with all four choices of sign.

Function	Domain (x)	Image (y)	Monotonicity	Inverse
$y = f_1(x)$	$(-\infty, -1]$	$[0, \infty)$	decreasing	$x = -\sqrt{1 + \sqrt{y}}$
$y = f_2(x)$	$[-1, 0]$	$[0, 1]$	increasing	$x = -\sqrt{1 - \sqrt{y}}$
$y = f_3(x)$	$[0, 1]$	$[0, 1]$	decreasing	$x = \sqrt{1 - \sqrt{y}}$
$y = f_4(x)$	$[1, \infty)$	$[0, \infty)$	increasing	$x = \sqrt{1 + \sqrt{y}}$

Ex. 10.3.4. Answer: Both are false.

Ex. 10.3.5. By hypothesis, f is defined near x_0.

(a) Subtracting and adding $f(x_0)$ in the numerator gives

$$\lim_{h \to 0} \frac{f(x_0 + h) - f(x_0 - h)}{2h}$$
$$= \lim_{h \to 0} \frac{\left(f(x_0 + h) - f(x_0)\right) + \left(f(x_0) - f(x_0 - h)\right)}{2h} = f'(x_0).$$

(b) We can deduce nothing. For example, if f is "even about x_0," the difference quotient is identically 0, but f might be discontinuous everywhere.

Ex. 10.3.7. Hint: Let $h(x) = f(x)/g(x)$ and calculate h'.

Ex. 10.3.8. Hint: Apply the mean value theorem to the function $h : [a, b] \to \mathbf{R}$ defined by
$$h(x) = f(x)\left(g(b) - g(a)\right) - g(x)\left(f(b) - f(a)\right).$$

Ex. 10.3.9. Hints: For (a), setting $f(c) = 0$ and $g(c) = 0$ extends f and g continuously. Pick δ in $(0, r)$. For each x in $(c, c + \delta)$, use the generalized mean value theorem of Exercise 10.3.8 on $[c, x]$ to write $(f/g)(x) = (f'/g')(x_0)$ for some x_0 in (c, x).

For (b), if $x > R$, define $F(x) = f(1/x^2)$ and $G(x) = g(1/x^2)$, then show part (a) can be applied at 0.

Exercises for Section 10.4

Ex. 10.4.2. We have

$$f'(x) = 4x(x^2 - 1) = 4x^3 - 4x = 4x(x - 1)(x + 1),$$
$$f''(x) = 12x^2 - 4 = 4(3x^2 - 1) = 4(\sqrt{3}x - 1)(\sqrt{3}x + 1).$$

Since $f' > 0$ for large x and changes sign at each root, f is decreasing on $(-\infty, -1]$ and $[0, 1]$ and increasing on $[-1, 0]$ and $[1, \infty)$. There are four maximal intervals of monotonicity.

Since $f'' > 0$ for large x and changes sign at each root, f is concave on $[-\sqrt{3}, \sqrt{3}]$ and convex on $(-\infty, -\sqrt{3}]$ and $[\sqrt{3}, \infty)$.

There are absolute minima at $x = -1$ and $x = 1$ (also apparent without calculus), a local maximum at $x = 0$, and inflection points above $x = -\sqrt{3}$ and $x = \sqrt{3}$.

Ex. 10.4.6. Hint: Intuitively, suppose f is defined on $[x_0, x_0 + h]$ for some positive h. By the mean value theorem applied to f on this interval, there exists a c such that $f'(c) = \big(f(x_0 + h) - f(x_0)\big)/h$. Now take the one-sided limit: Since $x_0 < c < x_0 + h$, we have $c \to x_0$ as $h \to 0^+$. The left-hand side approaches $f'(x_0^+)$, while the right-hand side is a one-sided limit for $f'(x_0)$ by definition.

If $h < 0$, we argue similarly, except applying the mean value theorem to f on $[x_0 + h, x_0]$ and letting $h \to 0^-$.

Ex. 10.4.7. Answer: Suitably patching parabolic arcs suffices. For convenience, we'll first construct a 2-periodic function ψ_2; the function $\psi(x) = \psi_2(2x)$ is 1-periodic.

The quadratic function $\psi_2(x) = x(1 - x)$ on $[0, 1]$ is smooth, non-constant, symmetric about the midpoint $1/2$, and its graph contains the origin, so its odd extension to $[-1, 1]$ is differentiable by patching, and the 2-periodic extension of *that* to $\mathbf{R}$ is differentiable by patching.

Ex. 10.4.12. Hints: For (a), assume x_0 is an interior point of I. Pick a positive r such that $B_r(x_0) \subseteq I$, and put $a = x_0 - r$ and $b = x_0 + r$. Use convexity to prove the graph of f on $[a, b]$ lies between the secant line through $\big(a, f(a)\big)$ and $\big(x_0, f(x_0)\big)$ and the secant line through $\big(x_0, f(x_0)\big)$ and $\big(b, f(b)\big)$. (Make a sketch to organize the necessary inequalities, and use the definition of convexity to establish these inequalities. "Between" refers to secant lines, not just secant segments.) For (b), show the one-point estimate of part (a) can be made local at an interior point of I.

Ex. 10.4.13. Notation slightly different from that in the text may be used to highlight parallel structure. As usual, ε and δ connote positive real numbers throughout.

(a) Continuous on $[a, b]$:

> For every ε and every x_0 in $[a, b]$,
>> there exists a $\delta(x_0)$ such that
>>> if $x \in [a, b]$ and $|x - x_0| < \delta(x_0)$, then $|f(x) - f(x_0)| < \varepsilon$.

Uniformly continuous on $[a, b]$:

> For every ε,
>> there exists a δ such that
>>> if $\{x, x_0\} \subseteq [a, b]$ and $|x - x_0| < \delta$, then $|f(x) - f(x_0)| < \varepsilon$.

Bounded stretch on $[a, b]$:

> For every ε,
>> there exists a positive M such that
>>> if $\{x, x_0\} \subseteq [a, b]$ and $|x - x_0| < \varepsilon/M$, then $|f(x) - f(x_0)| < \varepsilon$.

(b) Loosely, bounded stretch says we can pick $\delta = \varepsilon/M$; not just independently of x_0, but *proportional to* ε. The square root function on $[0, 1]$ is uniformly continuous by Proposition 9.3.4 as a continuous function on a closed, bounded interval. The square root does not have locally bounded stretch at 0: For every δ there exists no M such that $0 \le x \le \delta$ implies $\sqrt{x} \le Mx$.

If ψ is a smooth, non-constant, periodic function on $\mathbf{R}$, the differentiable function $f(x) = x^2 \psi(1/x^2)$ on $\mathbf{R} \setminus \{0\}$ has a differentiable extension to 0, hence is uniformly continuous on, say, $[-1, 1]$. Because the derivative is unbounded, however, for every δ and every real M, there exist points x and x_0 in $[-1, 1]$ such that $|x - x_0| < \delta$ and $|f(x) - f(x_0)| > M|x - x_0|$.

(c) Suppose f is continuous at x_0. For each ε, let

$$J(x_0, \varepsilon) = \sup\{r : x \in B_r(x_0) \cap [a, b] \text{ implies } |f(x) - f(x_0)| < \varepsilon\}$$

and put $\delta(x_0) = \sup J(x_0)$. This response is largest in the following sense: If $|x - x_0| < \delta$, then $|x - x_0| < r$ for some r in $J(x_0)$, so $|f(x) - f(x_0)| < \varepsilon$. Inversely, for every r greater than $\delta(x_0)$, if $[a, b] \not\subseteq B_r(x_0)$, then then there exists an x in $[a, b]$ such that $|x - x_0| < r$ and $|f(x) - f(x_0)| \ge \varepsilon$.

If $0 < \varepsilon' < \varepsilon$, then $J(x_0, \varepsilon') \subseteq J(x_0, \varepsilon)$ by transitivity of inequality. By Lemma 4.2.10, $\delta(x_0, \varepsilon') \le \delta(x_0, \varepsilon)$. Loosely, "the largest winning response is non-decreasing in ε."

(d) Uniform continuity is equivalent to "$\delta(\varepsilon) := \delta(x_0, \varepsilon)$ is independent of x_0."
Bounded stretch means "there exists a positive M such that $\varepsilon/M \leq \delta(\varepsilon)$."

(e) Bounded stretch implies $f(x) \approx f(x_0)+O(x-x_0)$ for every x_0 by definition, with "the same M at each x_0." The converse is not true. For example, the squaring function on $\mathbf{R}$, or the square root function on the *open* interval $(0, \infty)$, satisfy $f(x) \approx f(x_0) + O(x - x_0)$ for every x_0, but there is no upper bound to the bounding slopes.

(f) The points are, (i) We can "gauge" continuity by comparing $\delta(\varepsilon)$ with non-decreasing functions $\eta(\varepsilon)$), and smaller functions correspond to *weaker continuity*, qualitatively since for a given ε we must take $\delta \leq \eta(\varepsilon)$; (ii) If $0 < s < r < 1$ and M, M_r, and M_s are positive, then for sufficiently small positive ε, we have $(\varepsilon/M_s)^{1/s} < (\varepsilon/M_r)^{1/r} < \varepsilon/M$.

(g) If f is "1.0001-continuous" on an interval, then f is constant: For each x_0 in (a, b), $\Delta f(x_0, h) \approx O(h^{0.0001})$, so $f'(x_0) = 0$.

Solutions: The Fundamental Theorems of Calculus

Exercises for Section 11.1

Ex. 11.1.1. If $x \geq 0$, we have $F(x) = \int_0^x dt = x$; if instead $x < 0$, we have $F(x) = \int_0^x 0 \, dt = 0$. The graph is the union of the negative x-axis and the line $y = x$ if $x \geq 0$.

Ex. 11.1.3. Answers:

(a) $G'(x) = 2xf(x^2)$.   (b) $H'(x) = 2xf(x^2) - f(x)$.

(c) $\Phi'(x) = f\big(\phi(x)\big)\phi'(x) - f\big(\psi(x)\big)\psi'(x)$.

Ex. 11.1.5. (Existence). The function $F(x) = c + \int_{x_0}^x f \left[= c + \int_{x_0}^x f(t) \, dt \right]$ is a solution by Theorem 11.1.1.

   (Uniqueness). If G is another solution, the difference $H = F - G$ satisfies $H' = F' - G' = f - f = 0$ in I and $H(x_0) = 0$. By the identity theorem, H is constant. Thus, for all x in I, $F(x) - G(x) = H(x) = H(0) = 0$, so $F = G$.

Ex. 11.1.6. No, there does not. Theorem 11.1.1 implies

$$f(x) = -\frac{x}{\sqrt{1 - x^2}}, \quad -1 < x < 1,$$

and this function is not bounded, hence has no integrable extension.

Ex. 11.1.8. Our goal is to construct a "second primitive" of f.

(a) Immediately from Theorem 11.1.1, we have $g'(x) = xf(x)$. For h', factor x from the integral and use the product rule and Theorem 11.1.1 to get

$$h(x) = x \int_0^x f(t)\, dt, \qquad h'(x) = xf(x) + \int_0^x f(t)\, dt.$$

(b) Put

$$F(x) = \int_0^x (x - t)f(t)\, dt = h(x) - g(x), \qquad G(x) = \int_0^x \left[\int_0^s f(t)\, dt \right] ds.$$

By (a) and Theorem 11.1.1,

$$F'(x) = h'(x) - g'(x) = xf(x) + \int_0^x f(t)\, dt - xf(x) = \int_0^x f(t)\, dt = G'(x).$$

By the identity theorem, $F - G$ is constant. Since $F(0) = 0 = G(0)$, we have $F - G = 0$, or $F = G$. Two applications of differentiating an integral give $F'' = f$.

Ex. 11.1.9. Hint: Exercise 11.1.8 may help.

Ex. 11.1.10. Hint: First multiply the differential equation by $2y'$ and integrate.

Ex. 11.1.11. Hint: Consider the function $p_n(t) = t^n$ if $0 < t$ and 0 otherwise. First show p_n is of class $\mathscr{C}^{n-1}$.

Ex. 11.1.13. Hint: The derivative $F' = f$ is continuous, and therefore may be written as a difference of non-negative continuous functions by Exercise 8.2.6.

Exercises for Section 11.2

Ex. 11.2.4. By Exercise 11.2.3, the nth-degree germ of $f(-x)$ is $p^n(-x)$. Consequently, nth-degree germs of the even and odd parts of f are the even and odd parts of p^n, namely, the sum of the even-degree terms and the sum of the odd-degree terms.

Ex. 11.2.5. By hypothesis, $f(x) \approx p^n(x) + O(x^{n+1})$. Substituting x^2 for x gives $g(x) = f(x^2) \approx p^n(x^2) + O(x^{2(n+1)})$. The polynomial q^{2n} has degree at most $2n$, and approximates g to order $(2n + 2)$ near x_0. By Exercise 11.2.3, q^{2n} is the $(2n + 1)$th-degree germ of g at x_0.

Ex. 11.2.7. Hint: The $(n-1)$th-degree remainder is equal to the $(n-1)$th-degree term plus the nth-degree remainder. Apply the mean value theorem to $f^{(n)}$ on the interval between x_0 and z_n, and use $f^{(n+1)}(x_0) = \lim(f^{(n+1)}, x_0) \neq 0$.

Ex. 11.2.9. Hint: If $[\alpha, \beta] \subseteq [a, b]$ and $\overline{x} = \frac{1}{2}(\alpha + \beta)$ is the midpoint, use the second-degree germ with $x_0 = \overline{x}$ to prove

$$\left| \int_\alpha^\beta \left(f(x) - f(\overline{x}) \right) dx \right| \leq \frac{K_2(\beta - \alpha)^3}{24}.$$

Then apply this estimate to each piece of an equal-length splitting.

Ex. 11.2.10. Hints: Follow the general outline in Exercise 11.2.9: First establish the $n = 1$ estimate on an arbitrary subinterval. For that, let $T : [\alpha, \beta] \to \mathbf{R}$ be the trapezoid error for f on $[\alpha, x]$:

$$T(x) = \int_\alpha^x f(t)\, dt - \frac{1}{2}\left(f(\alpha) + f(x)\right)(x - \alpha),$$

and put

$$E(x) = T(x) - \left[\frac{x - \alpha}{\beta - \alpha} \right]^3 T(\beta).$$

Conceptually, interpolate T on $[\alpha, \beta]$ by a cubic $M(x - \alpha)^3$, the trapezoid error if f'' were constant, and let E be the difference. Apply the mean value theorem to E on $[\alpha, \beta]$ to prove there exists a z' in (α, β) such that $E'(z') = 0$. Then apply the mean value theorem to E' on $[\alpha, z']$ to prove there is a z in (α, z') such that $E''(z) = 0$. Conclude that $T(\beta) = -\frac{1}{12} f''(z)(\beta - \alpha)^3$.

Ex. 11.2.11. Hints: Follow the general outline in Exercise 11.2.9: First establish the $n = 1$ estimate on an arbitrary subinterval. For that, let $r = \frac{1}{2}(\beta - \alpha)$ be the radius of $[\alpha, \beta]$. Expand

$$F(x) = \int_{\overline{x}}^{\overline{x}+x} f(t)\, dt, \quad |x| \leq r,$$

about 0 as fourth-degree germ with integral remainder. Use this to prove the error of the parabolic sum is

$$F(r) - F(-r) - \frac{1}{6}\left(F'(-r) + 4F'(0) + F'(r)\right)(2r)$$

$$= \frac{1}{72} \int_0^r \left(F^{(5)}(s) + F^{(5)}(-s) \right)\left(3(r - s) - 4r\right)(r - s)^3\, dt.$$

Upon careful substitution, all the non-remainder terms in the error expression cancel. To handle the remainders (there are four integrals, two from the F terms and two from the F' terms), change variables so all are evaluated over $[0, r]$, then bring out common factors in the integrands.

Then prove that if $|f^{(4)}| \leq K_4$, the absolute value of the right-hand side is no larger than $(1/90)K_4 r^5 = (1/2880)K_4(\beta - \alpha)^5$. (Note: If $|s| \leq r$, then $F^{(5)}(s) = f^{(4)}(\overline{x} + s)$.)

Ex. 11.2.12. Writing $f(t) = (1 + t^4)^{1/2}$, we have $f'(t) = 2t^3(1 + t^4)^{-1/2}$, and (after a bit of algebra)

$$f''(t) = 2(1 + t^4)^{-3/2}(3t^2 + t^6).$$

To find an upper bound K_2 of $|f''|$, it suffices to find the absolute maximum of f'' on $[0, 1]$. By Proposition 10.1.10, the extrema occur at $t = 0$ or $t = 1$, which are endpoints of $[0, 1]$, or at points where $f'''(t) = 0$, which turns out to give the same two candidates. Alternatively, it suffices to maximize the square,

$$|f''(t)|^2 = \frac{4t^4(3 + t^4)^2}{(1 + t^4)^3} = \frac{4u(3 + u)^2}{(1 + u)^3}\bigg|_{u=t^4},$$

whose derivative is easier to calculate. Either way, the minimum is at $t = 0$ and the maximum at $t = 1$; thus $0 = f''(0) \leq f''(t) \leq f''(1) = 2\sqrt{2}$ for all t in $[0, 1]$.

Since $2\sqrt{2} < 3$, we may take $K_2 = 3$ in the midpoint method. The error bound of Exercise 11.2.9 says that with n equal-length intervals on $[0, x]$, the midpoint error is at most

$$\frac{K_2(x - 0)^3}{24n^2} < \frac{x^3}{8n^2}.$$

This is at most 0.5×10^{-4} if and only if $20,000x^3 \leq 8n^2$, or $n = 50x^{3/2}$. For example, if $x = \frac{1}{2}$, then 18 intervals suffice.

Ex. 11.2.13. Hint: Direct calculation of derivatives is tedious. Instead, write $f(t) = (1 + t^4)^{1/2}$, so $f^{(n)}(t) = (1 + t^4)^{(1/2)-n}p_n(t)$ for some polynomial p_n, find the recursion relation for p_n, and use that to calculate the derivatives.

Ex. 11.2.15. By Exercise 10.2.6, if $x > -1$ then

$$f^{(k)}(x) = (-1)^{k-1}\left[\prod_{j=1}^{k-1}\frac{2j - 1}{2}\right](1 + x)^{(1-2k)/2}, \quad \text{for every natural number } k.$$

Fix z in $(-1, \infty)$ arbitrarily, and put $R = \frac{1}{2}(1 + z)$, so $B_{2R}(z) \subseteq (-1, \infty)$. Substituting $k = n + 1$,

$$\left|\frac{f^{(n+1)}(z)R^{n+1}}{(n + 1)!}\right| = \frac{(1 + z)^{-(1+2n)/2}R^{n+1}}{n + 1}\prod_{j=1}^{n}\frac{2j - 1}{2j} \leq \frac{R}{(1 + z)^{1/2}(n + 1)},$$

which is bounded. By Exercise 11.2.14, f is real-analytic in $(-1, \infty)$.

Exercises for Section 11.3

Ex. 11.3.1. It suffices to split the first and fourth at 0.

(a) $\int_{-\infty}^{0} + \int_{0}^{\infty} = 2 \int_{0}^{\infty}$. Converges by comparison with $\max(1, 1/x^2)$.

(b) Converges since $(1 - x^2)^{-1/2} = [(1 - x)(1 + x)]^{-1/2} \leq (1 - x)^{-1/2}$.

(c) Diverges by comparison with $\max(1, 1/(2x))$.

(d) $\int_{-1}^{0} + \int_{0}^{1}$. Diverges by direct computation.

Ex. 11.3.4. The partial fractions decomposition of the integrand is

$$\frac{1}{1 - t^2} = \frac{1}{2}\left[\frac{1}{1 - t} + \frac{1}{1 + t}\right].$$

Change of variables gives

$$\begin{aligned}
\int_{0}^{x} \frac{dt}{1 - t^2} &= \frac{1}{2} \int_{0}^{x} \left[\frac{dt}{1 - t} + \frac{dt}{1 - t}\right] \\
&= \frac{1}{2}\left[-\int_{1}^{1-x} \frac{du}{u} + \int_{1}^{1+x} \frac{du}{u}\right] \\
&= \frac{1}{2}\left[\log(1 + x) - \log(1 - x)\right] = \frac{1}{2} \log \frac{1 + x}{1 - x}.
\end{aligned}$$

The integral is infinite at $x = \pm 1$, so converges if $|x| < 1$.

Ex. 11.3.5. Answer: The requested values are:

(a) $\displaystyle \int_{1}^{\infty} \frac{dt}{t^m(t + a)} = \frac{1}{(-a)^m}\left[-\log(1 + a) - \sum_{k=1}^{m-1} \frac{(-a)^k}{k}\right].$

(b) $\displaystyle \sum_{k=1}^{\infty} \frac{(-1)^{k-1}}{k} = \log 2.$

Solutions: Exponential Functions

Exercises for Section 12.1

Ex. 12.1.1. It does matter: $(a^b)^c \neq a^{(b^c)}$ in general, even if $a = b = c$. Since $(a^b)^c = a^{bc}$ can be written unambiguously without parentheses, it's more reasonable to define $a^{b^c} = a^{(b^c)}$.

Ex. 12.1.3. It's true mathematically—in a suitable model with no limitations on time—but laughably false in reality given what is observed about cosmology. In an alphabet of C characters, there are C^n strings of length n. The exponential growth of these sets motivates the question.

To make a quantitative model, let's assume there are one hundred characters (letters, numbers, punctuation), and one monkey types ten characters per second without rest. In those conditions, about how long do we expect to wait before we see Hamlet's "To be, or not to be;"? How does this estimate increase if we want to see the entire line, "To be, or not to be; that is the question." See also Munroe, [22].

Ex. 12.1.4. By the chain rule, $(\exp \circ u)' = (\exp' \circ u)u' = (\exp \circ u)u'$ and $(\log \circ |u|)' = (\log' \circ |u|)(\operatorname{sgn} u)u' = u'/u$.

Ex. 12.1.5. Answer: We have $f_{\text{even}}(t) = \dfrac{t}{2} \cdot \dfrac{e^{t/2} + e^{-t/2}}{e^{t/2} - e^{-t/2}}$ and $f_{\text{odd}}(t) = \dfrac{t}{2}$.

Ex. 12.1.7. Hint: The calculations can be done *ad hoc* but are streamlined with a lemma: By the product and chain rules, if p is a polynomial and $g(x) = p(x)e^{-x}$, then $g'(x) = p'(x)e^{-x} - p(x)e^{-x} = \big(p'(x) - p(x)\big)e^{-x}$.

Ex. 12.1.12. We first show $\log_{10} n$ is essentially the number of digits of n.

(a) The inequalities are equivalent because $\log_{10}$ is strictly increasing. Since 10^n in decimal is a one followed by n zeros and is the smallest such integer, an integer N has $n + 1$ digits if and only if $10^n \leq N < 10^{n+1}$.

(b) One googolplex, $10^{10^{100}}$, has $10^{100} + 1$ digits by (a). The other number is

$$N = 2^{2^{2^{2^{2^2}}}} = 2^{2^{2^{2^4}}} = 2^{2^{2^{16}}} = 2^{2^{65536}}.$$

The number of digits of N is $D = \lceil \log_{10} N \rceil = \lceil 2^{65536} \log_{10} 2 \rceil$. This is still too large to evaluate directly with a calculator. We have, however, $D = 10^{\log_{10} D}$, and

$$\log_{10} D \approx 65536 \log_{10} 2 + \log_{10}(\log_{10} 2) > 19728.$$

That is, $D > 10^{19728}$, and therefore $N > 10^{10^{19728}}$. This means N is so much larger than one googolplex that N divided by one googolplex is still essentially N. (When we divide powers, we *subtract* the exponents.)

Ex. 12.1.15. Hint for part (d): Apply the conclusion of part (c) to the numbers $A_k = a_k / \big[\sum_k |a_k|^p\big]^{1/p}$ and $B_k = b_k / \big[\sum_k |b_k|^q\big]^{1/q}$. Sum over k, and remember $1/p + 1/q = 1$.

For (e), if $\int_a^b |f|^p = 0$, then $\int_a^b |f| = 0$ (Why?), and the stated inequality is automatic. It therefore suffices to assume $\|f\|_p$ and $\|g\|_q$ are positive. Proceed as in (d), with $A = |f(x)|/\|f\|_p$ and $B = |g(x)|/\|g\|_q$, and integrate over $[a, b]$.

Exercises for Section 12.2

Ex. 12.2.1.

(a) Substituting $t = -\log x$ gives $\lim_{x \to 0^+} x \log x = \lim_{t \to \infty} -te^{-t} = \lim_{t \to \infty} -t/e^t$. Since $1 + t^2/2 < e^t$ if $t > 0$, the rightmost limit is 0.

(b) Since exp is continuous, part (a) implies

$$\lim_{x \to 0^+} x^x = \lim_{x \to 0^+} \exp(x \log x) = \exp(\lim_{x \to 0^+} x \log x) = \exp(0) = 1.$$

(c) Substituting $t = 1/x$ and using (b), $\lim_{x \to \infty} x^{1/x} = \lim_{t \to 0^+} 1/t^t = 1/[\lim_{t \to 0^+} t^t] = 1$.

Ex. 12.2.5. All three series converge.

(a) By Corollary 12.2.5,

$$\frac{\log n}{n^{3/2}} = \frac{\log n}{n^{1/4}} \cdot \frac{1}{n^{5/4}} \leq \frac{1}{n^{5/4}}$$

for sufficiently large n. Since the series with larger terms converges by the p-series test, the original series converges (absolutely in fact).

(b) The terms are eventually decreasing in absolute value and converge to 0, so the series converges by the alternating series test.

(c) The terms are positive and decreasing. By the integral test the series converges if and only if

$$\int_2^\infty \frac{dx}{x(\log x)^2} = \int_{\log 2}^\infty \frac{dt}{t^2}$$

converges, which it does.

Exercises for Section 12.3

Ex. 12.3.3. For all real x, $\cosh^2 x - \sinh^2 x = 1$.

(a) The function sinh is bijective, hence invertible. Algebraically, for each real t the equation $t = \sinh x$ may be written equivalently as $x = \sinh^{-1} t$. Since $\cosh x = \sqrt{\sinh^2 x + 1}$, substitution gives

$$\cosh(\sinh^{-1} t) = \sqrt{\sinh^2(\sinh^{-1} t) + 1} = \sqrt{t^2 + 1}.$$

(b) The function cosh is bijective, hence invertible, on each of $(-\infty, 0]$ and $[0, \infty)$. If $t \geq 1$, the equation $t = \cosh x$ may be written $x = \cosh^{-1} t$ for some unique non-negative real x. Since $\sinh x = \sqrt{\cosh^2 x - 1}$ if $x \geq 0$, substitution gives

$$\sinh(\cosh^{-1} t) = \sqrt{\cosh^2(\cosh^{-1} t) - 1} = \sqrt{t^2 - 1}.$$

The equation $t = \cosh x$ may also be written $x = \cosh^{-1} t$ for some unique non-positive real x. Since $\sinh x = -\sqrt{\cosh^2 x - 1}$ if $x \leq 0$, here we have

$$\sinh(\cosh^{-1} t) = -\sqrt{\cosh^2(\cosh^{-1} t) - 1} = -\sqrt{t^2 - 1}.$$

Particularly, the choice of sign is determined by the branch of $\cosh^{-1}$.

Ex. 12.3.5. By Proposition 12.3.3,

$$
\begin{aligned}
\int_0^x \sqrt{u^2 + 1}\, du &= \int_0^{\sinh^{-1} x} \sqrt{\sinh^2 t + 1}\, \cosh t\, dt && u = \sinh t, \quad du = \cosh t\, dt \\
&= \int_0^{\sinh^{-1} x} \cosh^2 t\, dt && \sinh^2 t + 1 = \cosh^2 t \\
&= \frac{1}{2} \int_0^{\sinh^{-1} x} (1 + \cosh 2t)\, dt && \sinh^2 t + \cosh^2 t = \cosh 2t \\
&= \tfrac{1}{2}\left(t + \tfrac{1}{2}\sinh 2t\right)\Big|_0^{\sinh^{-1} x} \\
&= \tfrac{1}{2}\left(t + \sinh t \cosh t\right)\Big|_0^{\sinh^{-1} x} && 2\sinh t \cosh t = \sinh 2t \\
&= \tfrac{1}{2}\left[\log(x + \sqrt{x^2 + 1}) + x\sqrt{1 + x^2}\right] && \cosh \sinh^{-1} x = \sqrt{x^2 + 1}.
\end{aligned}
$$

Ex. 12.3.7. Dropping a perpendicular from $(\cosh t, \sinh t)$ to the horizontal axis describes a right triangle of base $\cosh t$ and height $\sinh t$, hence area $\frac{1}{2}\cosh t \sinh t$. The complement of the shaded area in this triangle is described by inequalities $0 \leq y \leq \sqrt{x^2 - 1}$ and $1 \leq x \leq \cosh t$, so its area is

$$
\begin{aligned}
\int_1^{\cosh t} \sqrt{x^2 - 1}\, dx &= \int_0^t \sinh^2 u\, du && x = \cosh u, \quad dx = \sinh u\, du \\
&= \int_0^t \tfrac{1}{2}\left(\cosh(2u) - 1\right) du && \text{Proposition 12.3.5} \\
&= \tfrac{1}{2}\left(\tfrac{1}{2}\sinh(2u) - u\right)\Big|_{u=0}^t \\
&= \tfrac{1}{2}\left(\cosh t \sinh t - t\right) && \text{Proposition 12.3.5.}
\end{aligned}
$$

The shaded area is the difference, $t/2$.

Ex. 12.3.9. Hint: This can be done "naively," but a hyperbolic identity and abstract reasoning give a short, elegant proof.

Exercises for Section 12.4

Ex. 12.4.1. The first restates Proposition 12.2.3. The second restates Proposition 12.4.5 (ii).

Ex. 12.4.4. Since $-t^2 \leq -t$ if $t \geq 1$ and $e^{-t^2} \leq 1$ for all real t, we have, for every R greater than 1,

$$\int_0^R e^{-t^2}\, dt \leq 1 + \int_1^R e^{-t}\, dt < 1 + \int_0^R e^{-t}\, dt = 2 - e^{-R}.$$

Since the integral is bounded and increasing in R, it converges.

Ex. 12.4.6. By factorial growth rate,

$$e^{7/8}\left[\frac{2n}{e}\right]^{2n}\sqrt{2n} < (2n)! < e\left[\frac{2n}{e}\right]^{2n}\sqrt{2n},$$

$$e^{7/4}\left[\frac{n}{e}\right]^{2n} n < (n!)^2 < e^2\left[\frac{n}{e}\right]^{2n} n.$$

Neglecting the constant powers of e in front for the moment,

$$\frac{(2n/e)^{2n}\sqrt{2n}}{(n/e)^{2n}n} = \frac{2^{2n}\sqrt{2}}{\sqrt{n}}.$$

To get a lower bound, use the lower bound in the numerator and upper bound in the denominator. For an upper bound, use the upper bound in the numerator and lower bound in the denominator. Thus

$$\frac{2^{2n}\sqrt{2}}{e^{9/8}\sqrt{n}} \leq \binom{2n}{n} = \frac{(2n)!}{(n!)^2} \leq \frac{2^{2n}\sqrt{2}}{e^{3/4}\sqrt{n}}.$$

Incidentally, the sum over k of $\binom{2n}{k}$ is 2^{2n}. The result here says the middle (largest) binomial coefficient is about $1/\sqrt{n}$ of the total. When tossing a fair coin $2n$ times, the chance of getting precisely n heads is $\approx 1/\sqrt{n}$.

By contrast, if $\varepsilon > 0$, the chance of getting a fraction f of heads such that $|f - (1/2)| < \varepsilon$ turns out to approach 1 as $n \to \infty$. The chance of exactly half the tosses being heads is vanishingly small, but the chance the *proportion* of heads *differs* from one-half is also vanishingly small.

Ex. 12.4.9. Suggestion: Prove $\log \beta(\cdot, y)$ is convex along the same lines as log-convexity of Γ. Integrate $\beta(x + 1, y)$ by parts, and obtain an expression in terms of $\beta(x, y)$. Show $f(x) := \Gamma(x+y)\beta(x, y)/\Gamma(y)$ satisfies conditions (i)–(iii) of Proposition 12.4.5.

Ex. 12.4.10. Hint: For all positive x,

$$\frac{1}{e^x - 1} = \frac{e^{-x}}{1 - e^{-x}} = e^{-x} \sum_{m=0}^{\infty} e^{-mx}.$$

First prove the following converge and are equal:

$$\int_0^{\infty} \sum_{m=0}^{\infty} x^{s-1} e^{-x} e^{-mx} \, dx = \sum_{m=0}^{\infty} \int_0^{\infty} x^{s-1} e^{-x} e^{-mx} \, dx,$$

then do a suitable change of variables on the right-hand side. To show the two are equal, which is most of the work, assume $\delta > 0$ and $N \geq 1$. Separately show

$$\int_0^{\infty} \sum_{m=0}^{N-1} = \sum_{m=0}^{N-1} \int_0^{\infty},$$

and that with estimates customized to the structure of each term,

$$\left| \int_0^{\infty} \sum_{m=N}^{\infty} - \sum_{m=N}^{\infty} \int_0^{\infty} \right| = \left| \left[\int_0^{\delta} + \int_{\delta}^{\infty} \right] \sum_{m=N}^{\infty} - \sum_{m=N}^{\infty} \int_0^{\infty} \right|$$

can be made arbitrarily small.

Solutions: Circular Functions

Exercises for Section 13.1

Ex. 13.1.1. Differentiating termwise,

$$\cos' x = \sum_{k=1}^{\infty} \frac{(-1)^k x^{2k-1}}{(2k-1)!} = 0 - x + \frac{x^3}{3!} - \frac{x^5}{5!} + \frac{x^7}{7!} + \cdots$$

$$\sin' x = \sum_{k=0}^{\infty} \frac{(-1)^k x^{2k}}{(2k)!} = 1 - \frac{x^2}{2!} + \frac{x^4}{4!} - \frac{x^6}{6!} + \frac{x^8}{8!} - \cdots,$$

or $\cos' x = -\sin x$ and $\sin' x = \cos x$.

Exercises for Section 13.2

Ex. 13.2.1. The fundamental circular identity and double-angle formula for cos,

$$\left. \begin{array}{l} 1 = \cos^2 x + \sin^2 x \\ \cos(2x) = \cos^2 x - \sin^2 x \end{array} \right\} \quad \text{for all real } x$$

imply, for all real x,

$$\cos^2 x = \tfrac{1}{2}\big(1 + \cos(2x)\big), \qquad \sin^2 x = \tfrac{1}{2}\big(1 - \cos(2x)\big).$$

Writing $2x = \theta$ and taking square roots gives the *half-angle formulas*

$$\cos(\tfrac{1}{2}\theta) = \sqrt{\tfrac{1}{2}(1 + \cos\theta)}, \qquad \sin(\tfrac{1}{2}\theta) = \sqrt{\tfrac{1}{2}(1 - \cos\theta)},$$

which are true if $0 \le x \le \pi/2$ (the first quadrant), namely, for θ in $[0, \pi]$.

If $\theta = \pi/4$, we have $\cos\theta = \sqrt{2}/2$, so $\tfrac{1}{2}(1 \pm \cos\theta) = \tfrac{1}{4}(2 \pm \sqrt{2})$. Consequently,

$$\cos\tfrac{\pi}{8} = \sqrt{\tfrac{1}{4}(2 + \sqrt{2})} = \tfrac{1}{2}\sqrt{2 + \sqrt{2}}, \qquad \sin\tfrac{\pi}{8} = \sqrt{\tfrac{1}{4}(2 - \sqrt{2})} = \tfrac{1}{2}\sqrt{2 - \sqrt{2}}.$$

Ex. 13.2.3. Hint: Bound $\sin x$ below by a positive multiple of x to prove the integral converges, then use the substitution $x = 2u$ and the double angle formula for $\sin 2u$ to obtain an algebraic equation for the integral, and solve.

Ex. 13.2.7. Hint: Take $x = y = 1/2$ in Proposition 13.2.10.

Ex. 13.2.8. Hint: Exercises 10.1.9, 10.4.9, and 12.1.14 may be of interest.

Ex. 13.2.9. Hint: The summands of each function are individually in $[-1, 1]$. A "peak" of a sinusoid is a point where the value is 1. A "trough" is a point where the value is -1. The inequalities $-2 \le f(x) \le 2$ and $-2 \le g(x) \le 2$ are immediate. Further, the value 2 is achieved if and only if two peaks coincide, while -2 is achieved if and only if two troughs coincide.

Intuitively, because the periods of the summands have irrational ratio, there is no exact repetition of behavior, but the peaks of one summand "line up arbitrarily closely" with the peaks of the other, and similarly for troughs. The result of Exercise 5.3.10 may be helpful in making this intuition rigorous.

Exercises for Section 13.3

Ex. 13.3.1. Formally, since $\sec x = 1/\cos x$, the chain rule gives

$$\sec' x = \frac{-\cos' x}{\cos^2 x} = \frac{\sin x}{\cos^2 x} = \frac{1}{\cos x} \cdot \frac{\sin x}{\cos x} = \sec x \tan x.$$

Both sec and sec$'$ = sec tan are defined everywhere except the zero set of cos.

Ex. 13.3.3. Hint: The half-angle formulas in the solution of Exercise 13.2.1 may help.

Ex. 13.3.4. Each part refers to $\mathrm{SI}(x) = \int_0^x \frac{\sin t\, dt}{t}$.

(a) Using the sine series, we have

$$\frac{\sin t}{t} = \sum_{k=0}^{\infty} (-1)^k \frac{t^{2k}}{(2k+1)!} = 1 - \frac{t^2}{3!} + \frac{t^4}{5!} - \cdots$$

for all non-zero real t. The radius is ∞, just as for the sine series, so we may view s as a real-analytic (continuous in particular) function on $\mathbf{R}$ satisfying $s(0) = 1$. Integrating term by term,

$$\mathrm{SI}(x) = \int_0^x \frac{\sin t\, dt}{t} = \sum_{k=0}^{\infty} \frac{(-1)^k x^{2k+1}}{(2k+1)(2k+1)!} = x - \frac{x^3}{3 \cdot 3!} + \frac{x^5}{5 \cdot 5!} - \cdots.$$

(b) By Theorem 11.1.1, points where $\mathrm{SI}' = 0$ are the zeros of s, namely the non-zero integer multiples of π. The corresponding values are, for each positive integer n,

$$\mathrm{SI}(n\pi) = \int_0^{n\pi} \frac{\sin t\, dt}{t} = \sum_{k=0}^{n-1} \int_{k\pi}^{(k+1)\pi} \frac{\sin t\, dt}{t}.$$

(c) The quotient rule gives

$$\mathrm{SI}''(x) = s'(x) = \frac{x \cos x - \sin x}{x^2} \quad \text{if } x \neq 0,$$

and a short series calculation shows $s'(x) \approx O(x)$ if $x \approx 0$, so $\mathrm{SI}''(0) = 0$ if and only if $x \cos x - \sin x = 0$. At each non-zero solution, $\cos x \neq 0$. Consequently, $s'(x) = 0$ if and only if $x = \tan x$. By Corollary 13.3.8, there is exactly one solution x_k satisfying $(k - \frac{1}{2})\pi < x_k < (k + \frac{1}{2})\pi$. Since the sign of SI'' changes at x_k, x_k is an inflection point of SI, see Proposition 10.4.11.

(d) The function s "oscillates," but the size of the oscillations decreases with $|t|$. Precisely,

$$s(t + 2\pi) = \frac{\sin t}{t + 2\pi} < s(t) \quad \text{for all non-negative } t.$$

Since

$$\frac{|\sin t|}{(k+1)\pi} \leq \frac{|\sin t|}{t} \leq \frac{|\sin t|}{k\pi} \quad \text{on } [k\pi, (k+1)\pi],$$

we have

$$\frac{2}{(k+1)\pi} \leq \left| \int_{k\pi}^{(k+1)\pi} \frac{\sin t}{t} \, dt \right| \leq \frac{2}{k\pi} \quad \text{for all } k \text{ in } \mathbf{Z}^{+}.$$

Let a_k denote the integral. The signs of the a_k alternate, since the sine function is π-anti-periodic, and $|a_{k+1}| < |a_k|$ for all k because the denominator strictly increases from one interval to the next. The absolute maximum value of SI is therefore

$$\text{SI}(\pi) = \int_0^{\pi} \frac{\sin t \, dt}{t}.$$

(e) By the alternating series test, $\lim\big(\text{SI}(n\pi), \infty\big)$ exists, and since the partial sums bound the partial integrals, the improper integral

$$\lim_{x \to \infty} \text{SI}(x) = \int_0^{\infty} \frac{\sin t \, dt}{t}$$

converges. However, the integral is not absolutely convergent, since

$$\int_0^{\infty} \left| \frac{\sin t}{t} \right| dt = \sum_{k=0}^{\infty} \int_{k\pi}^{(k+1)\pi} \left| \frac{\sin t}{t} \right| dt \geq \sum_{k=0}^{\infty} \frac{2}{(k+1)\pi}$$

is (a multiple of) the harmonic series.

Ex. 13.3.6. Substituting $x = r\cos\theta$ and $y = r\sin\theta$ into the equation of the hyperbola gives $r^2 \cos^2\theta - r^2 \sin^2\theta = 1$, or, by the double angle formula for cos, $r^2 \cos(2\theta) = 1$. This equation has real solutions (r, θ) if and only if $0 < \cos(2\theta)$, in which case the hyperbola is the polar graph $r = \sqrt{\sec(2\theta)}$. To get a domain geometrically, consider the angles of rays from $(0,0)$ that cross the hyperbola. We may take $-\pi/4 < \theta < \pi/4$ or $3\pi/4 < \theta < 5\pi/4$, namely $I = (-\pi/4, \pi/4) \cup (3\pi/4, 5\pi/4)$.

Ex. 13.3.8. The hyperbola has polar equation $r^2 \cos(2\theta) = 1$ by Exercise 13.3.6. Inversion sends r to its reciprocal, which sends the hyperbola equation to $(1/r^2) \cos(2\theta) = 1$, or $\cos(2\theta) = r^2$, or $r = \sqrt{\cos(2\theta)}$. Because $\cos(2\theta) = 0$ gives $r = 0$, we may take the domain to be the closure of the intervals in Exercise 13.3.6, namely $I = [-\pi/4, \pi/4] \cup [3\pi/4, 5\pi/4]$.

To convert to a rectangular equation, one general idiom is to multiply both sides by enough powers of r to convert circular functions to rectangular coordinates. This does not always work, but does here because both circular functions are squared. Multiplying both sides by r^2 gives $r^2 \cos(2\theta) = r^4$, or $x^2 - y^2 = (x^2 + y^2)^2$. This is equivalent because the origin is on the polar graph $r = \sqrt{\cos(2\theta)}$, and the only introduced solution when multiplying by r^2 is $r = 0$.

Exercises for Section 13.4

Ex. 13.4.3. By Corollary 13.2.16, every point (x, y) such that $x > 0$ or $y \neq 0$ may be written uniquely in the form $(x, y) = (r \cos \theta, r \sin \theta)$ with r positive and θ in $(-\pi, \pi)$. If $x > 0$, then $-\pi/2 < \theta < \pi/2$, so x and y are both positive, and

$$y/x = \frac{r \sin \theta}{r \cos \theta} = \tan \theta, \qquad x/y = \frac{r \cos \theta}{r \sin \theta} = \cot \theta.$$

Since $|\arctan| < \pi/2$, we have $\arctan(y/x) = \theta$ if $x > 0$. If $y > 0$ in addition, then since $0 < \operatorname{arccot} < \pi$ we also have $\operatorname{arccot}(x/y) = \theta$. Consequently, the first and second formulas agree in the open first quadrant, where both are defined.

If instead $y < 0$, then $-\pi < \theta < 0$ and $\cot \theta = x/y$, so $\operatorname{arccot}(x/y) = \pi + \theta$, or $\operatorname{arccot}(x/y) - \pi = \theta$. Consequently, the first and third formulas agree in the open fourth quadrant, where both are defined.

Solutions: Complex Numbers

Exercises for Section 14.1

Ex. 14.1.1. Multiplying and dividing by $\overline{\alpha}$, we obtain $\dfrac{1}{\alpha} = \dfrac{\overline{\alpha}}{\alpha \overline{\alpha}} = \dfrac{a - bi}{a^2 + b^2}$.

Ex. 14.1.4. If z and w are complex numbers and n is a natural number, then

$$(z + w)^n = \sum_{k=0}^{n} \binom{n}{k} z^{n-k} w^k = \sum_{k \in \mathbf{Z}} \binom{n}{k} z^{n-k} w^k,$$

with the terms of the rightmost sum understood to be 0 unless $0 \leq k \leq n$.

Each proof we gave for the real case (operational and inductive) formally goes through verbatim. The crucial properties are the field axioms, and existence of binomial coefficients. (Some fields "do not contain a copy of the integers.")

Ex. 14.1.6. Answers: For all real x and y,

(a) $(x + iy)^2 = (x^2 - y^2) + (2xy)i$.

(b) $(x + iy)^3 = (x^3 - 3xy^2) + (3x^2y - y^3)i$.

(c) $(x + iy)^4 = (x^4 - 6x^2y^2 + y^4) + (4x^3y - 4xy^3)i$.

Ex. 14.1.9. Let m, m', n, and n' denote arbitrary integers.

(a) If $m + ni$ and $m' + n'i$ are in A, then

$$(m + ni)(m' + n'i) = (mm' - nn') + (mn' + m'n)i.$$

The real and imaginary parts are integers, so the product is in A.

(b) Every non-zero element of A has magnitude at least 1, so as a complex number has reciprocal of magnitude at most 1. The invertible elements of A are therefore elements of magnitude 1, namely ± 1 and $\pm i$.

Exercises for Section 14.2

Ex. 14.2.1. In polar form, $\omega_8^k = e^{k(2\pi i/8)} = e^{k\pi i/4}$. In rectangular form,

$$\omega_8^0 = 1, \qquad \omega_8^1 = \tfrac{1}{2}\sqrt{2}(1 + i), \qquad \omega_8^2 = i, \qquad \omega_8^3 = \tfrac{1}{2}\sqrt{2}(-1 + i),$$
$$\omega_8^4 = -1, \qquad \omega_8^5 = -\tfrac{1}{2}\sqrt{2}(1 + i), \qquad \omega_8^6 = -i, \qquad \omega_8^7 = \tfrac{1}{2}\sqrt{2}(\ 1 - i).$$

Ex. 14.2.8. Let x and y denote arbitrary real numbers. We have

$$x + iy = \exp(u + iv) = e^u(\cos v + i \sin v)$$

if and only if $(x, y) = (e^u \cos v, e^u \sin v)$. By Corollary 13.2.16, every point (x, y) other than $(0, 0)$ has a set of polar coordinates (r, θ) such that $r = \sqrt{x^2 + y^2} > 0$. Since every positive real number r may be written $r = e^u$ for a unique real u, and every polar angle θ is represented by a unique v in $(-\pi, \pi]$, exp maps H bijectively to $\mathbf{C}^\times$.

To calculate the principal logarithm of a complex number $z = x + iy$, we write $z = |z|e^{i\theta}$ in principal polar form, from which we read off the principal logarithm $\operatorname{Log} z = \log|z| + i\theta$. If α is real and positive, then $\operatorname{Log} \alpha$ is real, so $\alpha^\beta = \exp(\beta \operatorname{Log} \alpha) = e^{\beta \log \alpha}$, the definition in Chapter 12.

For i, $-1 = i^2$, and $-i = i^{-1} = i^3$, the principal polar forms are

$$i = \exp(i\pi/2), \qquad\qquad -1 = \exp(i\pi), \qquad\qquad -i = \exp(-i\pi/2),$$
$$\operatorname{Log} i = i\pi/2, \qquad\qquad \operatorname{Log}(-1) = i\pi, \qquad\qquad \operatorname{Log}(-i) = -i\pi/2.$$

Consequently, $i^i = \exp(i \operatorname{Log} i) = e^{-\pi/2}$, $(-1)^i = e^{-\pi}$, and $(-i)^i = e^{\pi/2}$. (Note and caution: Elsewhere you may see complex exponential expressions α^β taking multiple values, in general infinitely many. Only one of those values is the principal value defined here.)

Since $-i = (-1)i$ but $(-i)^i \neq (-1)^i i^i$, the formula $(z_1 z_2)^\beta = z_1^\beta z_2^\beta$ is not an identity.

Since $i^3 = -i$, we have $(i^3)^i = (-i)^i = e^{\pi/2}$, but $(i^i)^3 = (2^{-\pi/2})^3 = e^{-3\pi/2}$. Particularly, the formula $z^{\beta\beta'} = (z^\beta)^{\beta'}$ is not an identity.

For all non-zero complex z and all β, β', however, we do have

$$z^{\beta+\beta'} = \exp[(\beta + \beta') \operatorname{Log} z] = \exp(\beta \operatorname{Log} z) \cdot \exp(\beta' \operatorname{Log} z) = z^\beta \cdot z^{\beta'}.$$

Finally, the principal logarithms give

$$i^{1/2} = e^{i\pi/4} = (\sqrt{2}/2)(1 + i),$$
$$(-1)^{1/2} = e^{i\pi/2} = i,$$
$$(-i)^{1/2} = e^{-i\pi/4} = (\sqrt{2}/2)(1 - i).$$

Note that $(-i)^{1/2} = [(-1) \cdot i]^{1/2} \neq (-1)^{1/2} \cdot i^{1/2}$, see also [12].

Ex. 14.2.9. Answer: $\displaystyle\sum_{k=0}^{\infty} b_{2k} \frac{(-1)^k 4^k t^{2k}}{(2k)!}$.

Solutions: Linear Spaces

Exercises for Section 15.1

Ex. 15.1.1. A mapping $f : X \to Y$ is, by definition, a subset f of the ordered product $X \times Y$ with the property that for every x in X, there exists a unique y in Y such that $(x, y) \in f$. That is, the set of mappings from X to Y contains a "separate copy" of Y for each x in X. Selecting a specific mapping amounts to choosing, for each x in X, a unique y in Y, resulting in an indexed list $\left(f(x)\right)_{x \in X}$, analogous to how we view real sequences. The space of mappings is therefore analogous to an ordered product of Ys indexed by X, symbolically Y^X.

Ex. 15.1.3. Hint: To establish that every uniformly continuous function is in the image of R, prove that if f is uniformly continuous and (x_k) converges to a, the image sequence is condensing, hence has a limit, and this limit defines a continuous extension of f to a. The same argument shows f extends continuously to b.

Exercises for Section 15.2

Ex. 15.2.1. Answer: $e_0(x) = 1$, $e_1(x) = \sqrt{3}x$, $e_2(x) = \frac{1}{2}\sqrt{5}(3x^2 - 1)$,

$$e_3(x) = \frac{1}{2}\sqrt{7}(5x^2 - 3x), \qquad e_4(x) = \frac{3}{8}(35x^4 - 30x^2 + 3).$$

Ex. 15.2.2. By definition, $\mathbf{1} = (1, 1, \dots, 1)$.

(a) Since $\langle \mathbf{1}, \mathbf{e}_1 \rangle = 1$ and $\|\mathbf{1}\| = \sqrt{n}$, the angle between $\mathbf{1}$ and $\mathbf{e}_1$ is $\arccos(1/\sqrt{n})$. For $n = 2$, this is $\pi/4$, as expected from the diagonal of a square. For $n = 3$ the angle is $\arccos(1/\sqrt{3})$, a bit larger. Remarkably, if $n = 4$ the angle is $\arccos(1/2) = \pi/3$. As $n \to \infty$, the angle approaches $\pi/2$.

(b) In $\mathbf{R}^{10,000}$, the diagonal of a unit cube has length $\sqrt{10,000} = 100$, namely, one meter if the sides are 1cm. In $\mathbf{R}^{30,000} = \mathbf{R}^{10,000} \times \mathbf{R}^{10,000} \times \mathbf{R}^{10,000}$, there are three mutually perpendicular one-meter diagonals in a unit cube, whose ordered product is a 3-dimensional cube of side length 1m.

Ex. 15.2.7. Interpreting vectors in $\mathbf{R}^n$ as column matrices, the standard inner product $\langle \mathbf{u}, \mathbf{v} \rangle$ may be interpreted as the matrix product $\mathbf{u}^\mathsf{T}\mathbf{v}$. Consequently, $\langle A\mathbf{u}, A\mathbf{v} \rangle = \langle \mathbf{u}, \mathbf{v} \rangle$ for all $\mathbf{u}$ and $\mathbf{v}$ if and only if

$$\mathbf{u}^\mathsf{T} A^\mathsf{T} A\mathbf{v} = A\mathbf{u}^\mathsf{T} A\mathbf{v} = \mathbf{u}^\mathsf{T}\mathbf{v} = \mathbf{u}^\mathsf{T} I_n \mathbf{v} \quad \text{for all } \mathbf{u} \text{ and } \mathbf{v}.$$

Taking $\mathbf{u}$ and $\mathbf{v}$ to be arbitrary standard basis vectors shows $A^\mathsf{T} A$ has the same entries as I_n.

Ex. 15.2.9. Suggestion: First extend results about ℓ^2 to complex square-summable ("singly-infinite") sequences, then show a doubly infinite square-summable complex sequence is in effect a pair of square-summable complex sequences.

Ex. 15.2.10. Hints: For (a), apply the parallelogram law to $\mathbf{v} - \mathbf{u}_1$ and $\mathbf{v} - \mathbf{u}_2$ and isolate $\|\mathbf{u}_2 - \mathbf{u}_1\|^2$. Estimate the other side, noting that $\frac{1}{2}(\mathbf{u}_1 + \mathbf{u}_2) \in U$ and $\|2\mathbf{v} - (\mathbf{u}_1 + \mathbf{u}_2)\|^2 = 4\|\mathbf{v} - \frac{1}{2}(\mathbf{u}_1 + \mathbf{u}_2)\|^2$.

For (b), if $\mathbf{u}$ is an arbitrary vector in U, then $\|\mathbf{v} - (\mathbf{u}_0 + t\mathbf{u})\|^2$, a quadratic in t, has a minimum at $t = 0$.

Exercises for Section 15.3

Ex. 15.3.1. If c is real, then $1 = |c\mathbf{v}| = |c|\,|\mathbf{v}|$ if and only if $|c| = 1/|\mathbf{v}|$, if and only if $c = \pm 1/|\mathbf{v}|$. The two unit vectors proportional to $\mathbf{v}$ are $\pm\mathbf{v}/|\mathbf{v}|$.

Ex. 15.3.3. Hint: The proof for the real reverse triangle inequality, Proposition 3.2.11 (ii), carries through with natural modifications.

Ex. 15.3.4. Because the defining inequalities involve functions that are even in both variables, each ball is symmetric under reflection about either axis. It suffices to sketch the ball in the first quadrant and reflect across the axes.

(a) In the first quadrant, the closed unit ball is defined by the following inequalities:

$$\| \ \|_1 : \ v_1 + v_2 \le 1; \quad \| \ \|_2 : \ \sqrt{v_1^2 + v_2^2} \le 1; \quad \| \ \|_\infty : \ \max(v_1, v_2) \le 1.$$

The solution sets are in Figure A.6.

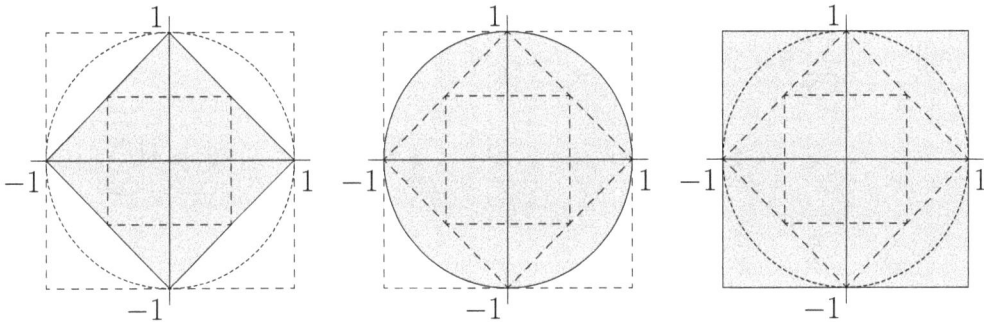

FIGURE A.6
The closed unit ball in the 1-, 2-, and ∞-norms.

(b) For all $\mathbf{v} = (v_1, v_2)$ in the plane,

$$\max(|v_1|, |v_2|)^2 \le v_1^2 + v_2^2 \quad \text{non-negative summands}$$
$$\le v_1^2 + 2|v_1 v_2| + v_2^2 = \left(|v_1| + |v_2| \right)^2$$
$$\le \left(2\max(|v_1|, |v_2|) \right)^2.$$

Since the square root function is increasing, taking square roots preserves the inequalities. Each inequality corresponds to an inclusion of sets, but the inclusion may be opposite to initial impression. An upper bound on a *larger* quantity bounds a smaller quantity. For example, if $\|\mathbf{v}\|_1 \le 1$, namely, if $\mathbf{v}$ is in the closed unit ball for the 1-norm, then $\|\mathbf{v}\|_2 \le \|\mathbf{v}\|_1 \le 1$, so $\mathbf{v}$ is in the closed unit ball for the 2-norm. The innermost dashed square in Figure A.6 is the closed ball of radius $1/2$ in the ∞-norm.

Ex. 15.3.6. This reiterates the triangle inequality: If $\|\mathbf{u}\| < 1$ and $\|\mathbf{v}\| < 1$, then since $0 \le t \le 1$,

$$\|(1-t)\mathbf{u} + t\mathbf{v}\| \le \|(1-t)\mathbf{u}\| + \|t\mathbf{v}\| = |1-t| \, \|\mathbf{u}\| + |t| \, \|\mathbf{v}\| = (1-t)\|\mathbf{u}\| + t\|\mathbf{v}\| < 1.$$

Assume $n \geq 2$. To prove $\| \ \|_p$ is not a norm on $\mathbf{R}^n$ if $0 < p < 1$, it suffices to consider the unit vectors $\mathbf{e}_1 = (1,0)$ and $\mathbf{e}_2 = (0,1)$, for which $\|\mathbf{e}_j\|_p = 1$, but for whose midpoint $\mathbf{v} = \frac{1}{2}(1,1)$ we have $\|\mathbf{v}\|_p = 2^{(1/p)-1} > 1$. Geometrically, the closed unit ball $\{\mathbf{v} \text{ in } \mathbf{R}^n : \|\mathbf{v}\|_p \leq 1\}$ is not convex.

Ex. 15.3.10. Put $M = \max(\|\mathbf{e}_k\|)_{k=0}^{n-1}$. By a straightforward induction,

$$\left\| \sum_{k=0}^{n-1} t_k \mathbf{e}_k \right\| \leq \sum_{k=0}^{n-1} |t_k| \, \|\mathbf{e}_k\| \leq M \sum_{k=0}^{n-1} |t_k| \quad \text{for all } (t_k)_{k=0}^{n-1}.$$

By the triangle and reverse triangle inequalities, Lemma 15.3.3, we have, for all $(t_k)_{k=0}^{n-1}$,

$$\|\mathbf{v}\| - M \sum_{k=0}^{n-1} |t_k| \leq \|\mathbf{v}\| - \left\| \sum_{k=0}^{n-1} t_k \mathbf{e}_k \right\| \leq \left\| \|\mathbf{v}\| - \left\| \sum_{k=0}^{n-1} t_k \mathbf{e}_k \right\| \right\|$$

$$\leq N_{\mathbf{v}}\left((t_k)_{k=0}^{n-1} \right) \leq \|\mathbf{v}\| + \left\| \sum_{k=0}^{n-1} t_k \mathbf{e}_k \right\| = \|\mathbf{v}\| + M \sum_{k=0}^{n-1} |t_k|,$$

or

$$\left| N_{\mathbf{v}}\left((t_k)_{k=0}^{n-1} \right) - N_{\mathbf{v}}(\mathbf{0}) \right| \leq M \sum_{k=0}^{n-1} |t_k|.$$

That is, $N_{\mathbf{v}}$ has bounded stretch at $\mathbf{0}$ relative to the 1-norm on $\mathbf{R}^n$, hence with respect to the 2-norm, which is equivalent.

Ex. 15.3.12. This rephrases properties of summable sequences from Chapter 7: If $\mathbf{v} = (v_k)_{k=0}^{\infty}$ is in ℓ^p, then

$$\|\mathbf{v}\|_p = \left[\sum_{k=0}^{\infty} |v_k|^p \right]^{1/p} < \infty.$$

If $\varepsilon > 0$, use convergence of the series inside the brackets to pick N such that

$$\sum_{k=N}^{\infty} |v_k|^p < \varepsilon^p.$$

The *truncation* of $\mathbf{v}$ after N terms, namely the sequence $\mathbf{u}$ defined by $u_k = v_k$ if $0 \leq k < N$ and $u_k = 0$ if $N \leq k$, is in $\mathbf{R}^{\infty}$ and $\|\mathbf{v} - \mathbf{u}\|_p < \varepsilon$.

Ex. 15.3.13. Answer: The functions $f(x) = 1$ and $g(x) = x - (1/2)$ serve for both parts.

Ex. 15.3.15. The hint, the triangle inequality, and summing over k give

$$\sum_{k=0}^{n-1} |a_k + b_k|^p \leq \sum_{k=0}^{n-1} |a_k| \, |a_k + b_k|^{p/q} + \sum_{k=0}^{n-1} |b_k| \, |a_k + b_k|^{p/q}.$$

By Exercise 12.1.15 (d) applied to each term, this is no larger than

$$\left[\sum_{k=0}^{n-1}|a_k|^p\right]^{1/p}\left[\sum_{k=0}^{n-1}|a_k+b_k|^p\right]^{1/q} + \left[\sum_{k=0}^{n-1}|b_k|^p\right]^{1/p}\left[\sum_{k=0}^{n-1}|a_k+b_k|^p\right]^{1/q}$$

$$= \left[\left[\sum_{k=0}^{n-1}|a_k|^p\right]^{1/p} + \left[\sum_{k=0}^{n-1}|b_k|^p\right]^{1/p}\right] \times \left[\sum_{k=0}^{n-1}|a_k+b_k|^p\right]^{1/q}.$$

Since $1 - (1/q) = 1/p$, dividing both sides of

$$\sum_{k=0}^{n-1}|a_k+b_k|^p \le \left[\left[\sum_{k=0}^{n-1}|a_k|^p\right]^{1/p} + \left[\sum_{k=0}^{n-1}|b_k|^p\right]^{1/p}\right] \times \left[\sum_{k=0}^{n-1}|a_k+b_k|^p\right]^{1/q}.$$

by $\left[\sum_k|a_k+b_k|^p\right]^{1/q}$ gives the stated conclusion.

Ex. 15.3.16. Hints: For positivity, continuity of f guarantees positivity of the integral of $|f|^p$ if f is non-zero, compare the proof of Proposition 9.4.4 (iii).

For the triangle inequality, proceed as in the solution to Exercise 15.3.15 using the result of Exercise 12.1.15 (e).

Ex. 15.3.17. Hint: Use Exercise 15.3.15 and proceed as in the proof of Corollary 15.1.22.

Solutions: Metric Spaces

Exercises for Section 16.1

Ex. 16.1.2. Since $B_{r_x}(x) \subseteq A$ for each x, the union is contained in A. Conversely, for each x in A we have $x \in B_{r_x}(x)$, so A is contained in the union.

Ex. 16.1.4. By the triangle inequality,

$$d(x,y) \le d(x,x') + d(x',y) \le d(x,x') + d(x',y') + d(y',y),$$
$$d(x',y') \le d(x,x') + d(x,y') \le d(x,x') + d(x,y) + d(y',y).$$

(Each these might be called a *quadrangle inequality*.) Subtracting the "mixed" term from each gives

$$d(x,y) - d(x',y') \le d(x,x') + d(y',y),$$
$$d(x',y') - d(x,y) \le d(x,x') + d(y',y).$$

As in Proposition 3.2.10 (ii), these imply $|d(x, y) - d(x', y')| \leq d(x, x') + d(y', y)$.

Ex. 16.1.6. The inclusion $\overline{\overline{A}} \supseteq \overline{A}$ is immediate.

To prove the other inclusion, we argue contrapositively from the partition of X into interior points of A, exterior points, isolated points, and border points: Every point of X that is not in $\overline{A}$ is exterior to A, and hence exterior to $\overline{A}$, hence not in $\overline{\overline{A}}$.

Ex. 16.1.8. (i). Since every open d-ball is d-open and every open d'-ball is d'-open, equivalence of d and d' immediately implies every d-open ball is d'-open and *vice versa*.

Conversely, assume every d'-open ball is d open, and O is a d'-open set. If x_0 is an arbitrary point of O, there exists an r' such that the d'-ball $B_{r'}^{d'}(x_0)$ is contained in O. By hypothesis this ball is d-open, so there exists an r such that $B_r^d(x_0) \subseteq B_{r'}^{d'}(x_0) \subseteq O$. Since x_0 was arbitrary, O is d-open; thus, every d'-open set is d-open. Reversing roles, every d-open set is d'-open.

(ii). Assume d and d' are equivalent, and that $(x_k) \to x_\infty$ with respect to d'. Fix ε arbitrarily. Because the d-ball $B_\varepsilon^d(x_\infty)$ is d'-open, there exists an r such that $B_r^{d'}(x_\infty) \subseteq B_\varepsilon^d(x_\infty)$. Since $(x_k) \to x_\infty$ in (X, d'), there exists an N such that if $k \geq N$ then $d'(x_k, x_\infty) < r$; that is, $x_k \in B_r^{d'}(x_\infty) \subseteq B_\varepsilon^d(x_\infty)$, which implies $d(x_k, x_\infty) < \varepsilon$. Since ε was arbitrary, $(x_k) \to x_\infty$ in (X, d). Reversing the roles of d and d', if $(x_k) \to x_\infty$ in (X, d), then $(x_k) \to x_\infty$ in (X, d').

Ex. 16.1.10. Hint for (a): If f is concave on $[0, \infty)$ and a, b are non-negative, the concave function $g(t) = f(t) - [f(a+b)/(a+b)]t$ is convenient for proving $f(a+b) \leq f(a) + f(b)$.

Ex. 16.1.12. By Exercise 15.3.9, for each x in $\mathbf{R}^n$ we have

$$|f_k(x)| \leq \|f(x)\|_2 \leq \|f(x)\|_1 = \sum_{k=0}^{m-1} |f_k(x)| \quad \text{for all } k.$$

Fix ε and x in $\mathbf{R}^n$ arbitrarily. If f is continuous at x, then there exists a δ such that if $\|x - x'\|_2 < \delta$ in $\mathbf{R}^n$, then $\|f(x) - f(x')\|_2 < \varepsilon$ in $\mathbf{R}^m$. The first inequality guarantees $|f_k(x) - f_k(x')| < \varepsilon$ for all k, so each f_k is continuous at x. Since x was arbitrary, f_k is continuous on $\mathbf{R}^n$ for all k.

Conversely, if f_k is continuous for all k, then for each x in $\mathbf{R}^n$, there exists a δ such that if $\|x - x'\|_2 < \delta$, then $|f_k(x) - f_k(x')| < \varepsilon/m$ for each k. The second inequality above implies f is continuous at x.

Ex. 16.1.14. Each component function is a rational function in two variables, and the denominator is non-vanishing, so f is continuous by Exercises 16.1.12 and 16.1.13. Following the suggestion, note that

$$(2u)^2 + (2v)^2 + (u^2 + v^2 - 1)^2 = 2u^2 + 4v^2 + (u^2 + v^2)^2 - 2(u^2 + v^2) + 1 = (u^2 + v^2 + 1)^2,$$

so the image of f is contained in the unit sphere. Further, $z < 1$ on the image of f, so the image of f is contained in the unit sphere with $(0, 0, 1)$ removed.

Since

$$z = \frac{u^2 + v^2 - 1}{u^2 + v^2 + 1} = 1 - \frac{2}{u^2 + v^2 + 1}, \qquad 1 - z = \frac{2}{u^2 + v^2 + 1},$$

we have $(u, v) = \frac{1}{1-z}(x, y)$. Let X be the unit sphere with the point $(0, 0, 1)$ removed, and define $g : X \to \mathbf{R}^2$ by $g((x, y, z)) = \frac{1}{1-z}(x, y)$. The preceding calculation shows $(g \circ f)(u, v) = (u, v)$ for all (u, v) in $\mathbf{R}^2$. Further, for all (x, y, z) in X, we have

$$u^2 + v^2 + 1 = \left[\frac{x}{1-z}\right]^2 + \left[\frac{y}{1-z}\right]^2 + \left[\frac{1-z}{1-z}\right]^2 = \frac{x^2 + y^2 + 1 - 2z + z^2}{(1-z)^2} = \frac{2}{1-z},$$

from which we immediately deduce $(f \circ g)(x, y, z) = (x, y, z)$. Since $f : \mathbf{R}^2 \to X$ and $g : X \to \mathbf{R}^2$ are inverse mappings, each is bijective; particularly, f is injective, and its image is X, the unit sphere with $(0, 0, 1)$ removed.

Exercises for Section 16.2

Ex. 16.2.7. Suppose $A \subseteq X$. A point x of X is exterior to A if and only if there exists a positive r such that $B_r(x) \cap A = \varnothing$, if and only if $0 < d(x, A)$.

Consequently, A is dense in X if and only if the exterior of A in X is empty, if and only if $d(x, A) = 0$ for all x in X.

Ex. 16.2.6. Let x_0 be an arbitrary element of X.

If A is bounded, pick x in A arbitrarily and put $r = d(x_0, x) + 2 \operatorname{diam} A$. If x' is an arbitrary element of A, the triangle inequality for d implies

$$d(x_0, x') \leq d(x_0, x) + d(x, x') \leq d(x_0, x) + \operatorname{diam} A < r.$$

Since x' was an arbitrary element of A, $A \subseteq B_r(x_0)$.

Conversely, if $A \subseteq B_r(x_0)$ for some r, then $\operatorname{diam} A \leq \operatorname{diam} B_r(x_0) \leq 2r$ by Exercise 16.2.2 (d).

Exercises for Section 16.3

Ex. 16.3.1. This is false. For instance $V^+ = \mathbf{R}^2 \setminus \{(0, y) : y \geq 0\}$ and $V^- = \mathbf{R}^2 \setminus \{(0, y) : y \leq 0\}$ are connected, but their intersection is the disconnected set A of Exercise 16.1.5 (c).

Ex. 16.3.4. Hint: Use Exercise 16.3.3 and the hub lemma, Exercise 16.3.2.

Ex. 16.3.6. If $K_N = \varnothing$ for some N, then because the sets are nested inward, intersection $\bigcap_n K_n \subseteq K_N$ is empty.

Conversely, assume the intersection is empty. We wish to show $K_N = \varnothing$ for some N. Since compact sets are closed, the sets $O_n = X \setminus K_n$ are open. Further, $O_n \subseteq O_{n+1}$ for each n. By the complement law, $\{O_n\}_{n=0}^{\infty}$ is an open-cover of $K_1 \setminus \bigcap_n K_n$. If $\bigcap_n K_n$ is empty, the $\{O_n\}$ cover K_1. By compactness, some finite subcollection covers, and since the O_n are nested outward, $K_1 \subseteq O_N = X \setminus K_N$ for some N. Since $K_N \subseteq K_1 \subseteq X \setminus K_N$, we have $K_N = \varnothing$.

Exercises for Section 16.4

Ex. 16.4.1. The singleton $\{c\}$ is closed in flat m-space, so if f is continuous, then the preimage $f^*(\{c\})$ is closed in flat n-space by Propositions 16.1.38 and 5.3.12. By Theorem 16.4.4, the level of f is compact.

Ex. 16.4.2. Hint: A uniformly continuous, non-negative function whose zero set is C can be written briefly and explicitly.

Ex. 16.4.4. Using subscripts to denote column indices and superscripts to denote row indices, we may write $A = [A_j^i]$ and $B = [B_j^i]$. The definition of matrix multiplication gives

$$(A^\mathsf{T} B)_j^i = \sum_{k=0}^{n-1} A_k^j B_k^i, \qquad \langle A, B \rangle = \sum_{j,k=0}^{n-1} A_k^j B_k^j.$$

This is the standard inner product on $\mathbf{R}^{n \times n}$: Treat an array as a list of n^2 entries, multiply corresponding entries and sum over all entries. (If this is not entirely clear, writing out small cases explicitly may help.)

Ex. 16.4.5. The determinant function on $\mathbf{R}^{n \times n}$ is polynomial, hence continuous in the flat metric by Exercise 16.1.13. The preimage of $\{1\}$, namely $SL(n, \mathbf{R})$, is therefore closed. On the other hand, the matrices $\mathrm{diag}[e^t, e^{-t}, 1, \ldots, 1]$ are in $SL(n, \mathbf{R})$ for all real t, so $SL(n, \mathbf{R})$ is not bounded, hence not compact.

Ex. 16.4.9. Hint for (c): If $\mathrm{osc}_c(f) < r$, then there exists a δ such that $U_c f(\delta) - L_c f(\delta) < r$. Prove $B_\delta(c)$ is disjoint from D_r, which implies the complement of D_r is open in I.

Ex. 16.4.10. Hints: For each positive r, Exercise 16.4.9 (c) implies the set $D_r = \{c \text{ in } I : \mathrm{osc}_c f \geq r\}$ is closed in $[a, b]$, hence compact. If the set of discontinuities has measure zero, then by Exercise 16.4.8 (c), D_r can be covered by finitely many closed intervals of total length at most $\varepsilon/(4M)$.

Inversely, if the set of discontinuities does not have measure zero, first prove there exists a positive integer n such that $D_{1/n}$ does not have measure zero.

Exercises for Section 16.5

Ex. 16.5.1. Because f is ℓ-periodic, $f([0, \ell]) = f(\mathbf{R})$; in fact, f maps every closed interval of length ℓ onto its image. Since $[0, \ell]$ is compact, its image under f is compact by Theorem 16.5.1.

Ex. 16.5.7. Let I and D denote integration and differentiation operators.

(a) By linearity and the triangle inequality for integrals,

$$\|I(f) - I(g)\|_\infty = \sup_{x \in [a,b]} |I(f)(x) - I(g)(x)| \leq \sup_{x \in [a,b]} \left| \int_a^x (f - g) \right|$$

$$\leq \sup_{x \in [a,b]} \int_a^x \|f - g\|_\infty \leq (b - a)\|f - g\|_\infty.$$

(b) If f is smooth, consider the sequence (g_n) of smooth functions defined by $g_n(x) = f(x) + (1/n)\sin(nx)$. We have $\|g_n - f\|_\infty = 1/n \to 0$, so $(g_n) \to f$ in $(\mathscr{C}^\infty(I), d)$. By direct calculation, $Dg_n(x) = g_n'(x) = f'(x) + \cos(nx)$ for all n. Consequently, $\|Dg_n - Df\|_\infty = 1$ for all n. (We can arrange "worse" behavior, for example, by taking $h_n(x) = f(x) + (1/n)\sin(n^2x)$.)

Ex. 16.5.8. Hint: Use a basis to identify V with $\mathbf{R}^n$ for some n. By Exercise 15.3.10, $\|\ \|$ is continuous. Restrict to the unit sphere in flat n-space and use Theorem 16.5.1.

Ex. 16.5.9. Hint for (c): Start by using Exercise 15.2.10 to prove every $\mathbf{v}$ in V decomposes uniquely as a sum $\mathbf{v} = \mathbf{v}_0 + \mathbf{v}^\perp$ such that $\mathbf{v}_0$ in U and $\mathbf{v}^\perp$ in $U^\perp$.

Ex. 16.5.10. Hint: Since $\int_{-1}^1 \phi' = \phi(1) - \phi(-1) = 0$, g is orthogonal to the orthogonal complement of the constants. We cannot directly apply Exercise 16.5.9 (c) because the topological hypotheses are not satisfied, but inner product geometry may suggest an explicit approach.

Ex. 16.5.11. Assume i_1 and i_2 are isometries of (X, d). The composition $i_2 i_1$ is an isometry of (X, d): For all $\mathbf{x}$ and $\mathbf{x}'$ in X,

$$d\big(i_2 i_1(\mathbf{x}), i_2 i_1(\mathbf{x}')\big) = d\big(i_1(\mathbf{x}), i_1(\mathbf{x}')\big) = d(\mathbf{x}, \mathbf{x}').$$

Every distance-preserving mapping i is injective: If $i(\mathbf{x}) = i(\mathbf{x}')$ for some $\mathbf{x}$ and $\mathbf{x}'$ in X, then $0 = d\big(i(\mathbf{x}), i(\mathbf{x}')\big) = d(\mathbf{x}, \mathbf{x}')$, so $\mathbf{x} = \mathbf{x}'$. Since an isometry is a surjection by definition, every isometry is a bijection, hence invertible.

The inverse mapping i^{-1} is an isometry: If $\mathbf{y}$ and $\mathbf{y}'$ are arbitrary elements of X, then since i is a bijection, there exist unique $\mathbf{x}$ and $\mathbf{x}'$ in X such that $\mathbf{y} = i(\mathbf{x})$ and $\mathbf{y}' = i(\mathbf{x}')$. By definition, $\mathbf{x} = i^{-1}(\mathbf{y})$ and $\mathbf{x}' = i^{-1}(\mathbf{y}')$, so

$$d\big(i^{-1}(\mathbf{y}), i^{-1}(\mathbf{y}')\big) = d(\mathbf{x}, \mathbf{x}') = d\big(i(\mathbf{x}), i(\mathbf{x}')\big) = d(\mathbf{y}, \mathbf{y}').$$

Ex. 16.5.12. Hint: If f is an isometry of flat n-space, prove we can compose f with a translation and an orthogonal linear transformation to get an isometry i that fixes the origin and the standard basis vectors, see Exercise 15.2.7. Then use properties of the standard inner product to prove i is the identity mapping.

Ex. 16.5.14. Hint for (b): Use the quotient mapping to interpret $\overline{d}$ on the circle, where the triangle inequality may be easier to see geometrically.

Ex. 16.5.15. Hints: Thinking geometrically throughout is all but essential. For part (c), Exercises 5.3.10 and 13.2.9 are likely to be of interest. The author hopes the path components assertion in (d) is intuitively plausible once parts (a)–(c) are complete, though finding a proof may be vexing without further guidance. The author first proved the *path-lifting* property of the mapping f. Precisely, if $\gamma : [0,1] \to S^1 \times S^1$ is a continuous mapping, and if O is a point of $\mathbf{R}^2$ such that $f(O) = \gamma(0)$, then there exists a unique continuous "lift" $\tilde{\gamma} : [0,1] \to \mathbf{R}^2$ such that $\gamma = f \circ \tilde{\gamma}$ and $\tilde{\gamma}(0) = O$. The proof may be accomplished using interval induction and a special *covering* property of f: For every point (s_0, t_0) of the plane, there exists an open neighborhood V of $f(s_0, t_0)$ such that the preimage $f^*(V)$ is partitioned into components that are *mapped homeomorphically to V by f*.

Solutions: Approximation Theorems

Exercises for Section 17.1

Ex. 17.1.3. Hints: Exercise 16.1.4 will be useful in various places. For completeness of $\overline{d}$: Show that a condensing sequence $(\mathbf{x}_j) = (x_{j,k})$ in $\overline{X}$, namely, a condensing sequence of condensing sequences in X, converges to (the equivalence class of) the "diagonal" sequence $(x_{k,k})$.

Ex. 17.1.4. Suggestion: Establish as a lemma that if $f : (X, d) \to (Y, e)$ is uniformly continuous and (x_k) is condensing in (X, d), then $\big(f(x_k)\big)$ is condensing in (Y, e).

Ex. 17.1.5. Hints: Parts (i) and (ii) follow from results of Chapter 6 and are routine (indeed, fairly tedious). For (iii): If $A \subseteq \mathbf{R}$ is bounded above, pick a rational a_0 that is not an upper bound of A and a rational upper bound b_0. Define rational sequences by recursive bisection. In detail, set $c_0 = \frac{1}{2}(a_0 + b_0)$. Inductively, if c_k is an upper bound of A, define $a_{k+1} = a_k$ and $b_{k+1} = c_k$ (keep the same lower bound and reduce the upper bound); if c_k is not an upper bound of A, define $a_{k+1} = c_k$ and $b_{k+1} = b_k$ (keep the same upper bound and raise the lower bound); and put $c_{k+1} = \frac{1}{2}(a_{k+1} + b_{k+1})$.

Prove that (a_k) and (b_k) are condensing and equivalent, represent an upper bound of A, and no smaller real is an upper bound of A.

Exercises for Section 17.2

Ex. 17.2.3. Hint: Prove that a continuous bijection from $[0, 1]$ to $[0, 1]^2$ is a homeomorphism. Then show $[0, 1]$ is not homeomorphic to $[0, 1]^2$ by showing $[0, 1]$ can be disconnected by removing one point, while $[0, 1]^2$ cannot be.

Ex. 17.2.4. Hint: First prove inductively that there exists a continuous surjection from $[0, 1]^{2^n}$ to $[0, 1]^{2^{n+1}}$.

Ex. 17.2.5. Partition $\mathbf{R}^n$ into a countable list of unit cubes $(C_k)_{k=0}^{\infty}$ with vertices in $\mathbf{Z}^n$ and such that $C_0 = [0, 1]^n$. For convenience, let C_k^- be the corner whose coordinates are all smallest, and C_k^+ the corner whose coordinates are all largest. Fix a continuous surjection c_0 from $[0, 1] \to [0, 1]^n$ starting at $C_0^- = \mathbf{0}$ and ending at $\mathbf{1} = C_0^+$. Because c_0 is continuous on a compact metric space, it is uniformly continuous.

Construct a uniformly continuous mapping $c : \mathbf{R} \to \mathbf{R}^n$ as follows: Define $c(t) = \mathbf{0}$ if $t < 0$. If $0 \leq t \leq 1$, define $c(t) = c_0(t)$. Assume inductively that C has been constructed on some interval $[0, b_n]$ so its image is the union $\bigcup_{k=0}^{n-1} C_k$ and $c(b_n) = C_{n-1}^+$. Pick a unit-speed path of segments parallel to the coordinate axes proceeding "monotonically" from C_{n-1}^+ to C_n^-. If L_n is the length of this path, define c on $[b_n, b_n + L_n]$ by parameter shifting. Then define $c(t) = C_n^- + c_0(t - b_n - L_n)$ if $b_n + L_n \leq t \leq b_n + L_n + 1$. In words, use the "next" unit interval to cover C_n, ending at C_n^+. This inductive procedure defines a uniformly continuous surjection from $\mathbf{R}$ to $\mathbf{R}^n$. Naturally, there are many, many other ways to proceed.

Finally, the function $f(x) = 1/(1 - x) - 1/x$ is a continuous surjection from $(0, 1)$ to $\mathbf{R}$. The composition $(c \circ f) : (0, 1) \to \mathbf{R}^n$ is therefore continuous and surjective.

Exercises for Section 17.3

Ex. 17.3.4. Hints: Are there properties of polynomials on $\mathbf{R}$ not shared by continuous and/or uniformly continuous functions?

Exercises for Section 17.4

Ex. 17.4.1. Hint: It is logically permissible to use the chain rule and Theorem 11.1.1 without verifying hypotheses if doing so leads to verifiable conditions that justify the use of the theorems.

Exercises for Section 17.5

Ex. 17.5.1. The integral $\|c_0\|^2$ is 1, as it true for the standard inner product on $\mathscr{C}([a, b])$ regardless of a and b. For each m, we have

$$\|c_m\|^2 = \frac{1}{2\pi} \int_{-\pi}^{\pi} 2\cos^2(mx)\, dx = \frac{1}{2\pi} \int_{-\pi}^{\pi} \left(1 + \cos(2mx)\right) dx = 1,$$

$$\|s_m\|^2 = \frac{1}{2\pi} \int_{-\pi}^{\pi} 2\sin^2(mx)\, dx = \frac{1}{2\pi} \int_{-\pi}^{\pi} \left(1 - \cos(2mx)\right) dx = 1.$$

It remains to prove all other inner products are 0. By Corollary 13.1.6, for all real x and all natural numbers m and n, we have

$$\cos(mx)\cos(nx) = \tfrac{1}{2}\left(\cos(m-n)x + \cos(m+n)x\right),$$

$$\sin(mx)\sin(nx) = \tfrac{1}{2}\left(\cos(m-n)x - \cos(m+n)x\right),$$

$$\sin(mx)\cos(nx) = \tfrac{1}{2}\left(\sin(m+m)x + \sin(m-n)x\right).$$

If $m \neq n$ all three families integrate to 0 over $[-\pi, \pi]$.

Ex. 17.5.2. Answers: The series obtained are

(a) $\displaystyle\sum_{m=0}^{\infty} \frac{1}{(2m+1)^2} = \frac{\pi^2}{8}$,   (b) $\displaystyle\sum_{m=0}^{\infty} \frac{1}{(2m+1)^4} = \frac{\pi^4}{96}$,   (c) $\displaystyle\sum_{m=1}^{\infty} \frac{1}{m^4} = \frac{\pi^4}{90}$.

Ex. 17.5.3. Conceptually, if g is a step function, the triangle inequality gives $\|f - f_n\|_2 \leq \|f - g\|_2 + \|g - f_n\|_2$, and each term on the right can be made as small as we like.

Bibliography

[1] Ahlfors, Lars V. 1979. *Complex Analysis*. 3rd ed. McGraw-Hill.

[2] Basile, Jonathan 2015. *Universal Library Modeled after Borges's*.
`https://libraryofbabel.info/`

[3] Black, Sammy 2015. *Is there a function whose inverse is exactly the reciprocal of the function, that is $f^{-1} = \frac{1}{f}$?*.
`https://math.stackexchange.com/q/1585414` (version: 2015-12-22).

[4] Borges, Jorge Luis 1941. *The Library of Babel*. See:
`https://en.wikipedia.org/wiki/The_Library_of_Babel`

[5] Robin Chapman. 2010. *Proof of $\int_0^\infty (\frac{\sin x}{x})^2 \, dx = \frac{\pi}{2}$*.
`https://math.stackexchange.com/q/13362` (version: 2010-12-07).

[6] Dym, Harry and KcKean, Henry P. 1972. *Fourier Series and Integrals*.
Academic Press.

[7] Hardy, Godfrey Harold. 1921. *A Course of Pure Mathematics*. 3rd ed.
Cambridge University Press.
`http://www.gutenberg.org/ebooks/38769`.

[8] Hwang, Andrew D. 2013. *What are some common pitfalls when squaring both sides of an equation?*.
`https://math.stackexchange.com/q/445756` (version: 2013-07-17).

[9] ——— 2013. *Differentiating both sides of a non-differential equation*.
`https://math.stackexchange.com/q/529103` (version: 2013-10-16).

[10] ——— 2013. *If f' is differentiable at a then f' is continuous at $(a-\delta, a+\delta)$*.
`https://math.stackexchange.com/q/617575` (version: 2013-12-24).

[11] ——— 2015. *Why $y = e^x$ is not an algebraic curve?*.
`https://math.stackexchange.com/q/1336754` (version: 2015-06-23).

[12] ——— 2015. *For which complex a, b, c does $(a^b)^c = a^{bc}$ hold?*.
`https://math.stackexchange.com/q/1347587` (version: 2015-07-02).

[13] ———— 2015. *Is there a function whose inverse is exactly the reciprocal of the function, that is $f^{-1} = \frac{1}{f}$?*.
https://math.stackexchange.com/q/1585904 (version: 2015-12-22).

[14] ———— 2016. *Is $\lim_{x \to x_0} f'(x) = f'(x_0)$?*.
https://math.stackexchange.com/q/1833167 (version: 2016-06-20).

[15] ———— 2017. *What is the purpose of a function being surjective?*.
https://math.stackexchange.com/q/2325332 (version: 2017-06-16).

[16] ———— 2021. *Two metrics on $\mathbb{P}^n(\mathbf{C})$.*
https://math.stackexchange.com/q/4087053 (version: 2021-04-02).

[17] ———— 2022. *What are some strategies to write a proof that can be easily comprehended?*.
https://math.stackexchange.com/q/4457779 (version: 2022-05-28).

[18] Klein, Felix. 1893. *The Evanston Colloquium: Lectures on Mathematics.* MacMillan/Company.
http://www.gutenberg.org/ebooks/36154.

[19] Krishnan, Gautam Gopal. 2016. *Continued Fractions: Notes for a short course at the Ithaca High School Senior Math Seminar.*
https://pi.math.cornell.edu/~gautam/ContinuedFractions.pdf

[20] Lang, Serge. 1973. *A First Course in Calculus*, 3rd ed. Addison-Wesley.

[21] Munkres, James R. 1999. *Topology*, 2nd ed. Prentice-Hall.

[22] Munroe, Randall. Feb. 26, 2013. *How many unique English tweets are possible? . . . .*
https://what-if.xkcd.com/34/

[23] Preston, Richard. 1992. *The Mountains of Pi*, in *The New Yorker*.
https://www.newyorker.com/magazine/1992/03/02/the-mountains-of-pi

[24] Priestley, William McGowen. 1998. *Calculus: A Liberal Art*, 2nd ed. Springer-Verlag.

[25] Rudin, Walter. 1976. *Principles of Mathematical Analysis.* 3rd ed. McGraw-Hill.

[26] Simmons, George F. 1991. *Differential Equations with Applications and Historical Notes*, 2nd ed. McGraw-Hill.

[27] Spivak, Michael D. 1994. *Calculus*, 3rd ed. Publish or Perish.

Index

For Product Safety Concerns and Information please contact our EU
representative GPSR@taylorandfrancis.com
Taylor & Francis Verlag GmbH, Kaufingerstraße 24, 80331 München, Germany

www.ingramcontent.com/pod-product-compliance
Lightning Source LLC
Chambersburg PA
CBHW080125220326
41598CB00032B/4962

* 9 7 8 1 0 3 2 9 8 7 1 4 9 *